Springer Series in Reliability Engineering

Series Editor

Professor Hoang Pham
Department of Industrial Engineering
Rutgers
The State University of New Jersey
96 Frelinghuysen Road
Piscataway, NJ 08854-8018
USA

Other titles in this series

Warranty Management and Product Manufacture
D.N.P. Murthy and W. Blischke

Maintenance Theory of Reliability
T. Nakagawa
Publication due August 2005

Gregory Levitin

The Universal Generating Function in Reliability Analysis and Optimization

With 142 Figures

Gregory Levitin, PhD
The Israel Electric Corporation Ltd, P.O.B. 10, Haifa 31000, Israel

British Library Cataloguing in Publication Data
Levitin, Gregory
Universal generating function in reliability analysis and optimization. — (Springer series in reliability engineering)
1. Reliability (Engineering) — Mathematical models
I. Title
620′.00452′015118

Springer Series in Reliability Engineering ISSN 1614-7839
ISBN-13: 978-1-84996-962-8
e-ISBN-13: 978-1-84628-245-4
Springer Science+Business Media
springeronline.com

Printed in the United States of America (TB/IBT)
69/3830-543210 Printed on acid-free paper

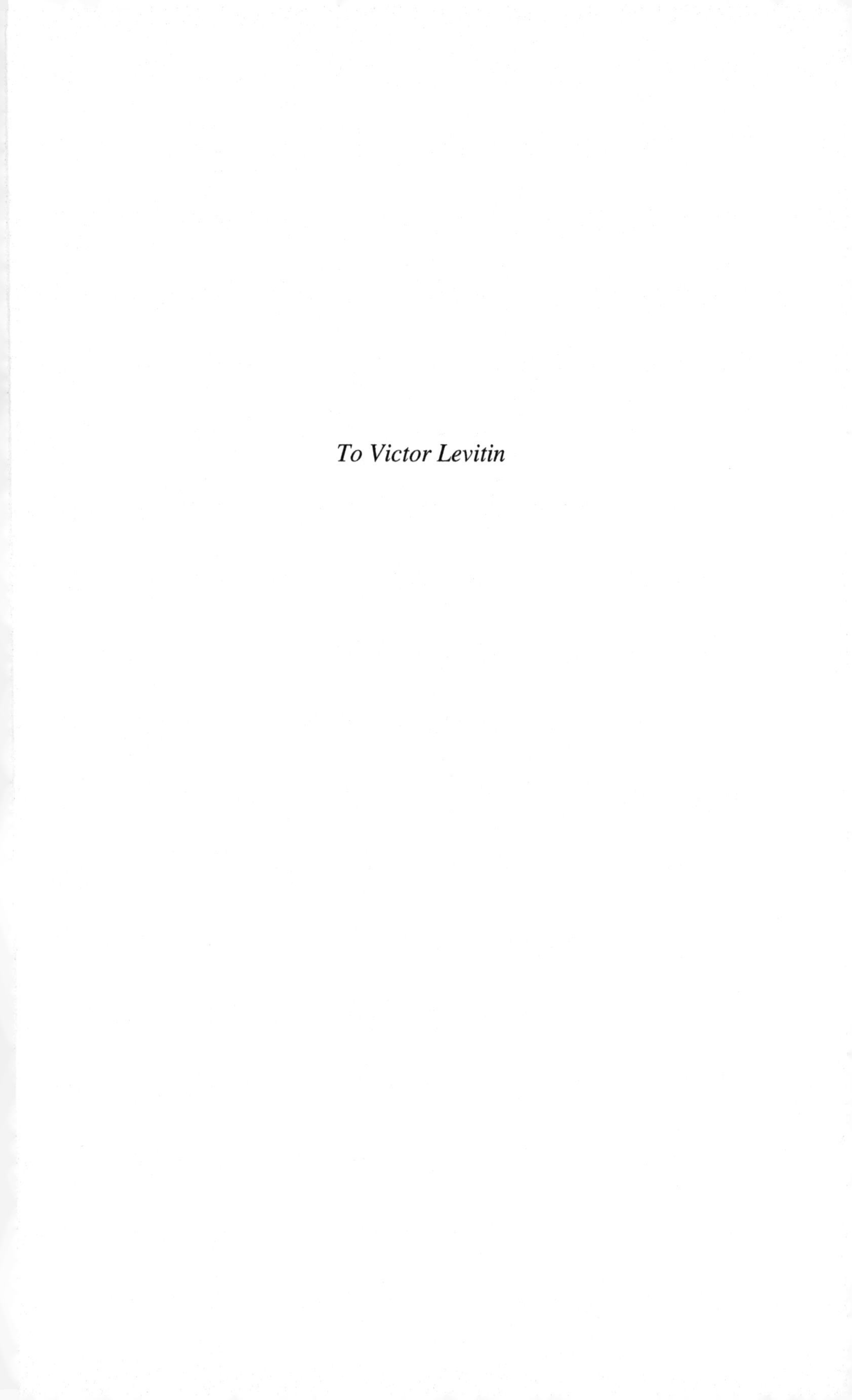

To Victor Levitin

Preface

Most books on reliability theory are devoted to traditional binary reliability models allowing only two possible states for a system and its components: perfect functionality and complete failure. Many real-world systems are composed of multi-state components, which have different performance levels and several failure modes with various effects on the system's entire performance. Such systems are called multi-state systems (MSSs). Examples of MSSs are power systems or computer systems where the component performance is respectively characterized by the generating capacity or the data processing speed. For MSSs, the outage effect will be essentially different for units with different performance rates. Therefore, the reliability analysis of MSSs is much more complex when compared with binary-state systems. In real-world problems of MSS reliability analysis, the great number of system states that need to be evaluated makes it difficult to use traditional binary reliability techniques.

The recently emerged universal generating function (UGF) technique allows one to find the entire MSS performance distribution based on the performance distributions of its elements by using algebraic procedures. This technique (also called the method of generalized generating sequences) generalizes the technique that is based on using a well-known ordinary generating function. The basic ideas of the method were introduced by Professor I. Ushakov in the mid 1980s [1, 2]. Since then, the method has been considerably expanded.

The UGF approach is straightforward. It is based on intuitively simple recursive procedures and provides a systematic method for the system states' enumeration that can replace extremely complicated combinatorial algorithms used for enumerating the possible states in some special types of system (such as consecutive systems or networks).

The UGF approach is effective. Combined with simplification techniques, it allows the system's performance distribution to be obtained in a short time. The computational burden is the crucial factor when one solves optimization problems where the performance measures have to be evaluated for a great number of possible solutions along the search process. This makes using the traditional methods in reliability optimization problematic. On the contrary, the UGF technique is fast enough to be implemented in optimization procedures.

The UGF approach is universal. An analyst can use the same recursive procedures for systems with a different physical nature of performance and different types of element interaction.

The first brief description of the UGF method appeared in our recent book (Lisnianski A, Levitin G. Multi-state system reliability. Assessment, optimization and applications, World Scientific 2003), where three basic approaches to MSS reliability analysis were presented: the extended Boolean technique, the random processes methods, and the UGF. Unlike the previous book that contained only a chapter devoted to the universal generating function, this book is the first to include a comprehensive up-to-date presentation of the universal generating function method and its application to analysis and optimization of different types of binary and multi-state system. It describes the mathematical foundations of the method, provides a generalized view of the performance-based reliability measures, and presents a number of new topics not included in the previous book, such as: UGF for analysis of binary systems, systems with dependent elements, simplified analysis of series-parallel systems, controllable series-parallel systems, analysis of continuous-state systems, optimal multistage modernization, incorporating common cause failures into MSS analysis, systems with multilevel protection, vulnerability importance, importance of multi-state elements in MSSs, optimization of MSS topology, asymmetric weighted voting systems, decision time of voting systems, multiple sliding window systems, fault-tolerant software systems, *etc*. It provides numerous examples of applications of the UGF method for a variety of technical problems.

In order to illustrate applications of the UGF to numerous optimization problems, the book also contains a description of a universal optimization technique called the genetic algorithm (GA). The main aim of the book is to show how the combination of the two universal tools (UGF and GA) helps in solving various practical problems of reliability and performance optimization.

The book is suitable for different types of reader. It primarily addresses practising reliability engineers and researchers who have an interest in reliability and performability analysis. It can also be used as a textbook for senior undergraduate or graduate courses in several departments: industrial engineering, nuclear engineering, electrical engineering, and applied mathematics.

The book is divided into eight chapters.

Chapter 1 presents two basic universal tools used in the book for MSS reliability assessment and optimization. It introduces the UGF as a generalization of the moment generating function and the z-transform; it defines the generic composition operator and describes its basic properties, and shows how the operator can be used for the determination of the probabilistic distribution of complex functions of discrete random variables. The chapter also shows how the combination of recursive determination of the functions with simplification techniques based on the like terms collection allows one to reduce considerably the computational burden associated with evaluating the probabilistic distribution of complex functions. This chapter also presents the GAs and discusses the basic steps in applying them to a specific optimization problem.

Chapter 2 describes the application of the UGF approach for the reliability evaluation of several binary reliability models.

Chapter 3 introduces the MSSs as an object of study. It defines the generic model and describes the basic properties of an MSS. This chapter also introduces some reliability indices used in MSSs and presents examples of different MSS models.

Chapter 4 is devoted to the application of the UGF method to reliability analysis of the most widely used series-parallel MSSs. It describes the extension of the reliability block diagram method to series-parallel MSS, presents methods for evaluating the influence of common cause failures on the entire MSS reliability, and discusses methods for evaluating element reliability importance in MSSs.

Chapter 5 describes the application of the UGF in the optimization of series-parallel MSSs. It contains definitions and solutions related to various application problems of structure optimization for different types of series-parallel MSS. It shows that, by optimizing the MSS maintenance policy, one can achieve the desired level of system reliability requiring minimal cost. It also considers the problems of survivability maximization for MSSs that are subject to common cause failures. The optimal separation and protection problems are discussed.

Chapter 6 is devoted to the adaptation of the UGF technique to different special types of MSS. It presents the UGF-based algorithms for evaluating the reliability of MSSs with bridge topology, MSSs with two failure modes, weighted voting systems and classifiers, and sliding window systems. For each algorithm it describes the methods for computational complexity reduction. The chapter also considers the problems of structure optimization subject to the reliability and survivability constraints for different types of system.

Chapter 7 is devoted to the adaptation of the UGF technique to several types of network. It presents the UGF-based algorithms for evaluating the reliability of linear multi-state consecutively connected systems and multi-state acyclic networks. The connectivity model and the transmission delay model are considered. The structure optimization problems subject to reliability and survivability constraints are presented.

Chapter 8 is devoted to the application of the UGF technique for software reliability. The multi-state nature of fault-tolerant programs is demonstrated in this chapter and the methods for obtaining the performance distribution of such programs is presented. The reliability of combined software-hardware systems is analyzed. Optimal software modules sequencing problem and the software structure optimization problem are formulated and solved using the techniques presented in the book.

I would like to express my sincere appreciation to Professor Hoang Pham from Rutgers University, Editor-in-Chief of the Springer Series in Reliability, for providing me with the chance to include this book in the series. I thank my colleagues Professor Igor Ushakov from the Canadian Training Group, San Diego, Professor Min Xie and Professor Kim Leng Poh from the National University of Singapore, Dr Yuanshun Dai from Purdue University, USA, Professor Enrico Zio and Dr Luca Podofillini from the Polytechnic of Milan, Dr Anatoly Lisnianski, Dr David Elmakis, and Dr Hanoch Ben Haim from The Israel Electric Corporation for collaboration in developing the UGF method. My special thanks to Dr Edward Korczak from the Telecommunications Research Institute, Warsaw, for his friendly support, correcting my mistakes, and discussions that benefited this book.

I am also indebted to the many researchers who have developed the underlying concepts of this book. Although far too numerous to mention, I have tried to recognize their contributions in the bibliographical references.

It was a pleasure working with the Springer-Verlag editors Michael Koy, Anthony Doyle and Oliver Jackson.

Haifa, Israel Gregory Levitin

Contents

General Notation and Acronyms

Notation

$\Pr\{e\}$	probability of event e
$E(X)$	expected value of random variable X
$1(x)$	unity function: $1(\boldsymbol{x}) = \begin{cases} 1, \boldsymbol{x} \text{ is true} \\ 0, \boldsymbol{x} \text{ is false} \end{cases}$
$\sum_{i=m}^{n} x_i$	sum of values x_i with indices running from m to n. If $n<m$ $\sum_{i=m}^{n} x_i = 0$
$\lfloor x \rfloor$	greatest integer not exceeding x
$\lceil x \rceil$	least integer not less than x
$\text{mod}_a(x)$	module function: $\text{mod}_a(x) = x - a\lfloor x/a \rfloor$
$F(G,W)$	system acceptability function
G	random system performance rate
W	random system demand
O	system performance measure (expected value of some function of G and W)
R	system reliability (expected acceptability)
A	system availability (expected acceptability)
ε	mean system performance
$\tilde{\varepsilon}$	conditional expected system performance
Δ	expected system performance deviation
Δ^-	expected system performance deficiency
$\boldsymbol{g}_j$	vector of possible performance levels for element j
$\boldsymbol{p}_j$	vector of probabilities corresponding to possible performance levels for element j
$\boldsymbol{w}$	vector of possible system demand levels
$\boldsymbol{q}$	vector of probabilities corresponding to possible demand levels
$u(z)$	u-function: polynomial-like structure representing probabilistic distribution of a random mathematical object
$d(z)$	d-function (double u-function) set of two related u-functions
$u_j(z)$	u-function representing performance distribution of individual element j
$U_m(z)$	u-function representing performance distribution of subsystem m
$U(z)$	u-function representing performance distribution of the entire system

$\underset{\phi}{\otimes}$ composition operator over u-functions

$\underset{\phi_{ser}}{\otimes}$ composition operator over u-functions of elements connected in series

$\underset{\phi_{br}}{\otimes}$ composition operator over u-functions of elements composing a bridge

$\underset{\phi_{par}}{\otimes}$ composition operator over u-functions of elements connected in parallel

Acronyms

ATB acceptance test block
CCF common cause failure
CCG common cause group
CMCCS circular multi-state consecutively connected system
GA genetic algorithm
LMCCS linear multi-state consecutively connected system
MAN multi-state acyclic network
ME multi-state element
MPS main producing system
MR minimal repair
MSS multi-state system
MSWS multiple sliding window system
NVP n-version programming
PD performance distribution
PM performance measure
PMA preventive maintenance action
p.m.f. probability mass function for discrete random value
PR preventive replacement
RBS recovery block scheme
RGS resource-generating subsystem
RGT reliability growth testing
SWS sliding window system
UGF universal generating function (u-function)
UOD unit output distribution
VWD voting weight distribution
WVC weighted voting classifier
WVS weighted voting system

1. Basic Tools and Techniques

1.1 Moment-generating Function and z-transform

Consider a discrete random variable X that can take on a finite number of possible values. The probabilistic distribution of this variable can be represented by the finite vector $\boldsymbol{x} = (x_0, \ldots, x_k)$ consisting of the possible values of X and the finite vector $\boldsymbol{p}$ consisting of the corresponding probabilities $p_i = \Pr\{X = x_i\}$.
The mapping $x_i \rightarrow p_i$ is usually called the *probability mass function* (p.m.f.)

X must take one of the values x_i. Therefore

$$\sum_{i=0}^{k} p_i = 1 \tag{1.1}$$

Example 1.1

Suppose that one performs k independent trials and each trial can result either in a success (with probability π) or in a failure (with probability $1-\pi$). Let random variable X represent the number of successes that occur in k trials. Such a variable is called a binomial random variable. The p.m.f. of X takes the form

$$x_i = i, \; p_i = \binom{k}{i} \pi^i (1-\pi)^{k-i}, \quad 0 \le i \le k$$

According to the binomial theorem it can be seen that

$$\sum_{i=0}^{k} p_i = \sum_{i=0}^{k} \binom{k}{i} \pi^i (1-\pi)^{k-i} = [\pi + (1-\pi)]^k = 1$$

The expected value of X is defined as a weighted average of the possible values that X can take on, where each value is weighted by the probability that X assumes that value:

$$E(X)=\sum_{i=0}^{k} x_i p_i \tag{1.2}$$

Example 1.2

The expected value of a binomial random variable is

$$E(X)=\sum_{i=0}^{k} x_i p_i = \sum_{i=0}^{k} i \binom{k}{i} \pi^i (1-\pi)^{k-i}$$

$$= k\pi \sum_{i=0}^{k-1} \binom{k-1}{i} \pi^i (1-\pi)^{k-i-1} = k\pi[\pi+(1-\pi)]^{k-1} = k\pi$$

The moment-generating function $m(t)$ of the discrete random variable X with p.m.f. $\boldsymbol{x}, \boldsymbol{p}$ is defined for all values of t by

$$m(t)=E(e^{tX})=\sum_{i=0}^{k} e^{tx_i} p_i \tag{1.3}$$

The function $m(t)$ is called the moment-generating function because all of the moments of random variable X can be obtained by successively differentiating $m(t)$. For example:

$$m'(t)=\frac{\mathrm{d}}{\mathrm{d}t}\left(\sum_{i=0}^{k} e^{tx_i} p_i\right)=\sum_{i=0}^{k} x_i e^{tx_i} p_i. \tag{1.4}$$

Hence

$$m'(0)=\sum_{i=0}^{k} x_i p_i = E(X) \tag{1.5}$$

Then

$$m''(t)=\frac{\mathrm{d}}{\mathrm{d}t}(m'(t))=\frac{\mathrm{d}}{\mathrm{d}t}\left(\sum_{i=0}^{k} x_i e^{tx_i} p_i\right)=\sum_{i=0}^{k} x_i^2 e^{tx_i} p_i \tag{1.6}$$

and

$$m''(0) = \sum_{i=0}^{k} x_i^2 p_i = E(X^2) \tag{1.7}$$

The nth derivative of $m(t)$ is equal to $E(X^n)$at $t = 0$.

Example 1.3

The moment-generating function of the binomial distribution takes the form

$$m(t) = E(e^{tX}) = \sum_{i=0}^{k} e^{ti} \binom{k}{i} \pi^i (1-\pi)^{k-i}$$

$$= \sum_{i=0}^{k} (\pi\ e^t)^i \binom{k}{i} (1-\pi)^{k-i} = (\pi\ e^t + 1 - \pi)^k$$

Hence

$$m'(t) = k(\pi\ e^t + 1 - \pi)^{k-1} \pi e^t \quad \text{and} \quad E(X) = m'(0) = k\pi.$$

The moment-generating function of a random variable uniquely determines its p.m.f. This means that a one-to-one correspondence exists between the p.m.f. and the moment-generating function.

The following important property of moment-generating function is of special interest for us. The moment-generating function of the sum of the independent random variables is the product of the individual moment-generating functions of these variables. Let $m_X(t)$ and $m_Y(t)$ be the moment-generating functions of random variables X and Y respectively. The p.m.f. of the random variables are represented by the vectors

$$\boldsymbol{x} = (x_0, \ldots, x_{k_X}), \quad \boldsymbol{p}_X = (p_{X0}, \ldots, p_{Xk_X})$$

and

$$\boldsymbol{y} = (y_0, \ldots, y_{k_Y}), \quad \boldsymbol{p}_Y = (p_{Y0}, \ldots, p_{Yk_Y})$$

respectively. Then $m_{X+Y}(t)$, the moment-generating function of $X+Y$, is obtained as

$$m_{X+Y}(t) = m_X(t) m_Y(t) = \sum_{i=0}^{k_X} e^{tx_i} p_{X_i} \sum_{j=0}^{k_Y} e^{ty_i} p_{Y_j}$$

$$= \sum_{i=0}^{k_X} \sum_{j=0}^{k_Y} e^{tx_i} e^{ty_j} p_{X_i} p_{Y_j} = \sum_{i=0}^{k_X} \sum_{j=0}^{k_Y} e^{t(x_i+y_j)} p_{X_i} p_{Y_j} \tag{1.8}$$

The resulting moment-generating function $m_{X+Y}(t)$ relates the probabilities of all the possible combinations of realizations $X = x_i$, $Y = y_j$, for any i and j, with the values that the random function $X + Y$ takes on for these combinations.

In general, for n independent discrete random variables $X_1, \ldots, X_n$

$$m_{\sum_{i=1}^{n} X_i}(t) = \prod_{i=1}^{n} m_{X_i}(t) \tag{1.9}$$

By replacing the function e^t by the variable z in Equation (1.3) we obtain another function related to random variable X that uniquely determines its p.m.f.:

$$\omega(z)=E(z^X)=\sum_{i=0}^{k} z^{x_i} p_i \tag{1.10}$$

This function is usually called the z-transform of discrete random variable X. The z-transform preserves some basic properties of the moment-generating functions. The first derivative of $\omega(z)$ is equal to $E(X)$ at $z = 1$. Indeed:

$$\omega'(z) = \frac{\mathrm{d}}{\mathrm{d}t}\left(\sum_{i=0}^{k} z^{x_i} p_i\right) = \sum_{i=0}^{k} x_i z^{x_i - 1} p_i \tag{1.11}$$

Hence

$$\omega'(1) = \sum_{i=0}^{k} x_i p_i = E(X) \tag{1.12}$$

The z-transform of the sum of independent random variables is the product of the individual z-transforms of these variables:

$$\begin{aligned}\omega_{X+Y}(z) &= \omega_X(z)\omega_Y(z) = \sum_{i=0}^{k_X} z^{x_i} p_{X_i} \sum_{j=0}^{k_Y} z^{y_i} p_{Y_j} \\ &= \sum_{i=0}^{k_X} \sum_{j=0}^{k_Y} z^{x_i} z^{y_j} p_{X_i} p_{Y_j} = \sum_{i=0}^{k_X} \sum_{j=0}^{k_Y} z^{(x_i+y_j)} p_{X_i} p_{Y_j}\end{aligned} \tag{1.13}$$

and in general

$$\omega_{\sum_{i=1}^{n} X_i} = \prod_{i=1}^{n} \omega_{X_i}(z) \tag{1.14}$$

The reader wishing to learn more about the generating function and z-transform is referred to the books by Grimmett and Stirzaker [3] and Ross [4].

Example 1.4

Suppose that one performs k independent trials and each trial can result either in a success (with probability π) or in a failure (with probability $1 - \pi$). Let random variable X_j represent the number of successes that occur in the jth trial.

The p.m.f. of any variable $X_j\,(1 \le j \le k)$ is

$$\Pr\{X_j = 1\} = \pi,\ \Pr\{X_j = 0\} = 1 - \pi.$$

The corresponding z-transform takes the form

$$\omega_{Xj}(z) = \pi z^1 + (1-\pi)z^0$$

The random number of successes that occur in k trials is equal to the sum of the numbers of successes in each trial

$$X = \sum_{j=1}^{k} X_j$$

Therefore, the corresponding z-transform can be obtained as

$$\omega_X(z) = \prod_{j=1}^{n} \omega_{X_j}(z) = [\pi\ z + (1-\pi)z^0]^k$$

$$= \sum_{i=0}^{k} \binom{k}{i} z^i \pi^i (1-\pi)^{k-i} = \sum_{i=0}^{k} z^i \binom{k}{i} \pi^i (1-\pi)^{k-i}$$

This z-transform corresponds to the binomial p.m.f:

$$x_i = i,\ p_i = \binom{k}{i} \pi^i (1-\pi)^{k-i},\ \ 0 \le i \le k$$

1.2 Mathematical Fundamentals of the Universal Generating Function

1.2.1 Definition of the Universal Generating Function

Consider n independent discrete random variables $X_1, \ldots, X_n$ and assume that each variable X_i has a p.m.f. represented by the vectors $\boldsymbol{x}_i$, $\boldsymbol{p}_i$. In order to evaluate the p.m.f. of an arbitrary function $f(X_1, \ldots, X_n)$, one has to evaluate the vector $\boldsymbol{y}$ of all of the possible values of this function and the vector $\boldsymbol{q}$ of probabilities that the function takes these values.

Each possible value of function f corresponds to a combination of the values of its arguments $X_1, \ldots, X_n$. The total number of possible combinations is

$$K = \prod_{i=1}^{n} (k_i + 1) \tag{1.15}$$

where $k_i + 1$ is the number of different realizations of random variable X_i. Since all of the n variables are statistically independent, the probability of each unique combination is equal to the product of the probabilities of the realizations of arguments composing this combination.

The probability of the jth combination of the realizations of the variables can be obtained as

$$q_j = \prod_{i=1}^{n} p_{ij_i} \tag{1.16}$$

and the corresponding value of the function can be obtained as

$$f_j = f(x_{1j_1}, \ldots, x_{nj_n}) \tag{1.17}$$

Some different combinations may produce the same values of the function. All of the combinations are mutually exclusive. Therefore, the probability that the function takes on some value is equal to the sum of probabilities of the combinations producing this value. Let A_h be a set of combinations producing the value f_h. If the total number of different realizations of the function $f(X_1, \ldots, X_n)$ is H, then the p.m.f. of the function is

$$\boldsymbol{y} = (f_h : 1 \le h \le H), \quad \boldsymbol{q} = \Big(\sum_{(x_{1j_1}, \ldots, x_{nj_n}) \in A_h} \prod_{i=1}^{n} p_{ij_i} : 1 \le h \le H \Big) \tag{1.18}$$

Example 1.5

Consider two random variables X_1 and X_2 with p.m.f. $\boldsymbol{x_1} = (1, 4)$, $\boldsymbol{p_1} = (0.6, 0.4)$ and $\boldsymbol{x_2} = (0.5, 1, 2)$, $\boldsymbol{p_2} = (0.1, 0.6, 0.3)$. In order to obtain the p.m.f. of the function $Y = X_1^{X_2}$ we have to consider all of the possible combinations of the values taken by the variables. These combinations are presented in Table 1.1.

The values of the function Y corresponding to different combinations of realizations of its random arguments and the probabilities of these combinations can be presented in the form

$$\boldsymbol{y} = (1, 2, 1, 4, 1, 16),\ \ \boldsymbol{q} = (0.06, 0.04, 0.36, 0.24, 0.18, 0.12)$$

Table 1.1. p.m.f. of the function of two variables

No of combination	Combination probability	Value of X_1	Value of X_2	Value of Y
1	0.6×0.1 = 0.06	1	0.5	1
2	0.4×0.1 = 0.04	4	0.5	2
3	0.6×0.6 = 0.36	1	1	1
4	0.4×0.6 = 0.24	4	1	4
5	0.6×0.3 = 0.18	1	2	1
6	0.4×0.3 = 0.12	4	2	16

Note that some different combinations produce the same values of the function Y. Since all of the combinations are mutually exclusive, we can obtain the probability that the function takes some value as being the sum of the probabilities of different combinations of the values of its arguments that produce this value:

$$\Pr\{Y = 1\} = \Pr\{X_1 = 1,\ X_2 = 0.5\} + \Pr\{X_1 = 1,\ X_2 = 1\}$$

$$+ \Pr\{X_1 = 1,\ X_2 = 2\} = 0.06 + 0.36 + 0.18 = 0.6$$

The p.m.f. of the function Y is

$$\boldsymbol{y} = (1, 2, 4, 16),\ \ \boldsymbol{q} = (0.\ 6, 0.04, 0.24, 0.12)$$

The z-transform of each random variable X_i represents its p.m.f. $(x_{i0}, \ldots, x_{ik_i})$, $(p_{i0}, \ldots, p_{ik_i})$ in the polynomial form

$$\sum_{j=0}^{k_i} p_{ij} z^{x_{ij}} \tag{1.19}$$

According to (1.14), the product of the z-transform polynomials corresponding to the variables $X_1, \ldots, X_n$ determines the p.m.f. of the sum of these variables.

In a similar way one can obtain the z-transform representing the p.m.f. of the arbitrary function f by replacing the product of the polynomials by a more general composition operator $\otimes_f$ over z-transform representations of p.m.f. of n independent variables:

$$\underset{f}{\otimes}\left(\sum_{j_i=0}^{k_i} p_{ij_i} z^{x_{ij_i}}\right) = \sum_{j_1=0}^{k_1} \sum_{j_2=0}^{k_2} \dots \sum_{j_n=0}^{k_n} \left(\prod_{i=0}^{n} p_{ij_i} z^{f(x_{ij_1},\dots,x_{nj_n})}\right) \tag{1.20}$$

The technique based on using z-transform and composition operators $\otimes_f$ is named the universal z-transform or universal (moment) generating function (UGF) technique. In the context of this technique, the z-transform of a random variable for which the operator $\otimes_f$ is defined is referred to as its u-function. We refer to the u-function of variable X_i as $u_j(z)$, and to the u-function of the function $f(X_1, \dots, X_n)$ as $U(z)$. According to this notation

$$U(z) = \underset{f}{\otimes}(u_1(z),\ u_2(z),\ \dots,\ u_n(z)) \tag{1.21}$$

where $u_i(z)$ takes the form (1.19) and $U(z)$ takes the form (1.20). For functions of two arguments, two interchangeable notations can be used:

$$U(z) = \underset{f}{\otimes}(u_1(z),\ u_2(z)) = u_1(z) \underset{f}{\otimes} u_2(z) \tag{1.22}$$

Despite the fact that the u-function resembles a polynomial, it is not a polynomial because:

- Its coefficients and exponents are not necessarily scalar variables, but can be other mathematical objects (*e.g.* vectors);
- Operators defined over the u-functions can differ from the operator of the polynomial product (unlike the ordinary z-transform, where only the product of polynomials is defined).

When the u-function $U(z)$ represents the p.m.f. of a random function $f(X_1,\dots, X_n)$, the expected value of this function can be obtained (as an analogy with the regular z-transform) as the first derivative of $U(z)$ at $z = 1$.

In general, the u-functions can be used not just for representing the p.m.f. of random variables. In the following chapters we also use other interpretations. However, in any interpretation the coefficients of the terms in the u-function represent the probabilistic characteristics of some object or state encoded by the exponent in these terms.

The u-functions inherit the essential property of the regular polynomials: they allow for collecting like terms. Indeed, if a u-function representing the p.m.f. of a random variable X contains the terms $p_h z^{x_h}$ and $p_m z^{x_m}$ for which $x_h = x_m$, the two terms can be replaced by a single term $(p_h + p_m)z^{x_m}$, since in this case $\Pr\{X = x_h\} = \Pr\{X = x_m\} = p_h+p_m$.

Example 1.6

Consider the p.m.f. of the function Y from Example 1.5, obtained from Table 1.1. The u-function corresponding to this p.m.f. takes the form:

$$U(z) = 0.06z^1 + 0.04z^2 + 0.36z^1 + 0.24z^4 + 0.18z^1 + 0.12z^{16}$$

By collecting the like terms in this u-function we obtain:

$$U(z) = 0.6z^1 + 0.04z^2 + 0.24z^4 + 0.12z^{16}$$

which corresponds to the final p.m.f. obtained in Example 1.5.

The expected value of Y can be obtained as

$$E(Y) = U'(1) = 0.6\times1 + 0.04\times2 + 0.24\times4 + 0.12\times16 = 3.56$$

The described technique of determining the p.m.f. of functions is based on an enumerative approach. This approach is extremely resource consuming. Indeed, the resulting u-function $U(z)$ contains K terms (see Equation (1.15)), which requires excessive storage space. In order to obtain $U(z)$ one has to perform $(n - 1)K$ procedures of probabilities multiplication and K procedures of function evaluation. Fortunately, many functions used in reliability engineering produce the same values for different combinations of the values of their arguments. The combination of recursive determination of the functions with simplification techniques based on the like terms collection allows one to reduce considerably the computational burden associated with evaluating the p.m.f. of complex functions.

Example 1.7

Consider the function

$$Y = f(X_1, \ldots, X_5) = (\max(X_1, X_2) + \min(X_3, X_4))\, X_5$$

of five independent random variables $X_1, \ldots, X_5$. The probability mass functions of these variables are determined by pairs of vectors $\boldsymbol{x}_i$, $\boldsymbol{p}_i$ $(0 \le i \le 5)$ and are presented in Table 1.2.

These p.m.f. can be represented in the form of u-functions as follows:

$$u_1(z) = p_{10}z^{x_{10}} + p_{11}z^{x_{11}} + p_{12}z^{x_{12}} = 0.6z^5 + 0.3z^8 + 0.1z^{12}$$

$$u_2(z) = p_{20}z^{x_{20}} + p_{21}z^{x_{21}} = 0.7z^8 + 0.3z^{10}$$

$$u_3(z) = p_{30}z^{x_{30}} + p_{31}z^{x_{31}} = 0.6z^0 + 0.4z^1$$

$$u_4(z) = p_{40}z^{x_{40}} + p_{41}z^{x_{41}} + p_{42}z^{x_{42}} = 0.1z^0 + 0.5z^8 + 0.4z^{10}$$

$$u_5(z) = p_{50}z^{x_{50}} + p_{51}z^{x_{51}} = 0.5z^1 + 0.5z^{1.5}$$

Using the straightforward approach one can obtain the p.m.f. of the random variable Y applying the operator (1.20) over these u-functions. Since $k_1 + 1 = 3$, $k_2 + 1 = 2$, $k_3 + 1 = 2$, $k_4 + 1 = 3$, $k_5 + 1 = 2$, the total number of term multiplication procedures that one has to perform using this equation is $3\times2\times2\times3\times2 = 72$.

Table 1.2. p.m.f. of random variables

X_1	p_1	0.6	0.3	0.1
	x_1	5	8	12
X_2	p_2	0.7	0.3	-
	x_2	8	10	-
X_3	p_3	0.6	0.4	-
	x_3	0	1	-
X_4	p_4	0.1	0.5	0.4
	x_4	0	8	10
X_5	p_5	0.5	0.5	-
	x_5	1	1.5	-

Now let us introduce three auxiliary random variables X_6, X_7 and X_8, and define the same function recursively:

$$X_6 = \max\{X_1, X_2\}$$

$$X_7 = \min\{X_3, X_4\}$$

$$X_8 = X_6 + X_7$$

$$Y = X_8\, X_5$$

We can obtain the p.m.f. of variable Y using composition operators over pairs of u-functions as follows:

$$u_6(z) = u_1(z) \underset{\max}{\otimes} u_1(z) = (0.6z^5 + 0.3z^8 + 0.1z^{12}) \underset{\max}{\otimes} (0.7z^8 + 0.3z^{10})$$

$$=0.42z^{\max\{5,8\}} + 0.21z^{\max\{8,8\}} + 0.07z^{\max\{12,8\}} + 0.18z^{\max\{5,10\}} + 0.09z^{\max\{8,10\}}$$

$$+ 0.03z^{\max\{12,10\}} = 0.63z^8 + 0.27z^{10} + 0.1z^{12}$$

$$u_7(z) = u_3(z) \underset{\min}{\otimes} u_4(z) = (0.6z^0 + 0.4z^2) \underset{\min}{\otimes} (0.1z^0 + 0.5z^3 + 0.4z^5)$$

$$= 0.06z^{\min\{0,0\}} + 0.04z^{\min\{2,0\}} + 0.3z^{\min\{0,3\}} + 0.2z^{\min\{2,3\}}$$

$$+0.24z^{\min\{0,5\}} + 0.16z^{\min\{2,5\}} = 0.64z^0 + 0.36z^2$$

$$u_8(z) = u_6(z) \underset{+}{\otimes} u_7(z) = (0.63z^8 + 0.27z^{10} + 0.1z^{12}) \underset{+}{\otimes} (0.64z^0 + 0.36z^2)$$

$$= 0.4032z^{8+0} + 0.1728z^{10+0} + 0.064z^{12+0} + 0.2268z^{8+2} + 0.0972z^{10+2}$$

$$+ 0.036z^{12+2} = 0.4032z^{8} + 0.3996z^{10} + 0.1612z^{12} + 0.036z^{14}$$

$$U(z) = u_8(z) \underset{\times}{\otimes} u_5(z)$$

$$= (0.4032z^{8} + 0.3996z^{10} + 0.1612z^{12} + 0.036z^{14})(0.5z^{1} + 0.5z^{1.5})$$

$$= 0.2016z^{8\times1} + 0.1998z^{10\times1} + 0.0806z^{12\times1} + 0.018z^{14\times1} + 0.2016z^{8\times1.5}$$

$$+ 0.1998z^{10\times1.5} + 0.0806z^{12\times1.5} + 0.018z^{14\times1.5} = 0.2016z^{8} + 0.1998z^{10}$$

$$+ 0.2822z^{12} + 0.018z^{14} + 0.1998z^{15} + 0.0806z^{18} + 0.018z^{21}$$

The final u-function $U(z)$ represents the p.m.f. of Y, which takes the form

$$\boldsymbol{y} = (8, 10, 12, 14, 15, 18, 21)$$

$$\boldsymbol{q} = (0.2016, 0.1998, 0.2822, 0.018, 0.1998, 0.0806, 0.018)$$

Observe that during the recursive derivation of this p.m.f. we used only 26 term multiplication procedures. This considerable computational complexity reduction is possible because of the like term collection in intermediate u-functions.

The problem of system reliability analysis usually includes evaluation of the p.m.f. of some random values characterizing the system's behaviour. These values can be very complex functions of a large number of random variables. The explicit derivation of such functions is an extremely complicated task. Fortunately, the UGF method for many types of system allows one to obtain the system u-function recursively. This property of the UGF method is based on the associative property of many functions used in reliability engineering. The recursive approach presumes obtaining u-functions of subsystems containing several basic elements and then treating the subsystem as a single element with the u-function obtained when computing the u-function of a higher level subsystem. Combining the recursive approach with the simplification technique reduces the number of terms in the intermediate u-functions and provides a drastic reduction of the computational burden.

1.2.2 Properties of Composition Operators

The properties of composition operator $\otimes_f$ strictly depend on the properties of the function $f(X_1, ..., X_n)$. Since the procedure of the multiplication of the probabilities in this operator is commutative and associative, the entire operator can also possess these properties if the function possesses them.

If

$$f(X_1, X_2, \ldots, X_n) = f(f(X_1, X_2, \ldots, X_{n-1}), X_n), \tag{1.23}$$

then:

$$U(z) = \underset{f}{\otimes}(u_1(z), u_2(z), \ \ldots, \ u_n(z))$$
$$= \underset{f}{\otimes}(\underset{f}{\otimes}(u_1(z), u_2(z), \ \ldots, \ u_{n-1}(z)), \ u_n(z)). \tag{1.24}$$

Therefore, one can obtain the u-function $U(z)$ assigning $U_1(z) = u_1(z)$ and applying operator $\underset{f}{\otimes}$ consecutively:

$$U_j(z) = \underset{f}{\otimes} \ (U_{j-1}(z), u_j(z)) \ \text{ for } 2 \le j \le n, \tag{1.25}$$

such that finally $U(z) = U_n(z)$.

If the function f possesses the associative property

$$f(X_1, \ldots, X_j, X_{j+1}, \ldots, X_n) = f(f(X_1, \ldots, X_j), \ f(X_{j+1}, \ldots, X_n)) \tag{1.26}$$

for any j, then the $\underset{f}{\otimes}$ operator also possesses this property:

$$\underset{f}{\otimes}(u_1(z), \ \ldots, \ u_n(z))$$
$$= \underset{f}{\otimes}(\underset{f}{\otimes}(u_1(z), \ \ldots, \ u_{j-1}(z)), \underset{f}{\otimes}(u_j(z), \ \ldots, \ u_n(z)) \tag{1.27}$$

If, in addition to the property (1.24), the function f is also commutative:

$$f(X_1, \ldots, X_j, X_{j+1}, \ldots, X_n) = f(X_1, \ldots, X_{j+1}, X_j, \ldots, X_n) \tag{1.28}$$

then for any j, which provides the commutative property for the $\underset{f}{\otimes}$ operator:

$$\underset{f}{\otimes}(u_1(z), \ldots, u_j(z), u_{j+1}(z), \ldots, u_n(z))$$
$$= \underset{f}{\otimes}(u_1(z), \ldots, u_{j+1}(z), u_j(z), \ldots, u_n(z)) \tag{1.29}$$

the order of arguments in the function $f(X_1, \ldots, X_n)$ is inessential and the u-function $U(z)$ can be obtained using recursive procedures (1.23) and (1.25) over any permutation of u-functions of random arguments $X_1, \ldots, X_n$.

If a function takes the recursive form

$$f(f_1(X_1, \ldots, X_j), \ f_2(X_{j+1}, \ldots, X_h), \ \ldots, \ f_m(X_l, \ldots, X_n)) \tag{1.30}$$

then the corresponding u-function $U(z)$ can also be obtained recursively:

$$\underset{f}{\otimes}(\underset{f_1}{\otimes}(u_1(z),\ldots,u_j(z)),\underset{f_2}{\otimes}(u_{j+1}(z),\ldots,u_h(z)),\ldots,\underset{f_m}{\otimes}(u_l(z),\ldots,u_n(z)). \quad (1.31)$$

Example 1.8

Consider the variables X_1, X_2, X_3 with p.m.f. presented in Table 1.1. The u-functions of these variables are:

$$u_1(z) = 0.6z^5 + 0.3z^8 + 0.1z^{12}$$

$$u_2(z) = 0.7z^8 + 0.3z^{10}$$

$$u_3(z) = 0.6z^0 + 0.4z^1$$

The function $Y = \min(X_1, X_2, X_3)$ possesses both commutative and associative properties. Therefore

$$\min(\min(X_1, X_2), X_3) = \min(\min(X_2, X_1), X_3) = \min(\min(X_1, X_3), X_2)$$

$$= \min(\min(X_3, X_1), X_2) = \min(\min(X_2, X_3), X_1) = \min(\min(X_3, X_2), X_1).$$

The u-function of Y can be obtained using the recursive procedure

$$u_4(z) = u_1(z)\underset{\min}{\otimes} u_2(z) = (0.6z^5 + 0.3z^8 + 0.1z^{12})\underset{\min}{\otimes}(0.7z^8 + 0.3z^{10})$$

$$= 0.42z^{\min\{5,8\}} + 0.21z^{\min\{8,8\}} + 0.07z^{\min\{12,8\}}$$

$$+ 0.18z^{\min\{5,10\}} + 0.09z^{\min\{8,10\}} + 0.03z^{\min\{12,10\}} = 0.6z^5 + 0.37z^8 + 0.03z^{10}$$

$$U(z) = u_4(z)\underset{\min}{\otimes} u_3(z) = (0.6z^5 + 0.37z^8 + 0.03z^{10})\underset{\min}{\otimes}(0.6z^0 + 0.4z^1)$$

$$= 0.36z^{\min\{5,0\}} + 0.222z^{\min\{8,0\}} + 0.018z^{\min\{12,0\}}$$

$$+ 0.24z^{\min\{5,1\}} + 0.148z^{\min\{8,1\}} + 0.012z^{\min\{12,1\}} = 0.6z^0 + 0.4z^1$$

The same u-function can also be obtained using another recursive procedure

$$u_4(z) = u_1(z)\underset{\min}{\otimes} u_3(z) = (0.6z^5 + 0.3z^8 + 0.1z^{12})\underset{\min}{\otimes}(0.6z^0 + 0.4z^1)$$

$$= 0.36z^{\min\{5,0\}} + 0.18z^{\min\{8,0\}} + 0.06z^{\min\{12,0\}}$$

$$+0.24z^{\min\{5,1\}} + 0.12z^{\min\{8,1\}} + 0.04z^{\min\{12,1\}} = 0.6z^0 + 0.4z^1;$$

$$U(z) = u_3(z)\underset{\min}{\otimes} u_2(z) = (0.6z^0 + 0.4z^1)\underset{\min}{\otimes}(0.7z^8 + 0.3z^{10})$$

$$= 0.42z^{\min\{0,8\}} + 0.28z^{\min\{1,8\}} + 0.18z^{\min\{0,10\}} + 0.12z^{\min\{1,10\}} = 0.6z^0 + 0.4z^1$$

Observe that while both recursive procedures produce the same u-function, their computational complexity differs . In the first case, 12 term multiplication operations have been performed; in the second case, only 10 operations have been performed.

Consider a random variable X with p.m.f. represented by u-function $u_X(z)=\sum_{j=0}^{k} p_j z^{x_j}$. In order to obtain the u-function representing the p.m.f. of function $f(X, c)$ of the variable X and a constant c one can apply the following simplified operator:

$$U(z) = u_X(z) \underset{f}{\otimes} c = (\sum_{j=0}^{k} p_j z^{x_j}) \underset{f}{\otimes} c = \sum_{j=0}^{k} p_j z^{f(x_j, c)} \tag{1.32}$$

This can be easily proved if we represent the constant c as the random variable C that can take the value of c with a probability of 1. The u-function of such a variable takes the form

$$u_c(z) = z^c \tag{1.33}$$

Applying the operator $\otimes_f$ over the two u-functions $u_X(z)$ and $u_c(z)$ we obtain Equation (1.32).

1.3 Introduction to Genetic Algorithms

An abundance of optimization methods have been used to solve various reliability optimization problems. The algorithms applied are either heuristics or exact procedures based mainly on modifications of dynamic programming and nonlinear programming. Most of these methods are strongly problem oriented. This means that, since they are designed for solving certain optimization problems, they cannot be easily adapted for solving other problems. In recent years, many studies on reliability optimization use a universal optimization approach based on metaheuristics. These metaheuristics hardly depend on the specific nature of the problem that is solved and, therefore, can be easily applied to solve a wide range of optimization problems. The metaheuristics are based on artificial reasoning rather than on classical mathematical programming. Their important advantage is that they do not require any information about the objective function besides its values corresponding to the points visited in the solution space. All metaheuristics use the idea of randomness when performing a search, but they also use past knowledge in order to direct the search. Such search algorithms are known as randomized search techniques.

Genetic algorithms (GAs) are one of the most widely used metaheuristics. They were inspired by the optimization procedure that exists in nature, the biological phenomenon of evolution. A GA maintains a population of different solutions allowing them to mate, produce offspring, mutate, and fight for survival. The principle of survival of the fittest ensures the population's drive towards optimization. The GAs have become the popular universal tool for solving various optimization problems, as they have the following advantages:

- they can be easily implemented and adapted;
- they usually converge rapidly on solutions of good quality;
- they can easily handle constrained optimization problems;
- they produce variety of good quality solutions simultaneously, which is important in the decision-making process.

The GA concept was developed by John Holland at the University of Michigan and first described in his book [5]. Holland was impressed by the ease with which biological organisms could perform tasks, which eluded even the most powerful computers. He also noted that very few artificial systems have the most remarkable characteristics of biological systems: robustness and flexibility. Unlike technical systems, biological ones have methods for self-guidance, self-repair and reproducing these features. Holland's biologically inspired approach to optimization is based on the following analogies:

- As in nature, where there are many organisms, there are many possible solutions to a given problem.
- As in nature, where an organism contains many genes defining its properties, each solution is defined by many interacting variables (parameters).
- As in nature, where groups of organisms live together in a population and some organisms in the population are more fit than others, a group of possible solutions can be stored together in computer memory and some of them are closer to the optimum than others.
- As in nature, where organisms that are fitter have more chances of mating and having offspring, solutions that are closer to the optimum can be selected more often to combine their parameters to form new solutions.
- As in nature, where organisms produced by good parents are more likely to be better adapted than the average organism because they received good genes, offspring of good solutions are more likely to be better than a random guess, since they are composed of better parameters.
- As in nature, where survival of the fittest ensures that the successful traits continue to get passed along to subsequent generations, and are refined as the population evolves, the survival-of-the-fittest rule ensures that the composition of the parameters corresponding to the best guesses continually get refined.

GAs maintain a population of individual solutions, each one represented by a finite string of symbols, known as the *genome*, encoding a possible solution within a given problem space. This space, referred to as the *search space*, comprises all of the possible solutions to the problem at hand. Generally speaking, a GA is applied to spaces, which are too large to be searched exhaustively.

GAs exploit the idea of the survival of the fittest and an interbreeding population to create a novel and innovative search strategy. They iteratively create new populations from the old ones by ranking the strings and interbreeding the fittest to create new strings, which are (hopefully) closer to the optimum solution for the problem at hand. In each generation, a GA creates a set of strings from pieces of the previous strings, occasionally adding random new data to keep the population from stagnating. The result is a search strategy that is tailored for vast, complex, multimodal search spaces.

The idea of survival of the fittest is of great importance to genetic algorithms. GAs use what is termed as the fitness function in order to select the fittest string to be used to create new, and conceivably better, populations of strings. The fitness function takes a string and assigns it a relative fitness value. The method by which it does this and the nature of the fitness value do not matter. The only thing that the fitness function must do is rank the strings in some way by producing their fitness values. These values are then used to select the fittest strings.

GAs use the idea of randomness when performing a search. However, it must be clearly understood that the GAs are not simply random search algorithms. Random search algorithms can be inherently inefficient due to the directionless nature of their search. GAs are not directionless. They utilize knowledge from previous generations of strings in order to construct new strings that will approach the optimal solution. GAs are a form of a randomized search, and the way that the strings are chosen and combined comprise a stochastic process.

The essential differences between GAs and other forms of optimization, according to Goldberg [6], are as follows.

GAs usually use a coded form of the solution parameters rather than their actual values. Solution encoding in a form of strings of symbols (an analogy to chromosomes containing genes) provides the possibility of crossover and mutation. The symbolic alphabet that was used was initially binary, due to certain computational advantages purported in [5]. This has been extended in recent years to include character-based encodings, integer and real-valued encodings, and tree representations [7].

GAs do not just use a single point on the problem space, rather they use a set, or population, of points (solutions) to conduct a search. This gives the GAs the power to search noisy spaces littered with local optimum points. Instead of relying on a single point to search through the space, GAs look at many different areas of the problem space at once, and use all of this information as a guide.

GAs use only payoff information to guide themselves through the problem space. Many search techniques need a range of information to guide themselves. For example, gradient methods require derivatives. The only information a GA needs to continue searching for the optimum is some measure of fitness about a point in the space.

GAs are probabilistic in nature, not deterministic. This is a direct result of the randomization techniques used by GAs.

GAs are inherently parallel. Herein lies one of their most powerful features. GAs, by their nature, are very parallel, dealing with a large number of solutions simultaneously. Using schemata theory, Holland has estimated that a GA, processing n strings at each generation, in reality processes n^3 useful substrings [6].

Two of the most common GA implementations are "generational" and "steady state", although recently the steady-state technique has received increased attention [8]. This interest is partly attributed to the fact that steady-state techniques can offer a substantial reduction in the memory requirements of a system: the technique abolishes the need to maintain more than one population during the evolutionary process, which is necessary in the generational GA. In this way, genetic systems have greater portability for a variety of computer environments because of the reduced memory overhead. Another reason for the increased interest in steady-state techniques is that, in many cases, a steady-state GA has been shown to be more effective than a generational GA [9, 10]. This improved performance can be attributed to factors such as the diversity of the population and the immediate availability of superior individuals.

A comprehensive description of a generational GA can be found in [6]. Here, we present the structure of a steady-state GA.

1.3.1 Structure of Steady-state Genetic Algorithms

The steady-state GA (see Figure 1.1) proceeds as follows [11]: an initial population of solutions is generated randomly or heuristically. Within this population, new solutions are obtained during the genetic cycle by using the *crossover* operator. This operator produces an offspring from a randomly selected pair of parent solutions (the parent solutions are selected with a probability proportional to their relative fitness), facilitating the inheritance of some basic properties from the parents to the offspring. The newly obtained offspring undergoes *mutation* with the probability p_{mut}.

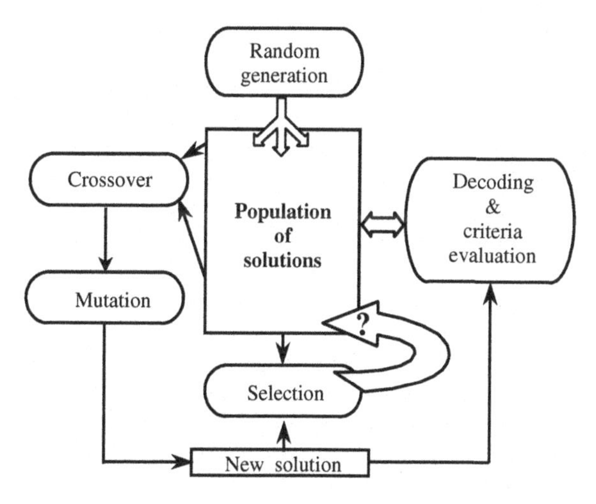

Figure 1.1. Structure of a steady-state GA

Each new solution is decoded and its objective function (fitness) values are estimated. These values, which are a measure of quality, are used to compare different solutions. The comparison is accomplished by a selection procedure that determines which solution is better: the newly obtained solution or the worst solution in the population. The better solution joins the population, while the other is discarded. If the population contains equivalent solutions following selection, then redundancies are eliminated and the population size decreases as a result.

A genetic cycle terminates when N_{rep} new solutions are produced or when the number of solutions in the population reaches a specified level. Then, new randomly constructed solutions are generated to replenish the shrunken population, and a new genetic cycle begins. The whole GA is terminated when its termination condition is satisfied. This condition can be specified in the same way as in a generational GA. The following is the steady-state GA in pseudo-code format.

```
begin STEADY STATE GA
 Initialize population Π
 Evaluate population Π {compute fitness values}
 while GA termination criterion is not satisfied do
       {GENETIC CYCLE}
    while genetic cycle termination criterion is not satisfied do
         Select at random Parent Solutions S1, S2 from Π
         Crossover: (S1, S2) → SO {offspring}
         Mutate offspring SO → S*O with probability pmut
         Evaluate S*O
              Replace SW {the worst solution in Π with S*O } if S*O is
              better than SW
         Eliminate identical solutions in Π
    end while
    Replenish Π with new randomly generated solutions
 end while
end GA
```

Example 1.9

In this example we present several initial stages of a steady-state GA, that maximizes the function of six integer variables $x_1, \ldots, x_6$ taking the form

$$f(x_1,\ldots,x_6)=1000\ [(x_1-3.4)^2+(x_2-1.8)^2+(x_3-7.7)^2$$
$$+(x_4-3.1)^2+(x_5-2.8)^2+(x_6-8.8)^2]^{-1}$$

The variables can take values from 1 to 9. The initial population, consisting of five solutions ordered according to their fitness (value of function f), is:

No.	x_1	x_2	x_3	x_4	x_5	x_6	$f(x_1,\ldots,x_6)$
1	4	2	4	1	2	5	297.8
2	3	7	7	7	2	7	213.8
3	7	5	3	5	3	9	204.2
4	2	7	4	2	1	4	142.5
5	8	2	3	1	1	4	135.2

Using the random generator that produces the numbers of the solutions, the GA chooses the first and third strings, *i.e.* (4 2 4 1 2 5) and (7 5 3 5 3 9) respectively. From these strings, it produces a new one by applying a crossover procedure that takes the three first numbers from the better parent string and the last three numbers from the inferior parent string. The resulting string is (4 2 4 5 3 9). The fitness of this new solution is $f(x_1, \ldots, x_6) = 562.4$. The new solution enters the population, replacing the one with the lowest fitness. The new population is now

No.	x_1	x_2	x_3	x_4	x_5	x_6	$f(x_1,\ldots,x_6)$
1	4	2	4	5	3	9	562.4
2	4	2	4	1	2	5	297.8
3	3	7	7	7	2	7	213.8
4	7	5	3	5	3	9	204.2
5	2	7	4	2	1	4	142.5

Choosing at random the third and fourth strings, (3 7 7 7 2 7) and (7 5 3 5 3 9) respectively, the GA produces the new string (3 7 7 5 3 9) using the crossover operator. This string undergoes a mutation that changes one of its numbers by one (here, the fourth element of the string changes from 5 to 4). The resulting string (3 7 7 4 3 9) has a fitness of $f(x_1, \ldots, x_6) = 349.9$. This solution is better than the inferior one in the population; therefore, the new solution replaces the inferior one. Now the population takes the form

No.	x_1	x_2	x_3	x_4	x_5	x_6	$f(x_1,\ldots,x_6)$
1	4	2	4	5	3	9	562.4
2	3	7	7	4	3	9	349.9
3	4	2	4	1	2	5	297.8
4	3	7	7	7	2	7	213.8
5	7	5	3	5	3	9	204.2

A new solution (4 2 4 4 3 9) is obtained by the crossover operator over the randomly chosen first and second solutions, *i.e.* (4 2 4 5 3 9) and (3 7 7 4 3 9) respectively. After the mutation this solution takes the form (4 2 4 5 3 9) and has the fitness $f(x_1, \ldots, x_6) = 1165.5$. The population obtained after the new solution joins it is

No.	x_1	x_2	x_3	x_4	x_5	x_6	$f(x_1,\ldots,x_6)$
1	4	2	5	4	3	9	1165.5
2	4	2	4	5	3	9	562.4
3	3	7	7	4	3	9	349.9
4	4	2	4	1	2	5	297.8
5	3	7	7	7	2	7	213.8

Note that the mutation procedure is not applied to all the solutions obtained by the crossover. This procedure is used with some prespecified probability p_{mut}. In our example, only the second and the third newly obtained solutions underwent the mutation.

The actual GAs operate with much larger populations and produce thousands of new solutions using the crossover and mutation procedures. The steady-state GA with a population size of 100 obtained the optimal solution for the problem presented after producing about 3000 new solutions. Note that the total number of possible solutions is $9^6 = 531441$. The GA managed to find the optimal solution by exploring less than 0.6% of the entire solution space.

Both types of GA are based on the crossover and mutation procedures, which depend strongly on the solution encoding technique. These procedures should preserve the feasibility of the solutions and provide the inheritance of their essential properties.

1.3.2 Adaptation of Genetic Algorithms to Specific Optimization Problems

There are three basic steps in applying a GA to a specific problem.

In the first step, one defines the solution representation (encoding in a form of a string of symbols) and determines the decoding procedure, which evaluates the fitness of the solution represented by the arbitrary string.

In the second step, one has to adapt the crossover and mutation procedures to the given representation in order to provide feasibility for the new solutions produced by these procedures as well as inheriting the basic properties of the parent solutions by their offspring.

In the third step, one has to choose the basic GA parameters, such as the population size, the mutation probability, the crossover probability (generational GA) or the number of crossovers per genetic cycle (in the steady-state GA), and formulate the termination condition in order to provide the greatest possible GA efficiency (convergence speed).

The strings representing GA solutions are randomly generated by the population generation procedure, modified by the crossover and mutation procedures, and decoded by the fitness evaluation procedure. Therefore, the solution representation in the GA should meet the following requirements:

- It should be easily generated (the sophisticated complex solution generation procedures reduce the GA speed).
- It should be as compact as possible (using very long strings requires excessive computational resources and slows the GA convergence).
- It should be unambiguous (i.e. different solutions should be represented by different strings).
- It should represent feasible solutions (if not any randomly generated string represents a feasible solution, then the feasibility should be provided by simple string transformation).
- It should provide feasibility inheritance of new solutions obtained from feasible ones by the crossover and mutation operators.

The field of reliability optimization includes the problems of finding optimal parameters, optimal allocation and assignment of different elements into a system, and optimal sequencing of the elements. Many of these problems are combinatorial by their nature. The most suitable symbol alphabet for this class of problems is integer numbers. The finite string of integer numbers can be easily generated and stored. The random generator produces integer numbers for each element of the string in a specified range. This range should be the same for each element in order to make the string generation procedure simple and fast. If for some reason different string elements should belong to different ranges, then the string should be transformed to provide solution feasibility.

In the following sections we show how integer strings can be interpreted for solving different kinds of optimization problems.

1.3.2.1 Parameter Determination Problems

When the problem lies in determining a vector of H parameters $(x_1, x_2, ..., x_H)$ that maximizes an objective function $f(x_1, x_2, ..., x_H)$ one always has to specify the ranges of the parameter variation:

$$x_j^{min} \le x_j \le x_j^{max} \text{ for } 1 \le j \le H \tag{1.34}$$

In order to facilitate the search in the solution space determined by inequalities (1.34), integer strings $\boldsymbol{a} = (a_1\ a_2\ ...\ a_H)$ should be generated with elements ranging from 0 to N and the values of parameters should be obtained for each string as

$$x_j = x_j^{min} + a_j(x_j^{max} - x_j^{min})/N. \tag{1.35}$$

Note that the space of the integer strings just approximately maps the space of the real-valued parameters. The number N determines the precision of the search. The search resolution for the jth parameter is $(x_j^{max} - x_j^{min})/N$. Therefore, the increase of N provides a more precise search. On the other hand, the size of the search space of integer strings grows drastically with the increase of N, which slows the GA convergence. A reasonable compromise can be found by using a multistage GA search. In this method, a moderate value of N is chosen and the GA is run to obtain a "crude" solution. Then the ranges of all the parameters are corrected to accomplish the search in a small vicinity of the vector of parameters obtained and the GA is started again. The desired search precision can be obtained by a few iterations.

Example 1.10

Consider a problem in which one has to minimize a function of seven parameters. Assume that following a preliminary decision the ranges of the possible variations of the parameters are different.

Let the random generator provide the generation of integer numbers in the range of 0 -100 (N = 100). The random integer string and the corresponding values of the parameters obtained according to (1.35) are presented in Table 1.3.

Table 1.3. Example of parameters encoding

No. of variable	1	2	3	4	5	6	7
x_j^{min}	0.0	0.0	1.0	1.0	1.0	0.0	0.0
x_j^{max}	3.0	3.0	5.0	5.0	5.0	5.0	5.0
Random integer string	21	4	0	100	72	98	0
Decoded variable	0.63	0.12	1.0	5.0	3.88	4.9	0.0

1.3.2.2 Partition and Allocation Problems

The partition problem can be considered as a problem of allocating Y items belonging to a set Φ in K mutually disjoint subsets Φ_i, *i.e.* such that

$$\bigcup_{i=1}^{K} \Phi_i = \Phi, \quad \Phi_i \bigcap \Phi_j = \varnothing, \; i \neq j \tag{1.36}$$

Each set can contain from 0 to Y items. The partition of the set Φ can be represented by the Y-length string $\boldsymbol{a} = (a_1\ a_2\ \ldots\ a_{Y-1}\ a_Y)$ in which a_j is a number of the set to which item j belongs. Note that, in the strings representing feasible solutions of the partition problem, each element can take a value in the range (1, K).

Now consider a more complicated allocation problem in which the number of items is not specified. Assume that there are H types of different items with an unlimited number of items for each type h. The number of items of each type allocated in each subset can vary. To represent an allocation of the variable number of items in K subsets one can use the following string encoding $\boldsymbol{a} = (a_{11}\ a_{12}\ \ldots a_{1K}\ a_{21}\ a_{22}\ \ldots\ a_{2K} \ldots\ a_{H1}\ a_{H2} \ldots\ a_{HK})$, in which a_{ij} corresponds to the number of items of type i belonging to subset j. Observe that the different subsets can contain identical elements.

Example 1.11

Consider the problem of allocating items of three different types in two disjoint subsets. In this problem, H=3 and K=2. Any possible allocation can be represented by an integer string using the encoding described above. For example, the string (2 1 0 1 1 1) encodes the solution in which two type 1 items are allocated in the first subset and one in the second subset, one item of type 2 is allocated in the second subset, one item of type 3 is allocated in each of the two subsets.

When K = 1, one has an assignment problem in which a number of different items should be chosen from a list containing an unlimited number of items of K different types. Any solution of the assignment problem can be represented by the string $\boldsymbol{a} = (a_1\ a_2\ \ldots\ a_K)$, in which a_j corresponds to the number of chosen items of type j.

The range of variance of string elements for both allocation and assignment problems can be specified based on the preliminary estimation of the characteristics of the optimal solution (maximal possible number of elements of the same type included into the single subset). The greater the range, the greater the solution space to be explored (note that the minimal possible value of the string element is always zero in order to provide the possibility of not choosing any element of the given type to the given subset). In many practical applications, the total number of items belonging to each subset is also limited. In this case, any string representing a solution in which this constraint is not met should be transformed in the following way:

$$a_{ij}^* = \begin{cases} \left\lfloor a_{ij}N_j / \sum_{h=1}^{H} a_{hj} \right\rfloor, & \text{if } N_j < \sum_{h=1}^{H} a_{hj} \\ a_{ij}, & \text{otherwise} \end{cases} \quad \text{for } 1 \le i \le H, 1 \le j \le K \tag{1.37}$$

where N_j is the maximal allowed number of items in subset j.

Example 1.12

Consider the case in which the items of three types should be allocated into two subsets. Assume that it is prohibited to allocate more than five items of each type to the same subset. The GA should produce strings with elements ranging from 0 to 5. An example of such a string is (4 2 5 1 0 2).

Assume that for some reason the total numbers of items in the first and in the second subsets are restricted to seven and six respectively. In order to obtain a feasible solution, one has to apply the transform (1.37) in which $N_1 = 7$, $N_2 = 6$:

$$\sum_{h=1}^{3} a_{h1} = 4+5+0=9, \quad \sum_{h=1}^{3} a_{h2} = 2+1+2=5$$

The string elements take the values

$$a_{11} = \lfloor 4\times7/9 \rfloor = 3,\ a_{21} = \lfloor 5\times7/9 \rfloor = 3,\ a_{31} = \lfloor 0\times7/9 \rfloor = 0$$

$$a_{12} = \lfloor 2\times6/5 \rfloor = 2,\ a_{22} = \lfloor 1\times6/5 \rfloor = 1,\ a_{32} = \lfloor 2\times6/5 \rfloor = 2$$

After the transformation, one obtains the following string: (3 2 3 1 0 2).

When the number of item types and subsets is large, the solution representation described above results in an enormous growth of the length of the string. Besides, to represent a reasonable solution (especially when the number of items belonging to each subset is limited), such a string should contain a large fraction of zeros because only a few items should be included in each subset. This redundancy causes an increase in the need of computational resources and lowers the efficiency of the GA. To reduce the redundancy of the solution representation, each inclusion of m items of type h into subset k is represented by a triplet (m h k). In order to preserve the constant length of the strings, one has to specify in advance a maximal reasonable number of such inclusions I. The string representing up to I inclusions takes the form (m_1 h_1 k_1 m_2 h_2 k_2 ... m_I h_I k_I). The range of string elements should be (0, max{M, H, K}), where M is the maximal possible number of elements of the same type included into a single subset. An arbitrary string generated in this range can still produce infeasible solutions. In order to provide the feasibility, one has to apply the transform $a_j^* = \mathrm{mod}_{x+1}\, a_j$, where x is equal to M, H and K for the string elements corresponding to m, h and k respectively. If one of the elements of the triplet is equal to zero, then this means that no inclusion is made.

For example, the string (3 *1* 2 1 2 *3 2 1* 1 2 2 2 *3 2*) represents the same allocation as string (3 2 3 1 0 2) in Example 1.12. Note that the permutation of triplets, as well as an addition or reduction of triplets containing zeros, does not change the solution. For example, the string (4 0 1 *2 3 2* 2 1 2 *3 1 1* 1 2 2 *3 2 1*) also represents the same allocation as that of the previous string.

1.3.2.3 Mixed Partition and Parameter Determination Problems

Consider a problem in which Y items should be allocated in K subsets and a value of a certain parameter should be assigned to each item. The first option of representing solutions of such a problem in the GA is by using a $2Y$-length string which takes the form $\boldsymbol{a} = (a_{11}\ a_{12}\ a_{21}\ a_{22}\ \dots\ a_{Y1}\ a_{Y2})$. In this string, a_{j1} and a_{j2} correspond respectively to the number of the set the item j belongs to and to the value of the parameter associated with this item. The elements of the string should be generated in the range $(0, \max\{K, N\})$, where N is chosen as described in Section 1.3.2.1. The solution decoding procedure should transform the odd elements of the string as follows:

$$a_{j1}^{*} = 1 + \operatorname{mod}_K a_{j1} \tag{1.38}$$

in order to obtain the class number in the range from 1 to K. The even elements of the string should be transformed as follows:

$$a_{j2}^{*} = \operatorname{mod}_{N+1} a_{j2} \tag{1.39}$$

in order to obtain the parameter value encoded by the integer number in the range from 0 to N. The value of the parameter is then obtained using Equation (1.35).

Example 1.13

Consider a problem in which seven items (N = 7) should be allocated to three separated subsets (K = 3) and a value of a parameter associated with each item should be chosen. The solution should encode both items' distribution among the subsets and the parameters. Let the range of the string elements be (0, 100) (N = 100). The string

(*99* 21 *22* 4 *75* 0 *14* 100 *29* 72 *60* 98 *1* 0)

(in which elements corresponding to the numbers of the subsets are marked in italics) represents the solution presented in Table 1.4. The values corresponding to the numbers of the groups are obtained using Equation (1.38) as

$$a_{11}^{*} = 1 + \operatorname{mod}_K a_{11} = 1 + \operatorname{mod}_3 99 = 1$$

$$a_{21}^{*} = 1 + \operatorname{mod}_K a_{21} = 1 + \operatorname{mod}_3 22 = 2$$

and so on. The numbers that determine the units' weights are obtained using Equation (1.39) as

$$a_{12}^* = \text{mod}_{101}\, a_{12} = \text{mod}_{101}\, 21 = 21$$

$$a_{22}^* = \text{mod}_{101}\, a_{22} = \text{mod}_{101}\, 4 = 4$$

and so on. Observe that, in this solution, items 1, 3, and 6 belong to the first subset, items 2 and 7 belong to the second subset, and items 4 and 5 belong to the third subset. The parameters are identical to those in Example 1.10.

Table 1.4. Example of the solution encoding for the mixed partition and parameter determination problem

No. of unit	1	2	3	4	5	6	7
No. of subset	1	2	1	3	3	1	2
Integer code parameter value	21	4	0	100	72	98	0

This encoding scheme has two disadvantages:

- A large number of different strings can represent an identical solution. Indeed, when K is much smaller than N, many different values of a_{ji} produce the same value of $1+\text{mod}_K\, a_{ji}$ (actually, this transform maps any value $mK+n$ for $n<K$ and $m = 1, 2, \ldots, \lfloor (N-n)/K \rfloor$ into the same number $n+1$). Note for example that the string

(*3* 21 *76* 4 *27* 0 *29* 100 *89* 72 *18* 98 *70*)

represents the same solution as the string presented above. This causes a situation where the GA population is overwhelmed with different strings corresponding to the same solution, which misleads the search process.

- The string is quite long, which slows the GA process and increases need for computational resources.

In order to avoid these problems, another solution representation can be suggested that lies in using a Y-length string in which element a_j represents both the number of the set and the value of the parameter corresponding to item j. To obtain such a compound representation, the string elements should be generated in the range $(0, K(N+1)-1)$. The number of the subset that element j belongs to should be obtained as

$$1+\lfloor a_j/(N+1) \rfloor \quad (1.40)$$

and the number corresponding to the value of jth parameter should be obtained as

$$\text{mod}_{N+1}\, a_j \quad (1.41)$$

Consider the example presented above with $K = 3$ and $N = 100$. The range of the string elements should be (0, 302). The string

(21 105 0 302 274 98 101)

corresponds to the same solution as the strings in the previous example (Table 1.4).

1.3.2.4 Sequencing Problems

The sequencing problem lies in ordering a group of unique items. It can be considered as a special case of the partition problem in which the number of items Y is equal to the number of subsets K and each subset should not be empty. As in the partition problem, the sequences of items can be represented by Y-length strings ($a_1\ a_2\ ...\ a_{Y-1}\ a_Y$) in which a_j is a number of a set to which item j belongs. However, in the case of the sequencing problem, the string representing a feasible solution should be a permutation of Y integer numbers, *i.e.* it should contain all the numbers from 1 to Y and each number in the string should be unique. While the decoding of such strings is very simple (it just explicitly represents the order of item numbers), the generation procedure should be more sophisticated to satisfy the above-mentioned constraints.

The simplest procedure for generating a random string permutation is as follows:

1. Fill the entire string with zeros.
2. For i from 1 to Y in the sequence:
   2.1. Generate a random number j in the range (1, Y).
   2.2. If $a_j = 0$ assign $a_j = i$ or else find the closest zero element to the right of a_j and assign i to this element (treat the string as a circle, *i.e.* consider a_0 to be the closest element to the right of a_Y).

Like the generation procedures for the partition problem, this one also requires the generation of Y random numbers.

1.3.2.5 Determination of Solution Fitness

Having a solution represented in the GA by an integer string $\boldsymbol{a}$ one then has to estimate the quality of this solution (or, in terms of the evolution process, the fitness of the individual). The GA seeks solutions with the greatest possible fitness. Therefore, the fitness should be defined in such a way that its greatest values correspond to the best solutions.

For example, when optimizing the system reliability R (which is a function of some of the parameters represented by $\boldsymbol{a}$) one can define the solution fitness equal to this index, since one wants to maximize it. On the contrary, when minimizing the system cost C, one has to define the solution fitness as $M - C$, where M is a constant number. In this case, the maximal solution fitness corresponds to its minimal cost.

In the majority of optimization problems, the optimal solution should satisfy some constraints. There are three different approaches to handling the constraints in GA [7]. One of these uses penalty functions as an adjustment to the fitness function; two other approaches use "decoder" or "repair" algorithms to avoid building illegal solutions or repair them respectively. The "decoder" and "repair" approaches suffer from the disadvantage of being tailored to the specific problems and thus are not sufficiently general to handle a variety of problems. On the other hand, the penalty approach based on generating potential solutions without considering the constraints and on decreasing the fitness of solutions, violating the constraints, is suitable for problems with a relatively small number of constraints. For heavily constrained problems, the penalty approach causes the GA to spend most of its time evaluating solutions violating the constraints. Fortunately, the reliability optimization problems usually deal with few constraints.

Using the penalty approach one transforms a constrained problem into an unconstrained one by associating a penalty with all constraint violations. The penalty is incorporated into the fitness function. Thus, the original problem of maximizing a function $f(\boldsymbol{a})$ is transformed into the maximization of the function

$$f(\boldsymbol{a}) - \sum_{j=1}^{J} \pi_j \eta_j \tag{1.42}$$

where J is the total number of constraints, π_j is a penalty coefficient related to the jth constraint ($j = 1, \ldots, J$) and η_j is a measure of the constraint violation. Note that the penalty coefficient should be chosen in such a way as to allow the solution with the smallest value of $f(\boldsymbol{a})$ that meets all of the constraints to have a fitness greater than the solution with the greatest value of $f(\boldsymbol{a})$ but violating at least one constraint.

Consider, for example, a typical problem of maximizing the system reliability subject to cost constraint: $R(\boldsymbol{a}) \rightarrow \max$ subject to $C(\boldsymbol{a}) \leq C^*$.

The system cost and reliability are functions of parameters encoded by a string $\boldsymbol{a}$: $C(\boldsymbol{a})$ and $R(\boldsymbol{a})$ respectively. The system cost should not be greater than C^*. The fitness of any solution $\boldsymbol{a}$ can be defined as

$$M+R(\boldsymbol{a})-\pi\eta(C^*, \boldsymbol{a})$$

where (1.43)

$$\eta(C^*, \boldsymbol{a})=(1+C(\boldsymbol{a})-C^*)1(C(\boldsymbol{a})>C^*)$$

The coefficient π should be greater than one. In this case the fitness of any solution violating the constraint is smaller than M (the smallest violation of the constraint $C(\boldsymbol{a}) \leq C^*$ produces a penalty greater than π) while the fitness of any solution meeting the constraint is greater than M. In order to keep the fitness of the solutions positive, one can choose $M > \pi(1+C_{max}-C^*)$, where C_{max} is the maximal possible system cost.

Another typical optimization problem is minimizing the system cost subject to the reliability constraint: $C(\boldsymbol{a}) \rightarrow \min$ subject to $R(\boldsymbol{a}) \geq R^*$.

The fitness of any solution $\boldsymbol{a}$ of this problem can be defined as

$$M-C(\boldsymbol{a})-\pi\eta(R^*,\boldsymbol{a})$$

where (1.44)

$$\eta(A^*, \boldsymbol{a})=(1+R^*-R(\boldsymbol{a}))1(R(\boldsymbol{a})<R^*)$$

The coefficient π should be greater than C_{max}. In this case, the fitness of any solution violating the constraint is smaller than $M - C_{max}$ whereas the fitness of any solution meeting the constraint is greater than $M - C_{max}$. In order to keep the fitness of the solutions positive, one can choose $M > C_{max} + 2\pi$.

1.3.2.6 Basic Genetic Algorithm Procedures and Parameters

The crossover procedures create a new solution as the offspring of a pair of existing ones (parent solutions). The offspring should inherit some useful properties of both parents in order to facilitate their propagation throughout the population. The mutation procedure is applied to the offspring solution. It introduces slight changes into the solution encoding string by modifying some of the string elements. Both of these procedures should be developed in such a way as to provide the feasibility of the offspring solutions given that parent solutions are feasible.

When applied to parameter determination, partition, and assignment problems, the solution feasibility means that the values of all of the string elements belong to a specified range. The most commonly used crossover procedures for these problems generate offspring in which every position is occupied by a corresponding element from one of the parents. This property of the offspring solution provides its feasibility. For example, in the *uniform crossover* each string element is copied either from the first or second parent string with equal probability.

The commonly used mutation procedure changes the value of a randomly selected string element by 1 (increasing or decreasing this value with equal probability). If after the mutation the element is out of the specified range, it takes the minimal or maximal allowed value.

When applied to the sequencing problems, the crossover and mutation operators should produce the offspring that preserve the form of permutations. This means that the offspring string should contain all of the elements that appear in the initial strings and each element should appear in the offspring only once. Any omission or duplication of the element constitutes an error. For example, in the *fragment crossover* operator all of the elements from the first parent string are copied to the same positions of the offspring. Then, all of the elements belonging to a randomly chosen set of adjacent positions in the offspring are reallocated within this set in the order that they appear in the second parent string. It can be seen that this operator provides the feasibility of the permutation solutions.

The widely used mutation procedure that preserves the permutation feasibility swaps two string elements initially located in two randomly chosen positions.

There are no general rules in order to choose the values of basic GA parameters for solving specific optimization problems. The best way to determine the proper combination of these values is by experimental comparison between GAs with different parameters.

A detailed description of a variety of different crossover and mutation operators and recommendations concerning the choice of GA parameters can be found in the GA literature.

2. The Universal Generating Function in Reliability Analysis of Binary Systems

While the most effective applications of the UGF method lie in the field of the MSS reliability, it can also be used for evaluating the reliability of binary systems. The theory of binary systems is well developed. Many algorithms exist for evaluating the reliability of different types of binary system. However, no universal systematic approach has been suggested for the wide range of system types. This chapter demonstrates the ability of the UGF approach to handle the reliability assessment problem for different types of binary system.

Since very effective specialized algorithms were developed for each type of system, the UGF-based procedures may not appear to be very effective in comparison with the best known algorithms (these algorithms can be found in the comprehensive book of Kuo and Zuo [12]). The aim of this chapter is to demonstrate how the UGF technique can be adapted for solving a variety of reliability evaluation problems.

2.1 Basic Notions of Binary System Reliability

System reliability analysis considers the relationship between the functioning of the system's elements and the functioning of the system as a whole. An element is an entity in a system that is not further subdivided. This does not imply that an element cannot be made of parts; rather, it means that, in a given reliability study, it is regarded as a self-contained unit and is not analyzed in terms of the functioning of its constituents.

In binary system reliability analysis it is assumed that each system element, as well as the entire system, can be in one of two possible states, *i.e.* working or failed. Therefore, the state of each element or the system can be represented by a binary random variable such that X_j indicates the state of element j: $X_j = 1$ if element j is in working condition and $X_j = 0$ if element j is failed; X indicates the state of the entire system: $X = 1$ if the system works, $X = 0$ if the system is failed.

The states of all n elements composing the system are represented by the so-called element state vector $(X_1, \ldots, X_n)$. It is assumed that the states of the system elements (the realization of the element state vector) unambiguously determine the state of the system. Thus, the relationship between the element state vector and the system state variable X can be expressed by the deterministic function

$$X = \phi(X_1,..., X_n) \tag{2.1}$$

This function is called the system structure function.

Example 2.1

Consider an air conditioning system that consists of two air conditioners supplied from a single power source. The system fails if neither air conditioner works.

The two air conditioners constitute a subsystem that fails if and only if all of its elements are failed. Such subsystems are called parallel.

Assume that the random binary variables X_1 and X_2 represent the states of the air conditioners and the random binary variable X_c represents the state of the subsystem. The structure function of the subsystem can be expressed as

$$X_c = \phi_{par}(X_1, X_2) = \max(X_1, X_2) = 1-(1-X_1)(1-X_2)$$

The entire system fails either if the power source fails or if the subsystem of the conditioners fails. The system that works if and only if all of its elements work is called a series system. Assume that the random binary variable X_3 represents the state of the power source. The structure function of the entire system takes the form

$$X = \phi_{ser}(X_3, X_c) = \min(X_3, X_c) = X_3 X_c$$

Combining the two expressions one can obtain the structure function of the entire system:

$$\begin{aligned} X &= \phi_{ser}(X_3, X_c) = \phi_{ser}(X_3, \phi_{par}(X_1, X_2)) \\ &= \min(X_3, \max(X_1, X_2)) = X_3(1-(1-X_1)(1-X_2)) \end{aligned}$$

In order to represent the nature of the relationship among the elements in the system, reliability block diagrams are usually used. The reliability block diagram of the system considered is presented in Figure 2.1.

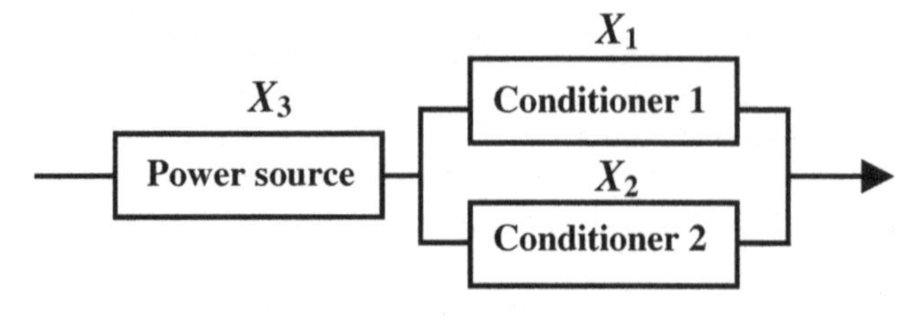

Figure 2.1. Reliability block diagram of an air conditioning system

The reliability is a property of any element or the entire system to be able to perform its intended task. Since we represent the system state by random binary variable X and $X=1$ corresponds to the system state in which it performs its task, the measures of the system reliability should express its ability to be in the state $X=1$. Different reliability measures can be defined in accordance with the conditions of the system's functioning.

When the system has a fixed mission time (for example, a satellite that should supply telemetric information during the entire preliminarily defined time of its mission), the reliability of such a system (and its elements) is defined to be the probability that it will perform its task during the mission time under specified working conditions. For any system element j its reliability p_j is

$$p_j = \Pr\{X_j = 1\} \tag{2.2}$$

and for the entire system its reliability R is

$$R = \Pr\{X = 1\} \tag{2.3}$$

Observe that the reliability can be expressed as the expected value of the state variable:

$$p_j = E(X_j), \quad R = E(X) \tag{2.4}$$

The reliabilities of the elements compose the element reliability vector $\boldsymbol{p} = (p_1, \ ..., \ p_n)$. Usually this vector is known and we are interested in obtaining the system reliability as a function of $\boldsymbol{p}$:

$$R = R(\boldsymbol{p}) = R(p_1, ..., p_n) \tag{2.5}$$

In systems with independent elements, such functions exist and depend on the system structure functions.

Example 2.2

Consider the system from Example 2.1 and assume that the reliabilities of the system elements p_1, p_2 and p_3 are known. Since the elements are independent, we can obtain the probability of each realization of the element state vector $(X_1, \ X_2, \ X_3) = (x_1, \ x_2, \ x_3)$ as

$$\Pr\{X_1 = x_1 \bigcap X_2 = x_2 \bigcap X_3 = x_3\}$$
$$= p_1^{x_1}(1-p_1)^{1-x_1} \, p_2^{x_2}(1-p_2)^{1-x_2} \, p_3^{x_3}(1-p_3)^{1-x_3}$$

Having the system structure function

$$X = \min(X_3,\ \max(X_1,\ X_2))$$

and the probability of each realization of the element state vector, we can obtain the probabilities of each system state that defines the p.m.f. of the system state variable X. This p.m.f. is presented in Table 2.1.

Table 2.1. p.m.f. of the system structure function

Realization of (X_1, X_2, X_3)	Realization probability	Realization of X
0,0,0	$(1-p_1)(1-p_2)(1-p_3)$	0
0,0,1	$(1-p_1)(1-p_2)p_3$	0
0,1,0	$(1-p_1)p_2(1-p_3)$	0
0,1,1	$(1-p_1)p_2p_3$	1
1,0,0	$p_1(1-p_2)(1-p_3)$	0
1,0,1	$p_1(1-p_2)p_3$	1
1,1,0	$p_1p_2(1-p_3)$	0
1,1,1	$p_1p_2p_3$	1

The system reliability can now be defined as the expected value of the random variable X (which is equal to the sum of the probabilities of states corresponding to $X = 1$):

$$R = E(X) = (1-p_1)p_2p_3+p_1(1-p_2)p_3+p_1p_2p_3$$

$$= [(1-p_1)p_2+p_1(1-p_2)+p_1p_2]\ p_3 = (p_1+p_2-p_1p_2)p_3$$

When the system operates for a long time and no finite mission time is specified, we need to know how the system's ability to perform its task changes over time. In this case, a dynamic measure called the reliability function is used. The reliability function of element j $p_j(t)$ or the entire system $R(t)$ is defined as the probability that the element (system) will perform its task beyond time t, while assuming that at the beginning of the mission the element (system) is in working condition: $p_j(0) = R(0) = 1$.

Having the reliability functions of independent system elements $p_j(t)$ $(1\le j\le n)$ one can obtain the system reliability function $R(t)$ using the same relationship $R(\boldsymbol{p})$ that was defined for the fixed mission time and by substituting p_j with $p_j(t)$.

Example 2.3

Consider the system from Example 2.2 and assume that the reliability functions of the system elements are

$$p_1(t) = e^{-\lambda_1 t}, \quad p_2(t) = e^{-\lambda_2 t}, \quad p_3(t) = e^{-\lambda_3 t}$$

The system reliability function takes the form

$$R(t)=E(X(t))=(p_1(t)+p_2(t)-p_1(t)p_2(t))p_3(t)$$

$$=(\mathrm{e}^{-\lambda_1 t}+\mathrm{e}^{-\lambda_2 t}-\mathrm{e}^{-(\lambda_1+\lambda_2)t})\mathrm{e}^{-\lambda_3 t}$$

In many practical cases the failed system elements can be repaired. While the failures bring the elements to a non-working state, repairs performed on them bring them back to a working state. Therefore, the state of each element and the state of the entire system can change between 0 and 1 several times during the system's mission. The probability that the element (system) is able to perform its task at a given time t is called the element (system) availability function:

$$a_j(t)=\Pr\{X_j=1\} \tag{2.6}$$

$$A(t)=\Pr\{X=1\}$$

For a repairable system, $X_j = 1$ ($X = 1$) indicates that the element (system) can perform its task at time t regardless of the states experienced before time t.

While the reliability reflects the internal properties of the element (system), the availability reflects both the ability of the element (system) to work without failures and the ability of the system's environment to bring the failed element (system) to a working condition. The same system working in a different maintenance environment has a different availability.

As a rule, the availability function is difficult to obtain. Instead, the steady-state system availability is usually used. It is assumed that enough time has passed since the beginning of the system operation so that the system's initial state has practically no influence on its availability and the availabilities of the system elements become constant

$$a_j=\lim_{t\to\infty} a_j(t) \tag{2.7}$$

Having the long-run average (steady-state) availabilities of the system elements one can obtain the steady-state availability A of the system by substituting in Equation (2.5) R with A and p_j with a_j.

One can see that since all of the reliability measures presented are probabilities, the same procedure of obtaining the system reliability measure from the reliability measures of its elements can be used in all cases. This procedure presumes:

- Obtaining the probabilities of each combination of element states from the element reliability vector.
- Obtaining the system state (the value of the system state variable) for each combination of element states (the realization of the element state vector) using the system structure function.

- Calculating the expected value of the system state variable from its p.m.f. defined by the element state combination probabilities and the corresponding values of the structure function.

This procedure can be formalized by using the UGF technique. In fact, the element reliability vector (p_1, …, p_n) determines the p.m.f. of each binary element that can be represented in the form of u-functions

$$u_j(z) = (1-p_j)\, z^0 + p_j z^1 \text{ for } 1 \le j \le n \tag{2.8}$$

Having the u-functions of system elements that represent the p.m.f. of discrete random variables $X_1, \ldots, X_n$ and having the system structure function $X = \phi(X_1, \ldots, X_n)$ we can obtain the u-function representing the p.m.f. of the system state variable X using the composition operator over u-functions of individual system elements:

$$U(z) = \underset{\phi}{\otimes}(u_1(z),\ldots,u_n(z)) \tag{2.9}$$

The system reliability measure can now be obtained as $E(X) = U'(1)$.

Note that the same procedure can be applied for any reliability measure considered. The system reliability measure (the fixed mission time reliability, the value of the reliability function at a specified time or availability) corresponds to the reliability measures used to express the state probabilities of elements. Therefore, we use the term reliability and presume that any reliability measure can be considered in its place (if some specific measure is not explicitly specified).

Example 2.4

The u-functions of the system elements from Example 2.2 are

$$u_1(z) = (1-p_1)z^0 + p_1 z^1, \quad u_2(z) = (1-p_2)z^0 + p_2 z^1, \quad u_3(z) = (1-p_3)z^0 + p_3 z^1$$

The system structure function is

$$X = \phi(X_1, X_2, X_3) = \min(X_3, \max(X_1, X_2))$$

Using the composition operator we obtain the system u-function representing the p.m.f. of the random variable X:

$$U(z) = \underset{\phi}{\otimes}(u_1(z), u_2(z), u_3(z))$$
$$= \underset{\phi}{\otimes}\left(\sum_{i=0}^{1} p_1^i (1-p_1)^{1-i} z^i, \sum_{k=0}^{1} p_2^k (1-p_2)^{1-k} z^k, \sum_{m=0}^{1} p_3^m (1-p_3)^{1-m} z^m\right)$$
$$= \sum_{i=0}^{1} \sum_{k=0}^{1} \sum_{m=0}^{1} p_1^i (1-p_1)^{1-i} p_2^k (1-p_2)^{1-k} p_3^m (1-p_3)^{1-m} z^{\min(\max(i,k),m)}$$

The resulting u-function takes the form

$$U(z) = (1-p_1)(1-p_2)(1-p_3)z^{\min(\max(0,0),0)}$$
$$+ (1-p_1)(1-p_2)p_3 z^{\min(\max(0,0),1)} + (1-p_1)p_2(1-p_3)z^{\min(\max(0,1),0)}$$
$$+ (1-p_1)p_2 p_3 z^{\min(\max(0,1),1)} + p_1(1-p_2)(1-p_3)z^{\min(\max(1,0),0)}$$
$$+ p_1(1-p_2)p_3 z^{\min(\max(1,0),1)} + p_1 p_2 (1-p_3) z^{\min(\max(1,1),0)}$$
$$+ p_1 p_2 p_3 z^{\min(\max(1,1),1)}$$
$$= (1-p_1)(1-p_2)(1-p_3)z^0 + (1-p_1)(1-p_2)p_3 z^0$$
$$+ (1-p_1)p_2(1-p_3)z^0 + (1-p_1)p_2 p_3 z^1 + p_1(1-p_2)(1-p_3)z^0$$
$$+ p_1(1-p_2)p_3 z^1 + p_1 p_2(1-p_3)z^0 + p_1 p_2 p_3 z^1$$

After collecting the like terms we obtain

$$U(z) = [(1-p_1)(1-p_2)(1-p_3) + (1-p_1)(1-p_2)p_3$$
$$+ (1-p_1)p_2(1-p_3) + p_1(1-p_2)(1-p_3) + p_1 p_2(1-p_3)]z^0$$
$$+ [p_1(1-p_2)p_3 + (1-p_1)p_2 p_3 + p_1 p_2 p_3]z^1$$

The system reliability is equal to the expected value of variable X that has the p.m.f. represented by the u-function $U(z)$. As we know, this expected value can be obtained as the derivative of $U(z)$ at $z = 1$:

$$R = E(X) = U'(1) = p_1(1-p_2)p_3 + (1-p_1)p_2 p_3 + p_1 p_2 p_3$$
$$= (p_1 + p_2 - p_1 p_2)p_3$$

It can easily be seen that the total number of combinations of states of the elements in the system with n elements is equal to 2^n. For systems with a great number of elements, the technique presented is associated with an enormous number of evaluations of the structure function value (the u-function of the system state variable X before the like term collection contains 2^n terms). Fortunately, the structure function can usually be defined recursively and the p.m.f. of intermediate variables corresponding to some subsystems can be obtained. These p.m.f. always consist of two terms. Substituting all the combinations of the elements composing the subsystem with its two-term p.m.f. (obtained by collecting the like terms in the u-function corresponding to the subsystem) allows one to achieve considerable reduction of the computational burden.

Example 2.5

Consider a series-parallel system consisting of five binary elements (Figure 2.2). The structure function of this system is

$$X = \phi(X_1, X_2, X_3, X_4, X_5) = \max(\max(X_1, X_2)\, X_3,\ X_4\, X_5)$$

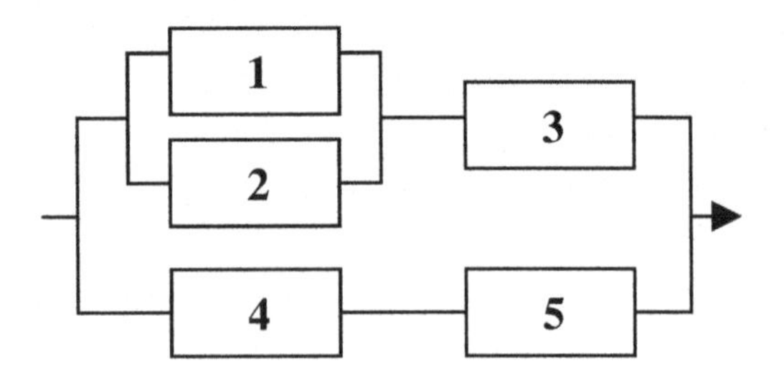

Figure 2.2. Reliability block diagram of series-parallel binary system

The u-functions of the elements take the form

$$u_j(z) = (1-p_j)\, z^0 + p_j z^1, \quad \text{for} \quad 1 \le j \le 5$$

Direct application of the operator $\underset{\phi}{\otimes}(u_1(z), u_2(z), u_3(z), u_4(z), u_5(z))$ requires $2^5 = 32$ evaluations of the system structure function.

The system structure function can be defined recursively:

$$X_6 = \max(X_1, X_2)$$

$$X_7 = X_6\, X_3$$

$$X_8 = X_4\, X_5$$

$$X = \max(X_7, X_8)$$

where X_6 is the state variable corresponding to the subsystem consisting of elements 1 and 2, X_7 is the state variable corresponding to the subsystem consisting of elements 1, 2 and 3, X_8 is the state variable corresponding to the subsystem consisting of elements 4 and 5.

The u-functions corresponding to variables X_6, X_7 and X_8 consist of two terms (after collecting the like terms) as well as u-functions corresponding to variables X_1, …, X_5. The number of evaluations of the structure functions representing the p.m.f. of variables X_6, X_7, X_8, and X is four. Therefore, the total number of such evaluations is 16. Note that the structure functions evaluated for the intermediate variables are much simpler than the structure function of the entire system that must be evaluated when applying the direct approach.

The process of obtaining the system reliability using the recursive approach is as follows:

$$U_6(z) = u_1(z) \underset{\max}{\otimes} u_2(z) = [p_1 z^1 + (1-p_1)z^0] \underset{\max}{\otimes} [p_2 z^1 + (1-p_2)z^0]$$
$$= p_1 p_2 z^{\max(1,1)} + p_1(1-p_2)z^{\max(1,0)} + (1-p_1)p_2 z^{\max(0,1)}$$
$$+ (1-p_1)(1-p_2)z^{\max(0,0)} = p_1 p_2 z^1 + p_1(1-p_2)z^1 + (1-p_1)p_2 z^1$$
$$+ (1-p_1)(1-p_2)z^0 = (p_1 + p_2 - p_1 p_2)z^1 + (1-p_1)(1-p_2)z^0$$

$$U_7(z) = U_6(z) \underset{\times}{\otimes} u_3(z) = [(p_1 + p_2 - p_1 p_2)z^1$$
$$+ (1-p_1)(1-p_2)z^0] \underset{\times}{\otimes} [p_3 z^1 + (1-p_3)z^0] = (p_1 + p_2 - p_1 p_2)p_3 z^{1\times 1}$$
$$+ (1-p_1)(1-p_2)p_3 z^{0\times 1} + (p_1 + p_2 - p_1 p_2)(1-p_3)z^{1\times 0}$$
$$+ (1-p_1)(1-p_2)(1-p_3)z^{0\times 0} = (p_1 + p_2 - p_1 p_2)p_3 z^1$$
$$+ [(1-p_1)(1-p_2) + (p_1 + p_2 - p_1 p_2)(1-p_3)]z^0$$

$$U_8(z) = u_4(z) \underset{\times}{\otimes} u_5(z) = [p_4 z^1 + (1-p_4)z^0] \underset{\times}{\otimes} [p_5 z^1 + (1-p_5)z^0]$$
$$= p_4 p_5 z^{1\times 1} + p_4(1-p_5)z^{1\times 0} + (1-p_4)p_5 z^{0\times 1} + (1-p_4)(1-p_5)z^{0\times 0}$$
$$= p_4 p_5 z^1 + (1 - p_4 p_5)z^0$$

$$U(z) = U_7(z) \underset{\max}{\otimes} U_8(z) = \{(p_1 + p_2 - p_1 p_2)p_3 z^1 + [(1-p_1)(1-p_2)$$
$$+ (p_1 + p_2 - p_1 p_2)(1-p_3)]z^0\} \underset{\max}{\otimes} [p_4 p_5 z^1 + (1 - p_4 p_5)z^0]$$

$$= (p_1 + p_2 - p_1 p_2) p_3 p_4 p_5 z^{\max(1,1)}$$
$$+ (p_1 + p_2 - p_1 p_2) p_3 (1 - p_4 p_5) z^{\max(1,0)}$$
$$+ [(1 - p_1)(1 - p_2) + (p_1 + p_2 - p_1 p_2)(1 - p_3)] p_4 p_5 z^{\max(0,1)}$$
$$+ [(1 - p_1)(1 - p_2) + (p_1 + p_2 - p_1 p_2)(1 - p_3)](1 - p_4 p_5) z^{\max(0,0)}$$
$$= \{(p_1 + p_2 - p_1 p_2) p_3 + [(1 - p_1)(1 - p_2)$$
$$+ (p_1 + p_2 - p_1 p_2)(1 - p_3)] p_4 p_5\} z^1$$
$$+ [(1 - p_1)(1 - p_2) + (p_1 + p_2 - p_1 p_2)(1 - p_3)](1 - p_4 p_5) z^0$$

The system reliability (availability) can now be obtained as

$$R = E(X) = U'(1) = (p_1 + p_2 - p_1 p_2) p_3 + [(1 - p_1)(1 - p_2)$$
$$+ (p_1 + p_2 - p_1 p_2)(1 - p_3)] p_4 p_5$$

In order to reduce the number of arithmetical operations in the term multiplication procedures performed when obtaining the u-functions of the system variable, the u-function of the binary elements that takes the form

$$u_j(z) = p_j z^1 + (1 - p_j) z^0 \quad (2.10)$$

can be represented in the form

$$u_j(z) = p_j (z^1 + q_j z^0) \quad (2.11)$$

where

$$q_j = p_j^{-1} - 1 \quad (2.12)$$

Factoring out the probability p_j from $u_j(z)$ results in fewer computations associated with performing the operators $U(z) \underset{\phi}{\otimes} u_j(z)$ for any $U(z)$ because the multiplications by 1 are implicit.

Example 2.6

In this example we obtain the reliability of the series-parallel system from Example 2.5 numerically for $p_1 = 0.8,\ p_2 = 0.9,\ p_3 = 0.7,\ p_4 = 0.9,\ p_5 = 0.7$.

The u-functions of the elements take the form

$$u_1(z) = 0.8z^1 + 0.2z^0,\ u_2(z) = u_4(z) = 0.9z^1 + 0.1z^0,\ u_3(z) = u_5(z) = 0.7z^1 + 0.3z^0$$

Following the procedure presented in Example 2.5 we obtain:

$$U_6(z) = u_1(z) \underset{\max}{\otimes} u_2(z) = (0.8z^1 + 0.2z^0) \underset{\max}{\otimes} (0.9z^1 + 0.1z^0)$$

$$= 0.8 \cdot 0.9z^1 + 0.8 \cdot 0.1z^1 + 0.2 \cdot 0.9z^1 + 0.2 \cdot 0.1z^0 = 0.98z^1 + 0.02z^0$$

$$U_7(z) = U_6(z) \underset{\times}{\otimes} u_3(z) = (0.98z^1 + 0.02z^0) \underset{\times}{\otimes} (0.7z^1 + 0.3z^0)$$

$$= 0.98 \times 0.7z^1 + 0.98 \times 0.3z^0 + 0.02 \times 0.7z^0 + 0.02 \times 0.3z^0$$
$$= 0.686z^1 + 0.314z^0$$

$$U_8(z) = u_4(z) \underset{\times}{\otimes} u_5(z) = (0.9z^1 + 0.1z^0) \underset{\times}{\otimes} (0.7z^1 + 0.3z^0)$$

$$= 0.9 \times 0.7z^1 + 0.9 \times 0.3z^0 + 0.1 \times 0.7z^0 + 0.1 \times 0.3z^0 = 0.63z^1 + 0.37z^0$$

$$U(z) = U_7(z) \underset{\max}{\otimes} U_8(z) = (0.686z^1 + 0.314z^0) \underset{\max}{\otimes} (0.63z^1 + 0.37z^0)$$

$$= 0.686 \times 0.63z^1 + 0.686 \times 0.37z^1 + 0.314 \times 0.63z^1 + 0.314 \times 0.37z^0$$

$$= 0.88382z^1 + 0.11618z^0$$

And, finally:

$$R = U'(1) = 0.88382 \approx 0.884$$

Representing the u-functions of the elements in the form

$$u_1(z) = 0.8(z^1 + 0.25)z^0,\ u_2(z) = u_4(z) = 0.9(z^1 + 0.111)z^0$$

$$u_3(z) = u_5(z) = 0.7(z^1 + 0.429)z^0$$

we can obtain the same result by fewer calculations:

$$U_6(z) = u_1(z) \underset{\max}{\otimes} u_2(z) = 0.8(z^1 + 0.25z^0) \underset{\max}{\otimes} 0.9(z^1 + 0.111z^0)$$

$$= 0.8 \cdot 0.9(z^1 + 0.111z^1 + 0.25z^1 + 0.25 \times 0.111z^0)$$
$$= 0.72(1.361z^1 + 0.0278)z^0$$

$$U_7(z) = U_6(z) \underset{\times}{\otimes} u_3(z) = 0.72(1.361z^1 + 0.028z^0) \underset{\times}{\otimes} 0.7(z^1 + 0.429z^0)$$

$$= 0.72 \times 0.7(1.361z^1 + 0.028z^0 + 1.361 \times 0.429z^0 + 0.028 \times 0.429z^0)$$
$$= 0.504(1.361z^1 + 0.623z^0)$$

$$U_8(z) = u_4(z) \underset{\times}{\otimes} u_5(z) = 0.9(z^1 + 0.111z^0) \underset{\times}{\otimes} 0.7(z^1 + 0.429z^0)$$

$$= 0.9 \times 0.7(z^1 + 0.111z^0 + 0.429z^0 + 0.111 \times 0.429z^0) = 0.63(z^1 + 0.588z^0)$$

$$U(z) = U_7(z) \underset{\max}{\otimes} U_8(z) = 0.504(1.361z^1 + 0.623z^0) \underset{\max}{\otimes} 0.63(z^1 + 0.588z^0)$$

$$= 0.504 \times 0.63(1.361z^1 + 0.623z^1 + 1.361 \times 0.588z^1 + 0.623 \times 0.588z^0)$$

$$= 0.3175(2.784z^1 + 0.366z^0)$$

$$R = U'(1) = 0.3175 \cdot 2.784 \approx 0.884$$

The simplification method presented is efficient in numerical procedures. In future examples we do not use it in order to preserve their clarity.

There are many cases where estimating the structure function of the binary system is a very complicated task. In some of these cases the structure function and the system reliability can be obtained recursively, as in the case of the complex series-parallel systems. The following sections of this chapter are devoted to such cases.

2.2 *k*-out-of-*n* Systems

Consider a system consisting of n independent binary elements that can perform its task (is "good") if and only if at least k of its elements are in working condition. This type of system is called a k-out-of-n:G system. The system that fails to perform its task if and only if at least k of its elements fail is called a k-out-of-n:F system. It can be seen that a k-out-of-n:G system is equivalent to an $(n-k+1)$-out-of-n:F system. Therefore, we consider only k-out-of-n:G systems and omit G from their denomination.

The pure series and pure parallel systems can be considered to be special cases of k-out-of-n systems. Indeed, the series system works if and only if all of its elements work. This corresponds to an n-out-of-n system. The parallel system works if and only if at least one of its elements works, which corresponds to a 1-out-of-n system.

The k-out-of-n systems are widely used in different technical applications. For example, an airplane survives if no more than two of its four engines are destroyed. The power generation system can meet its demand when at least three out of five of its generators function.

Consider the k-out-of-n system consisting of identical elements with reliability p. It can be seen that the number of working elements in the system follows the binomial distribution: the probability R_j that exactly j out of n elements work ($1 \le j \le n$) takes the following form:

$$R_j = \binom{n}{j} p^j (1-p)^{n-j} \tag{2.13}$$

Since the system reliability is equal to the probability that the number of working elements is not less than k, the overall system reliability can be found as

$$R = \sum_{j=k}^{n} R_j = \sum_{j=k}^{n} \binom{n}{j} p^j (1-p)^{n-j} \tag{2.14}$$

Using this equation one can readily obtain the reliability of the k-out-of-n system with independent identical binary elements. When the elements are not identical (have different reliabilities) the evaluation of the system reliability is a more complicated problem. The structure function of the system takes the form

$$\phi(X_1, \ ..., \ X_n) = 1(\sum_{i=1}^{n} X_i \geq k) \tag{2.15}$$

In order to obtain the probability R_j that exactly j out of n elements work ($1 \leq j \leq n$), one has to sum up the probabilities of all of the possible realizations of the element state vector $(X_1, \ ..., \ X_n)$ in which j state variables exactly take on the value of 1. Observe that, in such realizations, number i_1 of the first variable X_{i_1} from the vector that should be equal to 1 can vary from 1 to $n-j+1$. Indeed, if $X_{i_1} = 0$ for $1 \leq i_1 \leq n-j+1$, then the maximal number of variables taking a value of 1 is not greater than $j-1$. Using the same consideration, we can see that if the number of the first variable that is equal to 1 is i_1, the number of the second variable taking this value can vary from i_1+1 to $n-j+2$ and so on. Taking into account that $\Pr\{X_i = 1\} = p_i$ and $\Pr\{X_i = 0\} = 1 - p_i$ for any i: $1 \leq i \leq n$, we can obtain

$$R_j = \left[\prod_{i=1}^{n}(1-p_i)\right]\left[\sum_{i_1=1}^{n-j+1} \frac{p_{i_1}}{1-p_{i_1}} \sum_{i_2=i_1+1}^{n-j+2} \frac{p_{i_2}}{1-p_{i_2}} \cdots \sum_{i_j=i_{j-1}+1}^{n} \frac{p_{i_j}}{1-p_{i_j}}\right] \tag{2.16}$$

The reliability of the system is equal to the probability that j is greater than or equal to k. Therefore:

$$R = \sum_{j=k}^{n} R_j$$

$$= \left[\prod_{i=1}^{n}(1-p_i)\right] \sum_{j=k}^{n} \left[\sum_{i_1=1}^{n-j+1} \frac{p_{i_1}}{1-p_{i_1}} \sum_{i_2=i_1+1}^{n-j+2} \frac{p_{i_2}}{1-p_{i_2}} \cdots \sum_{i_j=i_{j-1}+1}^{n} \frac{p_{i_j}}{1-p_{i_j}}\right] \tag{2.17}$$

The computation of the system reliability based on this equation is very complicated. The UGF approach provides for a straightforward method of k-out-of-n system reliability computation that considerably reduces the computational complexity. The basics of this method were mentioned in the early Reliability Handbook by Kozlov and Ushakov [13]; the efficient algorithm was suggested by Barlow and Heidtmann [14].

Since the p.m.f. of each element state variable X_j can be represented by the u-function

$$u_j(z) = p_j z^1 + (1 - p_j) z^0 \tag{2.18}$$

the operator

$$U(z) = \underset{+}{\otimes}(u_1(z), \ldots, u_n(z)) \tag{2.19}$$

gives the distribution of the random variable X:

$$X = \sum_{i=1}^{n} X_i \tag{2.20}$$

which is equal to the total number of working elements in the system.

The resulting u-function representing the p.m.f. of the variable X takes the form

$$U(z) = \sum_{j=0}^{n} R_j z^j \tag{2.21}$$

where $R_j = \Pr\{X = j\}$ is the probability that exactly j elements work. By summing the coefficients of the u-function $U(z)$ corresponding to $k \le j \le n$, we obtain the system reliability.

Taking into account that the operator $\otimes_+$ possesses the associative property (Equation (1.27)) and using the structure function formalism we can define the following procedure that obtains the reliability of a k-out-of-n system:

1. Determine u-functions of each element in the form (2.18).
2. Assign $U_1(z) = u_1(z)$.
3. For j = 2, …, n obtain $U_j(z) = U_{j-1}(z) \otimes_+ u_j(z)$ (the final u-function $U_n(z)$ represents the p.m.f. of random variable X).
4. Obtain u-function $U(z)$ representing the p.m.f. of structure function (2.15) as $U(z) = U_n(z) \underset{\varphi}{\otimes} k$, where $\varphi(X,k) = 1(X \ge k)$.
5. Obtain the system reliability as $E(\varphi(X,k)) = U'(1)$.

Example 2.7

Consider a 2-out-of-4 system consisting of elements with reliabilities $p_1 = 0.8$, $p_2 = 0.6$, $p_3 = 0.9$, and $p_4 = 0.7$.

First, determine the u-functions of the elements:

$$u_1(z) = 0.8z^1 + 0.2z^0$$
$$u_2(z) = 0.6z^1 + 0.4z^0$$
$$u_3(z) = 0.9z^1 + 0.1z^0$$
$$u_4(z) = 0.7z^1 + 0.3z^0$$

Follow step 2 and assign

$$U_1(z) = u_1(z) = 0.8z^1 + 0.2z^0$$

Using the recursive equation (step 3 of the procedure) obtain

$$U_2(z) = (0.8z^1 + 0.2z^0)\underset{+}{\otimes}(0.6z^1 + 0.4z^0)$$
$$= (0.8z^1 + 0.2z^0)(0.6z^1 + 0.4z^0) = 10^{-2}(48z^2 + 44z^1 + 8z^0)$$

$$U_3(z) = 10^{-2}(48z^2 + 44z^1 + 8z^0)\underset{+}{\otimes}(0.9z^1 + 0.1z^0)$$
$$= 10^{-3}(48z^2 + 44z^1 + 8z^0)(9z^1 + 1z^0)$$
$$= 10^{-3}(432z^3 + 444z^2 + 116z^1 + 8z^0)$$

$$U_4(z) = 10^{-3}(432z^3 + 444z^2 + 116z^1 + 8z^0)\underset{+}{\otimes}(0.7z^1 + 0.3z^0)$$
$$= 10^{-4}(432z^3 + 444z^2 + 116z^1 + 8z^0)(7z^1 + 3z^0)$$
$$= 10^{-4}(3024z^4 + 4404z^3 + 2144z^2 + 404z^1 + 24z^0)$$

Following step 4 obtain

$$U(z) = U_4(z)\underset{\varphi}{\otimes}2 = 10^{-4}(3024z^1 + 4404z^1 + 2144z^1 + 404z^0 + 24z^0)$$
$$= 0.9572z^1 + 0.0428z^0$$

The system reliability can now be obtained as

$$U'(1) = 0.9572$$

Note that the UGF method requires less computational effort than simple enumeration of possible combinations of states of the elements. In order to obtain $U_2(z)$ we used four term multiplication operations. In order to obtain $U_3(z)$, six operations were used (because $U_2(z)$ has only three different terms after collecting the like terms). In order to obtain $U_4(z)$, eight operations were used (because $U_3(z)$ has only four different terms after collecting the like terms). The total number of the term multiplication operations used in the example is 18.

When the enumerative approach is used, one has to evaluate the probabilities of $2^4 = 16$ combinations of the states of the elements. For each combination the product of four element state probabilities should be obtained. This requires three multiplication operations. The total number of the multiplication operations is $16\times3 = 48$. The difference in the computational burden increases with the growth of n.

The computational complexity of this algorithm can be further reduced in its modification that avoids calculating the probabilities R_j. Note that it does not matter for the k-out-of-n system how many elements work if the number of the working elements is not less than k. Therefore, we can introduce the intermediate variable X^*:

$$X^* = \min\{k, \sum_{i=1}^{n} X_i\} \tag{2.22}$$

and define the system structure function as

$$\phi(X_1,..., X_n) = \eta(X^*,k) = 1(X^* = k) \tag{2.23}$$

In order to obtain the u-function of the variable X^* we introduce the following composition operator:

$$U(z) = \underset{\theta_k}{\otimes}(u_1(z),...,u_n(z)) \tag{2.24}$$

where

$$\theta_k(x_1,..., x_n) = \min\{k, \sum_{i=1}^{n} x_i\} \tag{2.25}$$

It can be easily seen that this operator possesses the associative and commutative properties and, therefore, the u-function of X^* can be obtained recursively:

$$U_1(z) = u_1(z) \tag{2.26}$$

$$U_j(z) = U_{j-1}(z) \underset{\theta_k}{\otimes} u_j(z) \text{ for } j = 2, \ldots, n \tag{2.27}$$

The u-functions $U_j(z)$ for $j<k$ do not contain terms with exponents equal to k. The first u-function that contains such a term is $U_k(z)$. This u-function can be represented as

$$U_k(z) = \sum_{i=0}^{k-1} \alpha_i z^i + \varepsilon_k z^k \tag{2.28}$$

Applying the operator $\underset{\theta_k}{\otimes}$ over $U_k(z)$ and $u_{k+1}(z)$ we obtain

$$\begin{aligned} U_{k+1}(z) &= (\sum_{i=0}^{k-1} \alpha_i z^i + \varepsilon_k z^k) \underset{\theta_k}{\otimes} [p_{k+1} z^1 + (1-p_{k+1}) z^0] \\ &= p_{k+1} \sum_{i=0}^{k-1} \alpha_i z^{\theta_k(i,1)} + (1-p_{k+1}) \sum_{i=0}^{k-1} \alpha_i z^{\theta_k(i,0)} \\ &+ \varepsilon_k p_{k+1} z^{\theta_k(k,1)} + \varepsilon_k (1-p_{k+1}) z^{\theta_k(k,0)} \\ &= p_{k+1} \sum_{i=1}^{k} \alpha_{i-1} z^i + (1-p_{k+1}) \sum_{i=0}^{k-1} \alpha_i z^i + \varepsilon_k [p_{k+1} + (1-p_{k+1})] z^k \\ &= \sum_{i=0}^{k-1} \beta_i z^i + \varepsilon_{k+1} z^k + \varepsilon_k z^k \end{aligned} \tag{2.29}$$

where

$$\varepsilon_{k+1} = p_{k+1} \alpha_{k-1} \tag{2.30}$$

Here, we used the important property of the function θ_k that for any $x \ge 0$

$$\theta_k(k, x) = k \tag{2.31}$$

It can be proven by induction that

$$U_n(z) = \sum_{i=0}^{k-1} \upsilon_i z^i + (\varepsilon_n + \varepsilon_{n-1} + \ldots + \varepsilon_{k+1} + \varepsilon_k) z^k \tag{2.32}$$

where ε_i (for $k \le i \le n$) is a product of the coefficient of the term with z^{k-1} from $U_{i-1}(z)$ and p_i. Observe that the coefficient of the term with z^k from $U_{i-1}(z)$ does not participate in calculating $\varepsilon_i, \ldots, \varepsilon_n$.

The u-function of the system structure function can now be determined as

$$U(z) = U_n(z) \underset{\eta}{\otimes} k = (\sum_{i=0}^{k-1} \upsilon_i) z^0 + (\varepsilon_n + \varepsilon_{n-1} + ... + \varepsilon_{k+1} + \varepsilon_k) z^1 \qquad (2.33)$$

and the system reliability can be calculated as

$$R = U'(1) = \varepsilon_n + \varepsilon_{n-1} + ... + \varepsilon_{k+1} + \varepsilon_k \qquad (2.34)$$

The considerations presented lie at the base of the following simplified algorithm of the system reliability determination:

1. Determine u-functions of each element in the form (2.18).
2. Assign $R = 0$, $U_1(z) = u_1(z)$.
3. For $j = 2, \ldots, n$:
   3.1. Obtain $U_j(z) = U_{j-1}(z) \underset{\theta_k}{\otimes} u_j(z)$.

   3.2. If the u-function $U_j(z)$ contains a term with z^k, remove this term from $U_j(z)$ and add its coefficient to R.

After termination of the algorithm, R is equal to the system reliability.

Note that in this algorithm the operator $\underset{\theta_k}{\otimes}$ can be replaced by the operator $\underset{+}{\otimes}$. Indeed, after removing the term with z^k from $U_j(z)$ the operators $U_j(z) \underset{+}{\otimes} u_{j+1}(z)$ and $U_j(z) \underset{\theta_k}{\otimes} u_{j+1}(z)$ become equivalent (since the function $\theta_k(x,1)$ for $x<k$ is equivalent to function $x+1$).

In the simplified algorithm, obtaining $U_j(z)$ for $j \le k$ requires $2j$ term multiplication operations and obtaining $U_j(z)$ for $k < j \le n$ requires $2k$ term multiplication operations. The total number of these operations is

$$\begin{aligned} & 2 \sum_{j=2}^{k} j + \sum_{j=k+1}^{n} 2k(n-k) = 2[0.5k(k+1) - 1] + 2k(n-k) \\ & = 2nk + k - k^2 - 2 \end{aligned} \qquad (2.35)$$

Example 2.8

Consider a 2-out-of-4 system from Example 2.7 and apply the technique described for the recursive derivation of the system reliability:

Assign $R = 0$;

$$U_1(z) = u_1(z) = 0.8z^1 + 0.2z^0$$

$$U_2(z) = (0.8z^1 + 0.2z^0) \underset{+}{\otimes} (0.6z^1 + 0.4z^0)$$

$$= (0.8z^1 + 0.2z^0)(0.6z^1 + 0.4z^0) = 10^{-2}(48z^2 + 44z^1 + 8z^0)$$

Remove the term with z^2 from $U_2(z)$ and add its coefficient to R:

$$R = 0.48$$

$$U_2(z) = 10^{-2}(44z^1 + 8z^0)$$

Further:

$$U_3(z) = 10^{-2}(44z^1 + 8z^0)\underset{+}{\otimes}(0.9z^1 + 0.1z^0)$$

$$= 10^{-3}(44z^1 + 8z^0)(9z^1 + 1z^0) = 10^{-3}(396z^2 + 116z^1 + 8z^0)$$

Remove the term with z^2 from $U_3(z)$ and add its coefficient to R:

$$R = 0.48 + 0.396 = 0.876$$

$$U_3(z) = 10^{-3}(116z^1 + 8z^0)$$

In the next step

$$U_4(z) = 10^{-3}(116z^1 + 8z^0)\underset{+}{\otimes}(0.7z^1 + 0.3z^0)$$

$$= 10^{-4}(116z^1 + 8z^0)(7z^1 + 3z^0) = 10^{-4}(812z^2 + 404z^1 + 24z^0)$$

Finally, after adding the coefficient of the term with z^2 from $U_4(z)$ to R we obtain:

$$R = 0.876 + 0.0812 = 0.9572$$

Observe that the total number of the term multiplication operations used in this example is 12.

2.3 Consecutive *k*-out-of-*n* Systems

Consider a system consisting of n independent binary elements that are linearly connected in such a way that the system fails if and only if at least k consecutive elements fail. Such a model named linear consecutive k-out-of-n:F system is used for evaluating system reliability in telecommunications, oil pipeline systems, spacecraft relay stations, *etc*. [15-20].

Example 2.9

Consider a set of n+2 radio relay stations with a transmitter allocated at the first station and a receiver allocated at the last station. Each one of n intermediate stations has retransmitters generating signals that cover the distance including the next k stations. The aim of the system is to provide propagation of a signal from transmitter to receiver. It is evident that the system fails if at least k consecutive retransmitters fail. An example of the radio relay system with $k = 3$ and $n = 8$ is shown in Figure 2.3. Whenever the number of consecutive failures is less than three, the signal flow is not interrupted and the signal reaches the receiver.

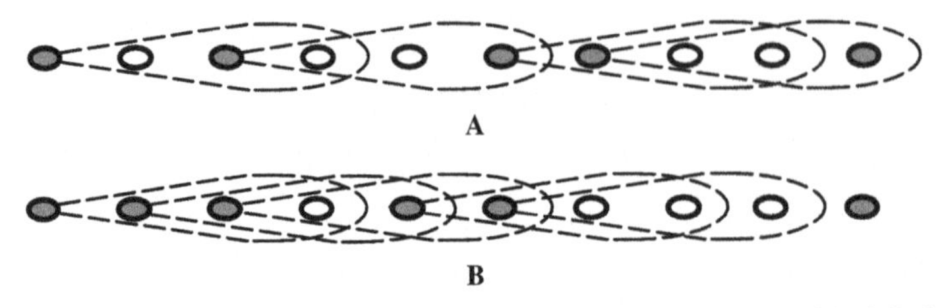

Figure 2.3. Linear consecutive 3-out-of-8 system in working state (A) and in failed state (B)

Example 2.10

In the pipeline systems transporting oil from a source point to a destination point the pressure is provided by n equally spaced pump stations. Each pump station provides pressure sufficient to transport oil to a distance that includes the k next stations. If m out of k stations following the working one fail ($m<k$), then the flow of oil is not interrupted because the remaining $k-m$ stations still carry the load. When the k adjacent stations fail, no working pumps remain in the part of the pipeline reached by the oil transported by the last working station. The oil flow is interrupted and the system fails.

The system in which n independent binary elements are linearly connected in such a way that the system works if and only if at least k consecutive elements are working is named the k-out-of-n:G system. The k-out-of-n:F and k-out-of-n:G systems are duals of each other [19]. This means that if the reliability of any element j in a k-out-of-n:F system is equal to the unreliability of element j a in k-out-of-n:G system (with the same k and n), then the reliability of the entire k-out-of-n:F system is equal to the unreliability of the entire k-out-of-n:G system. Therefore, the same algorithms for reliability evaluation can be applied to both types of system. In this chapter we consider only k-out-of-n:F systems and omit F from their denomination.

The linear consecutive k-out-of-n system was formally introduced by Chiang and Niu [15] but had been previously mentioned by Kontoleon [21]. The methods for evaluating the reliability of a linear consecutive k-out-of-n system with identical elements were suggested in [15, 16, 22-26]. The more complex case of systems with different elements was studied in [20, 21, 27].

The model in which the elements are circularly connected so that the first and the nth elements become adjacent to each other is named a circular consecutive k-out-of-n system. Examples of such a system can be found in monitoring, nuclear accelerators, *etc.* [12].

Example 2.11

For taking pictures of high-energy particles in a nuclear accelerator, n high-speed cameras are installed around the accelerator. If more then k adjacent cameras fail to take pictures, the particle behaviour cannot be analyzed.

Example 2.12

The vacuum system of an electronic accelerator consists of a large number of vacuum bulbs placed evenly along a ring. The vacuum system fails if at least k adjacent vacuum bulbs fail.

The circular consecutive k-out-of-n system was introduced by Derman *et al.* [22]. The algorithms for evaluating the system reliability were suggested in [22, 24, 25, 28–30] for a system with identical elements and in [20, 27, 31, 32] for a system with different elements.

The series and parallel systems can be considered as special cases of the consecutive k-out-of-n system. Indeed, when $k = 1$ the failure of any element causes the failure of the entire system, and the system becomes series one. When $k = n$ the entire system fails only if all of its elements fail, which corresponds to the parallel system.

2.3.1 Consecutive *k*-out-of-*n* Systems with Identical Elements

In this section we consider the consecutive k-out-of-n system in which all of its elements are identical, *i.e.* each individual element has the same reliability p. In the algorithms suggested by Lambiris and Papastavridis [24] and Goulden [25] for evaluating the reliability of such systems, the generating function approach is used in order to determine the number of ways to arrange j failed elements in a line with $n-j$ working elements such that no k or more failed elements are consecutive. Having this number $N(j, k, n)$ for any j we can obtain the reliability of linear consecutive k-out-of-n system $R_L(k, n)$ as

$$R_{\mathrm{L}}(k,n) = \sum_{j=0}^{n}(1-p)^{j} p^{n-j} N(j,k,n) \tag{2.36}$$

If the system contains exactly j failed elements, then the remaining $n-j$ working elements divide the system into $n-j+1$ segments (the first segment is to the left of the first working element, $n-j-1$ segments are between any two working elements that are close to each other, and the last segment is to the right of the last working element). Each segment may contain from 0 to j failed elements. The allocations in which no one segment contains k or more failed elements correspond to the system's success.

Let u-function $u_i(z)$ represent the distribution of the number of ways the failed elements can be allocated in section i such that the system does not fail:

$$u_i(z) = z^0 + z^1 + ... + z^{k-1} \tag{2.37}$$

This representation corresponds to the fact that from 0 to $k-1$ elements can be allocated in the section (which is expressed by exponents of the u-function) and only one way exists to allocate any number of failed elements in this single section (all the coefficients are equal to 1). The distributions of the number of ways the failed elements can be allocated in several sections can be obtained using the $\otimes_+$ operator.

Indeed, in the u-function

$$u_i(z)\underset{+}{\otimes}u_m(z) = (z^0 + z^1 + ... + z^{k-1})(z^0 + z^1 + ... + z^{k-1})$$
$$= \sum_{h=0}^{2k-2} a_h z^h \tag{2.38}$$

a_h is equal to the number of ways h elements can be distributed between sections i and m. Applying the $\otimes_+$ operator over $n-j+1$ identical u-functions we obtain the resulting u-function

$$U(z) = \underset{+}{\otimes}(u_1(z), u_2(z), ..., u_{n-j+1}(z)) = (z^0 + z^1 + ... + z^{k-1})^{n-j+1}$$
$$= \sum_{h=1}^{(k-1)(n-j+1)} \alpha_h z^h \tag{2.39}$$

that represents the number of ways different numbers of failed elements can be allocated in $n-j+1$ sections. The coefficient α_j of term $\alpha_j z^j$ represents the number of ways exactly j failed elements can be allocated in $n-j+1$ segments. Therefore, $N(j,k,n) = \alpha_j$.

When j is close to n, the j failed elements cannot be allocated among $n-j+1$ segments in a manner where any segment contains less than k failed elements. There exists a maximum number of element failures j_{max} that the system may experience without failing independently on the location of the failed elements:

$$N(j_{max}, k, n) > 0$$
$$N(j_{max} + 1, k, n) = 0 \tag{2.40}$$

Indeed, when j failed elements are distributed among $n-j+1$ segments, the minimal number of elements in each segment is achieved when the elements are distributed as evenly as possible. In this case, some segments contain

$\lceil j/(n-j+1) \rceil$ failed elements and some segments contain $\lfloor j/(n-j+1) \rfloor$ failed elements. The system succeeds if

$$\lceil j/(n-j+1) \rceil \le k-1 \tag{2.41}$$

This inequality holds when

$$j \le (k-1)(n-j+1) \tag{2.42}$$

From this expression we obtain

$$j \le n+1-\frac{n+1}{k} \tag{2.43}$$

which means that the maximum possible value of j is

$$j_{\max} = \left\lfloor n+1-\frac{n+1}{k} \right\rfloor \tag{2.44}$$

Observe also that $N(0,k,n)=1$ for any k and n. Indeed, only one way exists for allocating zero failed elements among n+1 segments. Therefore, expression (2.36) can be rewritten as

$$R_{\mathrm{L}}(k,n) = p^n + \sum_{j=1}^{j_{\max}} (1-p)^j p^{n-j} N(j,k,n) \tag{2.45}$$

Example 2.13

Consider a linear consecutive 2-out-of-5 system with identical elements. This system should not contain more than k–1 = 2–1 = 1 failed elements in each segment between the working elements. Therefore, the u-function corresponding to a single segment is $u(z)=z^0+z^1$. According to (2.44)

$$j_{\max} = \left\lfloor 5+1-\frac{5+1}{2} \right\rfloor = 3$$

For j = 1, according to (2.39), we obtain

$$U(z) = (z^0+z^1)^{5-1+1} = (z^0+z^1)^5 = z^0+5z^1+10z^2+10z^3+5z^4+z^5$$

$N(1,k,n)$ is equal to the coefficient of the term with exponent 1: $N(1,k,n)=5$.

For $j = 2$:

$$U(z) = (z^0 + z^1)^{5-2+1} = (z^0 + z^1)^4 = z^0 + 4z^1 + 6z^2 + 4z^3 + z^4$$

$N(2,k,n)$ is equal to the coefficient of the term with exponent 2: $N(2,k,n) = 6$.
For $j = 3$:

$$U(z) = (z^0 + z^1)^{5-3+1} = (z^0 + z^1)^3 = z^0 + 3z^1 + 3z^2 + z^3$$

$N(3,k,n)$ is equal to the coefficient of the term with exponent 3: $N(3,k,n) = 1$.
According to (2.45) the system reliability is

$$R_{\mathrm{L}}(2,5) = p^5 + 5p^4(1-p) + 6p^3(1-p)^2 + p^2(1-p)^3$$

The problem of evaluating the reliability of a circular system with identical elements can be reduced to the problem of evaluating the reliability of a linear system (Derman *et al.* [22]). Indeed, consider a point between two arbitrary adjacent elements in a circular consecutive k-out-of-n system and find two working elements clockwise and counter clockwise to this point (Figure 2.4). These two working elements divide the circle into two fragments. The system reliability is equal to the probability that the fragment, including the marked point, contains less than k elements and the fragment not including the marked point forms a working linear consecutive k-out-of-n system.

Let n_1 and n_2 indicate the number of failed elements between the marked point and the first working elements. It can be easily seen that for any $i < n-1$

$$\Pr\{n_1 = i\} = \Pr\{n_2 = i\} = p(1-p)^i \tag{2.46}$$

The probability that the fragment, including the marked point, contains exactly i failed elements is

$$\begin{aligned}\Pr\{n_1 + n_2 = i\} &= \sum_{j=0}^{i} \Pr\{n_1 = j\}\Pr\{n_2 = i-j\} \\ &= \sum_{j=0}^{i} p(1-p)^j p(1-p)^{i-j} = \sum_{j=0}^{i} p^2(1-p)^i = (i+1)p^2(1-p)^i\end{aligned} \tag{2.47}$$

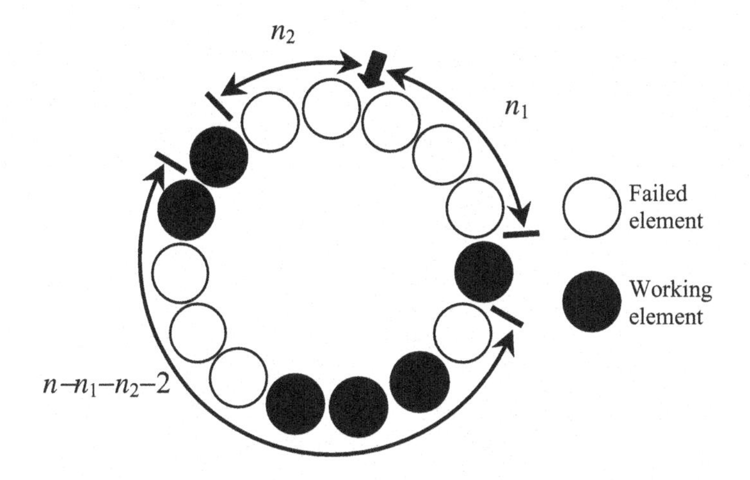

Figure 2.4. Circular consecutive *k*-out-of-*n* system with identical elements

The probability that the fragment contains less than the *k* failed elements is

$$\Pr\{n_1 + n_2 < k\} = \sum_{i=0}^{k-1} \Pr\{n_1 + n_2 = i\} = p^2 \sum_{i=0}^{k-1} (i+1)(1-p)^i \tag{2.48}$$

If the fragment, including the marked point, consists of *i* failed elements, then the second fragment consists of $n-i-2$ remaining elements. The reliability of the second fragment is $R_L(k, n-i-2)$. Therefore, the reliability of the entire circular system is

$$R_C(k,n) = p^2 \sum_{i=0}^{k-1} (i+1)(1-p)^i R_L(k, n-i-2) \tag{2.49}$$

Equation (2.49) is obtained on the assumption that $n>1$ and $k<n$. When $k = n$ we have a parallel system for which $R_C(n,n) = p^n$; when $n = 1$ we have a trivial case: $R_C(1,1) = p$.

2.3.2 Consecutive *k*-out-of-*n* Systems with Different Elements

This section considers consecutive *k*-out-of-*n* systems consisting of elements with different reliabilities. In order to describe the UGF-based algorithm for evaluating the reliability of this type of system, we represent the linear consecutive *k*-out-of-*n* system as a set of n+2 consecutively ordered nodes: 0, 1, 2, …, n+1 (see Figure 2.5). Each node j $(1 \le j \le n)$ has two states. In state $X_j = 1$ the arcs from the node j

to nodes $\alpha(j+1)$, $\alpha(j+2), \ldots$, $\alpha(j+k)$ exist, where $\alpha(x) = \min\{x, n+1\}$. In state $X_j = 0$ no arcs from node j exist. The probabilities of states 1 and 0 of the node j are respectively p_j and $1-p_j$. Node 0 is fully reliable ($X_0 \equiv 1$) and provides arcs to nodes 1, 2, …, k. Node $n+1$ is a dummy node and its state does not matter. The system reliability is the probability that a path exists between the nodes 0 and $n+1$.

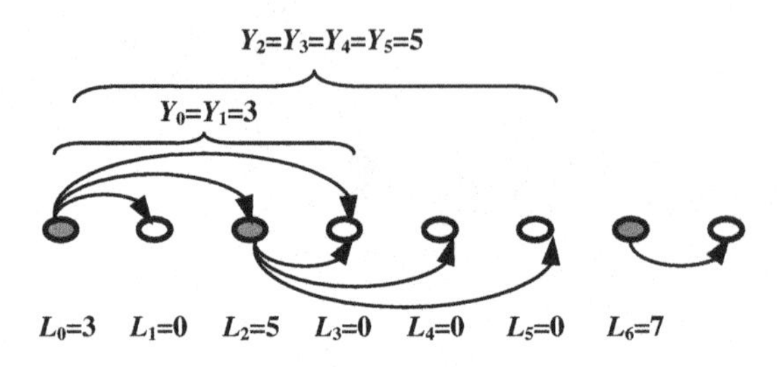

Figure 2.5. Example of state of linear consecutive 3-out-of-7 system

Let random value L_j be the number of the most remote node to which an arc from node j exists. It can be seen that, for $0 \le j \le n$, $L_j = X_j\alpha(j+k)$. The u-function $u_j(z)$ representing the p.m.f. of L_j takes the form

$$u_0(z) = z^k \text{ and } u_j(z) = p_j z^{\alpha(j+k)} + (1-p_j)z^0 \text{ for } 1 \le j \le n \tag{2.50}$$

Let random value Y_m be the number of the most remote node to which the path from node 0 provided by nodes 1, 2, …, m exists. It can be seen that if the path to the node $m+1$ provided by the nodes 1, 2, …, m exists ($Y_m \ge m+1$), then the path to node max($m+1$, L_{m+1}) also exists and the number of the most remote node connected with node 0 is equal to max(Y_m, L_{m+1}). If the path to the node $m+1$ does not exist ($Y_m < m+1$), then this node does not participate in prolonging the path and $Y_{m+1} = Y_m$. This consideration gives the recursive expression

$$Y_0 = L_0$$

$$Y_{m+1} = f(Y_m, L_{m+1}) = \begin{cases} \max\{Y_m, L_{m+1}\} & \text{if } Y_m > m \\ Y_m, & \text{if } Y_m < m+1 \end{cases} \quad \text{for } 0 \le m < n \tag{2.51}$$

The system structure function can be expressed as

$$\phi(X_1,...,X_n) = 1[Y_n(L_1,...,L_n) = n+1] \tag{2.52}$$

The u-function $U_m(z)$ representing the p.m.f. of each random variable Y_m can now be obtained as

$$U_0(z)=u_0(z)$$

$$U_{m+1}(z) = U_m(z) \underset{f}{\otimes} u_{m+1}(z), \quad 0 \le m < n \tag{2.53}$$

The term of the u-function $U_n(z)$ that has the exponent n+1 corresponds to the system state in which the path from node 0 to node n+1 exists. The system reliability $R = E(1(Y_n = n+1))$ is equal to the coefficient of this term.

When $Y_m < m+1$, the path from node 0 to nodes with numbers greater than m does not exist. Therefore, the states corresponding to $Y_m < m+1$ do not participate in the combinations of the node states that provide the success of the entire system. The terms corresponding to these states can be removed from the u-function $U_m(z)$. After removing the terms corresponding to $Y_m < m+1$, the function $f(Y_m, L_{m+1})$ takes the form $\max(Y_m, L_{m+1})$ and the simple operator $\underset{\max}{\otimes}$ can be used in Equation (2.53).

The following recursive procedure determines the reliability of the linear consecutive k-out-of-n system with different elements:

1. Define $u_0(z) = z^k$, $u_j(z) = p_j z^{\alpha(j+k)} + (1-p_j)z^0$ for $1 \le j \le n$.
2. Assign $U_0(z) = u_0(z)$.
3. For $0 \le j < n$: remove terms with z^s where $s \le j$ from $U_j(z)$ and obtain $U_{j+1}(z) = U_j(z) \underset{\max}{\otimes} u_{j+1}(z)$.
4. Obtain the system availability as the coefficient of the term with z^{n+1} in $U_n(z)$.

Example 2.14

Consider a linear consecutive 3-out-of-5 system with different elements. The u-functions of the individual elements are:

$$u_0(z) = z^3,\ u_1(z) = p_1 z^4 + q_1 z^0,\ u_2(z) = p_2 z^5 + q_2 z^0$$

$$u_3(z) = p_3 z^6 + q_3 z^0,\ u_4(z) = p_4 z^6 + q_4 z^0,\ u_5(z) = p_5 z^6 + q_5 z^0$$

where $q_j = 1-p_j$. Following the recursive procedure we obtain

$$U_0(z) = u_0(z) = z^3$$

$$U_1(z) = U_0(z) \underset{\max}{\otimes} u_1(z) = z^3 \underset{\max}{\otimes} (p_1 z^4 + q_1 z^0) = p_1 z^4 + q_1 z^3$$

$$U_2(z) = U_1(z) \underset{\max}{\otimes} u_2(z) = (p_1 z^4 + q_1 z^3) \underset{\max}{\otimes} (p_2 z^5 + q_2 z^0)$$

$$= p_2 z^5 + p_1 q_2 z^4 + q_1 q_2 z^3$$

$$U_3(z) = U_2(z) \underset{\max}{\otimes} u_3(z) = (p_2 z^5 + p_1 q_2 z^4 + q_1 q_2 z^3) \underset{\max}{\otimes} (p_3 z^6 + q_3 z^0)$$

$$= p_3 z^6 + p_2 q_3 z^5 + p_1 q_2 q_3 z^4 + q_1 q_2 q_3 z^3$$

After removing the term with z^3

$$U_3(z) = p_3 z^6 + p_2 q_3 z^5 + p_1 q_2 q_3 z^4$$

$$U_4(z) = U_3(z) \underset{\max}{\otimes} u_4(z) = (p_3 z^6 + p_2 q_3 z^5 + p_1 q_2 q_3 z^4) \underset{\max}{\otimes} (p_4 z^6 + q_4 z^0)$$

$$= (p_3 + p_2 q_3 + p_1 q_2 q_3) p_4 z^6 + p_3 q_4 z^6 + p_2 q_3 q_4 z^5 + p_1 q_2 q_3 q_4 z^4$$
$$= [p_4(1 - q_1 q_2 q_3) + p_3 q_4] z^6 + p_2 q_3 q_4 z^5 + p_1 q_2 q_3 q_4 z^4$$

After removing the term with z^4

$$U_4(z) = [p_4(1 - q_1 q_2 q_3) + p_3 q_4] z^6 + p_2 q_3 q_4 z^5$$

$$U_5(z) = U_4(z) \underset{\max}{\otimes} u_5(z)$$

$$= \{[p_4(1 - q_1 q_2 q_3) + p_3 q_4] z^6 + p_2 q_3 q_4 z^5\} \underset{\max}{\otimes} (p_5 z^6 + q_5 z^0)$$

$$= [p_4(1 - q_1 q_2 q_3) + p_3 q_4 + p_2 q_3 q_4 p_5] z^6 + p_1 q_2 q_3 q_4 q_5 z^5$$

The system reliability is equal to the coefficient of the term with z^6:

$$R = [p_4(1 - q_1 q_2 q_3) + p_3 q_4 + p_2 q_3 q_4 p_5]$$

The reduction of the problem of evaluating the reliability of the circular system with different independent elements into the problem of evaluating the reliability of the linear system was proposed by Hwang [27].

Let $R_L(k, \langle i, j \rangle)$ be the reliability of the linear consecutively connected k-out-of-(j–i+1) subsystem consisting of elements i, i+1, …, j–1, j.

Consider the circular system presented in Figure 2.6. Let in this system f and l be the numbers of the first and last working elements respectively in the sequence from 1 to n. The reliability of the circular system is equal to the probability that the fragment between elements l and f through element n contains less than k elements and the fragment between elements f and l, not including element n, forms a working linear consecutive k-out-of-n system. There are $n-l+f-1$ elements in the first fragment (including element n). The probability that all of the elements belonging to this fragment fail while elements f and l work is

$$p_f \prod_{i=1}^{f-1}(1-p_i)p_l \prod_{j=l+1}^{n}(1-p_j) \tag{2.54}$$

The reliability of the second fragment (not including element n) is $R_L(k,\langle f+1,\ l-1\rangle)$. The reliability of the circular consecutive k-out-of-n system $R_C(k,n)$ is equal to the sum of the probabilities that the system works for all possible combinations of f and l meeting the constraint $n-l+f-1<k$:

$$\begin{aligned} R_C(k,n) &= \sum_{n-l+f-1<k} p_f \prod_{i=1}^{f-1}(1-p_i)p_l \\ &\times \prod_{j=l+1}^{n}(1-p_j)R_L(k,\langle f+1,l-1\rangle) \\ &= \sum_{a=0}^{k-1}\sum_{b=0}^{a} p_{b+1}\prod_{i=1}^{b}(1-p_i)p_{n-a+b} \\ &\times \prod_{j=n-a+b+1}^{n}(1-p_j)R_L(k,\langle b+2,n-a+b-1\rangle) \end{aligned} \tag{2.55}$$

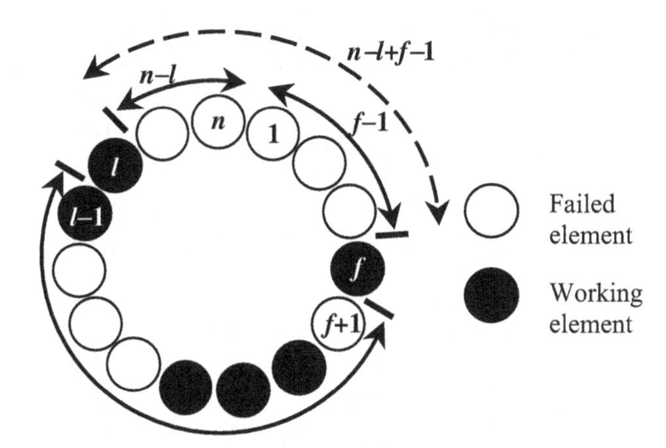

Figure 2.6. Circular consecutive k-out-of-n system with different elements

As in the case of a system with identical elements, Equation (2.55) is obtained on the assumption that $n>1$ and $k<n$. When $k=n$ we have a parallel system, for which $R_C(n,n)=\prod_{j=1}^{n} p_j$; when $n=1$, we have a trivial case: $R_C(1,1)=p_1$.

It should be mentioned that other recursive formulae for evaluating $R_C(k,n)$ were later suggested by Antonopoulou and Papastavridis [33], Korczak [34] and Chang *et al.* [35]. The last reference presents the most effective algorithm of the system reliability evaluation.

2.4 Consecutive *k*-out-of-*r*-from-*n* Systems

The linear consecutive *k*-out-of-*r*-from-*n*:*F* system has *n* ordered elements and fails if at least *k* out of any *r* consecutive elements fail. The system that works if at least *k* out of any *r* consecutive elements are working is called the consecutive *k*-out-of-*r*-from-*n*:*G* system. It can be seen that a *k*-out-of-*r*-from-*n*:*F* system is equivalent to an (*r*-*k*+1)-out-of-*r*-from-*n*:*G* system. Therefore, we consider only *k*-out-of-*n*:*F* systems and omit *F* from their denomination.

The linear consecutive *k*-out-of-*r*-from-*n* system was formally introduced by Griffith [36], but had been previously mentioned by Tong [37], Saperstein [38, 39], Naus [40] and Nelson [41] in connection with tests for non-random clustering, quality control and inspection procedures, service systems, and radar detection problems.

The models presented in the two previous sections can be considered as special cases of the linear consecutive *k*-out-of-*r*-from-*n* system. When $r=n$ one has the simple k-out-of-*n* system. When $k=r$ one has the consecutive *k*-out-of-*n* system.

Example 2.15

Consider a quality control system that randomly selects for a quality check *r* items produced consecutively by a manufacturing process. If within the selected sample at least *k* items are defective, then the system concludes that the process needs to be adjusted. If the process produces *n* items in a certain period of time, then we are interested in the probability that such a random quality check is able to detect a problem in the process.

Example 2.16

An outdoor industrial conveyor transports identical sealed containers (Figure 2.7). The containers are loaded onto pallets placed on the conveyor belt. The conveyor carries *r* pallets simultaneously. If the container sealing fails, its weight becomes greater due to humidity penetration. The maximum allowable load of the conveyor corresponds to *k*–1 containers with failed sealing. The system fails if more than *k* such containers are loaded on *r* consecutive pallets. Having the probability that each sealing fails, we obtain the system reliability as a probability that the system does not fail during the time when n containers are transported.

Figure 2.7. Industrial conveyor as an example of k-out-of-r-from-n system

The algorithms suggested for evaluating the reliability of linear consecutive k-out-of-r-from-n systems either consider the case of identical elements (elements with equal reliability) and a limited set of parameters [36, 40, 42] or provide bounds for system reliability [42-44] that are good enough only for element reliabilities very close to 1. Because of the difficulty in estimating the exact value of the system reliability, Psillakis [45] proposed a simulation approach and provided the error analysis. Malinowski and Preuss [46] suggested an enumerative algorithm for the exact evaluation of the system reliability based on recursive computation of conditional probabilities.

Let X_j be the binary state variable of element j. The u-function $u_j(z)$ that takes the form (2.8) represents the p.m.f. of X_j. The system succeeds if any group of r consecutive elements contains at least $r-k+1$ working elements. Therefore, the system reliability can be defined as

$$R = \Pr\{\bigcup_{h=1}^{n-r+1} (\sum_{j=h}^{h+r-1} X_j > r\text{-}k)\} \tag{2.56}$$

Let V_h be a group of r consecutive elements numbered from h to $h+r-1$. The state of this group can be represented by a random binary state vector $\boldsymbol{Y}_h = \{X_h, \ldots, X_{h+r-1}\}$.

Each realization $\boldsymbol{y}_{h,m}$ of vector $\boldsymbol{Y}_h$ constitutes a state m of V_h. Since the elements are independent, the probability of any state of the group V_h is equal to the product of the probabilities of the corresponding states of the individual elements. The p.m.f. of $\boldsymbol{Y}_h$ can be represented by the u-function $U_h(z)$. The total number of different states of the group of r elements is equal to 2^r. Therefore, the u-function $U_h(z)$ consists of 2^r different terms.

The u-function corresponding to the hth group of r consecutive elements V_h takes the form

$$U_h(z) = \sum_{x_h=0}^{1} \sum_{x_{h+1}=0}^{1} \dots \sum_{x_{h+r-1}=0}^{1} [\prod_{j=h}^{h+r-1} p_j^{x_j} (1-p_j)^{1-x_j}] z^{(x_h,\dots,x_{h+r-1})} \quad (2.57)$$

Simplifying this representation one obtains

$$U_h(z) = \sum_{m=1}^{2^r} Q_{h,m} z^{\boldsymbol{y}_{h,m}} \quad (2.58)$$

where $Q_{h,m}$ is the probability that the hth group is in state m and r-length binary vector $\boldsymbol{y}_{h,m}$ represents the states of the elements when the group is in state m. The u-function obtained defines all of the possible states of the group V_h.

Let random variable S_h represent the sum of random binary variables X_h, X_{h+1}, ..., X_{h+r-1} (which corresponds to the sum of the state variables of the elements belonging to the group V_h). According to definition (2.56) the system structure function takes the form

$$\phi(X_1,\dots,X_n) = \prod_{h=1}^{n-r+1} 1(S_h > r-k) \quad (2.59)$$

Having the vectors $\boldsymbol{y}_{h,m}$ representing states of elements belonging to V_h in any state m, one can obtain the realization of S_h in this state by summing the vector elements. Therefore, the p.m.f. of S_h can be represented by the u-function $\hat{U}_h(z)$, which is obtained from $U_h(z)$ by replacing the vectors $\boldsymbol{y}_{h,m}$ with sums of their elements. The u-function $\tilde{U}_h(z)$, representing p.m.f. of the binary function $1(S_h \le r-k)$, can be obtained by applying the operator $\tilde{U}_h(z) = \hat{U}_h(z) \underset{\varphi}{\otimes} (r-k)$, where $\varphi(Y, r-k) = 1(Y \le r-k)$. Calculating the expected value of the function $1(S_h \le r-k)$ one obtains the probability of failure of the hth group of r consecutive elements V_h:

$$\Pr\{S_h \le r-k\} = E(1(S_h \le r-k)) = \tilde{U}'_h(1) \quad (2.60)$$

Observe that this operation is equivalent to summing the coefficients of terms containing in their state vectors k or more zeros in the u-function $U_h(z)$. Therefore, the failure probability can also be obtained by applying the following operator Θ_k directly over the u-function $U_h(z)$:

$$\Pr\{S_h \le r-k\} = \Theta_k(U_h(z)) = \sum_{m=1}^{2^r} Q_{h,m} \times 1(\theta(\boldsymbol{y}_{h,m}) \ge k) \quad (2.61)$$

where $\theta(\boldsymbol{y})$ is a sum of zeros in vector $\boldsymbol{y}$.

Example 2.17

Consider a system with k = 2, n = 4 and r = 3. The binary state variables of the system elements are X_1, X_2, X_3 and X_4. The state of the second group of three variables V_2 is represented by the vector $\boldsymbol{Y}_2 = (X_2, X_3, X_4)$. If the reliability of each element is $p_i = \Pr\{X_i = 1\}$, then the probability of each possible realization of the vector $\boldsymbol{Y}_2$ is

$$\Pr\{\boldsymbol{Y}_2 = (x_2, x_3, x_4)\} = p_2^{x_2}(1-p_2)^{1-x_2}\, p_3^{x_3}(1-p_3)^{1-x_3}\, p_4^{x_4}(1-p_4)^{1-x_4}$$

The condition of failure of the group V_2 is $S_2 \le r-k$, where $S_2 = X_2+X_3+X_4$ and $r-k=3-2=1$. The u-function that represents the distribution of $\boldsymbol{Y}_2$ takes the form

$$U_2(z) = p_1p_2p_3z^{(1,1,1)} + p_1p_2q_3z^{(1,1,0)} + p_1q_2p_3z^{(1,0,1)} + p_1q_2q_3z^{(1,0,0)}$$
$$+ q_1p_2p_3z^{(0,1,1)} + q_1p_2q_3z^{(0,1,0)} + q_1q_2p_3z^{(0,0,1)} + q_1q_2q_3z^{(0,0,0)}$$

where $q_j = 1-p_j$. The u-functions $\hat{U}_2(z)$ and $\tilde{U}_2(z)$ that represent the distributions of the functions S_2 and $1(S_2 \le 1)$ respectively are

$$\hat{U}_2(z) = p_1p_2p_3z^3 + p_1p_2q_3z^2 + p_1q_2p_3z^2 + p_1q_2q_3z^1 + q_1p_2p_3z^2$$
$$+ q_1p_2q_3z^1 + q_1q_2p_3z^1 + q_1q_2q_3z^0 = p_1p_2p_3z^3 + (p_1p_2q_3 + p_1q_2p_3$$
$$+ q_1p_2p_3)z^2 + (p_1q_2q_3 + q_1p_2q_3 + q_1q_2p_3)z^1 + q_1q_2q_3z^0$$

and

$$\tilde{U}_2(z) = p_1p_2p_3z^0 + (p_1p_2q_3 + p_1q_2p_3 + q_1p_2p_3)z^0$$
$$+ (p_1q_2q_3 + q_1p_2q_3 + q_1q_2p_3)z^1 + q_1q_2q_3z^1$$

The failure probability is

$$\Pr\{S_2 \le 1\} = E(1(S_2 \le 1)) = \tilde{U}_2'(1) = p_1q_2q_3 + q_1p_2q_3 + q_1q_2p_3 + q_1q_2q_3$$

Note that the linear consecutive k-out-of-r-from-n system contains exactly $n-r+1$ groups of r consecutive elements and each element can belong to no more than r such groups. To obtain the u-functions corresponding to all the groups of r consecutive elements, the following definitions are introduced:

1. Define u-function $U_{1-r}(z)$ as follows:

$$U_{1-r}(z) = z^{y_0} \qquad (2.62)$$

where the vector $\boldsymbol{y}_0$ consists of r zeros.

2. Define the following shift operator over u-function $U_h(z)$:

$$\begin{aligned} U_h(z) \underset{\leftarrow}{\otimes} u_{h+r}(z) &= (\sum_{m=1}^{2^r} Q_{h,m} z^{y_{h,m}}) \underset{\leftarrow}{\otimes} [p_{h+r} z^1 + (1-p_{h+r}) z^0] \\ &= p_{h+r} \sum_{m=1}^{2^r} Q_{h,m} z^{y_{h,m} \leftarrow 1} + (1-p_{h+r}) \sum_{m=1}^{2^r} Q_{h,m} z^{y_{h,m} \leftarrow 0} \end{aligned} \tag{2.63}$$

where operator $\boldsymbol{y} \leftarrow x$ over arbitrary vector $\boldsymbol{y}$ and value x shifts all the vector elements one position left: $y(j-1) = y(j)$ for $1 < j \leq r$ and assigns the value x to the last element of $\boldsymbol{y}$: $y(r) = x$ (the first element of vector $\boldsymbol{y}$ disappears after applying the operator). The operator $\boldsymbol{y} \leftarrow x$ removes the state of the first element of the group and adds the state of the next (not yet considered) element to the group, preserving the order of the elements belonging to the group. Therefore, applying this operator over the u-function $U_h(z)$ that represents the state distribution of the group V_h, one obtains the u-function $U_{h+1}(z)$ representing the state distribution of the group V_{h+1}.

Using the operator $\underset{\leftarrow}{\otimes}$ in sequence as follows:

$$U_{j+1-r}(z) = U_{j-r}(z) \underset{\leftarrow}{\otimes} u_j(z) \tag{2.64}$$

for $j = 1, \ldots, n$ one obtains u-functions for all of the possible groups V_h: $U_1(z), \ldots, U_{n-r+1}(z)$. Note that the u-function $U_1(z)$ for the first group V_1 is obtained after applying the operator $\underset{\leftarrow}{\otimes}$ r times.

Consider a u-function $U_h(z)$ representing the distribution of the random vector $\boldsymbol{Y}_h = \{X_h, \ldots, X_{h+r-1}\}$. For each combination of values $X_{h+1}, \ldots, X_{h+r-1}$ it contains two terms corresponding to values 0 and 1 of X_h (states 0 and 1 of element h). After applying the operator $\underset{\leftarrow}{\otimes}$, X_h disappears from the vector $\boldsymbol{Y}_h$, being replaced with X_{h+1}. This produces two terms with the same state vector $\boldsymbol{y}_{h+1,m}$ for each state m of the group V_{h+1} in the u-function $U_{h+1}(z)$. The coefficients of the two terms with the same state vector $\boldsymbol{y}_{h+1,m}$ are equal to the probabilities that the group V_{h+1} is in state m while element h is in states 0 and 1 respectively. By summing these two coefficients (collecting the like terms in $U_{h+1}(z)$), one obtains a single term for each vector $\boldsymbol{y}_{h+1,m}$ with a coefficient equal to the overall probability that the group V_{h+1} is in state m. Therefore, the number of different terms in each u-function $U_h(z)$ is always equal to 2^r and

$$U_h(z) \underset{\leftarrow}{\otimes} u_{h+r}(z) = \sum_{m=1}^{2^r} Q_{h+1,m} z^{y_{h+1,m}} \tag{2.65}$$

Applying the operator Θ_k (2.61) over the u-functions $U_1(z)$, …, $U_{n-r+1}(z)$ one can obtain the failure probability for each group of r consecutive elements.

Example 2.18

Consider the system from Example 2.17 and obtain the u-functions for all of the possible groups of three consecutive elements using the recursive procedure described above. There are two such groups in the system: V_1 with element state vector (X_1, X_2, X_3) and V_2 with element state vector (X_2, X_3, X_4).
First define

$$U_{-2}(z) = z^{(0,0,0)}$$

Following (2.63) obtain

$$U_{-1}(z) = U_{-2}(z) \underset{\leftarrow}{\otimes} u_1(z) = z^{(0,0,0)} \underset{\leftarrow}{\otimes} (p_1 z^1 + q_1 z^0)$$

$$= p_1 z^{(0,0,1)} + q_1 z^{(0,0,0)}$$

$$U_0(z) = U_{-1}(z) \underset{\leftarrow}{\otimes} u_2(z) = (p_1 z^{(0,0,1)} + q_1 z^{(0,0,0)}) \underset{\leftarrow}{\otimes} (p_2 z^1 + q_2 z^0)$$

$$= p_1 p_2 z^{(0,1,1)} + q_1 p_2 z^{(0,0,1)} + p_1 q_2 z^{(0,1,0)} + q_1 q_2 z^{(0,0,0)}$$

$$U_1(z) = U_0(z) \underset{\leftarrow}{\otimes} u_3(z) = (p_1 p_2 z^{(0,1,1)} + q_1 p_2 z^{(0,0,1)} + p_1 q_2 z^{(0,1,0)}$$

$$+ q_1 q_2 z^{(0,0,0)}) \underset{\leftarrow}{\otimes} (p_3 z^1 + q_3 z^0) = p_1 p_2 p_3 z^{(1,1,1)} + q_1 p_2 p_3 z^{(0,1,1)}$$

$$+ p_1 q_2 p_3 z^{(1,0,1)} + q_1 q_2 p_3 z^{(0,0,1)} + p_1 p_2 q_3 z^{(1,1,0)} + q_1 p_2 q_3 z^{(0,1,0)}$$
$$+ p_1 q_2 q_3 z^{(1,0,0)} + q_1 q_2 q_3 z^{(0,0,0)}$$

The u-function $U_1(z)$ represents the distribution of the random vector (X_1, X_2, X_3) and contains $2^3 = 8$ terms. In order to obtain the failure probability of group V_1 for $k = 2$ we apply the operator $\Theta_2\,(U_1(z))$:

$$\Pr\{S_1 \le 1\} = \Theta_2(U_1(z)) = q_1 q_2 p_3 + p_1 q_2 q_3 + q_1 p_2 q_3 + q_1 q_2 q_3$$

In order to obtain the u-function $U_2(z)$ representing the distribution of the random vector (X_2, X_3, X_4) we apply the operator $\underset{\leftarrow}{\otimes}$ once more:

$$U_2(z) = U_1(z) \underset{\leftarrow}{\otimes} u_4(z) = (p_1 p_2 p_3 z^{(1,1,1)} + q_1 p_2 p_3 z^{(0,1,1)}$$

$$+ p_1 q_2 p_3 z^{(1,0,1)} + q_1 q_2 p_3 z^{(0,0,1)} + p_1 p_2 q_3 z^{(1,1,0)} + q_1 p_2 q_3 z^{(0,1,0)}$$

$$+ p_1 q_2 q_3 z^{(1,0,0)} + q_1 q_2 q_3 z^{(0,0,0)}) \underset{\leftarrow}{\otimes} (p_4 z^1 + q_4 z^0) = p_1 p_2 p_3 p_4 z^{(1,1,1)}$$

$$+ q_1 p_2 p_3 p_4 z^{(1,1,1)} + p_1 q_2 p_3 p_4 z^{(0,1,1)} + q_1 q_2 p_3 p_4 z^{(0,1,1)}$$

$$+ p_1 p_2 q_3 p_4 z^{(1,0,1)} + q_1 p_2 q_3 p_4 z^{(1,0,1)} + p_1 q_2 q_3 p_4 z^{(0,0,1)}$$

$$+ q_1 q_2 q_3 p_4 z^{(0,0,1)} + p_1 p_2 p_3 q_4 z^{(1,1,0)} + q_1 p_2 p_3 q_4 z^{(1,1,0)}$$

$$+ p_1 q_2 p_3 q_4 z^{(0,1,0)} + q_1 q_2 p_3 q_4 z^{(0,1,0)} + p_1 p_2 q_3 q_4 z^{(1,0,0)}$$

$$+ q_1 p_2 q_3 q_4 z^{(1,0,0)} + p_1 q_2 q_3 q_4 z^{(0,0,0)} + q_1 q_2 q_3 q_4 z^{(0,0,0)}$$

This *u*-function contains pairs of terms with the same state vectors. For example, both terms $p_1 p_2 p_3 p_4 z^{(1,1,1)}$ and $q_1 p_2 p_3 p_4 z^{(1,1,1)}$ correspond to the cases when $X_2 = X_3 = X_4 = 1$, but the first term corresponds to probability $\Pr\{X_2 = X_3 = X_4 = 1,\ X_1 = 1\}$ whereas the second term corresponds to probability $\Pr\{X_2 = X_3 = X_4 = 1,\ X_1 = 0\}$. The overall probability $\Pr\{X_2 = X_3 = X_4 = 1\}$ is equal to the sum of the probabilities. Therefore:

$$\Pr\{X_2 = X_3 = X_4 = 1\} = p_1 p_2 p_3 p_4 + q_1 p_2 p_3 p_4 = p_2 p_3 p_4$$

By summing the coefficients of the terms with the same state vectors (collecting the like terms) we obtain the probabilities of the state combinations of the group V_2:

$$U_2(z) = p_1 p_2 p_3 z^{(1,1,1)} + p_1 p_2 q_3 z^{(1,1,0)} + p_1 q_2 p_3 z^{(1,0,1)}$$

$$+ p_1 q_2 q_3 z^{(1,0,0)} + q_1 p_2 p_3 z^{(0,1,1)} + q_1 p_2 q_3 z^{(0,1,0)} + q_1 q_2 p_3 z^{(0,0,1)}$$

$$+ q_1 q_2 q_3 z^{(0,0,0)}$$

This *u*-function also contains $2^3 = 8$ terms. The failure probability of the group V_2 represented by the *u*-function $U_2(z)$ is

$$\Pr\{S_2 \le 1\} = \Theta_2(U_2(z)) = p_1 q_2 q_3 + q_1 p_2 q_3 + q_1 q_2 p_3 + q_1 q_2 q_3$$

The variables S_h are mutually dependent because different groups V_h contain the same elements. Therefore, the failure probability of the entire system cannot be obtained as a sum of the probabilities $\Pr\{S_h \le r-k\}$ for $1 \le h \le n-r+1$. However, excluding the terms corresponding to the failure states from the u-functions we can obtain the system failure probability. Indeed, if for some combination of element states a certain group fails, then the entire system fails independently of the states of the elements that do not belong to this group. Therefore, the terms corresponding to the group failure can be removed from the *u*-function since they should not participate in determining further state combinations that cause system failures. This consideration lies at the base of the following algorithm, which evaluates the system reliability using the enumerative technique in order to obtain all of the possible element state combinations leading to the system's failure.

1. Initialization.

   $F = 0$; $U_{1-r}(z) = z^{\mathbf{y}_0}$ ($\mathbf{y}_0$ consists of r zeros).

2. Main loop. Repeat the following for $j = 1, \ldots, n$:

   Obtain $U_{j+1-r}(z) = U_{j-r}(z) \underset{\leftarrow}{\otimes} u_j(z)$ and collect like terms in the *u*-function obtained.

   If $j \geq r$, then add value $\Theta_k(U_{j+1-r}(z))$ to F and remove all of the terms with state vectors containing more than k zeros from $U_{j+1-r}(z)$.

Obtain the system reliability as $R = 1-F$. Alternatively, the system reliability can be obtained as the sum of the coefficients of the last *u*-function $U_{n+1-r}(z)$.

Here, we omit the proof that this algorithm obtains the system reliability. The proof can be found in [47].

Example 2.19

Consider the system from Example 2.17 and obtain the system reliability applying the algorithm presented above. The u-functions $U_{-2}(z)$, …, $U_1(z)$ are obtained in the same way as in Example 2.18. After obtaining $U_1(z)$ in the form

$$U_1(z) = p_1p_2p_3z^{(1,1,1)} + q_1p_2p_3z^{(0,1,1)} + p_1q_2p_3z^{(1,0,1)} + q_1q_2p_3z^{(0,0,1)}$$
$$+ p_1p_2q_3z^{(1,1,0)} + q_1p_2q_3z^{(0,1,0)} + p_1q_2q_3z^{(1,0,0)} + q_1q_2q_3z^{(0,0,0)}$$

we add the value of $\Theta_2(U_1(z)) = q_1q_2p_3 + p_1q_2q_3 + q_1p_2q_3 + q_1q_2q_3$ to F and remove the terms with two or more zeros from this *u*-function. The remaining *u*-function $U_1(z)$ takes the form

$$U_1(z) = p_1p_2p_3z^{(1,1,1)} + q_1p_2p_3z^{(0,1,1)} + p_1q_2p_3z^{(1,0,1)} + p_1p_2q_3z^{(1,1,0)}$$

Now we obtain $U_2(z)$ as

$$U_2(z) = U_1(z) \underset{\leftarrow}{\otimes} u_4(z) = [p_1p_2p_3z^{(1,1,1)} + q_1p_2p_3z^{(0,1,1)}$$
$$+ p_1q_2p_3z^{(1,0,1)} + p_1p_2q_3z^{(1,1,0)}] \underset{\leftarrow}{\otimes} [p_4z^1 + q_4z^0]$$
$$= p_2p_3p_4z^{(1,1,1)} + p_1q_2p_3p_4z^{(0,1,1)} + p_1p_2q_3p_4z^{(1,0,1)}$$
$$+ p_2p_3q_4z^{(1,1,0)} + p_1q_2p_3q_4z^{(0,1,0)} + p_1p_2q_3q_4z^{(1,0,0)}$$

Having $U_2(z)$ we obtain $\Theta_2(U_2(z)) = p_1(q_2p_3 + p_2q_3)q_4$. This probability is added to F. Now F takes the form

$$F = q_1q_2p_3 + p_1q_2q_3 + q_1p_2q_3 + q_1q_2q_3 + p_1(q_2p_3 + p_2q_3)q_4$$

After removing the terms with two or more zeros from $U_2(z)$ it takes the form

$$U_2(z) = p_2p_3p_4z^{(1,1,1)} + p_1q_2p_3p_4z^{(0,1,1)}$$
$$+ p_1p_2q_3p_4z^{(1,0,1)} + p_2p_3q_4z^{(1,1,0)}$$

The system reliability is equal to the sum of the coefficients of $U_2(z)$:

$$R = p_2p_3p_4 + p_1(q_2p_3 + p_2q_3)p_4 + p_2p_3q_4$$

The same result can be obtained as $R = 1-F$. This can be verified by summing R and F:

$$R + F = p_2p_3p_4 + p_1(q_2p_3 + p_2q_3)p_4 + p_2p_3q_4$$
$$+ q_1q_2p_3 + p_1q_2q_3 + q_1p_2q_3 + q_1q_2q_3 + p_1(q_2p_3 + p_2q_3)q_4$$
$$= p_2p_3p_4 + p_1(q_2p_3 + p_2q_3) + p_2p_3q_4 + q_1q_2p_3 + p_1q_2q_3$$
$$+ q_1p_2q_3 + q_1q_2q_3 = p_2p_3 + p_2q_3 + q_2p_3 + q_2q_3 = 1$$

3. Introduction to Multi-state Systems

3.1 Main Definitions and Models

3.1.1 Basic Concepts of Multi-state Systems

All technical systems are designed to perform their intended tasks in a given environment. Some systems can perform their tasks with various distinguished levels of efficiency, usually referred to as *performance rates*. A system that can have a finite number of performance rates is called an MSS. Usually, an MSS is composed of elements that in their turn can be multi-state.

Actually, a binary system is the simplest case of an MSS having two distinguished states (perfect functioning and complete failure).

There are many different situations in which a system should be considered to be an MSS. Any system consisting of different units that have a cumulative effect on the entire system performance has to be considered as an MSS. Indeed, the performance rate of such a system depends on the availability of its units, as different numbers of the available units can provide different levels of task performance.

The simplest example of such a situation is the well-known k-out-of-n system (considered in Section 2.2). These systems consist of n identical binary units and can have $n+1$ states depending on the number of available units. The system performance rate is assumed to be proportional to the number of available units and the performance rates corresponding to more than $k-1$ available units are acceptable. When the contributions of different units to the cumulative system performance rate are different, the number of possible MSS states grows dramatically, as different combinations of k available units can provide different performance rates for the entire system. Such systems cannot be analyzed using the technique developed for the binary case.

The performance rate of elements composing a system can also vary as a result of their deterioration (fatigue, partial failures) or because of variable ambient conditions. Element failures can lead to the degradation of the entire MSS performance.

The performance rates of the elements can range from perfect functioning up to complete failure. Failures that lead to decrease in the element performance are called partial failures. After partial failure, elements continue to operate at reduced performance rates, and after complete failure the elements are totally unable to perform their tasks.

Example 3.1

In a power supply system consisting of generating and transmitting facilities, each generating unit can function at different levels of capacity. Generating units are complex assemblies of many parts. The failures of different parts may lead to situations in which the generating unit continues to operate, but at a reduced capacity. This can occur during the outages of several auxiliaries, such as pulverizers, water pumps, fans, *etc.* For example, Billinton and Allan [48] describe a three-state 50 MW generating unit. The performance rates (generating capacity) corresponding to these states and probabilities of the states are presented in Table 3.1.

Table 3.1. Capacity distribution of 50 MW generator

Number of state	Generating capacity (MW)	State probability
1	50	0.960
2	30	0.033
3	0	0.007

Example 3.2

The state of each retransmission station in the wireless communication system considered in Example 2.9 is defined by the number of subsequent stations covered in its range. This number depends not only on the availability of station amplifiers, but also on the conditions for signal propagation, which depend on weather, solar activity, *etc.* Therefore, each station can have a different and varying range and the number of subsequent stations connected to each station is a random discrete variable that can take more than two different integer values.

The theory of MSSs was developed since the mid 1970s, when the studies of Murchland [49], El-Neveihi *et al.* [50], Barlow and Wu [51], Ross [52] appeared. These studies formulated the basic concepts of MSS reliability. Griffith [53], Natvig [54] and Hudson and Kapur [55] subsequently generalized the results obtained in the earlier studies. Natvig [56], El-Neveihi and Prochan [57], Reinshke and Ushakov [58] summarized the achievements attained until the mid 1980s. The review of the modern state of the art in MSS reliability can be found in [59].

3.1.2 Generic Multi-state System Model

In order to analyze MSS behaviour one has to know the characteristics of its elements. Any system element j can have k_j different states corresponding to the performance rates, represented by the set $\boldsymbol{g}_j=\{g_{j0}, g_{j1},\ldots, g_{jk_j-1}\}$, where g_{ji} is the performance rate of element j in the state i, $i \in \{0, 1, \ldots, k_j-1\}$.

The performance rate G_j of element j at any time instant is a random variable that takes its values from $\boldsymbol{g}_j$: $G_j \in \boldsymbol{g}_j$. In some cases the element performance cannot be measured by just a single value; more complex mathematical objects are required, usually vectors. In these cases, the element performance is defined as a random vector $\boldsymbol{G}_j$. (Some examples of such special cases are considered in Chapter 7.)

The probabilities associated with the different states (performance rates) of the system element j can be represented by the set

$$\boldsymbol{p}_j=\{p_{j0}, p_{j1},\ldots, p_{jk_j-1}\} \tag{3.1}$$

where

$$p_{ji} = \Pr\{G_j = g_{ji}\} \tag{3.2}$$

As in the case of binary systems, the state probabilities of the MSS elements can be interpreted as the state probabilities during a fixed mission time, the state probabilities at a specified time, or the availabilities (in the case of binary elements). The system reliability measure corresponds to the reliability measures used to express the state probabilities of elements.

Note that, since the element states compose the complete group of mutually exclusive events (meaning that the element can always be in one and only in one of k_j states) $\sum_{i=0}^{k_j-1} p_{ji} = 1$.

Expression (3.2) defines the p.m.f. for a discrete random variable G_j. The collection of pairs g_{ji}, p_{ji}, $i = 0, 1,\ldots, k_j-1$, completely determines the probability distribution of performance (PD) of the element j.

Observe that the behaviour of binary elements (elements with only total failures) can also be represented by performance distribution. Indeed, consider a binary element i with a nominal performance (performance rate corresponding to a fully operable state) g^* and the probability that the element is in the fully operable state p. Assuming that the performance rate of the element in a state of complete failure is zero, one obtains its PD as follows: $\boldsymbol{g}_i=\{0, g^*\}$, $\boldsymbol{p}_i=\{1-p, p\}$.

The PDs can be represented graphically in the form of cumulative curves. In this representation, each value of performance x corresponds to the probability that the element provides a performance rate that is no less than this level: $\Pr\{G_j \geq x\}$.

For comparison, the graphs representing the PD of the binary element i and the element j with five different states are presented in Figure 3.1. Observe that the cumulative discrete PD is always a decreasing stepwise function.

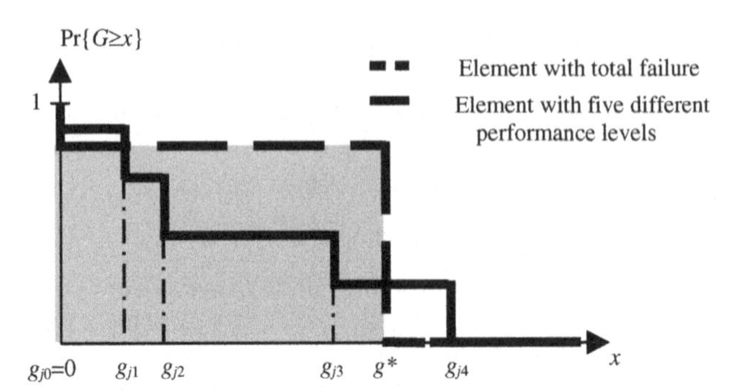

Figure 3.1. Cumulative performance curves of multi-state elements

When the MSS consists of n elements, its performance rates are unambiguously determined by the performance rates of these elements. At each moment, the system elements have certain performance rates corresponding to their states. The state of the entire system is determined by the states of its elements. Assume that the entire system has K different states and that g_i is the entire system performance rate in state i ($i \in \{0, ..., K-1\}$). The MSS performance rate is a random variable that takes values from the set $\{g_1, ..., g_{K-1}\}$.

Let $L^n = \{g_{10},...,g_{1k_1-1}\} \times \{g_{20},...,g_{2k_2-1}\} \times ... \times \{g_{n0},...,g_{nk_n-1}\}$ be the space of possible combinations of performance rates for all of the system elements and $M=\{g_0, ..., g_{K-1}\}$ be the space of possible values of the performance rate for the entire system. The transform $\phi(\mathrm{G}_1,...,\mathrm{G}_\mathrm{n})$: $L^n \rightarrow M$, which maps the space of the elements' performance rates into the space of system's performance rates, is named the system structure function. Note that the MSS structure function is an extension of a binary structure function. The only difference is in the definition of the state spaces: the binary structure function is mapped $\{0,1\}^n \rightarrow \{0,1\}$, whereas in the MSS one deals with much more complex spaces.

The set of random element performances $\{G_1, ..., G_n\}$ plays the same role in a MSS that the element state vector plays in binary systems.

Now we can define a generic model of the MSS. This model includes the p.m.f. of performances for all of the system elements and system structure function:

$$\mathbf{g}_j, \mathbf{p}_j, \ 1 \le j \le n \tag{3.3}$$

$$\phi(G_1,...,G_n) \tag{3.4}$$

It should be noted that this simple MSS model, while being satisfactory for many applications, fails to describe some important characteristics of MSSs, such as mean time to failure, mean number of failures during the operation period, *etc.* Analysis of these characteristics requires application of a random process approach and is beyond the scope of this book.

It does not matter how the structure function is defined. It can be represented in a table, in analytical form, or be described as an algorithm for unambiguously determining the system performance G for any given set $\{G_1, \ldots, G_n\}$.

Example 3.3

Consider an oil transmission system (Figure 3.2A) consisting of three pipes (elements). The oil flow is transmitted from point C to point E. The pipes' performance is measured by their transmission capacity (tons per minute). Elements 1 and 2 are binary. A state of total failure for both elements corresponds to a transmission capacity of zero and the operational state corresponds to the capacities of the elements 1.5 tons min^{-1} and 2 tons min^{-1} respectively, so that $G_1 \in \{0, 1.5\}$, $G_2 \in \{0, 2\}$. Element 3 can be in one of three states: a state of total failure corresponding to a capacity of zero, a state of partial failure corresponding to a capacity of 1.8 tons min^{-1} and a fully operational state with a capacity of 4 tons min^{-1} so that $G_3 \in \{0, 1.8, 4\}$. The system output performance rate is defined as the maximum flow that can be transmitted from C to E.

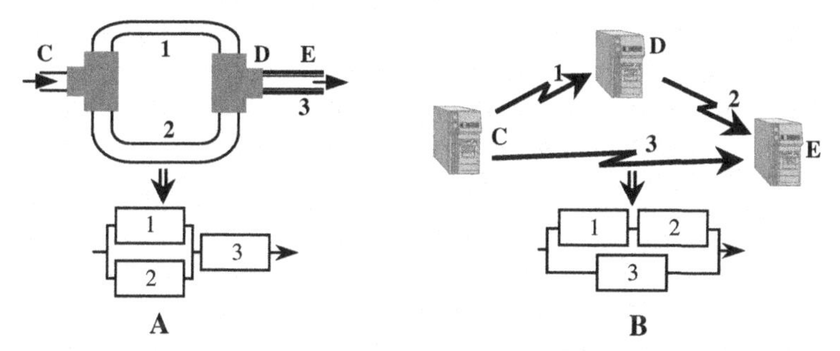

Figure 3.2. Two different MSSs with identical structure functions

The total flow between points C and D through the parallel pipes 1 and 2 is equal to the sum of the flows through each of these pipes. The flow from point D to point E is limited by the transmitting capacity of element 3. On the other hand, this flow cannot be greater than the flow between points C and D. Therefore, the flow between points C and E (the system performance) is

$$G = \phi(G_1, G_2, G_3) = \min\{G_1+G_2, G_3\}$$

The values of the system structure function $G = \phi(G_1, G_2, G_3)$ for all the possible system states are presented in Table 3.2.

Table 3.2. Possible states of an oil transmission system

G_1	0	0	0	0	0	0	1.5	1.5	1.5	1.5	1.5	1.5
G_2	0	0	0	2	2	2	0	0	0	2	2	2
G_3	0	1.8	4	0	1.8	4	0	1.8	4	0	1.8	4
$\phi(G_1, G_2, G_3)$	0	0	0	1	1.8	2	0	1.5	1.5	0	1.8	3.5

Example 3.4

Consider a data transmission system (Figure 3.2B) consisting of three fully reliable network servers and three data transmission channels (elements). The data can be transmitted from server C to server E through server D or directly. The time of data transmission between the servers depends on the state of the corresponding channel and is considered to be the channel performance rate. This time is measured in seconds.

Elements 1 and 2 are binary. They may be in a state of total failure when data transmission is impossible. In this case the data transmission time is formally defined as ∞. They may also be in a fully operational state, when they provide data transmission during 1.5 s and 2 s respectively: $G_1 \in \{\infty,1.5\}$, $G_2 \in \{\infty,2\}$. Element 3 can be in one of three states: a state of total failure, a state of partial failure with data transmission during 4 s and a fully operational state with data transmission during 1.8 s: $G_3 \in \{\infty,4,1.8\}$. The system performance rate is defined as the total time the data can be transmitted from server A to server C.

When the data are transmitted through server D, the total time of transmission is equal to the sum of times G_1 and G_2 it takes to transmit it from server C to server D and from server D to server E respectively. If either element 1 or 2 is in a state of total failure, then data transmission through server D is impossible. For this case we formally state that $(\infty+2) = \infty$ and $(\infty+1.5) = \infty$. When the data is transmitted from server C to server E directly, the transmission time is G_3. The minimum time needed to transmit the data from C to E directly or through D determines the system transmission time. Therefore, the MSS structure function takes the form

$$G = \phi(G_1, G_2, G_3,) = \min\{G_1+G_2, G_3\}$$

Note that the different technical systems in Examples 3.3 and 3.4, even when having different reliability block diagrams (Figure 3.2A and B), correspond to the identical MSS structure functions.

3.1.3 Acceptability Function

The MSS behaviour is characterized by its evolution in the space of states. The entire set of possible system states can be divided into two disjoint subsets corresponding to acceptable and unacceptable system functioning. The system entrance into the subset of unacceptable states constitutes a failure. The MSS reliability can be defined as its ability to remain in the acceptable states during the operation period.

Since the system functioning is characterized by its output performance G, the state acceptability depends on the value of this index. In some cases this dependency can be expressed by the binary acceptability function $F(G)$ that takes a value of 1 if and only if the MSS functioning is acceptable. This takes place when the efficiency of the system functioning is completely determined by its internal state (for example, only the states where a network preserves its connectivity are acceptable). In such cases, a particular set of MSS states is of interest to the customer. Usually, the unacceptable states (corresponding to $F(G) = 0$) are interpreted as system failure states, which, when reached, imply that the system should be repaired or discarded. The set of acceptable states can also be defined when the system functionality level is of interest at a particular point in time (such as at the end of the warranty period).

Much more frequently, the system state acceptability depends on the relation between the MSS performance and the desired level of this performance (demand) that is determined outside of the system. When the demand is variable, the MSS operation period T is often partitioned into M intervals T_m ($1 \le m \le M$) and a constant demand level w_m is assigned to each interval m. In this case the demand W can be represented by a random variable that can take discrete values from the set $\boldsymbol{w}=\{w_1,\dots,w_M\}$. The p.m.f. of the variable demand can be represented (in analogy with the p.m.f. of MSS performance) by two vectors ($\boldsymbol{w}$, $\boldsymbol{q}$), where $\boldsymbol{q}=\{q_1, \dots, q_M\}$ is the vector of probabilities of corresponding demand levels $q_j = \Pr\{W = w_j\}$. The desired relation between the system performance and the demand can also be expressed by the acceptability function $F(G, W)$. The acceptable system states correspond to $F(G, W) = 1$ and the unacceptable states correspond to $F(G, W) = 0$. The last equation defines the MSS failure criterion.

Example 3.5

An on-load tap changer control system is aimed at maintaining the voltage in the electric power distribution system between $u_{\min}$ and $u_{\max}$. The exit of the system voltage outside this range constitutes the system's failure. The system's output performance is the controlled voltage $G = U$ that can vary discretely. The acceptability function can be expressed as

$$F(G)=1(G \le u_{\max}) \times 1(G \ge u_{\min})$$

or

$$F(G)=1(|2G-(u_{\min}+u_{\max})| \le u_{\max}-u_{\min})$$

Example 3.6

A power generation system should supply the customers with variable demand W. If the cumulative power of the available generating units is much greater than the demand (usually at night) then some units can be disconnected and transferred to a standby state. If the cumulative power of all of the available units is not enough to meet the demand (either because of a sharp increase in demand or due to the outage of some of the units) then the system fails. The system's performance is the cumulative available power G, which should exceed the random demand W. In this case the acceptability function takes the form

$$F(G,W) = 1(G>W)$$

This type of acceptability function is used in many practical cases when the MSS performance should exceed the demand.

3.2. Types of Multi-state System

According to the generic model (3.3) and (3.4), one can define different types of MSS by determining the performance distribution of its elements and defining the system's structure function. It is possible to invent an infinite number of different structure functions in order to obtain different models of MSS. The question is whether or not the MSS model can be applied to real technical systems. This section presents different application-inspired MSS models that are most commonly used in reliability engineering.

3.2.1 Series Structure

The series connection of system elements represents a case where a total failure of any individual element causes an overall system failure. In the binary system the series connection has a purely logical sense. The topology of the physical connections among elements represented by a series reliability block diagram can differ, as can their allocation along the system's functioning process. The essential property of the binary series system is that it can operate only when all its elements are fully available.

When an MSS is considered and the system performance characteristics are of interest, the series connection usually has a "more physical" sense. Indeed, assuming that MSS elements are connected in a series means that some processes proceed stage by stage along a line of elements. The process intensity depends on the performance rates of the elements. Observe that the MSS definition of the series connection should preserve its main property: the total failure of any element (corresponding to its performance rate equal to zero) causes the total failure of the entire system (system performance rate equal to zero).

One can distinguish several types of series MSS, depending on the type of performance and the physical nature of the interconnection among the elements.

First, consider a system that uses the capacity (productivity or throughput) of its elements as the performance measure. The operation of these systems is associated with some media flow continuously passing through the elements. Examples of these types of system are power systems, energy or materials continuous transmission systems, continuous production systems, *etc.* The element with the minimal transmission capacity becomes the bottleneck of the system [51, 60]. Therefore, the system capacity is equal to the capacity of its "weakest" element. If the capacity of this element is equal to zero (total failure), then the entire system capacity is also zero.

Example 3.7

An example of the flow transmission (capacity-based) series system is a power station coal transportation unit (Figure 3.3) that continuously supplies the system of boilers and consists of five basic elements:

1. Primary feeder, which loads the coal from the bin to the primary conveyor.
2. Set of primary conveyors, which transport the coal to the stacker-reclaimer.
3. Stacker-reclaimer, which lifts the coal up to the secondary conveyor level.
4. Secondary feeder, which loads the set of secondary conveyors.
5. Set of secondary conveyors, which supplies the burner feeding system of the boilers.

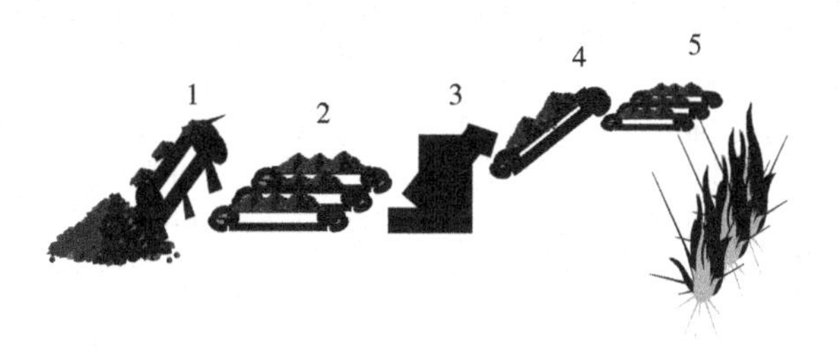

Figure 3.3. Example of flow transmission series system

The amount of coal supplied to the boilers at each time unit proceeds consecutively through each element. The feeders and the stacker-reclaimer can have two states: working with nominal throughput and total failure. The throughput of the sets of conveyors (primary and secondary) can vary depending on the availability of individual two-state conveyors. It can easily be seen that the throughput of the entire system is determined as the throughput of its elements having minimal transmission capacity. The system reliability is defined as its ability to supply a given amount of coal (demand) during a specified operation time.

Another category in the series systems is a task processing system, for which the performance measure is characterized by an operation time (processing speed). This category may include control systems, information or data processing systems, manufacturing systems with constrained operation time, *etc*. The operation of these systems is associated with consecutive discrete actions performed by the ordered line of elements. The total system operation time is equal to the sum of the operation times of all of its elements. When one measures the element (system) performance in terms of processing speed (reciprocal to the operation time), the total failure corresponds to a performance rate of zero. If at least one system element is in a state of total failure, then the entire system also fails completely. Indeed, the total failure of the element corresponds to its processing speed equal to zero, which is equivalent to an infinite operation time. In this case, the operation time of the entire system is also infinite.

Example 3.8

An example of the task processing series system is a manipulator control system (Figure 3.4) consisting of:

1. Visual image processor.
2. Multi-channel data transmission subsystem, which transmits the data from the image processor to main processing unit.
3. Main multi-processor unit, which generates control signals for manipulator actuators.
4. Manipulator.

The system performance is measured by the speed of its response to the events occurring. This speed is determined by the sum of the times needed for each element to perform its task (from initial detection of the event to the completion of the manipulator actuators performance). The time of data transmission also depends on the availability of channels, and the time of data processing depends on the availability of the processors as well as on the complexity of the image. The system reliability is defined as its ability to react within a specified time during an operation period.

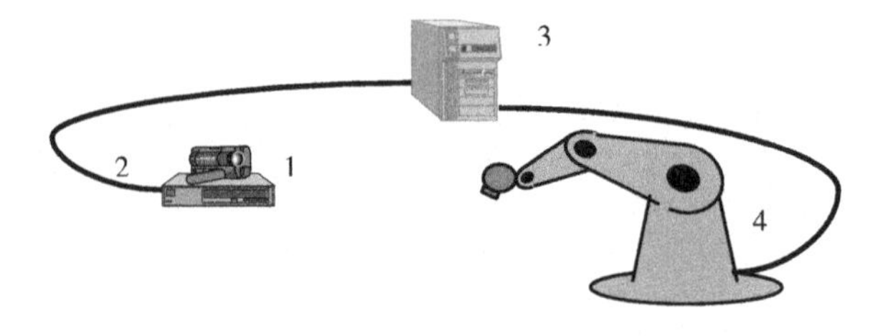

Figure 3.4. Example of task processing series system

3.2.2 Parallel Structure

The parallel connection of system elements represents a case where a system fails if and only if all of its elements fail. Two basic models of parallel systems are distinguished in binary reliability analysis. The first one is based on the assumption that all of the elements are active and work sharing. The second one represents a situation where only one element is operating at a time (active or standby redundancy without work sharing).

An MSS with a parallel structure inherits the essential property of the binary parallel system so that the total failure of the entire system occurs only when all of its elements are in total failure states. The assumption that MSS elements are connected in parallel means that some tasks can be performed by any one of the elements. The intensity of the task accomplishment depends on the performance rate of available elements.

For an MSS with work sharing, the entire system performance rate is usually equal to the sum of the performance rates of the parallel elements for both flow transmission and task processing systems. Indeed, the total flow through the former type of system is equal to the sum of flows through its parallel elements. In the latter type of MSS, the system processing speed depends on the rules of the work sharing. The most effective rule providing the minimal possible time of work completion shares the work among the elements in proportion to their processing speed. In this case, the processing speed of the parallel system is equal to the sum of the processing speeds of all of the elements.

Example 3.9

Consider a system of several parallel coal conveyors supplying the same system of boilers (Figure 3.5A) or a multi-processor control unit (Figure 3.5B), assuming that the performance rates of the elements in both systems can vary. In the first case the amount of coal supplied is equal to the sum of the amounts supplied by each one of the conveyors. In the second case the unit processing speed is equal to the sum of the processing speeds of all of its processors.

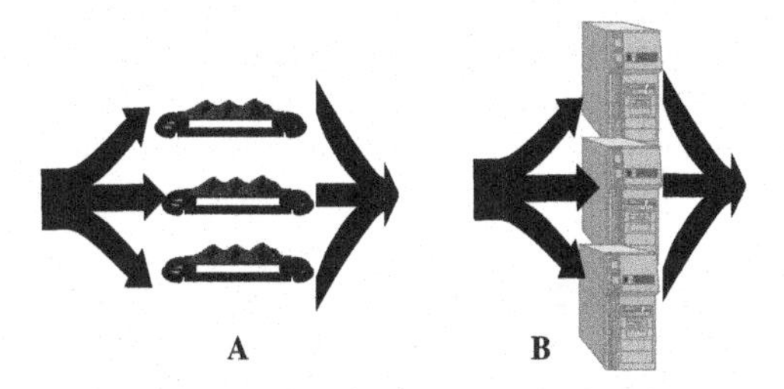

Figure 3.5. Examples of parallel systems with work sharing.
(A: flow transmission system; B: task processing system)

In an MSS without work sharing the system performance rate depends on the discipline of the elements' activation. Unlike binary systems, where all the elements have the same performance rate, the choice of an active element from the set of different ones affects the MSS performance. The most common policy in both flow transmission and task processing MSSs is to use an available element with the greatest possible performance rate. In this case, the system performance rate is equal to the maximal performance rate of the available parallel elements [51, 60].

Example 3.10

Consider a system with several generators and commutation equipment allowing only one generator to be connected to the electrical network (Figure 3.6A). If the system task is to provide the maximal possible power supply, then it keeps the most powerful generator from the set of those available in operation. The remainder of the generators can be either in an active state (hot redundancy), which means that they are rotating but are not connected to the network, or in a passive state (cold redundancy), where they do not rotate.

Another example is a multi-channel data transmission system (Figure 3.6B). When a message is sent simultaneously through all the channels, it reaches a receiver by the fastest channel and the transmission speeds of the rest of the channels do not matter.

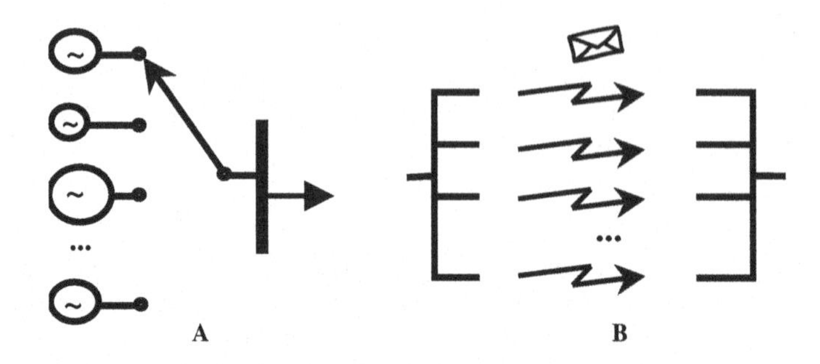

Figure 3.6. Examples of parallel systems without work sharing (A: flow transmission system; B: task processing system)

A hybrid combination of series and parallel structures results in series-parallel systems. The performance rates of these structures can be obtained by the consecutive evaluation of the performance rates of pure series or parallel subsystems and then considering these subsystems as single equivalent elements.

3.2.3 *k*-out-of-*n* Structure

The parallel MSS is not only a multi-state extension of the binary parallel structure, but it is also an extension of the binary k-out-of-*n* system. Indeed, the *k*-out-of-*n* system reliability is defined as a probability that at least *k* elements out of *n* are in operable condition (note that $k = n$ corresponds to the binary series system and $k = 1$ corresponds to the binary parallel one). The reliability of the parallel MSS with work sharing is defined as the probability that the sum of the elements' performance rates is not less than the demand. Assuming that the parallel MSS consists of *n* identical two-state elements having a capacity of 0 in a failure state and a capacity of 1 in an operational state and that the system demand is equal to *k*, one obtains the binary *k*-out-of-*n* system.

The first generalization *k*-out-of-*n* system to the multi-state case was suggested by Singh [61]. His model corresponds to the parallel flow transmission MSS with work sharing. Rushdi [62] and Wu and Chen in [63] suggested models in which the system elements have two states but can have different values of nominal performance rate. A review of the multi-state *k*-out-of-*n* models can be found in [64].

Huang *et al.* [65] suggested a multi-state generalization of the binary *k*-out-of-*n* model that cannot be considered as a parallel MSS. In this model, the entire system is in state *j* or above if at least k_j multi-state elements are in state *m*(*j*) or above.

Example 3.11

Consider a chemical reactor to which reagents are supplied by n interchangeable feeding subsystems consisting of pipes, valves, and pumps (Figure 3.7). Each feeding subsystem can provide a supply of the reagents under pressure depending on the technical state of the subsystem. Different technological processes require different numbers of reagents and different pressures. The system's state is determined by its ability to perform certain technological processes. For example, the first process requires a supply of $k_1 = 3$ reagents under pressure level $m(1) = 1$, the second process requires a supply of $k_2 = 2$ reagents under pressure level $m(2) = 2$, *etc.*

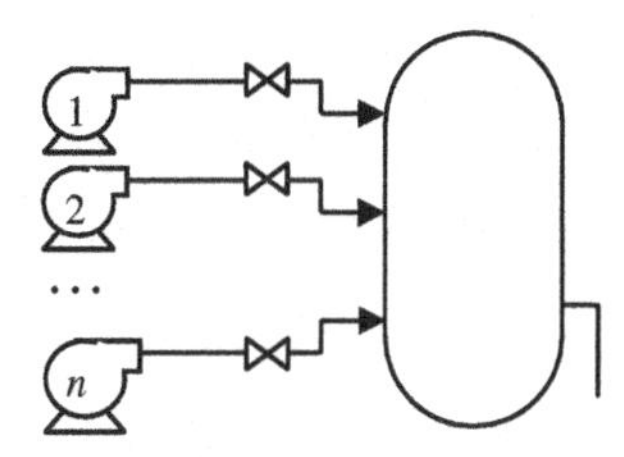

Figure 3.7. Example of multi-state k-out-of-n system that can be reduced to a parallel one

This multi-state model can be easily reduced to a set of binary k-out-of-n models. Indeed, for each system state j, every multi-state element i having the random performance G_i can be replaced with a binary element characterized by the binary state variable $X_i = 1(G_i \geq m(j))$ and the entire system can be considered as k_j-out-of-n.

3.2.4 Bridge Structure

Many reliability configurations cannot be reduced to a combination of series and parallel structures. The simplest and most commonly used example of such a configuration is a bridge structure (Figure 3.8). It is assumed that elements 1, 2 and 3, 4 of the bridge are elements of the same functionality separated from each other by some reason. The bridge structure is spread in spatially dispersed technical systems and in systems with vulnerable components separated to increase the entire system survivability. When the entire structure performance rate is of interest, it should be considered as an MSS.

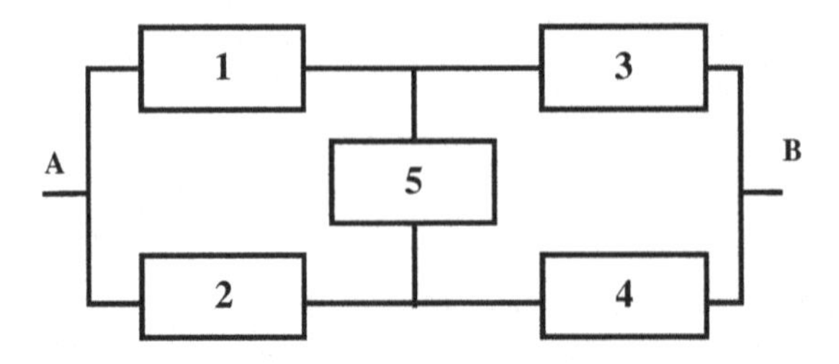

Figure 3.8 Bridge structure

Example 3.12

A local power supply system, presented in Figure 3.9, is aimed at supplying a common load. It consists of two spatially separated components containing generators and two spatially separated components containing transformers. Generators and transformers of different capacities within each component are connected by the common bus bar. To provide interchangeability of the components, bus bars of the generators are connected by a group of cables. The system output capacity (performance) must be not less than a specified load level (demand).

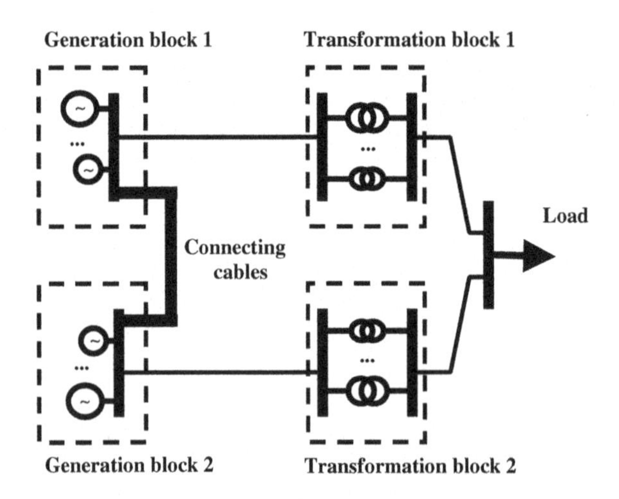

Figure 3.9. Example of MSS with bridge structure

Example 3.13

Consider a transportation task defined on a network of roads with different speed limitations (Figure 3.10). Each possible route from A to B consists of several different sections. The total travel time is determined by the random speed limitations at each section (depending on the traffic and the weather conditions) of the network and by the chosen route. This time characterizes the system performance and must be no less than some specified value (demand).

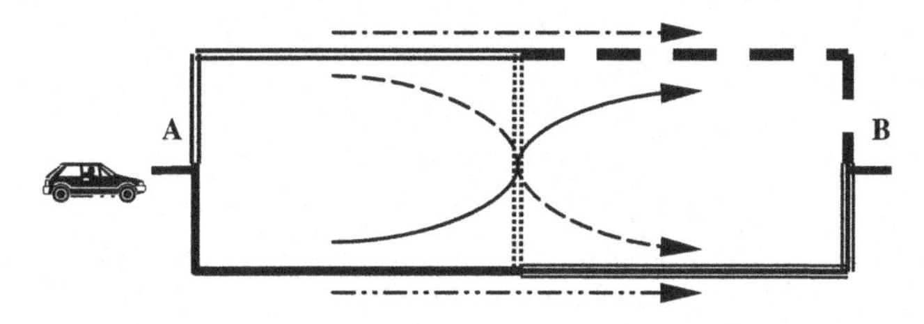

Figure 3.10. Bridge-shaped network of roads with different speed limitations

Note that the first example belongs to the flow transmission MSS. The overall power supplied to the load is equal to the total power flow through the bridge structure. The second example belongs to the task processing MSS, where the task of a vehicle is to go from point A to point B using one of four possible routes.

Determining the bridge performance rate based on its elements' performance rates is a more complicated problem than in the case of series-parallel systems. This will be addressed in the coming chapters.

3.2.5 Systems with Two Failure Modes

Systems with two failure modes consist of devices that can fail in either of two different modes. For example, switching systems not only can fail to close when commanded to close, but they can also fail to open when commanded to open. Typical examples of a switching device with two failure modes are a fluid flow valve and an electronic diode.

The binary reliability analysis considers only the reliability characteristics of elements composing the system. In many practical cases, measures of element (system) performance must be taken into account. For example, fluid-transmitting capacity is an important characteristic of a system containing fluid valves (flow transmission system), while operating time is crucial when a system of electronic switches (task processing system) is considered. The entire system with two failure modes can have different levels of output performance in both modes depending on the states of its elements at any given moment. Therefore, the system should be considered to be multi-state.

When applied to an MSS with two failure modes, reliability is usually considered to be a measure of the ability of a system to meet the demand in each mode (note that demands for the open and closed modes are different). If the probabilities of failures in open and closed modes are respectively Q_o and Q_c and the probabilities of both modes are equal to 0.5, then the entire system reliability can be defined as $R=1-0.5(Q_o+Q_c)$, since the failures in open and closed modes are mutually exclusive events.

An important property of systems with two failure modes is that redundancy, introduced into a system without any change in the reliability of the individual devices, may either increase or decrease the entire system's reliability.

Example 3.14

Consider an elevator that should be gently stopped at the upper end position by two end switches connected in a series within a circuit that activates the main engine (Figure 3.11). Assume that the operation times of the switches are T_1 and T_2 respectively in both the open and closed modes. When the switches are commanded to open (the elevator arrives at the upper end position), the first one that completes the command execution disconnects the engine. When the switches are commanded to close (an operator releases the elevator), both of them should complete the command execution in order to make the engine connected. Therefore, the slowest switch determines the execution time of the system.

The system performance (execution time) is equal to $\min\{T_1, T_2\}$ in the open mode and is equal to $\max\{T_1, T_2\}$ in the closed mode. It can be seen that if one of the two switches fails to operate ($T_j = \infty$) then the system is unable to connect the engine in the closed mode because $\max(T_1, \infty) = \max(\infty, T_2) = \infty$, whereas it remains operable in the open mode.

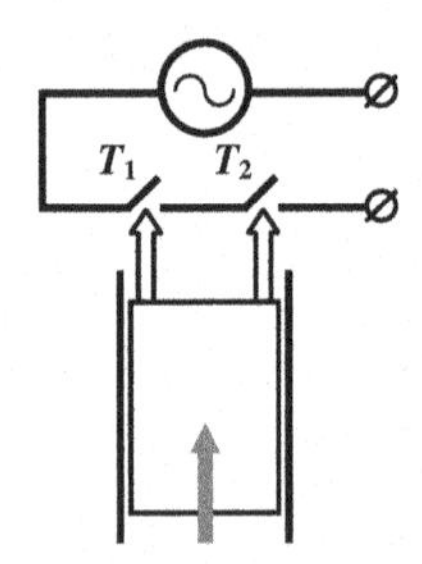

Figure 3.11. Series system with two failure modes

3.2.6 Weighted Voting Systems

Voting is widely used in human organizational systems, as well as in technical decision making systems. The use of voting for obtaining highly reliable data from multiple unreliable versions was first suggested in the mid 1950s by von Neumann. Since then the concept has been extended in many ways.

A voting system makes a decision about propositions based on the decisions of n independent individual voting units. The voting units can differ in the hardware or software used and/or by available information. Each proposition is *a priori* right or wrong, but this information is available for the units in implicit form. Therefore, the units are subject to the following three errors:

- Acceptance of a proposition that should be rejected (fault of being too optimistic).
- Rejection of a proposition that should be accepted (fault of being too pessimistic).
- Abstaining from voting (fault of being indecisive).

This can be modelled by considering the system input being either 1 (proposition to be accepted) or 0 (proposition to be rejected), which is supplied to each unit. Each unit j produces its decision (unit output), which can be 1, 0, or x (in the case of abstention). The decision made by the unit is wrong if it is not equal to the input. The errors listed above occur when:

- the input is 0, the decision is 1;
- the input is 1, the decision is 0;
- the decision is x without regard to the input.

Accordingly, the reliability of each individual voting unit can be characterized by the probabilities of its errors.

To make a decision about proposition acceptance, the system incorporates all unit decisions into a unanimous system output which is equal to x if all the voting units abstain, equal to 1 if at least k units produce decision 1, and otherwise equal to 0 (in the most commonly used majority voting systems $k = n/2$).

Note that the voting system can be considered as a special case of a k-out-of-n system with two failure modes. Indeed, if in both modes (corresponding to two possible inputs) at least k units out of n produce a correct decision, then the system also produces the correct decision. (Unlike the k-out-of-n system, the voting system can also abstain from voting, but the probability of this event can easily be evaluated as a product of the abstention probabilities of all units.)

Since the system output (number of 1-opting units) can vary, the voting systems can also be considered as the simplest case of an MSS. Such systems were intensively studied in [66-72].

A generalization of the voting system is a weighted voting system where each unit has its own individual weight expressing its relative importance within the system. The system output is x if all the units abstain. It is 1 if the cumulative weight of all 1-opting units is at least a prespecified fraction τ of the cumulative weight of all non-abstaining units. Otherwise the system output is 0.

Observe that the multi-state parallel system with two failure modes is a special case of the weighted voting system in which voting units never abstain. Indeed, in both modes (corresponding to two possible inputs) the total weight (performance) of units producing a correct decision should exceed some value (demand) determined by the system threshold.

The weighted voting systems have been suggested by Gifford [73] for maintaining the consistency and the reliability of the data stored with replication in distributed computer systems. The applications of these systems can be found in imprecise data handling, safety monitoring and self-testing, multi-channel signal processing, pattern recognition, and target detection. The reliability of weighted voting systems was studied in [73-75].

Example 3.15

An undersea target detection system consists of n electronic sensors each scanning the depths for an enemy target [76]. The sensors may both ignore a target and falsely detect a target when nothing is approaching. Each sensor has different technical characteristics and, therefore, different failure probabilities. Thus, each has a different output weight. It is important to determine a threshold level that maximizes the probability of making the correct decision.

A generalization of a weighted voting system is weighted voting classifiers consisting of n units where each one provides individual classification decisions. Each unit obtains information about some features of an object to be classified. Each object *a priori* belongs to one of the K classes, but the information about it is available to the units in implicit or incomplete form. Therefore, the units are subject to errors. The units can also abstain from making a decision (either because of unit unavailability or because of the uncertainty of information available to the unit). Obviously, some units will be highly reliable in recognizing objects of a certain class and much less reliable in recognizing objects of another class. This depends on unit specialization, as well as on the distance between objects in a space of parameters detected by the unit. The weights are used by the classifier to make unit decisions based on the analysis of some object parameters with greater influence on the system decision than ones based on other parameters.

The entire system output is based on tallying the weighted votes for each decision and choosing the winning one (plurality voting) or the one that has the total weight of supporting votes greater than some specified threshold (threshold voting). The entire system may abstain from voting if no decision ultimately wins.

The undersea target detection system from Example 3.15 becomes the weighted voting classifier if it has to detect not only a target but also to recognize the type of target (submarine, torpedo, surface vessel, *etc.*).

3.2.7 Multi-state Sliding Window Systems

The sliding window system model is a multi-state generalization of the binary consecutive k-out-of-r-from-n system, which has n ordered elements and fails if at least k out of any r consecutive elements fail (see Section 2.4). In this generalized model, the system consists of n linearly ordered multi-state elements. Each element can have a number of different states: from complete failure to perfect functioning. A performance rate is associated with each state. The system fails if an acceptability function of performance rates of any r consecutive elements is equal to zero. Usually, the acceptability function is formulated in such a manner that the system fails if the sum of the performance rates of any r consecutive elements is lower than the demand w. The special case of such a sliding window system in which all the n elements are identical and have two states with performance rates 0 and 1 is a k-out-of-r-from-n system where $w = r-k+1$.

As an example of the multi-state sliding window system, consider a conveyor-type service system that can process incoming tasks simultaneously according to a first-in-first-out rule and share a common limited resource. Each incoming task can have different states and the amount of the resource needed to process the task is different for each state of each task. The total resource needed to process r consecutive tasks should not exceed the available amount of the resource. The system fails if there is no available resource to process r tasks simultaneously.

Example 3.16

Consider a column of n vehicles crossing a bridge (Figure 3.12). The vehicles are loaded with a random number of concrete blocks (the load varies discretely from vehicle to vehicle). The maximum bridge load is w, the number of vehicles crossing the bridge simultaneously is r (this number is limited by the length of the bridge). The bridge collapses if the total load of any r consecutive vehicles is greater than w.

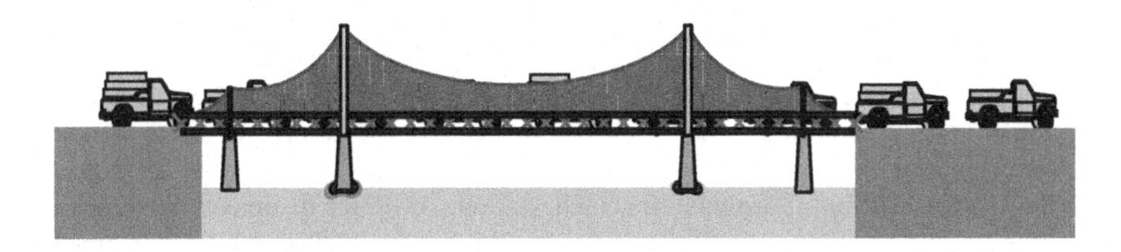

Figure 3.12. An example of a linear sliding window system

Example 3.17

In many quality control schemes the criterion for deciding when to initiate the search for a cause of a process change is based on the so-called zone tests. The process change is suspected whenever some warning limit of a measured process parameter is repeatedly or continuously violated by a sequence of points on the quality-control chart. If in k out of r consecutive tests the value of the parameter falls outside the warning limit, then the alarm search is initiated. A natural generalization of the single-zone test scheme is a scheme in which different levels of the parameter deviation are distinguished (multi-zone test). In this case, one can use the total overall parameter deviation during r consecutive tests as a search initiation criterion. For example, suppose that the parameter value can fall into M zones (the greater the zone number, the greater the parameter deviation). The alarm search should be initiated if the total sum of the numbers of zones the parameter falls in during r consecutive tests is greater than the specified value w.

3.2.8 Multi-state Consecutively Connected Systems

A linear consecutively connected system is a multi-state generalization of the binary linear consecutive k-out-of-n system that has n ordered elements and fails if at least k consecutive elements fail (see Section 2.3). In the multi-state model, the elements have different states, and when an element is in state i it is able to provide connection with i following elements (i elements following the one are assumed to be within its range). The linear multi-state consecutively connected system fails if its first and last elements are not connected (no path exists between these elements).

The first generalization of the binary consecutive k-out-of-n system was suggested by Shanthikumar [77, 78]. In his model, all of the elements can have two states, but in the working state different elements provide connection with different numbers of following elements. The multi-state generalization of the consecutive k-out-of-n system was first suggested by Hwang and Yao [79]. Algorithms for linear multi-state consecutive k-out-of-n system reliability evaluation were developed by Hwang & Yao [79], Kossow and Preuss [80], Zuo and Liang [81], and Levitin [82].

Example 3.18

Consider a set of radio relay stations with a transmitter allocated at the first station and a receiver allocated at the last station (Figure 3.13). Each station j has retransmitters generating signals that reach the next k_j stations. A more realistic model than the one presented in Example 2.9 should take into account differences in the retransmitting equipment for each station, different distances between the stations, and the varying weather conditions. Therefore, k_j should be considered to be a random value dependent on the power and availability of retransmitter amplifiers as well as on the signal propagation conditions. The aim of the system is to provide propagation of a signal from transmitter to receiver.

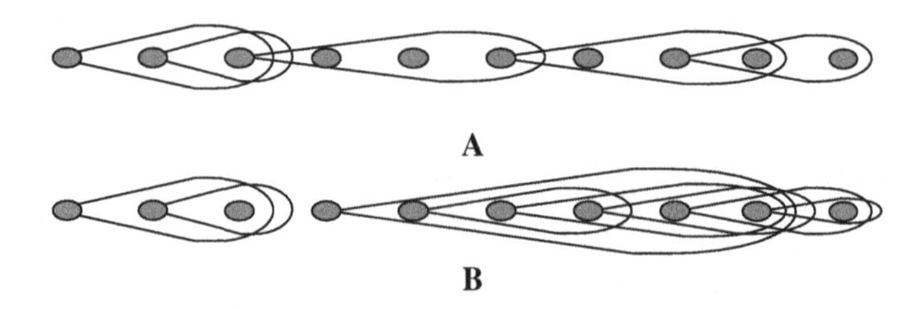

Figure 3.13. Linear consecutively connected MSS in states of successful functioning (A) and failure (B)

A circular consecutively connected system is a multi-state generalization of the binary circular consecutive k-out-of-n system. As in the linear system, each element can provide a connection to a different number of the following elements (nth element is followed by the first one). The system functions if at least one path exists that connects any pair of its elements; otherwise there is a system failure (Figure 3.14). Malinowski and Preuss [83] have shown that the problem of reliability evaluation for a circular consecutively connected system can be reduced to a set of problems of reliability evaluation for linear systems.

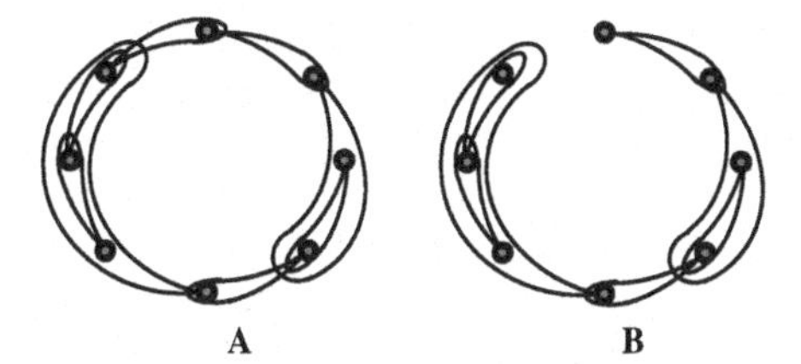

Figure 3.14. Circular consecutively connected MSS in states of successful functioning (A) and failure (B)

Example 3.19

An example of a circular consecutively connected system is a supervision system consisting of a set of detectors arranged in a circle. Each detector in state i can cover the interval between itself and the following i detectors. The state of each detector depends on the visibility conditions in its vicinity. The aim of the system is to cover the entire area.

In the examples discussed, the system reliability depends completely on the connectivity properties of the multi-state elements. In some applications, additional quantitative characteristics should be considered in order to evaluate system performance and its reliability characteristics. For example, consider a digital telecommunication system in which the signal retransmission process is associated with a certain delay. Since the delay is equal to the time needed for the digital retransmitter to processes the signal, it can be exactly evaluated and treated as a constant value for any given type of signal. When this is so, the total time of the signal propagation from transmitter to receiver can vary depending only on a combination of states of multi-state retransmitters. The entire system is considered to be in working condition if the time is not greater than a certain specified level, otherwise, the system fails.

In the more complex model, the retransmission delay of each multi-state element can also vary (depending on the load and the availability of processors). In this case each state of the multi-state element is characterized by a different delay and by a different set of following elements belonging to the range of the element.

The system's reliability for the multi-state consecutively connected systems can be defined as the probability that the system is connected or as the probability that the system's signal propagation time meets the demand. The expected system delay is also an important characteristic of its functioning.

3.2.9 Multi-state Networks

Networks are systems consisting of a set of vertices (nodes) and a set of edges that connect these vertices. Undirected and directed networks exist. While in the undirected network the edges merely connect the vertices without any consideration for direction, in the directed network the edges are ordered pairs of vertices. That is, each edge can be followed from one vertex to the next.

An acyclic network is a network in which no path (a list of vertices where each vertex has an edge from it to the next one) starts and ends at the same vertex. The directed networks considered in reliability engineering are usually acyclic.

The networks often have a single root node (source) and one or several terminal nodes (sinks). Examples of directed acyclic networks are presented in Figure 3.15A and B. The aim of the networks is the transmission of information or material flow from the source to the sinks. The transmission is possible only along the edges that are associated with the transmission media (lines, pipes, channels, *etc.*). The nodes are associated with communication centres (retransmitters, commutation, or processing stations, *etc.*)

The special case of the acyclic network is a three-structured network in which only a single path from the root node to any other node exists (Figure 3.15C). The three-structured network with a single terminal node is the linear consecutively connected system.

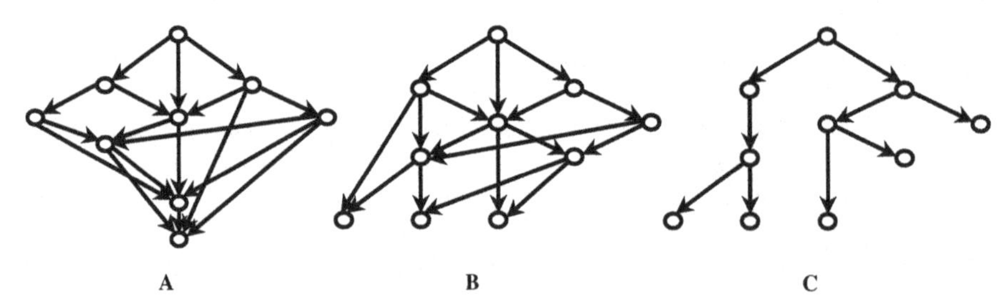

Figure 3.15. Examples of acyclic networks: A: a network with single terminal node; B: a network with several terminal nodes; C: a tree-structured network

Each network element can have its transmission characteristic, such as transmission capacity or transmission speed. The transmission process intensity depends on the transmission characteristics of the network elements and on the probabilistic properties of these elements. The most commonly used measures of the entire network performance are:

- The maximal flow between its source and sink (this measure characterizes the maximal amount of material or information that can be transmitted from the source to the sink through all of the network edges simultaneously).
- The flow of the single maximal flow path between the source and the sink (this measure characterizes the maximal amount of indivisible material or information that can be transmitted through the network by choosing a single path from the source to the sink).
- The time of transmission between the source and the sink (this measure characterizes the delivery delay in networks having edges and/or vertices with limited transmission speed).

In binary stochastic network theory, the network elements (usually edges) have a fixed level of the transmission characteristic in its working state and limited availability. The problem is to evaluate the probability that the sinks are connected to the source or the probabilic distribution of the network performance. There are several possible ways to extend the binary stochastic network model to the multi-state case.

In the multi-state edges models, the vertices are assumed fully reliable and edge transmission characteristics are random variables with a given distribution. The models correspond to:

- Communication systems with spatially distributed fully reliable stations and channels affected by environmental conditions or based on deteriorating equipment.
- Transportation systems in which the transmission delays are a function of the traffic.

In the multi-state vertices models, the edges are assumed fully reliable and the vertices are multi-state elements. Each vertex state can be associated with a certain delay, which corresponds to:

- Discrete production systems in which the vertices correspond to machines with variable productivity.
- Digital communication networks with retransmitters characterized by variable processing time.

These networks can be considered as an extension of task processing series-parallel reliability models to the case of the network structure.

The vertex states can also be associated with transmitting capacity, which corresponds to:

- Power delivery systems where vertices correspond to transformation substations with variable availability of equipment and edges to represent transmission lines.
- Continuous production systems in which vertices correspond to product processing units with variable capacity and edges represent the sequence of technological operations.

These networks can be considered as an extension of simple capacity-based series-parallel reliability models in the case of network structure (note that networks in which the maximal flow between its source and sink and the single maximal flow path between the source and the sink are of interest extend the series-parallel model with work sharing and without work sharing respectively).

In some models, each vertex state is determined by a set of vertices connected to the given one by edges. Such random connectivity models correspond mainly to wireless communication systems with spatially dispersed stations. Each station has retransmitters generating signals that can reach a set of the next stations. Note that the set composition for each station depends on the power and availability of the retransmitter amplifiers as well as on variable signal propagation conditions. The aim of the system is to provide propagation of a signal from an initial transmitter to receivers allocated at terminal vertices. (Note that it is not necessary for a signal to reach all the network vertices in order to provide its propagation to the terminal ones). This model can be considered as an extension of the multi-state linear consecutively connected systems in the case of the network structure.

The last model is generalized by assuming that the vertices can provide a connection to a random set of neighbouring vertices and can have random transmission characteristics (capacity or delay) at the same time.

In the most general mixed multi-state models, both the edges and the vertices are multi-state elements. For example, in computer networks the information transmission time depends on the time of signal processing in the node computers and the signal transmission time between the computers (depending on transmission protocol and channel loading).

The earliest studies devoted to the multi-state network reliability were by Doulliez, and Jamoulle [84], Evans [85] and Somers [86]. These models were intensively studied by Alexopoulos and Fishman [87-89], Lin [90-95], Levitin [96, 97], Yeh [98 - 101]. The three-structured networks were studied by Malinowski and Preuss [102, 103].

3.2.10 Fault-tolerant Software Systems

Software failures are caused by errors made in various phases of program development. When the software reliability is of critical importance, special programming techniques are used in order to achieve its fault tolerance. Two of the best-known fault-tolerant software design methods are n-version programming (NVP) and recovery block scheme (RBS) [104]. Both methods are based on the redundancy of software modules (functionally equivalent but independently developed) and the assumption that coincidental failures of modules are rare. The fault tolerance usually requires additional resources and results in performance penalties (particularly with regard to computation time), which constitutes a trade-off between software performance and reliability.

The NVP approach presumes the execution of n functionally equivalent software modules (called versions) that receive the same input and send their outputs to a voter, which is aimed at determining the system output. The voter produces an output if at least k out of n outputs agree (it is presumed that the probability that k wrong outputs agree is negligibly small). Otherwise, the system fails. Usually, majority voting is used in which n is odd and $k = (n+1)/2$.

In some applications, the available computational resources do not allow all of the versions to be executed simultaneously. In these cases, the versions are executed according to some predefined sequence and the program execution terminates either when k versions produce the same output (success) or after the execution of all the n versions when the number of equivalent outputs is less than k (failure). The entire program execution time is a random variable depending on the parameters of the versions (execution time and reliability), and on the number of versions that can be executed simultaneously.

Example 3.20

Consider an NVP system consisting of five versions (Figure 3.16). The system fails if the versions produce less than $k = 3$ coinciding (correct) outputs. Each version is characterized by its fixed running time and reliability. It can be seen that if the versions are executed consecutively (Figure 3.16A) the total system execution time can take the values of 28 (when the first three versions succeed), 46 (when any one of the first three versions fails and the fourth version succeeds) and 65 (when two of the first four versions fail and the fifth version succeeds). If two versions can be executed simultaneously and the versions start their execution in accordance with their numbers (Figure 3.16B) then the total system execution time can take on the values of 16 (when the first three versions succeed), 30 (when any one of the first three versions fails and the fourth version succeeds) and 35 (when two of the first four versions fail and the fifth version succeeds).

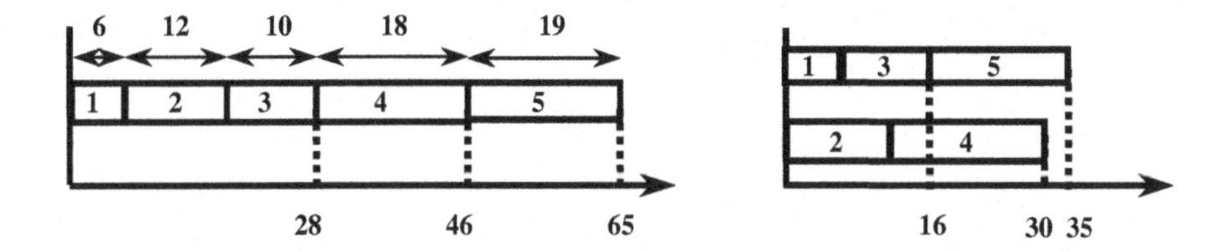

Figure 3.16. Execution of five-version fault-tolerant program with one (A) and two (B) versions executed simultaneously

In the RBS approach, after executing each version, its output is tested by an acceptance test block (ATB). If the ATB accepts the version output, then the process is terminated and the version output becomes the output of the entire system. If all of the n versions do not produce the accepted output, then the system fails. If the computational resources allow simultaneous execution of several versions, then the versions are executed according to some predefined sequence and the entire program terminates either when one of the versions produces the output accepted by the ATB (success) or after the execution of all the n versions if no output is accepted by the ATB (failure). If the acceptance test time is included in the execution time of each version, then the RBS performance model becomes identical to the performance model of the NVP with $k = 1$ (in this case k is the number of the correct outputs, but not the number of the outputs that agree).

The fault-tolerant programs can be considered as MSSs with the system performance defined as its total execution time. These systems are extensions of the binary k-out-of-n systems. Indeed, the fault-tolerant program in which all of its versions are executed simultaneously and have the same termination time is a simple k-out-of-n system.

Estimating the effect of the fault-tolerant programming on system performance is especially important for safety-critical real-time computer applications. In applications where the execution time of each task is of critical importance, the system reliability is defined as a probability that the correct output is produced within a specified time.

In applications where the average system productivity (the number of tasks executed) over a fixed mission time is of interest, the system reliability is defined as the probability that it produces correct outputs regardless of the total execution time, while the conditional expected system execution time is considered to be a measure of its performance. This index determines the system expected execution time given that the system does not fail.

Since the performance of fault-tolerant programs depends on the availability of computational resources, the impact of hardware availability should also be taken into account when the system's reliability is evaluated.

3.3 Measures of Multi-state System Performance and their Evaluation Using the UGF

To characterize MSS behaviour numerically from a reliability and performance point of view, one has to determine the MSS performance measures. Some of the measures are based on a consideration of the system's evolution in the time domain. In this case, the relation between the system's output performance and the demand represented by the two corresponding stochastic processes must be studied. This study is not within the scope of this book since the UGF technique allows one to determine only the measures based on performance distributions.

When a system is considered in the given time instant or in a steady state (when its output performance distribution does not depend on time) its behaviour is determined by its performance rate represented as a random variable G. Consider several measures of system output performance that can characterize any system state.

The first natural measure of a system's performance is its output performance rate G. This measure can be obtained by applying the system structure function over the performance rates of the system's elements. Each specific system state j is characterized by the associated system performance rate $G = g_j$, which determines the system's behaviour in the given state but does not reflect the acceptability of the state from the customer's point of view.

In order to represent the system state acceptability, we can use the acceptability function $F(G)$ or $F(G,W)$ defined in Section 3.1.3. The acceptability function divides the entire set of possible system states into two disjoint subsets (acceptable and unacceptable states). Therefore, if the system's behaviour is represented by an acceptability function, the system as a whole can be considered to be a binary one.

In many practical cases it is not enough to know whether the state is acceptable or not. The damage caused by an unacceptable state can be a function of the system's performance rate deviation from a demand. Usually, the one-sided performance deviation (performance deviation from a demand when the demand is not met) is of interest. For example, the cumulative generating capacity of available electric generators should exceed the demand. In this case the possible performance deviation (performance deficiency) takes the form

$$D^{-}(G,W) = \max(W - G, 0) \tag{3.5}$$

When the system's performance should not exceed demand (for example, the time needed to complete the assembling task in an assembly line should be less than a maximum allowable value in order to maintain the desired productivity), the performance redundancy is used as a measure of the performance deviation:

$$D^{+}(G,W)=\max(G-W,0) \tag{3.6}$$

Figure 3.17 shows an example of the behaviour of the MSS performance and the demand as the realizations of the discrete stochastic processes and the corresponding realizations of the measures of the system's output performance.

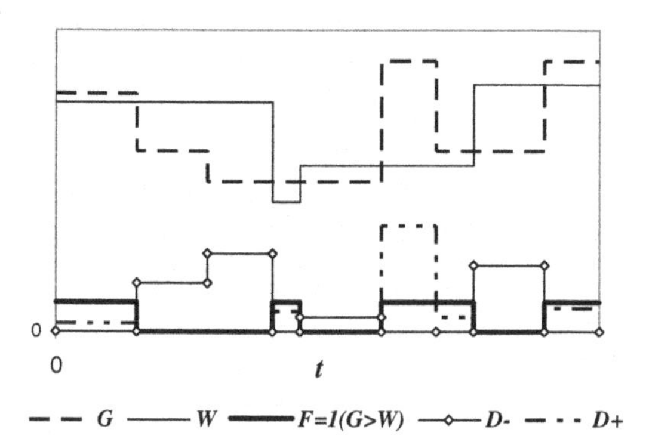

Figure 3.17. Example of a realization of the measures of system output performance

The expected system acceptability $E(F(G,W))$ determines the system reliability or availability (the probability that the MSS is in one of the acceptable states: $\Pr\{F(G,W)=1\}$). Depending on the meaning of the system and element state probabilities, it can be interpreted as $R(t)$, the MSS reliability at a specified time t, as $R(T)$, the MSS reliability during a fixed mission time T (for unrepairable systems), or as instantaneous (point) availability $A(t)$ or steady-state availability A (for repairable systems).

The expected system performance deviation $E(D^{-}(G,W))$ or $E(D^{+}(G,W))$can be interpreted as Δ_t, the expected instantaneous performance deviation at instant t, or as a mean steady-state performance deviation Δ.

In some cases we need to know the conditional expected performance of the MSS. This measure represents the mean performance of the MSS given that it is in acceptable states. In order to determine the conditional expected performance $\tilde{\varepsilon}$ we define the auxiliary function as $\tilde{G}(G,W)=GF(G,W)$. The measure $\tilde{\varepsilon}$ can be determined as follows:

$$\tilde{\varepsilon}=E(\tilde{G})/\Pr\{F(G,W)=1\}=E(GF(G,W))/E(F(G,W)) \tag{3.7}$$

Having the p.m.f. of the random MSS output performance G and the p.m.f. of the demand W in the form of u-functions $U_{\mathrm{MSS}}(z)$ and $u_w(z)$, one can obtain the u-functions representing the p.m.f. of the random functions $F(G,W)$,

$\tilde{G}(G,W)$, $D^{-}(G,W)$ or $D^{+}(G,W)$ using the corresponding composition operators over $U_{\text{MSS}}(z)$ and $u_w(z)$:

$$U_F(z) = U_{\text{MSS}}(z) \underset{F}{\otimes} u_w(z) \tag{3.8}$$

$$U_{\tilde{G}}(z) = U_{\text{MSS}}(z) \underset{\tilde{G}}{\otimes} u_w(z) \tag{3.9}$$

$$U_D(z) = U_{\text{MSS}}(z) \underset{D}{\otimes} u_w(z) \tag{3.10}$$

Since the expected values of the functions G, F, D and $\tilde{G}$ are equal to the derivatives of the corresponding u-functions $U_{\text{MSS}}(z)$, $U_F(z)$, $U_D(z)$ and $U_{\tilde{G}}(z)$ at $z = 1$, the MSS performance measures can now be obtained as

$$E(G) = U'_{\text{MSS}}(1) \tag{3.11}$$

$$E(F(G,W)) = U'_F(1) \tag{3.12}$$

$$E(D(G,W)) = U'_D(1) \tag{3.13}$$

$$E(\tilde{G}(G,F)) / E(F(G,W)) = U'_{\tilde{G}}(1) / U'_F(1) \tag{3.14}$$

Example 3.21

Consider two power system generators with a nominal capacity of 100 MW as two separate MSSs. In the first generator, some types of failure require its capacity G_1 to be reduced to 60 MW and other types lead to a complete outage. In the second generator, some types of failure require its capacity G_2 to be reduced to 80 MW, others lead to a capacity reduction to 40 MW, and others lead to a complete outage. The generators are repairable and each of their states has a steady-state probability.

Both generators should meet a variable two-level demand W. The high level (day) demand is 50 MW and has the probability 0.6; the low level (night) demand is 30 MW and has the probability 0.4.

The capacity and demand can be presented as a fraction of the nominal generator capacity. There are three possible relative capacity levels that characterize the performance of the first generator:

$$g_{10} = 0.0,\ \ g_{11} = 60/100 = 0.6,\ \ g_{12} = 100/100 = 1.0$$

and four relative capacity levels that characterize the performance of the second generator:

$$g_{20} = 0.0,\ \ g_{21} = 40/100 = 0.4,\ \ g_{22} = 80/100 = 0.8,\ \ g_{23} = 100/100 = 1.0$$

Assume that the corresponding steady-state probabilities are

$$p_{10} = 0.1,\ \ p_{11} = 0.6,\ \ p_{12} = 0.3$$

for the first generator and

$$p_{20} = 0.05,\ p_{21} = 0.35,\ p_{22} = 0.3,\ p_{23} = 0.3$$

for the second generator and that the demand distribution is

$$w_1 = 50/100 = 0.5,\ w_2 = 30/100 = 0.3,\ q_1 = 0.6,\ q_2 = 0.4$$

The u-functions representing the capacity distribution of the generators (the p.m.f. of random variables G_1 and G_2) take the form

$$U_1(z) = 0.1z^0+0.6z^{0.6}+0.3z^1,\quad U_2(z) = 0.05z^0+0.35z^{0.4}+0.3z^{0.8}+0.3z^1$$

and the u-function representing the demand distribution takes the form

$$u_w(z) = 0.6z^{0.5}+0.4z^{0.3}$$

The mean steady-state performance (capacity) of the generators can be obtained directly from these u-functions:

$$\varepsilon_1 = E(G_1) = U'_1(1) = 0.1\times 0 + 0.6\times 0.6 + 0.3\times 1.0 = 0.66$$

which means 66% of the nominal generating capacity for the first generator, and

$$\varepsilon_2 = E(G_2) = U'_2(1) = 0.05\times 0 + 0.35\times 0.4 + 0.3\times 0.8 + 0.3\times 1.0 = 0.68$$

which means 68% of the nominal generating capacity for the second generator.

The available generation capacity should be no less than the demand. Therefore, the system acceptability function takes the form

$$F(G,W) = 1(G \geq W)$$

and the system performance deficiency takes the form

$$D^-(G,W) = \max(W - G, 0)$$

The u-functions corresponding to the p.m.f. of the acceptability function are obtained using the composition operator $\underset{F}{\otimes}$:

$$U_{F1}(z) = U_1(z)\underset{F}{\otimes}u_w(z) = (0.1z^0 + 0.6z^{0.6} + 0.3z^1)\underset{F}{\otimes}(0.6z^{0.5} + 0.4z^{0.3})$$

$$= 0.06z^{1(0\geq 0.5)} + 0.36z^{1(0.6\geq 0.5)} + 0.18z^{1(1\geq 0.5)} + 0.04z^{1(0\geq 0.3)}$$

$$+ 0.24z^{1(0.6\geq 0.3)} + 0.12z^{1(1\geq 0.3)} = 0.06z^0 + 0.36z^1 + 0.18z^1 + 0.04z^0$$

$$+ 0.24z^1 + 0.12z^1 = 0.9z^1 + 0.1z^0$$

$$U_{F2}(z) = U_2(z) \underset{F}{\otimes} u_w(z)$$
$$= (0.05z^0 + 0.35z^{0.4} + 0.3z^{0.8} + 0.3z^1) \underset{F}{\otimes} (0.6z^{0.5} + 0.4z^{0.3})$$
$$= 0.03z^0 + 0.21z^0 + 0.18z^1 + 0.18z^1 + 0.02z^0 + 0.14z^1$$
$$+ 0.12z^1 + 0.12z^1 = 0.74z^1 + 0.26z^0$$

The system availability (expected acceptability) is

$$A_1 = E(1(G_1 \geq W) = U'_{F1}(1) = 0.9$$
$$A_2 = E(1(G_2 \geq W) = U'_{F2}(1) = 0.74$$

The u-functions corresponding to the p.m.f. of the performance deficiency function are obtained using the composition operator $\underset{D}{\otimes}$:

$$U_{D1}(z) = U_1(z) \underset{D}{\otimes} u_w(z)$$
$$= (0.1z^0 + 0.6z^{0.6} + 0.3z^1) \underset{D}{\otimes} (0.6z^{0.5} + 0.4z^{0.3})$$
$$= 0.06z^{\max(0.5-0,0)} + 0.36z^{\max(0.5-0.6,0)} + 0.18z^{\max(0.5-1,0)}$$
$$+ 0.04z^{\max(0.3-0,0)} + 0.24z^{\max(0.3-0.6,0)} + 0.12z^{\max(0.3-1,0)}$$
$$= 0.06z^{0.5} + 0.36z^0 + 0.18z^0 + 0.04z^{0.3} + 0.24z^0 + 0.12z^0$$
$$= 0.06z^{0.5} + 0.04z^{0.3} + 0.9z^0$$

$$U_{D2}(z) = U_2(z) \underset{D}{\otimes} u_w(z)$$
$$= (0.05z^0 + 0.35z^{0.4} + 0.3z^{0.8} + 0.3z^1) \underset{D}{\otimes} (0.6z^{0.5} + 0.4z^{0.3})$$
$$= 0.03z^{0.5} + 0.21z^{0.1} + 0.18z^0 + 0.18z^0 + 0.02z^{0.3} + 0.14z^0$$
$$+ 0.12z^0 + 0.12z^0 = 0.03z^{0.5} + 0.21z^{0.1} + 0.02z^{0.3} + 0.74z^0$$

The expected performance deficiency is

$$\Delta_1 = E(\max(W - G_1, 0)) = U'_{D1}(1)$$
$$= 0.06 \times 0.5 + 0.04 \times 0.3 + 0.9 \times 0 = 0.042$$

$$\Delta_2 = E(\max(W - G_2, 0)) = U'_{D2}(1)$$
$$= 0.03 \times 0.5 + 0.21 \times 0.1 + 0.02 \times 0.3 + 0.74 \times 0 = 0.042$$

In this case, Δ may be interpreted as expected electrical power unsupplied to consumers. The absolute value of this unsupplied demand is 4.2 MW for both generators. Multiplying this index by T, the system operating time considered, one can obtain the expected unsupplied energy.

Note that since the performance measures obtained have different natures they cannot be used interchangeably. For instance, in the present example the first generator performs better than the second one when availability is considered ($A_1>A_2$), the second generator performs better than the first one when the expected capacity is considered ($\varepsilon_1<\varepsilon_2$), and both generators have the same expected unsupplied demand ($\Delta_1 = \Delta_2$).

Now we determine the conditional expected system performance. The u-functions corresponding to the p.m.f. of the function $\tilde{G}$ are obtained using the composition operator $\underset{\tilde{G}}{\otimes} = \underset{GF}{\otimes}$:

$$U_{\tilde{G}1}(z) = U_1(z)\underset{\tilde{G}}{\otimes}u_w(z) = (0.1z^0 + 0.6z^{0.6} + 0.3z^1)\underset{GF}{\otimes}(1z^{0.5} + 0.4z^{0.3})$$

$$= 0.06z^{0\times1(0\geq0.5)} + 0.36z^{0.6\times1(0.6\geq0.5)} + 0.18z^{1\times1(1\geq0.5)} + 0.04z^{0\times1(0\geq0.3)}$$

$$+ 0.24z^{0.6\times1(0.6\geq0.3)} + 0.12z^{1\times1(1\geq0.3)} = 0.06z^0 + 0.36z^{0.6} + 0.18z^1$$

$$+ 0.04z^0 + 0.24z^{0.6} + 0.12z^1 = 0.3z^1 + 0.6z^{0.6} + 0.1z^0$$

$$U_{\tilde{G}2}(z) = U_2(z)\underset{\tilde{G}}{\otimes}u_w(z)$$

$$= (0.05z^0 + 0.35z^{0.4} + 0.3z^{0.8} + 0.3z^1)\underset{G\cdot F}{\otimes}(0.6z^{0.5} + 0.4z^{0.3})$$

$$= 0.03z^0 + 0.21z^0 + 0.18z^{0.8} + 0.18z^1 + 0.02z^0 + 0.14z^{0.4}$$

$$+ 0.12z^{0.8} + 0.12z^1 = 0.3z^1 + 0.3z^{0.8} + 0.14z^{0.4} + 0.26z^0$$

The system conditional expected performance is

$$\tilde{\varepsilon}_1 = U'_{\tilde{G}1}(1)/U'_{F1}(1) = (0.3\times1 + 0.6\times0.6 + 0.1\times0)/0.9$$
$$= 0.66/0.9 = 0.733$$

$$\tilde{\varepsilon}_2 = U'_{\tilde{G}2}(z)/U'_{F2}(1) = (0.3\times1 + 0.3\times0.8 + 0.14\times0.4 + 0.26\times0)/0.74$$
$$= 0.596/0.74 = 0.805$$

This means that generators 1 and 2 when they meet the variable demand, have average capacities 73.3 MW and 80.5 MW respectively.

Observe that the acceptability function $F(G,W)$ is a binary one. Therefore, if the demand is constant ($W \equiv w$), operator $U_F(z) = U_{\text{MSS}}(z) \otimes_F w$ produces a u-

function in which all of the terms corresponding to the acceptable states will have the exponent 1 and all of the terms corresponding to the unacceptable states will have the exponent 0. It is easily seen that $U'_F(1)$ is equal to the sum of the coefficients of the terms with exponents 1. Therefore, instead of obtaining the *u*-function of $F(G, w)$ and calculating its derivative at $z = 1$ for determining the system's reliability (availability), one can calculate the sum of the terms in $U_{\text{MSS}}(z)$ that correspond to the acceptable states. Introducing an operator $\delta_w(U_{\text{MSS}}(z))$ that produces the sum of the coefficients of those terms in $U_{\text{MSS}}(z)$ that have exponents g satisfying the condition $F(g,w) = 1$ (and correspond to the states acceptable for the demand level w) we obtain the following simple expression for the system's expected acceptability:

$$E(F(G,w)) = \delta_w(U_{\text{MSS}}(z)) \tag{3.15}$$

When the demand is variable, the system reliability can also be obtained as

$$\sum_{i=1}^{M} \Pr(W = w_i)E(F(G, w_i)) = \sum_{i=1}^{M} q_i E(F(G, w_i)) = \sum_{i=1}^{M} q_i \delta_{w_i}(U_{\text{MSS}}(z)) \tag{3.16}$$

Example 3.22

Consider the two power system generators presented in the previous example and obtain the system availability directly from the *u*-functions $U_{\text{MSS}}(z)$ using Equation (3.16). Since, in this example, $F(G,W) = 1(G \ge W)$, the operator $\delta_w(U(z))$ sums the coefficients of the terms having exponents not less than w in the *u*-function $U(z)$:

For the first generator with $U_1(z) = 0.1z^0+0.6z^{0.6}+0.3z^1$

$$E(F(G_1, w_1)) = \delta_{0.5}(0.1z^0 + 0.6z^{0.6} + 0.3z^1) = 0.6 + 0.3 = 0.9$$

$$E(F(G_1, w_2)) = \delta_{0.3}(0.1z^0 + 0.6z^{0.6} + 0.3z^1) = 0.6 + 0.3 = 0.9$$

$$A_1 = q_1 E(F(G_1, w_1)) + q_2 E(F(G_1, w_2)) = 0.6 \times 0.9 + 0.4 \times 0.9 = 0.9$$

For the second generator with $U_2(z) = 0.05z^0 + 0.35z^{0.4} + 0.3z^{0.8} + 0.3z^1$

$$E(F(G_2, 0.5)) = \delta_{0.5}(0.05z^0 + 0.35z^{0.4} + 0.3z^{0.8} + 0.3z^1) = 0.3+0.3 = 0.6$$

$$E(F(G_2, 0.3)) = \delta_{0.3}(0.05z^0 + 0.35z^{0.4} + 0.3z^{0.8} + 0.3z^1)$$
$$= 0.35 + 0.3 + 0.3 = 0.95$$

$$A_1 = q_1 E(F(G_2, w_1)) + q_2 E(F(G_2, w_2)) = 0.6 \times 0.6 + 0.4 \times 0.95 = 0.74$$

4. Universal Generating Function in Analysis of Series-Parallel Multi-state Systems

4.1 Reliability Block Diagram Method

Having a generic model of an MSS in the form of Equations (3.3) and (3.4) we can obtain the measures of MSS reliability by applying the following steps:

1. Represent the p.m.f. of the random performance of each system element j, Equations (3.1) and (3.2), in the form of the u-function

$$u_j(z) = \sum_{i=0}^{k_j-1} p_{ji} z^{g_{ji}}, \quad 1 \le j \le n \tag{4.1}$$

2. Obtain the u-function of the entire system (representing the p.m.f. of the random variable G) applying the composition operator that uses the system structure function.
3. Obtain the u-functions representing the random functions F, $\tilde{G}$ and D using operators (3.8)-(3.10).
4. Obtain the system reliability measures by calculating the values of the derivatives of the corresponding u-functions at $z = 1$ and applying Equations (3.11)-(3.14).

While steps 1, 3 and 4 are rather trivial, step 2 may involve complicated computations. Indeed, the derivation of a system structure function for various types of system is usually a difficult task.

As shown in Chapter 1, representing the functions in the recursive form is beneficial from both the derivation clarity and computation simplicity viewpoints. In many cases, the structure function of the entire MSS can be represented as the composition of the structure functions corresponding to some subsets of the system elements (MSS subsystems). The u-functions of the subsystems can be obtained separately and the subsystems can be further treated as single equivalent elements with the performance p.m.f. represented by these u-functions.

The method for distinguishing recurrent subsystems and replacing them with single equivalent elements is based on a graphical representation of the system structure and is referred to as the reliability block diagram method. This approach is usually applied to systems with a complex series-parallel configuration.

While the structure function of a binary series-parallel system is unambiguously determined by its configuration (represented by the reliability block diagram), the structure function of a series-parallel MSS also depends on the physical meaning of the system and of the elements' performance and on the nature of the interaction among the elements.

4.1.1 Series Systems

In the flow transmission MSS, where performance is defined as capacity or productivity, the total capacity of a subsystem containing n independent elements connected in series is equal to the capacity of a bottleneck element (the element with least performance). Therefore, the structure function for such a subsystem takes the form

$$\phi_{ser}(G_1,...,G_n) = \min\{G_1,...,G_n\} \quad (4.2)$$

In the task processing MSS, where the performance is defined as the processing speed (or operation time), each system element has its own operation time and the system's total task completion time is restricted. The entire system typically has a time resource that is larger than the time needed to perform the system's total task. But unavailability or deteriorated performance of the system elements may cause time delays, which in turn would cause the system's total task performance time to be unsatisfactory. The definition of the structure function for task processing systems depends on the discipline of the elements' interaction in the system.

When the system operation is associated with consecutive discrete actions performed by the ordered line of elements, each element starts its operation after the previous one has completed its operation. Assume that the random performances G_j of each element j is characterized by its processing speed. The random processing time T_j of any system element j is defined as $T_j = 1/G_j$. The total time of task completion for the entire system is

$$T = \sum_{j=1}^{n} T_j = \sum_{j=1}^{n} G_j^{-1} \quad (4.3)$$

The entire system processing speed is therefore

$$G = 1/T = (\sum_{j=1}^{n} G_j^{-1})^{-1} \quad (4.4)$$

Note that if for any j $G_j = 0$ the equation cannot be used, but it is obvious that in this case $G = 0$. Therefore, one can define the structure function for the series task processing system as

$$\phi_{\text{ser}}(G_1,\ldots,G_n)=\times(G_1,\ldots,G_n)=\begin{cases}1/\sum_{j=1}^{n}G_j^{-1} & \text{if } \prod_{j=1}^{n}G_j\neq 0\\ 0 & \text{if } \prod_{j=1}^{n}G_j=0\end{cases} \quad (4.5)$$

One can see that the structure functions presented above are associative and commutative (*i.e.* meet conditions (1.26) and (1.28)). Therefore, the u-functions for any series system of described types can be obtained recursively by consecutively determining the u-functions of arbitrary subsets of the elements. For example the u-function of a system consisting of four elements connected in a series can be determined in the following ways:

$$\begin{aligned}&[(u_1(z)\underset{\phi_{\text{ser}}}{\otimes}u_2(z))\underset{\phi_{\text{ser}}}{\otimes}u_3(z)]\underset{\phi_{\text{ser}}}{\otimes}u_4(z)\\&=(u_1(z)\underset{\phi_{\text{ser}}}{\otimes}u_2(z))\underset{\phi_{\text{ser}}}{\otimes}(u_3(z)\underset{\phi_{\text{ser}}}{\otimes}u_4(z))\end{aligned} \quad (4.6)$$

and by any permutation of the elements' u-functions in this expression.

Example 4.1

Consider a system consisting of n elements with the total failures connected in series. Each element j has only two states: operational with a nominal performance of g_{j1} and failure with a performance of zero. The probability of the operational state is p_{j1}. The u-function of such an element is presented by the following expression:

$$u_j(z)=(1-p_{j1})z^0+p_{j1}z^{g_{j1}},\quad j=1,\ldots,n$$

In order to find the u-function for the entire MSS, the corresponding $\otimes_{\phi_{\text{ser}}}$ operators should be applied. For the MSS with the structure function (4.2) the system u-function takes the form

$$U(z)=\underset{\min}{\otimes}(u_1(z),\ldots,u_n(z))=(1-\prod_{j=1}^{n}p_{j1})z^0+\prod_{j=1}^{n}p_{j1}z^{\min\{g_{11},\ldots,g_{n1}\}}$$

For the MSS with the structure function (4.5) the system u-function takes the form

$$U(z)=\underset{\times}{\otimes}\{u_1(z),\ldots,u_n(z)\}=(1-\prod_{j=1}^{n}p_{j1})z^0+\prod_{j=1}^{n}p_{j1}z^{(\sum_{j=1}^{n}g_{j1}^{-1})^{-1}}$$

Since the failure of each single element causes the failure of the entire system, the MSS can have only two states: one with the performance level of zero (failure of at least one element) and one with the performance level $\hat{g} = \min\{g_{11},..., g_{n1}\}$ for the flow transmission MSS and $\hat{g} = 1/\sum_{j=1}^{n} g_{j1}^{-1}$ for the task processing MSS.

The measures of the system performance $A(w) = \Pr\{G \geq w\}$, $\Delta^{-}(w) = E(\max(w-G,0))$ and $\varepsilon = E(G)$ are presented in the Table 4.1.

Table 4.1. Measures of MSS performance

w	$A(w)$	$\Delta^{-}(w)$	ε
$w > \hat{g}$	0	$w(1-\prod_{j=1}^{n} p_{j1}) + (w-\hat{g})\prod_{j=1}^{n} p_{j1} = w - \hat{g}\prod_{j=1}^{n} p_{j1}$	$\hat{g}\prod_{j=1}^{n} p_{j1}$
$0 < w \leq \hat{g}$	$\prod_{j=1}^{n} p_{j1}$	$w(1-\prod_{j=1}^{n} p_{j1})$	

The u-function of a subsystem containing n identical elements ($p_{j1}=p$, $g_{j1}=g$ for any j) takes the form

$$(1-p^n)z^0 + p^n z^g \tag{4.7}$$

for the system with the structure function (4.2) and takes the form

$$(1-p^n)z^0 + p^n z^{g/n} \tag{4.8}$$

for the system with the structure function (4.5).

4.1.2 Parallel Systems

In the flow transmission MSS, in which the flow can be dispersed and transferred by parallel channels simultaneously (which provides the work sharing), the total capacity of a subsystem containing n independent elements connected in parallel is equal to the sum of the capacities of the individual elements. Therefore, the structure function for such a subsystem takes the form

$$\phi_{\text{par}}(G_1,...,G_n) = +(G_1,...,G_n) = \sum_{j=1}^{n} G_j \tag{4.9}$$

In some cases only one channel out of n can be chosen for the flow transmission (no flow dispersion is allowed). This happens when the transmission is associated with the consumption of certain limited resources that does not allow simultaneous use of more than one channel. The most effective way for such a system to function is by choosing the channel with the greatest transmission capacity from the set of available channels. In this case, the structure function takes the form

$$\phi_{\text{par}}(G_1,...,G_n) = \max\{G_1,...,G_n\}. \tag{4.10}$$

In the task processing MSS, the definition of the structure function depends on the nature of the elements' interaction within the system.

First consider a system without work sharing in which the parallel elements act in a competitive manner. If the system contains n parallel elements, then all the elements begin to execute the same task simultaneously. The task is assumed to be completed by the system when it is completed by at least one of its elements. The entire system processing time is defined by the minimum element processing time and the entire system processing speed is defined by the maximum element processing speed. Therefore, the system structure function coincides with (4.10).

Now consider a system of n parallel elements with work sharing for which the following assumptions are made:

1. The work x to be performed can be divided among the system elements in any proportion.
2. The time required to make a decision about the optimal work sharing is negligible, the decision is made before the task execution and is based on the information about the elements state during the instant the demand for the task executing arrives.
3. The probability of the elements failure during any task execution is negligible.

The elements start performing the work simultaneously, sharing its total amount x in such a manner that element j has to perform x_j portion of the work and $x = \sum_{j=1}^{n} x_j$. The time of the work processing by element j is x_j/G_j. The system processing time is defined as the time during which the last portion of work is completed: $T = \max_{1 \le j \le n}\{x_j / G_j\}$. The minimal time of the entire work completion can be achieved if the elements share the work in proportion to their processing speed G_j: $x_j = xG_j / \sum_{k=1}^{n} G_k$. The system processing time T in this case is equal to $x / \sum_{k=1}^{n} G_k$ and its total processing speed G is equal to the sum of the processing speeds of its elements. Therefore, the structure function of such a system coincides with the structure function (4.9).

One can see that the structure functions presented also meet the conditions (1.26) and (1.28). Therefore, the u-functions for any parallel system of described types can be obtained recursively by the consecutive determination of u-functions of arbitrary subsets of the elements.

Example 4.2

Consider a system consisting of two elements with total failures connected in parallel. The elements have nominal performance g_{11} and g_{21} ($g_{11}<g_{21}$) and the

probability of operational state p_{11} and p_{21} respectively. The performances in the failed states are $g_{10} = g_{20} = 0$. The u-function for the entire MSS is

$$U(z) = u_1(z) \underset{\phi_{\text{par}}}{\otimes} u_2(z)$$

$$= [(1-p_{11})z^0 + p_{11}z^{g_{11}}] \underset{\phi_{\text{par}}}{\otimes} [(1-p_{21})z^0 + p_{21}z^{g_{21}}]$$

which for structure function (4.9) takes the form

$$U(z) = (1-p_{11})(1-p_{21})z^0 + p_{11}(1-p_{21})z^{g_{11}}$$

$$+ p_{21}(1-p_{11})z^{g_{21}} + p_{11}p_{21}z^{g_{11}+g_{21}}$$

and for structure function (4.10) takes the form

$$U(z) = (1-p_{11})(1-p_{21})z^0 + p_{11}(1-p_{21})z^{g_{11}} + p_{21}(1-p_{11})z^{g_{21}}$$

$$+ p_{11}p_{21}z^{\max(g_{11},g_{21})} = (1-p_{11})(1-p_{21})z^0 + p_{11}(1-p_{21})z^{g_{11}} + p_{21}z^{g_{21}}$$

The measures of the system output performance for MSSs of both types are presented in Tables 4.2 and 4.3.

Table 4.2. Measures of MSS performance for system with structure function (4.9)

w	$A(w)$	$\Delta^-(w)$	ε
$w>g_{11}+g_{21}$	0	$w-p_{11}g_{11}-p_{21}g_{21}$	
$g_{21}<w\leq g_{11}+g_{21}$	$p_{11}p_{21}$	$g_{11}p_{11}(p_{21}-1)+g_{21}p_{21}(p_{11}-1)+w(1-p_{11}p_{21})$	
$g_{11}<w\leq g_{21}$	p_{21}	$(1-p_{21})(w-g_{11}p_{11})$	$p_{11}g_{11}+p_{21}g_{21}$
$0<w\leq g_{11}$	$p_{11}+p_{21}-p_{11}p_{21}$	$(1-p_{11})(1-p_{21})w$	

Table 4.3. Measures of MSS performance for system with structure function (4.10)

w	$A(w)$	$\Delta^-(w)$	ε
$w>g_{21}$	0	$w-p_{11}g_{11}-p_{21}g_{21}+p_{11}p_{21}g_{11}$	
$g_{11}<w\leq g_{21}$	p_{21}	$(1-p_{21})(w-g_{11}p_{11})$	$p_{11}(1-p_{21})g_{11}+p_{21}g_{21}$
$0<w\leq g_{11}$	$p_{11}+p_{21}-p_{11}p_{21}$	$(1-p_{11})(1-p_{21})w$	

The u-function of a subsystem containing n identical parallel elements ($p_{j1} = p$, $g_{j1} = g$ for any j) can be obtained by applying the operator $\otimes_{\phi_{\text{par}}}(u(z),\ldots,u(z))$ over n identical u-functions $u(z)$ of an individual element. The u-function of this subsystem takes the form

$$\sum_{k=0}^{n} \frac{n!}{k!(n-k)!} p^k (1-p)^{n-k} z^{kg} \tag{4.11}$$

for the structure function (4.9) and

$$(1-p)^n z^0 + (1-(1-p)^n) z^g \tag{4.12}$$

for the structure function (4.10).

4.1.3 Series-Parallel Systems

The structure functions of complex series-parallel systems can always be represented as compositions of the structure functions of statistically independent subsystems containing only elements connected in a series or in parallel. Therefore, in order to obtain the *u*-function of a series-parallel system one has to apply the composition operators recursively in order to obtain *u*-functions of the intermediate pure series or pure parallel structures.

The following algorithm realizes this approach:

1. Find the pure parallel and pure series subsystems in the MSS.
2. Obtain *u*-functions of these subsystems using the corresponding $\otimes_{\phi_{\text{ser}}}$ and $\otimes_{\phi_{\text{par}}}$ operators.
3. Replace the subsystems with single elements having the *u*-function obtained for the given subsystem.
4. If the MSS contains more then one element return to step 1.

The resulting *u*-function represents the performance distribution of the entire system.

The choice of the structure functions used for series and parallel subsystems depends on the type of system. Table 4.4 presents the possible combinations of structure functions corresponding to the different types of MSS.

Table 4.4. Structure functions for a purely series and for purely parallel subsystems

No of MSS type	Description of MSS	Structure function for series elements (ϕ_{ser})	Structure function for parallel elements (ϕ_{par})
1	Flow transmission MSS with flow dispersion	(4.2)	(4.9)
2	Flow transmission MSS without flow dispersion	(4.2)	(4.10)
3	Task processing MSS with work sharing	(4.5)	(4.9)
4	Task processing MSS without work sharing	(4.5)	(4.10)

Example 4.3

In order to illustrate the recursive approach (the reliability block diagram method) consider the series-parallel system presented in Figure 4.1A.

First, one can find only one pure series subsystem consisting of elements with the u-functions $u_2(z)$, $u_3(z)$ and $u_4(z)$. By calculating the u-function $U_1(z) = u_2(z) \underset{\phi_{ser}}{\otimes} u_3(z) \underset{\phi_{ser}}{\otimes} u_4(z)$ and replacing the three elements with a single element with the u-function $U_1(z)$ one obtains a system with the structure presented in Figure 4.1B. This system contains a purely parallel subsystem consisting of elements with the u-functions $U_1(z)$ and $u_5(z)$, which in their turn can be replaced by a single element with the u-function $U_2(z) = U_1(z) \underset{\phi_{par}}{\otimes} u_5(z)$ (Figure 4.1C). The structure obtained has three elements connected in a series that can be replaced with a single element having the u-function $U_3(z) = u_1(z) \underset{\phi_{ser}}{\otimes} U_2(z) \underset{\phi_{ser}}{\otimes} u_6(z)$ (Figure 4.1D). The resulting structure contains two elements connected in parallel. The u-function of this structure representing the p.m.f. of the entire MSS performance is obtained as $U(z) = U_3(z) \underset{\phi_{par}}{\otimes} u_7(z)$.

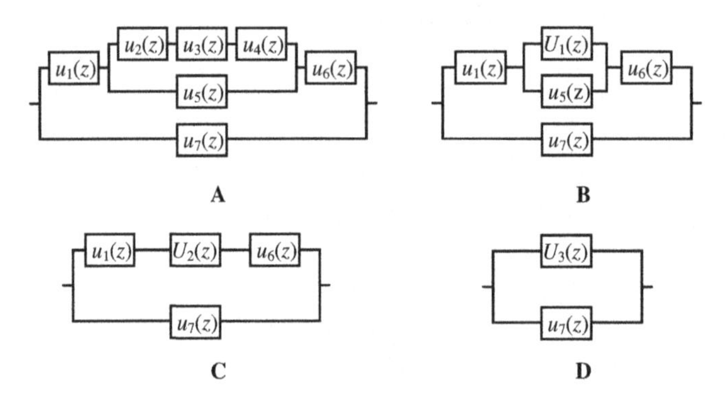

Figure 4.1. Example of recursive determination of the MSS u-function

Assume that in the series-parallel system presented in Figure 4.1A all of the system elements can have two states (elements with total failure) and have the parameters presented in Table 4.5. Each element j has a nominal performance rate g_{j1} in working state and performance rate of zero when it fails. The system is repairable and the steady-state probability that element j is in working state (element availability) is p_{j1}.

Table 4.5. Parameters of elements of series-parallel system

j	1	2	3	4	5	6	7
g_{j1}	5	3	5	4	2	6	3
p_{j1}	0.9	0.8	0.9	0.7	0.6	0.8	0.8

The process of calculating $U(z)$ for the flow transmission system with flow dispersion (for which ϕ_{ser} and ϕ_{par} functions are defined by Equations (4.2) and (4.9) respectively) is as follows:

$$u_2(z) \underset{\min}{\otimes} u_3(z) = (0.8z^3+0.2z^0) \underset{\min}{\otimes} (0.9z^5+0.1z^0) = 0.72z^3+0.28z^0$$

$$U_1(z) = (u_2(z) \underset{\min}{\otimes} u_3(z)) \underset{\min}{\otimes} u_4(z)$$

$$= (0.72z^3+0.28z^0) \underset{\min}{\otimes} (0.7z^4+0.3z^0) = 0.504z^3+0.496z^0$$

$$U_2(z) = U_1(z) \underset{+}{\otimes} u_5(z) = (0.504z^3+0.496z^0) \underset{+}{\otimes} (0.6z^3+0.4z^0)$$

$$= 0.3024z^6+0.4992z^3+0.1984z^0$$

$$u_1(z) \underset{\min}{\otimes} U_2(z) = (0.9z^5+0.1z^0) \underset{\min}{\otimes} (0.3024z^6+0.4992z^3+0.1984z^0)$$

$$= 0.27216z^5+0.44928z^3+0.27856z^0;$$

$$U_3(z) = (u_1(z) \underset{\min}{\otimes} U_2(z)) \underset{\min}{\otimes} u_6(z) = (0.27216z^5+0.44928z^3$$

$$+0.27856z^0) \underset{\min}{\otimes} (0.8z^6+0.2z^0) = 0.217728z^5+0.359424z^3+0.422848z^0$$

$$U(z) = U_3(z) \underset{+}{\otimes} u_7(z)$$

$$= (0.217728z^5+0.359424z^3+0.422848z^0) \underset{+}{\otimes} (0.8z^3+0.2z^0)$$

$$= 0.1741824z^8+0.2875392z^6+0.0435456z^5+0.4101632z^3+0.0845696z^0$$

Having the system u-function that represents its performance distribution one can easily obtain the system mean performance $\varepsilon = U'(1) = 4.567$. The system availability for different demand levels can be obtained by applying the operator δ_w (3.15) over the u-function $U(z)$:

$A(w) = 0.91543$ for $0<w\leq 3$

$A(w) = 0.50527$ for $3<w\leq 5$

$A(w) = 0.461722$ for $5<w\leq 6$

$A(w) = 0.174182$ for $6<w\leq 8$

$A(w) = 0$ for $w>8$

The process of calculating $U(z)$ for the task processing system without work sharing (for which ϕ_{ser} and ϕ_{par} functions are defined by Equations (4.5) and (4.10) respectively) is as follows:

$$u_2(z) \underset{\times}{\otimes} u_3(z) = (0.8z^3+0.2z^0) \underset{\times}{\otimes} (0.9z^5+0.1z^0) = 0.72z^{1.875}+0.28z^0;$$

$$U_1(z) = (u_2(z) \underset{\times}{\otimes} u_3(z)) \underset{\times}{\otimes} u_4(z)$$

$$= (0.72z^{1.875}+0.28z^0) \underset{\times}{\otimes} (0.7z^4+0.3z^0) = 0.504z^{1.277}+0.496z^0$$

$$U_2(z) = U_1(z) \underset{\max}{\otimes} u_5(z)) = (0.504z^{1.277}+0.496z^0) \underset{\max}{\otimes} (0.6z^2+0.4z^0)$$

$$= 0.6z^2+0.2016z^{1.277}+0.1984z^0$$

$$u_1(z) \underset{\times}{\otimes} U_2(z) = (0.9z^5+0.1z^0) \underset{\times}{\otimes} (0.6z^2+0.2016z^{1.277}+0.1984z^0)$$

$$= 0.54z^{1.429}+0.18144z^{1.017}+0.27856z^0$$

$$U_3(z) = (u_1(z) \underset{\times}{\otimes} U_2(z)) \underset{\times}{\otimes} u_6(z) = (0.54z^{1.429}+0.18144z^{1.017}$$

$$+0.27856z^0) \underset{\times}{\otimes} (0.8z^6+0.2z^0) = 0.432z^{1.154}+0.145152z^{0.87}+0.422848z^0$$

$$U(z) = U_3(z) \underset{\max}{\otimes} u_7(z) = (0.432z^{1.154}+0.145152z^{0.87}+0.422848z^0)$$

$$\underset{\max}{\otimes} (0.8z^3+0.2z^0) = 0.8z^3+0.0864z^{1.154}+0.0290304z^{0.87}$$

$$+0.08445696z^0$$

The main performance measures of this system are:

$$\varepsilon = U'(1) = 2.549$$

$$A(w) = 0.91543 \text{ for } 0<w\leq 0.87,\ A(w) = 0.8864 \text{ for } 0.87<w\leq 1.429$$

$$A(w) = 0.8 \text{ for } 1.429<w\leq 3,\ A(w) = 0 \text{ for } w>3$$

The procedure described above obtains recursively the same MSS u-function that can be obtained directly by operator $\underset{\phi}{\otimes}(u_1(z),u_2(z),u_3(z),u_4(z),u_5(z))$ using the following structure function:

$$\phi(G_1, G_2, G_3, G_4, G_5, G_6, G_7)$$

$$= \phi_{par}(\phi_{ser}(G_1, \phi_{par}(\phi_{ser}(G_2, G_3, G_4), G_5), G_6), G_7)$$

The recursive procedure for obtaining the MSS u-function is not only more convenient than the direct one, but, and much more important, it allows one to reduce the computational burden of the algorithm considerably. Indeed, using the direct procedure corresponding to Equation (1.20) one has to evaluate the system structure function for each combination of values of random variables G_1, ..., G_7 ($\prod_{j=1}^{7} k_j$ times, where k_j is the number of states of element j). Using the recursive algorithm one can take advantage of the fact that some subsystems have the same performance rates in different states, which makes these states indistinguishable and reduces the total number of terms in the corresponding u-functions.

In Example 4.3 the number of evaluations of the system structure function using the direct Equation (1.20) for the system with two-state elements is $2^7 = 128$. Each evaluation requires calculating a function of seven arguments. Using the reliability block diagram method one obtains the system u-function just by 30 procedures of structure function evaluation (each procedure requires calculating simple functions of just two arguments). This is possible because of the reduction in the lengths of intermediate u-functions by like terms collection. For example, it can be easily seen that in the subsystem of elements 2, 3 and 4 all eight possible combinations of the elements' states produce just two different values of the subsystem performance: 0 and $\min(g_{21}, g_{31}, g_{41})$ in the case of the flow transmission system, or 0 and $g_{21}g_{31}g_{41}/(g_{21}g_{31}+g_{21}g_{41}+g_{31}g_{41})$ in the case of the task processing system. After obtaining the u-function $U_1(z)$ for this subsystem and collecting like terms one gets a two-term equivalent u-function that is used further in the recursive algorithm. Such simplification is impossible when the entire expression (1.20) is used.

Example 4.4

Assume that in the series-parallel system presented in Figure 4.1A all of the system elements can have two states (elements with total failure). The system is unrepairable and the reliability of each element is defined by the Weibull hazard function

$$h(t) = \lambda^{\gamma}\gamma t^{\gamma-1}$$

The accumulated hazard function takes the form

$$H(t) = (\lambda t)^{\gamma}$$

The elements' nominal performance rates g_{j1}, the hazard function scale parameters λ_j and the shape parameters γ_j are presented in Table 4.6. One can see that some elements have increasing failure rates (γ >1) that correspond to their aging and some elements have constant failure rates (γ = 1).

Since the MSS reliability varies with the time, in order to obtain the performance measures of the system the reliability of its elements $\Pr\{G_j = g_{j1}\} = \exp(-H_j(t))$ should be calculated for each time instant. Then the entire system characteristics can be evaluated for the given demand w. Figures 4.2, 4.3 and 4.4

present ε, $R(w)$ and $\Delta^{-}(w)$ as functions of time for different types of system (numbered according to Table 4.4).

Table 4.6. Parameters of system elements

No of element	Nominal performance rate	Hazard function parameters	
	g	λ	γ
1	5	0.018	1.0
2	3	0.010	1.2
3	5	0.015	1.0
4	4	0.022	1.0
5	2	0.034	1.0
6	6	0.012	2.2
7	3	0.025	1.8

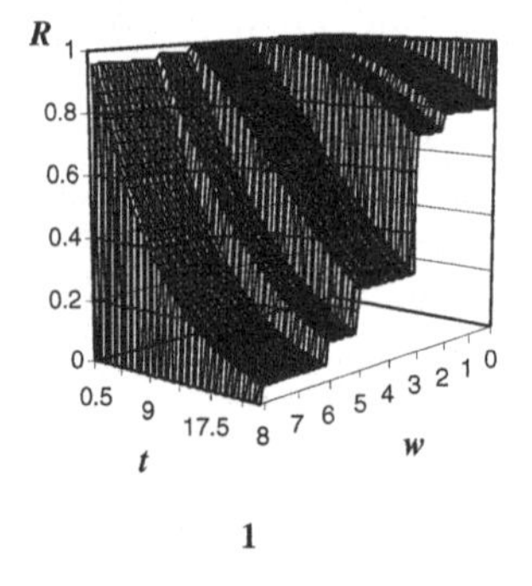

1

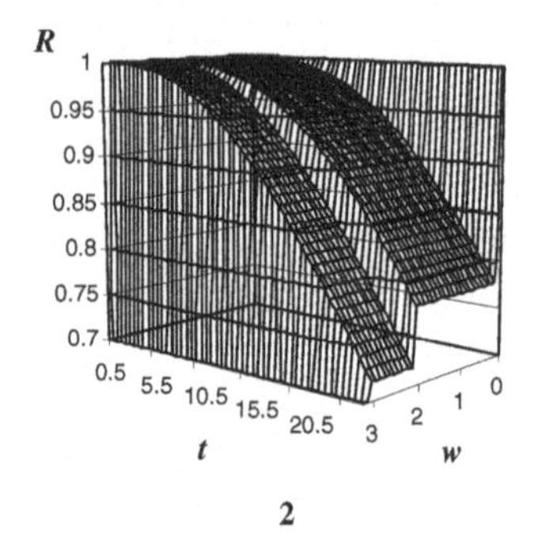

2

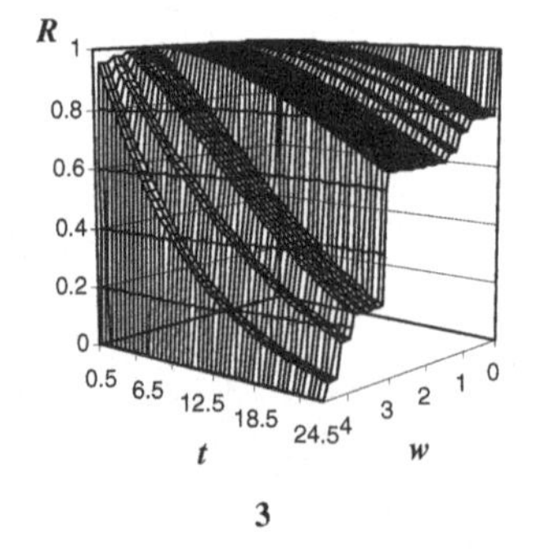

3

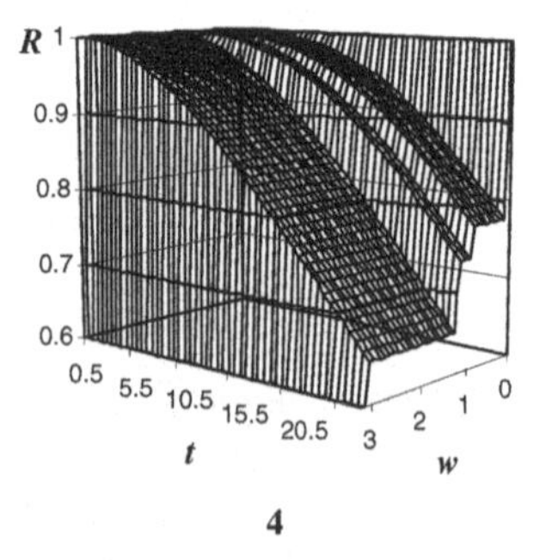

4

Figure 4.2. System reliability function for different types of MSS

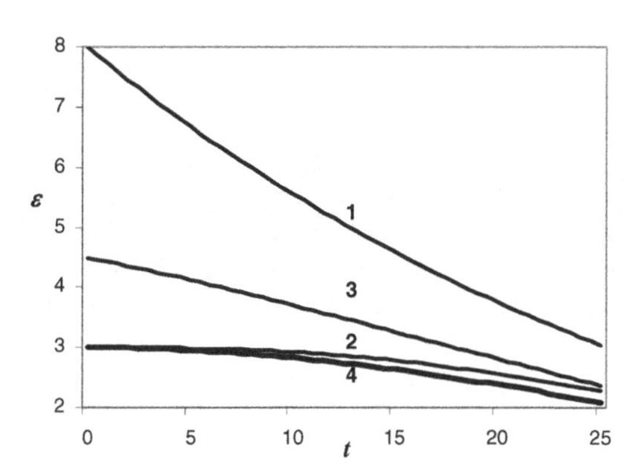

Figure 4.3. System mean performance for different types of MSS

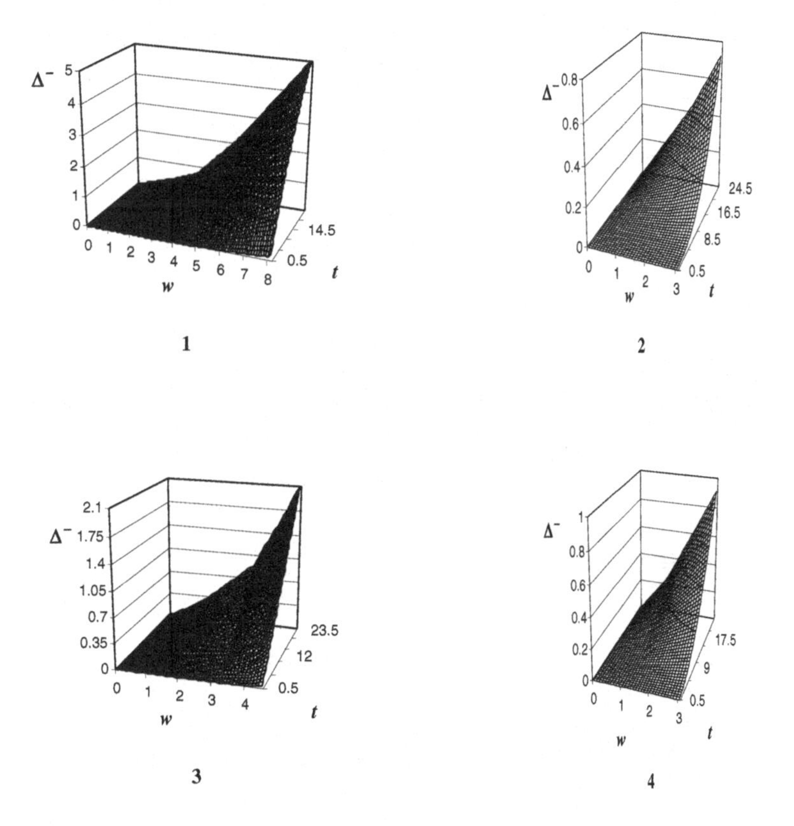

Figure 4.4. System performance deficiency for different types of MSS

4.1.4 Series-Parallel Multi-state Systems Reducible to Binary Systems

In some special cases the reliability (availability) of the entire system can be obtained without derivation of its u-function. In the final stage of reliability evaluation, such systems can be treated as binary systems.

Consider, for example, a flow transmission system consisting of n independent multi-state components connected in a series (each component in its turn can be a series-parallel subsystem). Let G_j be the random performance of component j. The structure function of the series flow transmission system is $G = \phi(G_1,\dots,G_n) = \min\{G_1,\dots,G_n\}$.

Assume that the system should meet a constant demand w. Therefore, the system acceptability function takes the form $F(G,w) = 1(G \geq w)$. It can be seen that in this special case

$$F(G,w) = 1(\min\{G_1,\dots,G_n\} \geq w) = \prod_{j=1}^{n} 1(G_j \geq w) \tag{4.13}$$

The system's reliability is defined as the probability that G is no less than w and takes the form

$$R(w) = \Pr\{F(G,w)=1\} = \Pr\{\prod_{j=1}^{n} 1(G_j \geq w) = 1\}$$
$$= \prod_{j=1}^{n} \Pr\{1(G_j \geq w) = 1\} = \prod_{j=1}^{n} \Pr\{F(G_j,w)=1\} \tag{4.14}$$

This means that the system's reliability is equal to the product of the reliabilities of its components.

Each component j can be considered to be a binary element with the state variable $X_j = F(G_j,w)$ and the entire system becomes the binary series system with the state variable X and the binary structure function φ:

$$F(G,w) = X = \varphi(X_1,\dots,X_n) = \prod_{j=1}^{n} X_j \tag{4.15}$$

The algorithm for evaluating the system reliability can now be simplified. It consists of the following steps:

1. Obtain the u-functions $U_j(z)$ of all of the series components.
2. Obtain the reliability of each component j as $R_j(w) = \delta_w(U_j(z))$.

3. Calculate the entire system reliability as $R=\prod_{j=1}^{n}R_j(w)=\prod_{j=1}^{n}\delta_w(U_j(z))$.

It can easily be seen that for the discrete random demand with p.m.f. $\boldsymbol{w}=\{w_1,\ldots,w_M\}$, $\boldsymbol{q}=\{q_1,\ldots,q_M\}$ the system reliability takes the form

$$R=\sum_{m=1}^{M}q_mR(w_m)=\sum_{m=1}^{M}q_m\prod_{j=1}^{n}R_j(w_m)=\sum_{m=1}^{M}q_m\prod_{j=1}^{n}\delta_{w_m}(U_j(z)) \tag{4.16}$$

Another example is a flow transmission system without flow dispersion consisting of n independent multi-state components connected in parallel. The structure function of such a system is $G=\phi(G_1,\ldots,G_n)=\max\{G_1,\ldots,G_n\}$. If the system should meet a constant demand w, its acceptability function also takes the form $F(G,w)=1(G\geq w)$. The probability of the system's failure is

$$\begin{aligned}&\Pr\{F(G,w)=0\}=\Pr\{G<w\}=\Pr\{\max\{G_1,\ldots,G_n\}<w\}\\&=\Pr\{\prod_{j=1}^{n}1(G_j<w)\}=\prod_{j=1}^{n}\Pr\{1(G_j<w)\}=\prod_{j=1}^{n}(1-\Pr\{1(G_j\geq w)\})\\&=\prod_{j=1}^{n}(1-\Pr\{F(G_j,w)=1\})\end{aligned} \tag{4.17}$$

The entire system reliability can now be determined as

$$\begin{aligned}&\Pr\{F(G,w)=1\}=1-\Pr\{F(G,w)=0\}\\&=1-\prod_{j=1}^{n}(1-\Pr\{F(G_j,w)=1\})\end{aligned} \tag{4.18}$$

This means that each component j can be considered to be a binary element with the state variable $X_j=F(G_j,w)$ and the entire system becomes the binary parallel system with the state variable X and the binary structure function σ

$$F(G,w)=X=\sigma(X_1,\ldots,X_n)=1-\prod_{j=1}^{n}(1-X_j) \tag{4.19}$$

After obtaining the u-functions $U_j(z)$ of all of the parallel components one can calculate the system reliability as

$$R(w)=1-\prod_{j=1}^{n}(1-R_j(w))=1-\prod_{j=1}^{n}(1-\delta_w(U_j(z))) \tag{4.20}$$

for the constant demand and as

$$R = \sum_{m=1}^{M} q_m \{1 - \prod_{j=1}^{n} [1 - R_j(w_m)]\} = \sum_{m=1}^{M} q_m \{1 - \prod_{j=1}^{n} [1 - \delta_{w_m}(U_j(z))]\} \quad (4.21)$$

for the discrete random demand.

Example 4.5

Consider the flow transmission series-parallel system presented in Figure 4.5. The system consists of three components connected in a series. The first component consists of two different elements and constitutes a subsystem without flow dispersion. The second and third components are subsystems with a flow dispersion consisting of two and three identical elements respectively. Each element j can have only two states: total failure (corresponding to a performance of zero) and operating with the nominal performance g_{j1}. The availability of element j is p_{j1}.

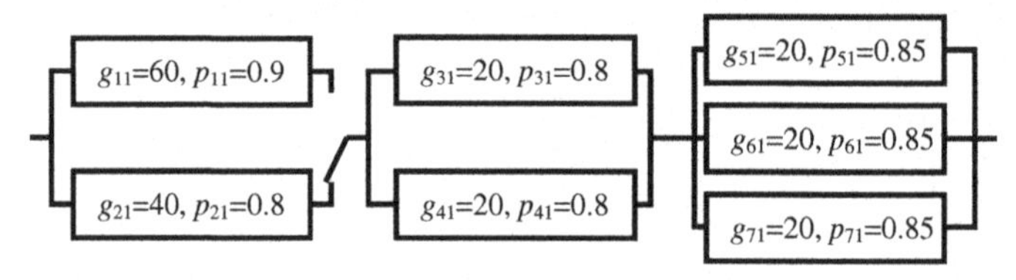

Figure 4.5. Example of a series-parallel system reducible to a binary system

The u-functions of the individual elements are:

$$u_1(z) = 0.9z^{60} + 0.1z^0, \; u_2(z) = 0.8z^{40} + 0.2z^0$$

$$u_3(z) = u_4(z) = 0.8z^{20} + 0.2z^0$$

$$u_5(z) = u_6(z) = u_7(z) = 0.85z^{20} + 0.15z^0$$

The u-functions of the components are obtained using the corresponding $\underset{\phi_{\text{par}}}{\otimes}$ operators:

$$U_1(z) = (0.9z^{60} + 0.1z^0) \underset{\max}{\otimes} (0.8z^{40} + 0.2z^0) = 0.9z^{60} + 0.08z^{40} + 0.02z^0$$

$$U_2(z) = (0.8z^{20} + 0.2z^0) \underset{+}{\otimes} (0.8z^{20} + 0.2z^0) = 0.64z^{40} + 0.32z^{20} + 0.04z^0$$

$$U_3(z) = (0.85z^{20} + 0.15z^0) \underset{+}{\otimes} (0.85z^{20} + 0.15z^0) \underset{+}{\otimes} (0.85z^{20} + 0.15z^0)$$

$$= (0.85z^{20} + 0.15z^0)^3 = 0.6141z^{60} + 0.3251z^{40} + 0.0574z^{20} + 0.0034z^0$$

For demand $w = 20$ we obtain

$$\delta_{20}(U_1(z)) = \delta_{20}(0.9z^{60} + 0.08z^{40} + 0.02z^0) = 0.98$$

$$\delta_{20}(U_2(z)) = \delta_{20}(0.64z^{40} + 0.32z^{20} + 0.04z^0) = 0.96$$

$$\delta_{20}(U_3(z)) = \delta_{20}(0.6141z^{60} + 0.3251z^{40} + 0.0574z^{20} + 0.0034z^0) = 0.9966$$

The entire system availability is

$$A(20) = \delta_{20}(U_1(z))\delta_{20}(U_2(z))\delta_{20}(U_3(z)) = 0.98 \times 0.96 \times 0.9966 = 0.9376$$

For demand $w = 40$ we obtain

$$\delta_{40}(U_1(z)) = \delta_{40}(0.9z^{60} + 0.08z^{20} + 0.02z^0) = 0.9$$

$$\delta_{40}(U_2(z)) = \delta_{40}(0.64z^{40} + 0.32z^{20} + 0.04z^0) = 0.64$$

$$\delta_{40}(U_3(z)) = \delta_{40}(0.6141z^{60} + 0.3251z^{40} + 0.0574z^{20} + 0.0034z^0) = 0.9392$$

The entire system availability is

$$A(40) = \delta_{40}(U_1(z))\delta_{40}(U_2(z))\delta_{40}(U_3(z)) = 0.9 \times 0.64 \times 0.9392 = 0.541$$

If the demand is the random variable W with p.m.f. $\boldsymbol{w} = \{20, 40\}$, $\boldsymbol{q} = \{0.7, 0.3\}$, the system availability is

$$A = 0.7A(20) + 0.3A(40) = 0.7 \times 0.9376 + 0.3 \times 0.541 = 0.8186$$

It should be noted that only the reliability (availability) of series-parallel systems can be evaluated using the MSS reduction to the binary system. The evaluation of the mean performance and the performance deviation measures still require the derivation of the u-function of the entire system.

4.2 Controllable Series-Parallel Multi-state Systems

Some series-parallel systems can change their configuration following certain rules aimed at achieving maximal system efficiency. Such systems belong to the class of controllable systems. If the rules that determine the system configuration depend on external factorsthen thesystemreliabilitymeasuresshould be determined for each possible configuration. If the rules are based on the states of the system elements then they can be incorporated into algorithms evaluating the system reliability measures. The application of simple operators $\otimes_{\phi_{\text{ser}}}$ and $\otimes_{\phi_{\text{par}}}$ over *u*-functions of the system elements is usually not enough in order to obtain the *u*-function of the entire system since its structure function is affected by the configuration control rules.

Examples of systems with controllable configuration are systems that contain elements with fixed resource consumption [105]. Many technical devices (processes) can work only if the available amount of some of the resources that they consume is not lower than the specified limits. If this requirement is not met, then the device (process) fails to work. An example of such a situation is a control system that stops the controlled process if a decrease in its computational resources does not allow the necessary information to be processed within the required cycle time. Another example is a metalworking machine that cannot perform its task if the flow of coolant supplied is less than required.

For a resource-consuming system that consists of several units, the amount of resource necessary to provide the normal operation of a given composition of the main producing units (controlled processes or machines) is fixed. Any deficit of the resource makes it impossible for all of the units from the composition to operate together (in parallel), because no unit can reduce the amount of resource it consumes. Therefore, any resource deficit leads to turning off some of the producing units.

Consider a system consisting of H resource-generating subsystems (RGSs) that supply different (not interchangeable) resources to the main producing system (MPS). RGSs can have an arbitrary series-parallel configuration, while the MPS consists of n elements connected in parallel (Figure 4.6). Each element of the MPS is an element with total failure and can perform in its working state only by consuming a fixed amount of resources. The MPS is the flow transmission system with flow dispersion. If, following failures, in any RGS there are not enough resources to allow all of the available producing elements to work, some of these elements should be turned off. We assume that the choice of the working MPS elements is made by a control system in such a way as to maximize the total performance rate of the MPS under the given resource constraints.

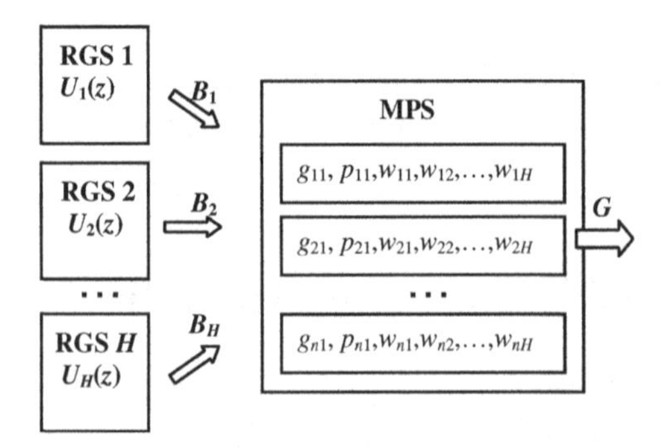

Figure 4.6. Structure of controllable system with fixed resource consumption

Assume that the RGS h produces a random amount B_h of the resource. The p.m.f. of B_h is represented by the u-function $U_h(z)=\sum_{i=0}^{k_h-1} q_{hi} z^{\beta_{hi}}$, where β_{hi} is the performance rate of RGS h in state i and q_{hi} = Pr{B_h = β_{hi}}. Each element j of the MPS has a nominal performance g_{j1} and availability p_{j1} and requires the amount w_{jh} of each resource h $(1 \le h \le H)$ for normal functioning (if different MPS elements consume different subsets of the set of H resources, this can be represented by assigning zero to w_{jh} for any resource h that is not required by element j). The p.m.f. of the random performance G_j of element j is represented by the u-function $u_j(z)=p_{j1}z^{g_{j1}}+(1-p_{j1})z^0$. The distribution of the available performance of the entire MPS G_{MPS} can be obtained as $U_{\text{MPS}}(z)=\otimes_{+}(u_1(z),\dots,u_n(z))$. Observe that the performance G_{MPS} represents the potential performance ability of the MPS. It does not always coincide with the output performance of the entire system G. $U_{\text{MPS}}(z)$ represents the conditional distribution of G corresponding to a situation when the resources are supplied without limitations. In order to take into account the possible deficiency of the resources supplied we have to incorporate the MPS control rule (the rule of turning the MPS elements off and on) into the derivation of the system u-function $U(z)$ representing the p.m.f. of G.

4.2.1 Systems with Identical Elements in the Main Producing System

If an MPS contains only identical elements with $g_{j1} = g$, $p_{j1} = p$ and $w_{jh} = w_h>0$ for any j and h, the number of elements that can work in parallel when the available amount of resource h is β_{hi} is $\lfloor \beta_{hi}/w_h \rfloor$, which corresponds to the total system performance $\gamma_{hi} = g \lfloor \beta_{hi}/w_h \rfloor$ (the remainder of the MPS elements must be turned off). It must be noted that γ_{hi} represents the total theoretical performance, which can be achieved by using the available resource h by an unlimited number of producing elements. In terms of the entire system output performance, the u-function of the RGS h can be obtained in the following form:

$$U_h^{\gamma}(z)=\sum_{i=0}^{k_h-1} q_{hi} z^{\gamma_{hi}} = \sum_{i=0}^{k_h-1} q_{hi} z^{g\lfloor \beta_{hi}/w_h \rfloor} \tag{4.22}$$

The RGS, which can provide the work of a minimal number of producing units, becomes the system's bottleneck. This RGS limits the total system performance. Therefore, the u-function for a system containing H different RGS in terms of system output performance can be obtained as

$$U_{\text{RGS}}(z)=\underset{\min}{\otimes}(U_1^{\gamma}(z),\dots,U_H^{\gamma}(z)) \tag{4.23}$$

Function $U_{\text{RGS}}(z)$ represents the entire system performance distribution in the case of an unlimited number of available elements in the MPS.

The entire system performance is equal to the minimum of the total theoretical performance, which can be achieved using available resources and the total performance of the available MPS elements. To obtain the u-function $U(z)$ of the entire system representing the p.m.f. of its performance G, the same operator $\underset{\min}{\otimes}$ should be applied over the u-functions $U_{\text{RGS}}(z)$ and $U_{\text{MPS}}(z)$:

$$U(z) = U_{\text{RGS}}(z) \underset{\min}{\otimes} U_{\text{MPS}}(z) = \underset{\min}{\otimes}(U_1^{\gamma}(z),...,U_H^{\gamma}(z),U_{\text{MPS}}(z))$$

$$= \underset{\min}{\otimes}(U_1^{\gamma}(z),...,U_H^{\gamma}(z),\underset{+}{\otimes}(u_1(z),...,u_n(z))) \quad (4.24)$$

4.2.2 System with Different Elements in the Main Producing System

If the MPS consists of n different elements, then it can be in one of 2^n possible states corresponding to the different combinations of the available elements. Let S be a random set of numbers of available MPS elements and S_k be a realization of S in state k $(1 \le k \le 2^n)$. The probability of state k can be evaluated as follows:

$$\tilde{P}_k = \prod_{j=1}^{n} p_{j1}^{1(j \in S_k)} (1 - p_{j1})^{1(i \notin S_k)} \quad (4.25)$$

The maximal possible performance of the MPS and the corresponding maximal resources consumption in state k are

$$g_k^{\max} = \sum_{j=1}^{n} (g_{j1})^{1(j \in S_k)} \quad (4.26)$$

and

$$w_{hk}^{\max} = \sum_{j=1}^{n} (w_{jh})^{1(j \in S_k)} \; (1 \le h \le H) \quad (4.27)$$

respectively.

Let us define a u-function representing the distribution of the random set of available elements. For a single element j this u-function takes the form

$$u_j(z) = p_{j1} z^{\{j\}} + (1 - p_{j1}) z^{\varnothing} \quad (4.28)$$

Using the union procedure $\cup$ in the composition operator $\underset{\cup}{\otimes}$ we can obtain the distribution of the random set of available elements in the system consisting of several elements. For example, if the MPS consists of two elements

$$u_1(z) = p_{11}z^{\{1\}} + (1-p_{11})z^{\varnothing},\ u_2(z) = p_{21}z^{\{2\}} + (1-p_{21})z^{\varnothing} \tag{4.29}$$

and the distribution of the set of available elements takes the form

$$\begin{aligned}
U_{\text{MPS}}(z) &= u_1(z)\underset{\cup}{\otimes}u_2(z) \\
&= [p_{11}z^{\{1\}} + (1-p_{11})z^{\varnothing}]\underset{\cup}{\otimes}[p_{21}z^{\{2\}} + (1-p_{21})z^{\varnothing}] \\
&= p_{11}p_{21}z^{\{1\}\cup\{2\}} + p_{11}(1-p_{21})z^{\{1\}\cup\varnothing} + (1-p_{11})p_{21}z^{\varnothing\cup\{2\}} \\
&+ (1-p_{11})(1-p_{21})z^{\varnothing\cup\varnothing} = p_{11}p_{21}z^{\{1,2\}} + p_{11}(1-p_{21})z^{\{1\}} \\
&+ (1-p_{11})p_{21}z^{\{2\}} + (1-p_{11})(1-p_{21})z^{\varnothing}
\end{aligned} \tag{4.30}$$

For an MPS consisting of n elements the u-function representing the distribution of a random set of available elements takes the form

$$U_{MPS}(z) = \underset{\cup}{\otimes}(u_1(z),\ldots,u_n(z)) = \sum_{k=1}^{2^n} \tilde{P}_k z^{S_k} \tag{4.31}$$

When each RGS h is in state i_h the amount β_{hi_h} of the resource generated by this RGS can be not enough to provide the maximal performance of the MPS at state k. In order to provide the maximum possible performance G of the MPS under the resource constraints one has to solve the following linear programming problem for any combination of states $i_1,\ldots,i_H$ of H RDSs and state k of the MPS:

$$\begin{aligned}
&\text{opt}(\beta_{1i_1}, \beta_{2i_2}, \ldots, \beta_{Hi_H}, S_k) = \max \sum_{j\in S_k} g_{j1}x_j \\
&\text{subject to} \\
&\sum_{j\in S_k} w_{jh}x_j \le \beta_{hi_h},\quad \text{for } 1\le h\le H \\
&x_j \in \{0,1\}
\end{aligned} \tag{4.32}$$

where $x_j = 1$ if the available element j is turned on (works providing performance rate g_{j1} and consuming w_{jh} of each resource $1 \le h \le H$) and $x_j = 0$ if the element is turned off.

The performance distribution of the entire system can be obtained by considering all of the possible combinations of the available resources generated by the RGS and the states of the MPS. For each combination, a solution of the above formulated optimization problem defines the system's performance. The u-function representing the p.m.f. of the entire system performance G can be defined as follows:

$$U(z) = \underset{\text{opt}}{\otimes}(U_1(z),...,U_H(z),U_{\text{MPS}}(z))$$

$$= \sum_{i_1=0}^{k_1-1}\sum_{i_2=0}^{k_2-1}...\sum_{i_H=0}^{k_H-1}\left\{(\prod_{h=1}^{H} q_{hi_h}) \sum_{k=1}^{2^n} \widetilde{P}_k z^{\text{opt}(\beta_{1i_1},\beta_{2i_2},...,\beta_{Hi_H},S_k)}\right\} \quad (4.33)$$

To obtain the system u-function, its optimal performance should be determined for each unique combination of available resources and for each unique state of the MPS. In general, the total number of linear programs to be solved in order to obtain $U(z)$ is $2^n \prod_{h=1}^{H} k_h$. In practice, the number of programs to be solved can be reduced drastically using the following rules:

1. If for the given vector $\beta_{1i_1},...,\beta_{Hi_H}$ and for each element j from the given set of MPS elements S_k there exists h for which $\beta_{hi_h} < w_{jh}$, then the system performance is equal to zero. In this case the system performance is equal to zero also for all combinations $(\beta_{1j_1},...,\beta_{Hj_H},S_m)$ such that $\beta_{1j_1} \le \beta_{1i_1},...,\beta_{Hj_H} \le \beta_{Hi_H}$ and $S_m \subseteq S_k$.

2. If element $j \in S_k$ exists for which $\beta_{hi_h} < w_{jh}$ for some h, this means that in the program (4.32) x_j must be zeroed. In this case, the integer program dimension can be reduced by removing all such elements from S_k.

3. If for the given vector $\beta_{1i_1},\beta_{2i_2}...,\beta_{Hi_H}$ and for the given set S_k the solution of the integer program (4.32) determines subset $\hat{S}_k$ of turned-on MPS elements ($j \in \hat{S}_k$ if x_j=1), then the same solution must be optimal for the MPS states characterized by any set S_m: $\hat{S}_k \subset S_m \subset S_k$. This allows one to avoid solving many integer programs by assigning the value of $\text{opt}(\beta_{1i_1},\beta_{2i_2}...,\beta_{Hi_H},S_k)$ to all the $\text{opt}(\beta_{1i_1},\beta_{2i_2}...,\beta_{Hi_H},S_m)$.

Example 4.6

Three different metalworking units (Figure 4.7) have the respective productivities and availabilities g_{11}=10, p_{11}=0.8, g_{21}=15, p_{21}=0.9 and g_{31}=20, p_{31}=0.85. The system productivity should be no less than a constant demand w. Each unit consumes two resources: electrical power and coolant.

The constant power consumption of the units is w_{11} = 5, w_{21} = 2, w_{31} = 3. The power is supplied by the system consisting of two transformers that work without load sharing (only one of the two transformers can work at any moment). The power of the transformers is 10 and 6. The availability of the transformers is 0.9 and 0.8 respectively. The constant coolant flow consumed by the units is w_{12} = 4,

$w_{22} = 5$ $w_{32} = 7$. Two identical pumps working in parallel supply the coolant (both pumps can work simultaneously). The nominal coolant flow provided by each pump is 9. The availability of each pump is 0.8.

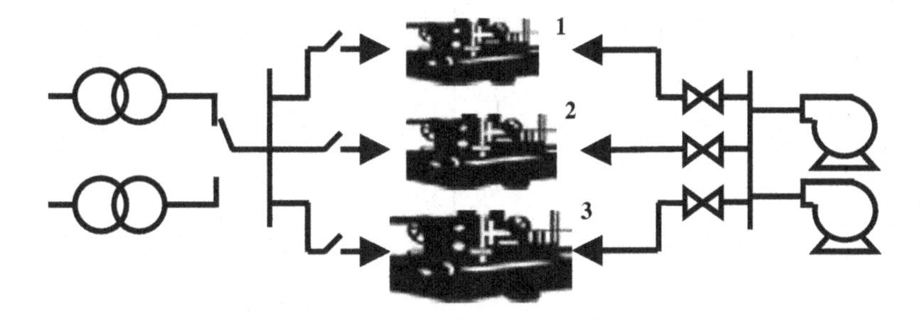

Figure 4.7. Example of controllable series-parallel system

The u-function representing the distribution of available power takes the form

$$U_1(z) = (0.9z^{10} + 0.1z^0) \underset{\max}{\otimes} (0.8z^6 + 0.2z^0) = 0.9z^{10} + 0.08z^6 + 0.02z^0$$

and the u-function representing the distribution of the available coolant takes the form

$$U_2(z) = (0.8z^9 + 0.2z^0) \underset{+}{\otimes} (0.8z^9 + 0.2z^0) = 0.64z^{18} + 0.32z^9 + 0.04z^0$$

The u-function representing the distribution of the set of available metalworking units takes the form

$$U_{\text{MPS}}(z) = (0.8z^{\{1\}} + 0.02z^{\varnothing}) \underset{\cup}{\otimes} (0.9z^{\{2\}} + 0.1z^{\varnothing}) \underset{\cup}{\otimes} (0.85z^{\{3\}}$$

$$+ 0.15z^{\varnothing}) = 0.003z^{\varnothing} + 0.012z^{\{1\}} + 0.027z^{\{2\}} + 0.017z^{\{3\}}$$
$$+ 0.108z^{\{1,2\}} + 0.068z^{\{1,3\}} + 0.153z^{\{2,3\}} + 0.612z^{\{1,2,3\}}$$

The values of the opt function obtained for all of the possible combinations of available metalworking units (realizations S_k of the random set S) and available resources (realizations of B_1 and B_2) are presented in Table 4.7. The table contains the maximal possible productivity of the MPS $g_k^{\max}$ and the corresponding maximal required resources $w_{hk}^{\max}$ for any set S_k that is not empty. It also contains the optimal system productivity G (values of the opt function) and the corresponding sets of turned-on elements $\hat{S}_k$.

Table 4.7. Solutions of a linear program for a system with different elements in an MPS

				B_1=6, B_2=9		B_1=6, B_2=18		B_1=10, B_2=9		B_1=10, B_2=18	
S_k	g_k^{max}	w_{1k}^{max}	w_{2k}^{max}	$\hat{S}_k$	G	$\hat{S}_k$	G	$\hat{S}_k$	G	$\hat{S}_k$	G
{1}	10	5	4	{1}	10	{1}	10	{1}	10	{1}	10
{2}	15	2	5	{2}	15	{2}	15	{2}	15	{2}	15
{3}	20	3	7	**{3}**	**20**	{3}	20	{3}	20	{3}	20
{1,2}	25	7	9	{2}	15	{2}	15	**{1,2}**	**25**	{1,2}	25
{1,3}	30	8	11	**{3}**	**20**	{3}	20	{3}	20	{1,3}	30
{2,3}	35	5	12	**{3}**	**20**	**{2,3}**	**35**	{3}	20	{2,3}	35
{1,2,3}	45	10	16	*{3}*	*20*	*{2,3}*	*35*	*{1,2}*	*25*	{1,2,3}	45

It is obvious that if $S_k = \varnothing$ then the entire system performance is equal to zero. If $B_1 = 0$ or $B_2 = 0$ then the entire system performance is also equal to zero according to rule 1 (these solutions are not included in the table). Note that the solutions marked in bold are obtained without solving the linear program (they were obtained using rule 3 from the solutions marked in italic).

The u-function of the entire system obtained in accordance with Table 4.7 after collecting the like terms takes the following form:

$$U(z) = \underset{\text{opt}}{\otimes}[(0.9z^{10} + 0.08z^{6} + 0.02z^{0}), (0.64z^{18} + 0.32z^{9} + 0.04z^{0}),$$
$$(0.003z^{\varnothing} + 0.012z^{\{1\}} + 0.027z^{\{2\}} + 0.017z^{\{3\}} + 0.108z^{\{1,2\}} + 0.068z^{\{1,3\}}$$
$$+ 0.153z^{\{2,3\}} + 0.612z^{\{1,2,3\}})] = 0.062z^{0} + 0.0113z^{10} + 0.0337z^{15}$$
$$+ 0.104z^{20} + 0.270z^{25} + 0.039z^{30} + 0.127z^{35} + 0.353z^{45}$$

Having the system u-function we can easily obtain its mean performance

$$\varepsilon = U'(1) = 0.0113 \times 10 + 0.0337 \times 15 + 0.104 \times 20 + 0.270 \times 25$$
$$+ 0.039 \times 30 + 0.127 \times 35 + 0.353 \times 45 = 30.94$$

and availability. For example, for system demand $w = 20$:

$$A(20) = \delta_{20}(U(z)) = \delta_{20}(0.062z^{0} + 0.0113z^{10} + 0.0337z^{15}$$
$$+ 0.104z^{20} + 0.270z^{25} + 0.039z^{30} + 0.127z^{35} + 0.353z^{45})$$
$$= 0.104 + 0.270 + 0.039 + 0.127 + 0.353 = 0.893$$

The system availability as a function of demand is presented in Figure 4.8.

Now consider the same system in which the MPS consists of three identical units with parameters g_{j1} = 20, p_{j1} = 0.85, w_{j1} = 3 and w_{j2} = 7. The reliability measures of such a system can be obtained in an easier manner by using the algorithm presented in Section 4.2.1.

From the u-functions of the RGSs $U_1(z)$ and $U_2(z)$ by applying Equation (4.22) we obtain for the first RGS:

$$U_1^{\gamma}(z) = 0.9z^{20\lfloor 10/3\rfloor} + 0.08z^{20\lfloor 6/3\rfloor} + 0.02z^{20\lfloor 0/5\rfloor}$$
$$= 0.9z^{60} + 0.08z^{40} + 0.02z^{0}$$

and for the second RGS:

$$U_2^{\gamma}(z) = 0.64z^{20\lfloor 18/7\rfloor} + 0.32z^{20\lfloor 9/7\rfloor} + 0.04z^{20\lfloor 0/7\rfloor}$$
$$= 0.64z^{40} + 0.32z^{20} + 0.04z^{0}$$

The u-function of the MPS is

$$U_{\text{MPS}}(z) = (0.85z^{20} + 0.15z^{0}) \underset{+}{\otimes} (0.85z^{20}$$

$$+ 0.15z^{0}) \underset{+}{\otimes} (0.85z^{20} + 0.15z^{0}) = (0.85z^{20} + 0.15z^{0})^3$$

$$= 0.6141z^{60} + 0.3251z^{40} + 0.0574z^{20} + 0.0034z^{0}$$

The u-function of the entire system after collecting the like terms takes the form:

$$U(z) = U_1^{\gamma}(z) \underset{\min}{\otimes} U_2^{\gamma}(z) \underset{\min}{\otimes} U_{\text{MPS}}(z)$$

$$= (0.9z^{60} + 0.08z^{40} + 0.02z^{0}) \underset{\min}{\otimes} (0.64z^{40} + 0.32z^{20} + 0.04z^{0})$$

$$\underset{\min}{\otimes} (0.6141z^{60} + 0.3251z^{40} + 0.0574z^{20} + 0.0034z^{0})$$

$$= 0.589z^{40} + 0.3486z^{20} + 0.0624z^{0}$$

From the system u-function we can obtain its mean performance

$$\varepsilon = U'(1) = 0.589 \times 40 + 0.3486 \times 20 = 30.532$$

and its availability. For example, for $w = 20$:

$$A(20) = \delta_{20}(U(z)) = \delta_{20}(0.589z^{40} + 0.3486z^{20} + 0.0624z^{0})$$
$$= 0.589 + 0.3486 = 0.9375$$

The system availability as a function of demand is presented in Figure 4.8.

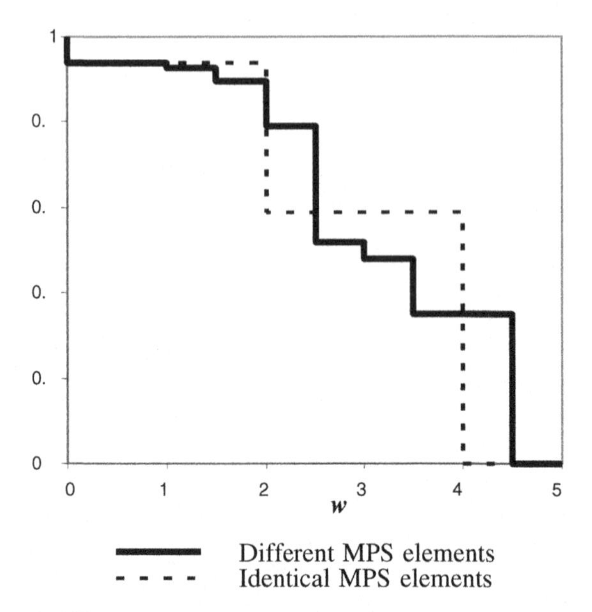

Figure 4.8. Availability of controllable series-parallel system as a function of demand

Since the system consists of three subsystems connected in a series and can be considered as a flow transmission system, its availability for any given demand can be obtained without derivation of the entire system u-function $U(z)$ using the simplified technique described in Section 4.1.4. The availability of the system for $w = 20$ is calculated using this simplified technique in Example 4.5.

The RGS-MPS model considered can easily be expanded to systems with a multilevel hierarchy. When analyzing multilevel systems, the entire RGS-MPS system (with its performance distribution represented by its u-function) may be considered in its turn as one of the RGSs for a higher level MPS (Figure 4.9).

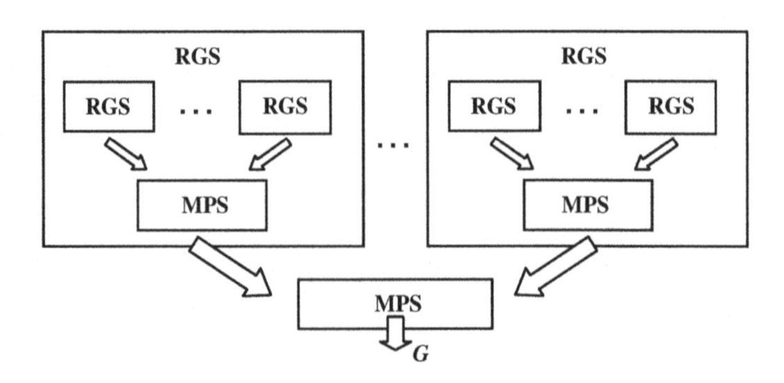

Figure 4.9. RGS-MPS system with hierarchical structure

4.3 Multi-state Systems with Dependent Elements

One of the main assumptions made in the previous sections is statistical independency of system elements. This assumption is not true for many technical systems. Fortunately, the UGF approach can be extended to cases when the performance distributions of some system elements are influenced by the states of other elements or subsystems [106].

4.3.1 *u*-functions of Dependent Elements

Consider a subsystem consisting of a pair of multi-state elements i and j in which the performance distribution of element j (p.m.f. of random performance G_j) depends on the state of element i. Since the states of the elements are distinguished by their corresponding performance rates, we can assume that the performance distribution of element j is determined by the performance rate of element i. Let $\boldsymbol{g}_i=\{g_{ih}:\ 1\le h\le k_i\}$ be the set of possible performance rates of element i. In general, this set can be separated into M mutually disjoint subsets $\boldsymbol{g}_i^m$ ($1\le m\le M$):

$$\bigcup_{m=1}^{M}\boldsymbol{g}_i^m=\boldsymbol{g}_i,\quad \boldsymbol{g}_i^m\cap\boldsymbol{g}_i^l=\varnothing,\ \text{if}\ m\neq l \tag{4.34}$$

such that when element i has the performance rate $g_{ih}\in\boldsymbol{g}_i^m$ the PD of element j is defined by the ordered sets $\boldsymbol{g}_{j|m}=\{g_{jc|m},\ 1\le c\le C_{j|m}\}$ and $\boldsymbol{q}_{j|m}=\{q_{jc|m},\ 1\le c\le C_{j|m}\}$, where

$$q_{jc|m}=\Pr\{G_j=g_{jc|m}\mid G_i=g_{ih}\in\boldsymbol{g}_i^m\} \tag{4.35}$$

If each performance rate of element i corresponds to a different PD of element j, then we have $M=k_i$ and $\boldsymbol{g}_i^m=\{g_{im}\}$.

We can define the set of all of the possible values of the performance rate of element j as $\boldsymbol{g}_j=\bigcup_{m=1}^{M}\boldsymbol{g}_{j|m}$ and redefine the conditional PD of element j when element i has the performance rate $g_{ih}\in\boldsymbol{g}_i^m$ using two ordered sets $\boldsymbol{g}_j=\{g_{jc},\ 1\le c\le C_j\}$ and $\boldsymbol{p}_{j|m}=\{p_{jc|m},\ 1\le c\le C_j\}$, where:

$$p_{jc|m}=\begin{cases}0, & g_{jc}\notin\boldsymbol{g}_{j|m}\\ q_{jc|m}, & g_{jc}\in\boldsymbol{g}_{j|m}\end{cases} \tag{4.36}$$

According to this definition

$$p_{jc|m} = \Pr\{G_j = g_{jc} \mid G_i = g_{ih} \in \boldsymbol{g}_i^m\} \tag{4.37}$$

for any possible realization of G_j and any possible realization of $G_i \in \boldsymbol{g}_i^m$.

Since the sets $\boldsymbol{g}_i^m$ ($1 \le m \le M$) are mutually disjoint, the unconditional probability that $G_j = g_{jc}$ can be obtained as

$$p_{jc} = \sum_{m=1}^{M} \Pr\{G_j = g_{jc} \mid G_i \in \boldsymbol{g}_i^m\} \Pr\{G_i \in \boldsymbol{g}_i^m\}$$
$$= \sum_{m=1}^{M} p_{jc|m} \sum_{h=1}^{k_i} p_{ih} 1(p_{ih} \in \boldsymbol{g}_i^m) \tag{4.38}$$

In the case when $\boldsymbol{g}_i^m = \{g_{im}\}$

$$p_{jc} = \sum_{m=1}^{k_i} p_{im} p_{jc|m} \tag{4.39}$$

The unconditional probability of the combination $G_i = g_{ih}$, $G_j = g_{jc}$ is equal to $p_{ih} p_{jc|\mu(h)}$, where $\mu(h)$ is the number of the set to which g_{ih} belongs: $g_{ih} \in \boldsymbol{g}_i^{\mu(h)}$.

Example 4.7

Assume that element 1 has the PD $\boldsymbol{g}_1 = \{0, 1, 2, 3\}$, $\boldsymbol{p}_1 = \{0.1, 0.2, 0.4, 0.3\}$ and the PD of element 2 depends on the performance rate of element 1 such that when $G_1 \le 2$ ($G_1 \in \boldsymbol{g}_1^1 = \{0,1,2\}$) element 2 has the PD $\boldsymbol{g}_{2|1} = \{0, 10\}$, $\boldsymbol{q}_{2|1} = \{0.3, 0.7\}$ while when $G_1 > 2$ ($G_1 \in \boldsymbol{g}_1^2 = \{3\}$) element 2 has the PD $\boldsymbol{g}_{2|2} = \{0, 5\}$, $\boldsymbol{q}_{2|2} = \{0.1, 0.9\}$. The conditional PDs of element 2 can be represented by the sets $\boldsymbol{g}_2 = \{0,5,10\}$ and $\boldsymbol{p}_{2|1} = \{0.3, 0, 0.7\}$, $\boldsymbol{p}_{2|2} = \{0.1, 0.9, 0\}$.

The unconditional probabilities p_{2c} are:

$$p_{21} = \Pr\{G_2 = 0\} = \Pr\{G_2 = 0 \mid G_1 \in \boldsymbol{g}_1^1\} \Pr\{G_1 \in \boldsymbol{g}_1^1\}$$
$$+ \Pr\{G_2 = 0 \mid G_1 \in \boldsymbol{g}_1^2\} \Pr\{G_1 \in \boldsymbol{g}_1^2\} = p_{21|1}(p_{11}+p_{12}+p_{13}) + p_{21|2}(p_{14})$$
$$= 0.3(0.1+0.2+0.4)+0.1(0.3) = 0.24$$

$$p_{22} = \Pr\{G_2 = 5\} = \Pr\{G_2 = 5 \mid G_1 \in \boldsymbol{g}_1^1\} \Pr\{G_1 \in \boldsymbol{g}_1^1\}$$
$$+ \Pr\{G_2 = 5 \mid G_1 \in \boldsymbol{g}_1^2\} \Pr\{G_1 \in \boldsymbol{g}_1^2\} = p_{22|1}(p_{11}+p_{12}+p_{13}) + p_{22|2}(p_{14})$$
$$= 0(0.1+0.2+0.4)+0.9(0.3) = 0.27$$

$$p_{23} = \Pr\{G_2 = 10\} = \Pr\{G_2 = 10 \mid G_1 \in \boldsymbol{g}_1^1\} \Pr\{G_1 \in \boldsymbol{g}_1^1\}$$
$$+ \Pr\{G_2 = 10 \mid G_1 \in \boldsymbol{g}_1^2\} \Pr\{G_1 \in \boldsymbol{g}_1^2\} = p_{23|1}(p_{11}+p_{12}+p_{13}) + p_{23|2}(p_{14})$$
$$= 0.7(0.1+0.2+0.4)+0(0.3) = 0.49$$

The probability of the combination $G_1 = 2$, $G_2 = 10$ is

$$p_{13}p_{23|\mu(3)} = p_{13}p_{23|1} = 0.4\times 0.7 = 0.28.$$

The probability of the combination $G_1 = 3$, $G_2 = 10$ is

$$p_{14}p_{23|\mu(4)} = p_{14}p_{23|2} = 0.3\times 0 = 0.$$

The sets $\boldsymbol{g}_j$ and $\boldsymbol{p}_{j|m}$ $1\le m\le M$ define the conditional PDs of element j. They can be represented in the form of the u-function with vector coefficients:

$$\bar{u}_j(z) = \sum_{c=1}^{C_j} \bar{\boldsymbol{p}}_{jc} z^{g_{jc}} \tag{4.40}$$

where

$$\bar{\boldsymbol{p}}_{jc} = (p_{jc|1}, p_{jc|2}, \ldots, p_{jc|M}) \tag{4.41}$$

Since each combination of the performance rates of the two elements $G_i = g_{ih}$, $G_j = g_{jc}$ corresponds to the subsystem performance rate $\phi(g_{ih}, g_{jc})$ and the probability of the combination is $p_{ih}p_{jc|\mu(h)}$, we can obtain the u-function of the subsystem as follows:

$$\begin{aligned} u_i(z) \underset{\phi}{\overset{\Rightarrow}{\otimes}} \bar{u}_j(z) &= \sum_{h=1}^{k_i} p_{ih} z^{g_{ih}} \underset{\phi}{\overset{\Rightarrow}{\otimes}} \sum_{c=1}^{C_j} \bar{\boldsymbol{p}}_{jc} z^{g_{jc}} \\ &= \sum_{h=1}^{k_i} p_{ih} \sum_{c=1}^{C_j} p_{jc|\mu(h)} z^{\phi(g_{ih}, g_{jc})} \end{aligned} \tag{4.42}$$

The function $\phi(g_{ih}, g_{jc})$ should be substituted by $\phi_{\text{par}}(g_{ih}, g_{jc})$ or $\phi_{\text{ser}}(g_{ih}, g_{jc})$ in accordance with the type of connection between the elements. If the elements are not connected in the reliability block diagram sense (the performance of element i does not directly affect the performance of the subsystem, but affects the PD of element j) the last equation takes the form

$$u_i(z) \overset{\Rightarrow}{\otimes} \bar{u}_j(z) = \sum_{h=1}^{k_i} p_{ih} z^{g_{ih}} \overset{\Rightarrow}{\otimes} \sum_{c=1}^{C_j} \bar{\boldsymbol{p}}_{jc} z^{g_{jc}} = \sum_{h=1}^{k_i} p_{ih} \sum_{c=1}^{C_j} p_{jc|\mu(h)} z^{g_{jc}} \tag{4.43}$$

Example 4.8

Consider two dependent elements from Example 4.7 and assume that these elements are connected in parallel in a flow transmission system (with flow dispersion). Having the sets $\boldsymbol{g}_1=\{0, 1, 2, 3\}$, $\boldsymbol{p}_1=\{0.1, 0.2, 0.4, 0.3\}$ and $\boldsymbol{g}_2=\{0,5,10\}$, $\boldsymbol{p}_{2|1}=\{0.3, 0, 0.7\}$, $\boldsymbol{p}_{2|2}=\{0.1, 0.9, 0\}$ we define the u-functions of the elements as

$$u_1(z) = 0.1z^0 + 0.2z^1 + 0.4z^2 + 0.3z^3$$

$$\overline{u}_2(z) = (0.3,0.1)z^0 + (0,0.9)z^5 + (0.7,0)z^{10}$$

The u-function representing the cumulative performance of the two elements is obtained according to (4.42):

$$u_1(z)\underset{+}{\overset{\Rightarrow}{\otimes}}\overline{u}_2(z) = \sum_{h=1}^{4} p_{1h} \sum_{c=1}^{3} p_{2c|\mu(h)} z^{g_{1h}+g_{2c}}$$

$$= 0.1(0.3z^{0+0} + 0z^{0+5} + 0.7z^{0+10}) + 0.2(0.3z^{1+0}$$

$$+ 0z^{1+5} + 0.7z^{1+10}) + 0.4(0.3z^{2+0} + 0z^{2+5} + 0.7z^{2+10})$$

$$+ 0.3(0.1z^{3+0} + 0.9z^{3+5} + 0z^{3+10}) = 0.03z^0 + 0.06z^1$$

$$+ 0.12z^2 + 0.03z^3 + 0.27z^8 + 0.07z^{10} + 0.14z^{11} + 0.28z^{12}$$

Now assume that the system performance is determined only by the output performance of the second element. The PD of the second element depends on the state of the first element (as in the previous example). According to (4.43) we obtain the u-function representing the performance of the second element:

$$u_1(z)\underset{+}{\overset{\Rightarrow}{\otimes}}\overline{u}_2(z) = \sum_{h=1}^{4} p_{1h} \sum_{c=1}^{3} p_{2c|\mu(h)} z^{g_{2c}}$$

$$= 0.1(0.3z^0 + 0z^5 + 0.7z^{10}) + 0.2(0.3z^0 + 0z^5 + 0.7z^{10})$$

$$+ 0.4(0.3z^0 + 0z^5 + 0.7z^{10}) + 0.3(0.1z^0 + 0.9z^5 + 0z^{10})$$

$$= 0.24z^0 + 0.27z^5 + 0.49z^{10}$$

4.3.2 *u*-functions of a Group of Dependent Elements

Consider a pair of elements e and j. Assume that both of these elements depend on the same element i and are mutually independent given the element i is in a certain state h. This means that the elements e and j are conditionally independent given the state of element i. For any state h of the element i ($g_{ih} \in \mathbf{g}_i^{\mu(h)}$) the PDs of the elements e and j are defined by the pairs of vectors $\mathbf{g}_e$, $\mathbf{p}_{e|\mu(h)}$ and $\mathbf{g}_j$, $\mathbf{p}_{j|\mu(h)}$, where $\mathbf{p}_{e|\mu(h)} = \{p_{ec|\mu(h)} \mid 1 \le c \le C_e\}$. Having these distributions, one can obtain the u-function corresponding to the conditional PD of the subsystem consisting of elements e and j by applying the operators

$$\sum_{c=1}^{C_e} p_{ec|\mu(h)} z^{g_{ec}} \underset{\phi}{\otimes} \sum_{s=1}^{C_j} p_{js|\mu(h)} z^{g_{js}}$$
$$= \sum_{c=1}^{C_e} \sum_{s=1}^{C_j} p_{ec|\mu(h)} p_{js|\mu(h)} z^{\phi(g_{ec}, g_{js})} \tag{4.44}$$

where the function $\phi(g_{ec}, g_{js})$ is substituted by $\phi_{\text{par}}(g_{ec}, g_{js})$ or $\phi_{\text{ser}}(g_{ec}, g_{js})$ in accordance with the type of connection between the elements. Applying the Equation (4.44) for any subset $\boldsymbol{g}_i^m$ ($1 \le m \le M$) we can obtain the u-function representing all of the subsystem's conditional PDs consisting of elements e and j using the following operator over the u-functions $\overline{u}_e(z)$ and $\overline{u}_j(z)$:

$$\overline{u}_e(z) \overset{\circ}{\underset{\phi}{\otimes}} \overline{u}_j(z) = \sum_{c=1}^{C_e} \overline{\boldsymbol{p}}_{ec} z^{g_{ec}} \overset{\circ}{\underset{\phi}{\otimes}} \sum_{s=1}^{C_j} \overline{\boldsymbol{p}}_{js} z^{g_{js}}$$
$$= \sum_{c=1}^{C_e} \sum_{s=1}^{C_j} \overline{\boldsymbol{p}}_{ec} \circ \overline{\boldsymbol{p}}_{js} z^{\phi(g_{ec}, g_{js})} \tag{4.45}$$

where

$$\overline{\boldsymbol{p}}_{ec} \circ \overline{\boldsymbol{p}}_{js} = (p_{ec|1} p_{js|1}, p_{ec|2} p_{js|2}, \ldots, p_{ec|M} p_{js|M}) \tag{4.46}$$

Example 4.9

A flow transmission system (with flow dispersion) consists of three elements connected in parallel. Assume that element 1 has the PD $\boldsymbol{g}_1 = \{0, 1, 3\}$, $\boldsymbol{p}_1 = \{0.2, 0.5, 0.3\}$.

The PD of element 2 depends on the performance rate of element 1 such that when $G_1 \le 1$ ($G_1 \in \{0, 1\}$) element 2 has the PD $\boldsymbol{g}_2 = \{0,3\}$, $\boldsymbol{q}_2 = \{0.3, 0.7\}$ while when $G_1 > 1$ ($G_1 \in \{3\}$) element 2 has the PD $\boldsymbol{g}_2 = \{0, 5\}$, $\boldsymbol{q}_2 = \{0.1, 0.9\}$.

The PD of element 3 depends on the performance rate of element 1 such that when $G_1 = 0$ ($G_1 \in \{0\}$) element 3 has the PD $\boldsymbol{g}_3 = \{0, 2\}$, $\boldsymbol{q}_3 = \{0.8, 0.2\}$ while when $G_1 > 0$ ($G_1 \in \{1, 3\}$) element 3 has the PD $\boldsymbol{g}_3 = \{0, 3\}$, $\boldsymbol{q}_3 = \{0.2, 0.8\}$.

The set $\boldsymbol{g}_1$ should be divided into three subsets corresponding to different PDs of dependent elements such that

for $G_1 \in \boldsymbol{g}_1^1 = \{0\}$ $\boldsymbol{g}_{2|1} = \{0, 3\}$, $\boldsymbol{q}_{2|1} = \{0.3, 0.7\}$ and $\boldsymbol{g}_{3|1} = \{0, 2\}$, $\boldsymbol{q}_{3|1} = \{0.8, 0.2\}$

for $G_1 \in \boldsymbol{g}_1^2 = \{1\}$ $\boldsymbol{g}_{2|2} = \{0, 3\}$, $\boldsymbol{q}_{2|2} = \{0.3, 0.7\}$ and $\boldsymbol{g}_{3|2} = \{0, 3\}$, $\boldsymbol{q}_{3|2} = \{0.2, 0.8\}$

for $G_1 \in \boldsymbol{g}_1^3 = \{3\}$ $\boldsymbol{g}_{2|3} = \{0, 5\}$, $\boldsymbol{q}_{2|3} = \{0.1, 0.9\}$ and $\boldsymbol{g}_{3|3} = \{0, 3\}$, $\boldsymbol{q}_{3|3} = \{0.2, 0.8\}$

The conditional PDs of elements 2 and 3 can be represented in the following form:

$$\boldsymbol{g}_2 = \{0,3,5\}, \boldsymbol{p}_{2|1} = \boldsymbol{p}_{2|2} = \{0.3, 0.7, 0\}, \ \boldsymbol{p}_{2|3} = \{0.1, 0, 0.9\}$$

$$g_3 = \{0,2,3\},\, p_{3|1} = \{0.8, 0.2, 0\},\, p_{3|2} = p_{3|3} = \{0.2, 0, 0.8\}$$

The *u*-functions $\overline{u}_1(z)$ and $\overline{u}_2(z)$ take the form

$$\overline{u}_2(z) = (0.3,0.3,0.1)z^0 + (0.7,0.7,0)z^3 + (0,0,0.9)z^5$$

$$\overline{u}_3(z) = (0.8,0.2,0.2)z^0 + (0.2,0,0)z^2 + (0,0.8,0.8)z^3$$

The *u*-function of the subsystem consisting of elements 2 and 3 according to (4.45) is

$$\overline{U}_4(z) = \overline{u}_2(z) \underset{+}{\overset{\circ}{\otimes}} \overline{u}_3(z) = [(0.3,0.3,0.1)z^0 + (0.7,0.7,0)z^3$$

$$+ (0,0,0.9)z^5] \underset{+}{\overset{\circ}{\otimes}} [(0.8,0.2,0.2)z^0 + (0.2,0,0)z^2 + (0,0.8,0.8)z^3]$$

$$= (0.24,0.06,0.02)z^0 + (0.06,0,0)z^2 + (0,0.24,0.08)z^3$$
$$+ (0.56,0.14,0)z^3 + (0.14,0,0)z^5 + (0,0.56,0)z^6$$
$$+ (0,0,0.18)z^5 + (0,0,0)z^7 + (0,0,0.72)z^8$$
$$= (0.24,0.06,0.02)z^0 + (0.06,0,0)z^2 + (0.56,0.38,0.08)z^3$$
$$+ (0.14,0,0.18)z^5 + (0,0.56,0)z^6 + (0,0,0.72)z^8$$

Now we can replace elements 2 and 3 by a single equivalent element with the *u*-function $\overline{U}_4(z)$ and consider the system as consisting of two elements with *u*-functions $u_1(z)$ and $\overline{U}_4(z)$. The *u*-function of the entire system according to (4.42) is:

$$U(z) = u_1(z) \underset{+}{\overset{\Rightarrow}{\otimes}} \overline{U}_4(z) = \sum_{h=1}^{3} p_{1h} \sum_{c=1}^{6} p_{4c|\mu(h)} z^{g_{1h}+g_{4c}}$$
$$= 0.2(0.24z^{0+0} + 0.06z^{0+2} + 0.56z^{0+3} + 0.14z^{0+5}) + 0.5(0.06z^{1+0}$$
$$+ 0.38z^{1+3} + 0.56z^{1+6}) + 0.3(0.02z^{3+0} + 0.08z^{3+3} + 0.18z^{3+5}$$
$$+ 0.72z^{3+8}) = 0.048z^0 + 0.03z^1 + 0.012z^2 + 0.118z^3 + 0.19z^4$$
$$+ 0.028z^5 + 0.024z^6 + 0.28z^7 + 0.054z^8 + 0.216z^{11}$$

Note that the conditional independence of two elements *e* and *j* does not imply their unconditional independence. The two elements are conditionally independent if for any states *c*, *s* and *h*

$$\Pr\{G_e = g_{ec},\, G_j = g_{js} \mid G_i = g_{ih}\}$$

$$= \Pr\{G_e = g_{ec} \mid G_i = g_{ih}\} \Pr\{G_j = g_{js} \mid G_i = g_{ih}\}$$

The condition of independence of elements e and j

$$\Pr\{G_e = g_{ec}, G_j = g_{js}\} = \Pr\{G_e = g_{ec}\}\Pr\{G_j = g_{js}\}$$

does not follow from the previous equation. In our example we have

$$\Pr\{G_2 = 3\} = p_{22|1}p_{11} + p_{22|2}p_{12} + p_{22|3}p_{13}$$

$$= 0.7\times0.2 + 0.7\times0.5 + 0\times0.3 = 0.49$$

$$\Pr\{G_3 = 3\} = p_{33|1}p_{11} + p_{33|2}p_{12} + p_{33|3}p_{13}$$

$$= 0\times0.2 + 0.8\times0.5 + 0.8\times0.3 = 0.64$$

Hence

$$\Pr\{G_2 = 3\}\Pr\{G_3 = 3\} = 0.49\times0.64 = 0.3136$$

while

$$\Pr\{G_2 = 3, G_3 = 3\} = p_{22|1}p_{33|1}p_{11} + p_{22|2}p_{33|2}p_{12} + p_{22|3}p_{33|3}p_{13}$$

$$= 0.7\times0\times0.2 + 0.7\times0.8\times0.5 + 0\times0.8\times0.3 = 0.28$$

4.3.3 *u*-functions of Multi-state Systems with Dependent Elements

Consecutively applying the operators $\otimes_{\varphi}$, $\overset{\Rightarrow}{\otimes}_{\varphi}$ and $\overset{\Rightarrow}{\otimes}$ and replacing pairs of elements by auxiliary equivalent elements, one can obtain the *u*-function representing the performance distribution of the entire system. The following recursive algorithm obtains the system *u*-function:

1. Define the *u*-functions for all of the independent elements.
2. Define the *u*-functions for all of the dependent elements in the form (4.40) and (4.41).
3. If the system contains a pair of mutually independent elements connected in parallel or in a series, replace this pair with an equivalent element with the *u*-function obtained by $\otimes_{\varphi_{\text{par}}}$ or $\otimes_{\varphi_{\text{ser}}}$ operator respectively (if both elements depend on the same external element, i.e. they are conditionally independent, operators $\overset{\circ}{\otimes}_{\varphi_{\text{par}}}$ or $\overset{\circ}{\otimes}_{\varphi_{\text{ser}}}$ (4.45) should be applied instead of $\otimes_{\varphi_{\text{par}}}$ or $\otimes_{\varphi_{\text{ser}}}$ respectively).
4. If the system contains a pair of dependent elements, replace this pair with an equivalent element with the *u*-function obtained by $\overset{\Rightarrow}{\otimes}_{\varphi_{\text{par}}}$, $\overset{\Rightarrow}{\otimes}_{\varphi_{\text{ser}}}$ or $\overset{\Rightarrow}{\otimes}$ operator.
5. If the system contains more than one element, return to step 3.

The performance distribution of the entire system is represented by the u-function of the remaining single equivalent element.

Example 4.10

Consider an information processing system consisting of three independent computing blocks (Figure 4.10). Each block consists of a high-priority processing unit and a low-priority processing unit that share access to a database. When the high-priority unit operates with the database, the low-priority unit waits for access. Therefore, the processing speed of the low-priority unit depends on the load (processing speed) of the high-priority unit. The processing speed distributions of the high-priority units (elements 1, 3 and 5) are presented in Table 4.8.

Table 4.8. Unconditional PDs of system elements 1, 3 and 5

g_1	50	40	30	20	10	0
p_1	0.2	0.5	0.1	0.1	0.05	0.05
g_3	60	20	0			
p_3	0.2	0.7	0.1			
g_5	100	80	0			
p_5	0.7	0.2	0.1			

The conditional distributions of the processing speed of the low-priority units (elements 2, 4 and 6) are presented in Table 4.9. The high- and low-priority units share their work in proportion to their processing speed.

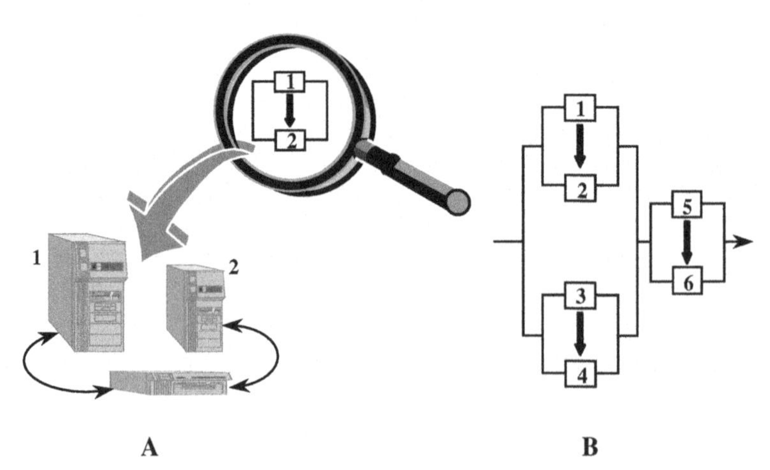

Figure 4.10. Information processing system
(A: structure of computing block; B: system logic diagram)

Table 4.9. Conditional PDs of system elements 2, 4 and 6

Condition for element 2	g_2:	30	15	0
$0\le G_1<15$	$p_{2\|m}$:	0.8	0.15	0.05
$15\le G_1<35$		0.4	0.55	0.05
$35\le G_1<70$		0	0.9	0.1
Condition for element 4	g_4:	30	15	0
$0\le G_3<15$	$p_{4\|m}$:	0.8	0.15	0.05
$15\le G_3<35$		0.6	0.35	0.05
$35\le G_3<70$		0	0.95	0.05
Condition for element 6	g_6:	50	30	0
$0\le G_5<30$	$p_{6\|m}$:	0.8	0.15	0.05
$30\le G_5<90$		0.5	0.4	0.1
$90\le G_5<150$		0.3	0.6	0.1

The first two computing blocks also share the computational load in proportion to their processing speed. The third block obtains the output of the first two blocks and starts processing when these blocks complete their work. The system fails if its processing speed is lower than the demand w.

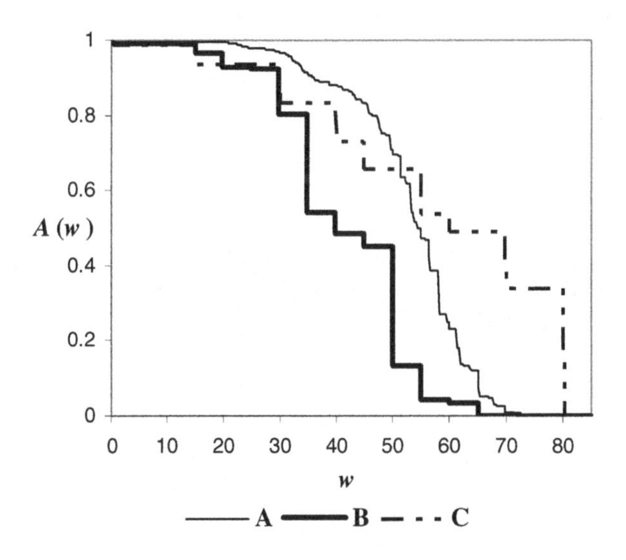

Figure 4.11. System availability as a function of demand w

The system belongs to the task processing type. In order to obtain the UGF representing the system PD, we first define the u-functions $u_1(z)$, $u_3(z)$, $u_5(z)$ from the unconditional PDs of the corresponding elements and the u-functions $\overline{u}_2(z), \overline{u}_4(z), \overline{u}_6(z)$ in accordance with (4.40) and (4.41):

$u_1(z) = 0.2z^{50}+0.5z^{40}+0.1z^{30}+0.1z^{20}+0.05z^{10}+0.05z^0$

$u_3(z) = 0.2z^{60}+0.7z^{20}+0.1z^0$, $u_5(z) = 0.7z^{100}+0.2z^{80}+0.1z^0$

$\bar{u}_2(z) = (0.8, 0.4, 0)z^{30}+(0.15, 0.55, 0.9)z^{15}+(0.05, 0.05, 0.1)z^0$

$\bar{u}_4(z) = (0.8, 0.6, 0)z^{30}+(0.15, 0.35, 0.95)z^{15}+(0.05, 0.05, 0.05)z^0$

$\bar{u}_6(z) = (0.8, 0.5, 0.3)z^{50}+(0.15, 0.4, 0.6)z^{30}+(0.05, 0.1, 0.1)z^0$

Then we apply the following operators producing the u-functions of the auxiliary equivalent elements:

$$U_7(z) = u_1(z) \overset{\Rightarrow}{\underset{+}{\otimes}} \bar{u}_2(z),\ U_8(z) = u_3(z) \overset{\Rightarrow}{\underset{+}{\otimes}} \bar{u}_4(z)$$

$$U_9(z) = u_5(z) \overset{\Rightarrow}{\underset{+}{\otimes}} \bar{u}_6(z)$$

The obtained u-functions represent the PD of the three computing blocks. The PD of the subsystem consisting of two parallel blocks (equivalent element 10) is represented by

$$U_{10}(z) = U_7(z) \underset{+}{\otimes} U_8(z)$$

The entire system can be represented as two elements with u-functions $u_{10}(z)$ and $u_9(z)$ connected in series. Since the system belongs to the task processing type, its u-function is obtained by the operator (4.5)

$$U(z) = U_{10}(z) \underset{\times}{\otimes} U_9(z)$$

The system availability can now be obtained by applying the operator δ_w over $U(z)$: $A(w)=\delta_w(U(z))$. The system availability, as a function of demand w, is presented in Figure 4.11 (curve A).

Example 4.11

A continuous production system (Figure 4.12) consists of two consecutive production blocks. Each block consists of a main production unit and an auxiliary production unit that share some preventive maintenance resources (cleaning, lubrication, *etc.*). When the main production unit is intensively loaded, the lack of resources prevents the auxiliary unit from being intensively loaded with high availability.

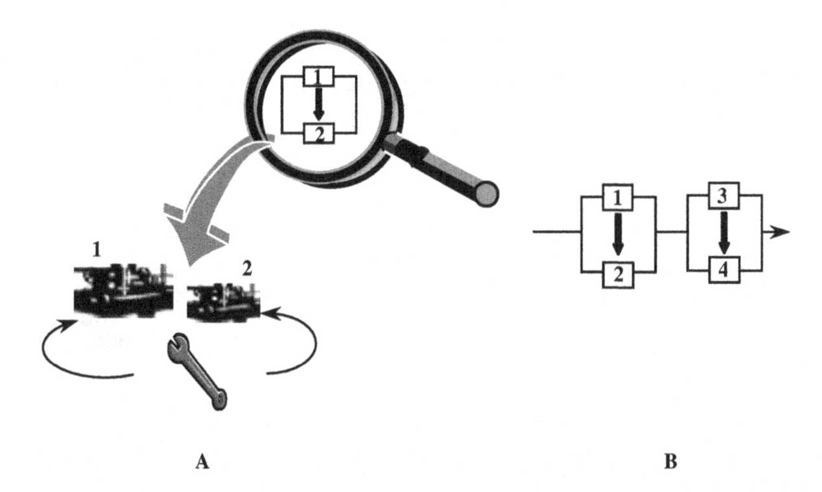

Figure 4.12. Continuous production system
(A: structure of production block; B: system logic diagram)

The productivity distributions of the main production units (elements 1 and 3) are presented in Table 4.8. The conditional distributions of the auxiliary units' productivities (elements 2 and 4) are presented in Table 4.9. The system fails if it does not meet the demand w.

The system belongs to the flow transmission type. In order to obtain the UGF representing the system PD, first we define the u-functions $u_1(z)$ and $u_3(z)$ from the unconditional PDs of the corresponding elements and the u-functions $\overline{u}_2(z)$ and $\overline{u}_4(z)$ in accordance with (4.40) and (4.41) (as in the previous example).

Then we apply the following operators producing the u-functions of auxiliary equivalent elements corresponding to the production blocks:

$$U_5(z) = u_1(z) \underset{+}{\overset{\Rightarrow}{\otimes}} \overline{u}_2(z), \; U_6(z) = u_3(z) \underset{+}{\overset{\Rightarrow}{\otimes}} \overline{u}_4(z)$$

The entire system can be represented as two elements with u-functions $U_5(z)$ and $U_6(z)$ connected in a series. Since the system belongs to the flow transmission type, its u-function takes the form:

$$U(z) = U_5(z) \underset{\min}{\otimes} U_6(z)$$

The system availability is obtained as $A(w) = \delta_w(U(z))$. The system availability as a function of demand w is presented in Figure 4.11 (curve B).

Example 4.12

Consider a system with indirect influence of part of the elements on its performance. A chemical reactor contains six heating elements and two identical mixers (Figure 4.13). Two heating elements have nominal heating power 8 and availability 0.9, four heating elements have heating power 5 and availability 0.85. The heating elements are powered by two independent power sources with nominal power 25 and availability 0.95 for each one. The heating power of the elements cannot exceed the total power of the available sources.

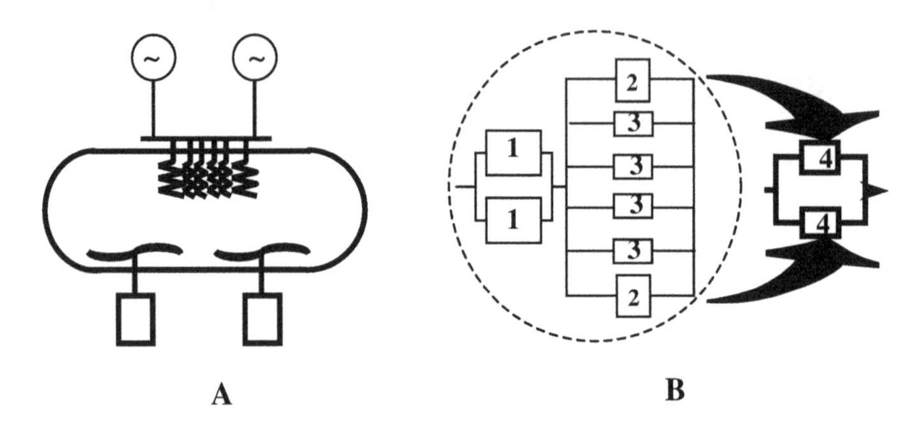

Figure 4.13. Chemical reactor (A: structure of reactor; B: system logic diagram)

The productivity distribution of each mixer depends on the cumulative power of the heaters. The greater the heating effect, the greater the productivity and availability of the mixers. The mixers are conditionally independent given the state of the heating subsystem. The conditional distributions of the mixers' productivities (element 4) are presented in Table 4.10. The total productivity of the reactor is equal to the cumulated productivity of the two mixers. The system fails if it does not meet the demand w.

Table 4.10. Conditional performance distributions of the mixers

Condition	g_4:	40	30	15	0
$0\leq G_h<10$		0	0	0.2	0.8
$10\leq G_h<20$	$p_{4\|m}$:	0	0	0.8	0.2
$20\leq G_h<25$		0	0.2	0.6	0.2
$25\leq G_h<30$		0.3	0.4	0.2	0.1
$30\leq G_h<40$		0.7	0.1	0.1	0.1

The heating subsystem is the series-parallel system of flow transmission type. In order to obtain the UGF representing the subsystem PD, first we define the u-functions $u_1(z)$, $u_2(z)$, $u_3(z)$ as

$$u_1(z) = 0.95z^{25}+0.05z^0,\ u_2(z) = 0.9z^8+0.1z^0,\ u_3(z)=0.85z^5+0.15z^0$$

and then obtain the u-function representing the PD of the subsystem by consecutively applying the composition operators. The u-function of the power supply system is

$$U_5(z) = u_1(z) \underset{+}{\otimes} u_1(z)$$

The u-function of the heaters is obtained as follows:

$$U_6(z) = u_2(z) \underset{+}{\otimes} u_2(z),\ U_7(z) = U_6(z) \underset{+}{\otimes} u_3(z),\ U_8(z) = U_7(z) \underset{+}{\otimes} u_3(z)$$

$$U_9(z) = U_8(z) \underset{+}{\otimes} u_3(z),\ U_{10}(z) = U_9(z) \underset{+}{\otimes} u_3(z)$$

Observe that this u-function can be obtained in a simpler manner by defining an auxiliary element with the u-function $U_7(z)$ equivalent to the u-function of two parallel elements 3:

$$U_6(z) = u_2(z) \underset{+}{\otimes} u_2(z), U_7(z) = u_3(z) \underset{+}{\otimes} u_3(z)$$

$$U_8(z) = U_7(z) \underset{+}{\otimes} U_7(z),\ U_{10}(z) = U_6(z) \underset{+}{\otimes} U_8(z)$$

The u-function of the entire heating system (power sources and heaters) is

$$U_h(z) = U_5(z) \underset{\min}{\otimes} U_{10}(z)$$

The mechanical system consists of two parallel mixers and belongs to the flow transmission type. Having the u-function $\overline{u}_4(z)$ of a single mixer defined in accordance with (4.40) and (4.41) as

$$\overline{u}_4(z) = (0, 0, 0, 0.3, 0.7)z^{40} + (0, 0, 0.2, 0.4, 0.1)z^{30}$$

$$+(0.2, 0.8, 0.6, 0.2, 0.1)z^{15} + (0.8, 0.2, 0.2, 0.1, 0.1)z^{0}$$

we obtain the u-function representing the conditional PDs of the system:

$$\overline{U}_{11}(z) = \overline{u}_4(z) \overset{\circ}{\underset{+}{\otimes}} \overline{u}_4(z)$$

Since the heating system affects the reactor's productivity only by influencing the PD of the mixers, we apply the $\overset{\Rightarrow}{\otimes}$ operator:

$$U(z) = U_h(z) \overset{\Rightarrow}{\otimes} \overline{U}_{11}(z)$$

The system availability can now be obtained as $A(w) = \delta_w(U(z))$. The system availability as a function of demand w is presented in Figure 4.11 (curve C).

4.4 Common Cause Failures in Multi-state Systems

Common cause (CC) failures (CCFs) are the failures of multiple elements due to a common cause (single occurrence or condition). The origin of CC events can be outside the system elements they affect (lightning or seismic events, sudden changes in the environment, a wide range of human interventions from maintenance errors, to intended enemy attacks), or they can originate from the elements themselves, causing other elements to fail (examples of such events are voltage surges caused by inappropriate switching in power systems leading to failure propagation, and pipe-whip events in high-pressure systems). The condition of a CCF occurring exists when some coupling factors affect a group of elements. These include the elements being:

- involved in the same process or procedure
- sharing a common resource
- having similar design or interface
- having the same manufacturer
- having the same or close location, *etc.*

CCFs increase joint-failure probabilities, thereby reducing the reliability of the technical systems.

It is assumed that all of the elements that can fail due to a certain CC belong to a corresponding CC group (CCG). There can be several CCGs in a system, since several factors can affect the functioning of its elements. Within each CCG, several failure processes can exist that cause the simultaneous failure of different subgroups of this CCG. In order to estimate the system's reliability, the characteristics of these failure processes should be included in the system model. The description of the methods for estimating the effect of CCFs on the reliability of the binary systems can be found in [107, 108].

4.4.1 Incorporating Common Cause Failures into Multi-state Systems Reliability Analysis

An algorithm presented in this section for incorporating the CCFs into the MSS reliability analysis is based on an implicit method suggested by Vaurio [109]. This implicit method uses formulas (derived by Chae and Clark [110]) for probabilities that specific elements subject to the same CCF remain in a working condition during a given time.

Consider an MSS consisting of two-state elements (elements with total failures). The elements are mutually independent (except for the elements belonging to the same CCG).

The system contains J CCGs such that each CCG j is defined by the set C_j of numbers of MSS elements belonging to this group.

Each element can belong to a single CCG (the CCGs are disjoint): $C_i \cap C_j = \varnothing$ if $i \neq j$. Each CCG j consists of L_j elements.

All of the elements subject to the same CC (belonging to the same CCG) have the same statistical characteristics (are statistically identical).

All elements belonging to the same CCG are subject to CCF by a number of different failure events. Each failure event ϑ_{jk} is independent and constitutes the simultaneous failures of a specific subset of k elements of CCG j. The probability of each failure event depends on the number of elements that fail, but it does not depend on the particular elements involved. Each particular element cannot individually affect the probability of the failure event it is involved in.

The implicit method for incorporating CCFs into the system reliability analysis suggested in [109] consists of the following three steps:

1. Assign the unique reliability p_j to all the basic system elements j.
2. Determine the expression for the system reliability in terms of the reliabilities of the basic elements without considering any CCF. This expression is in the form of an algebraic sum of the products (terms) of the basic element reliabilities.
3. In any term containing a product of k element reliabilities (*i.e.* $p_1p_2...p_k$) belonging to the same CCG j $(1 \le k \le L_j)$, replace that product with the probability $R_{j,L_j}^{(k)}$ that these specific k elements (which are subject to failure events $\vartheta_{j1},...,\vartheta_{jL_j}$) all remain in a working state.

This probability can be obtained recursively as follows [110]:

$$R_{j,n}^{(k)} = \prod_{i=n-k+1}^{n} R_{j,i}^{(1)} \tag{4.47}$$

$$R_{j,n}^{(1)} = \prod_{k=1}^{n} [\widetilde{P}_{jk}]^{\binom{n-1}{k-1}} \tag{4.48}$$

where $R_{j,n}^{(k)}$ is the probability that specific k elements belonging to CCG j, which contains a total of the n elements, all remain in working condition ($R_{j,n}^{(0)} = 1$ for any j and n by definition) and $\widetilde{P}_{jk}$ is the probability of the non-occurrence of the failed state caused by the event ϑ_{jk}.

The implicit method can be easily applied to an MSS if the final expression for its reliability is obtained in an explicit analytical form. Obtaining the analytical expressions for complex MSSs using the UGF method is an extremely time-consuming task. In contrast, the method provides simple numerical algorithms for computing the system's reliability for arbitrary time and demand without obtaining analytical expressions. To adapt the implicit method to the numerical algorithms, the modified u-function technique has been suggested [111].

In the u-function of the MSS subsystem e

$$U_e(z)=\sum_{i=1}^{k_e}\alpha_{ei}z^{g_{ei}} \tag{4.49}$$

the coefficients α_{ei} are products of the reliabilities of the individual elements. In order to keep track of the occurrence of different reliability functions in these coefficients, the u-function is modified as follows:

$$\widetilde{U}_e(z)=\sum_{i=1}^{k_e}\alpha_{ei}^{*}z^{g_{ei},\mathbf{s}_{ei}} \tag{4.50}$$

To obtain the system u-function in the form (4.50) from u-function (4.49), one has to perform the following steps for each term $\alpha_{ei}z^{g_{ei}}$:

1. Assign $\mathbf{0}$ to the vector $\mathbf{s}_{ei}$ that consists of J integer numbers.
2. Obtain coefficient α_{ei}^{*} by replacing in the product $\alpha_{ei}=p_1p_2...p_k$ all of the reliabilities of the individual elements belonging to any CCG with 1.
3. When replacing reliability p_h of element h belonging to CCG j, increment by 1 the corresponding element $s_{ei}(j)$ of the vector-indicator $\mathbf{s}_{ei}$. Finally each element $s_{ei}(j)$ of the vector-indicator $\mathbf{s}_{ei}$ contains a number of replaced reliabilities of elements belonging to CCG j.

Based on these steps one can obtain the u-function $\widetilde{u}_i(z)$ of a single two-state MSS element i not belonging to any CCG as

$$\widetilde{u}_i(z)=p_{i1}z^{g_{i1},\mathbf{0}}+(1-p_{i1})z^{f,\mathbf{0}}=p_{i1}z^{g_{i1},\mathbf{0}}+z^{f,\mathbf{0}}-p_{i1}z^{f,\mathbf{0}} \tag{4.51}$$

where g_{i1} and p_{i1} are the nominal performance and reliability of the element respectively, f is a performance rate in the failed state.

The u-function $\widetilde{u}_l(z)$ of the MSS element belonging to CCG j takes the form

$$\widetilde{u}_l(z)=z^{g_{l1},\mathbf{s}_l}+z^{f,\mathbf{0}}-z^{f,\mathbf{s}_l} \tag{4.52}$$

where $s_l(k)=1(k=j)$ for $1\le k\le J$.

The composition operators over u-functions (4.50) are the same as regular composition operators $\otimes_{\varphi}$ except for the rule that defines the treatment of vector-indicators:

$$\underset{\phi,+}{\otimes}(\widetilde{U}_1(z),\widetilde{U}_2(z))=\underset{\phi,+}{\otimes}\Big(\sum_{i=1}^{k_1}\alpha_{1i}^{*}z^{g_{1i},\mathbf{s}_{1i}},\sum_{j=1}^{k_2}\alpha_{2j}^{*}z^{g_{2j},\mathbf{s}_{2j}}\Big)$$
$$=\sum_{i=1}^{k_1}\sum_{j=1}^{k_2}\alpha_{1i}^{*}\alpha_{2j}^{*}z^{\phi(g_{1i},g_{2j}),\mathbf{s}_{1i}+\mathbf{s}_{2j}} \tag{4.53}$$

The vector-indicators are always summed independently on function ϕ chosen for a specific operator.

Consequently, applying the composition operators (4.53) in accordance with the reliability block diagram method (described in Section 4.1.3), one can obtain the u-function of the entire MSS in the form (4.50). In each term i of this sum, α_{ei}^{*} is a product of the reliabilities of the basic elements not belonging to any CCG, and g_{ei} is the total MSS output performance in state i of the system. Each element $s_{ei}(j)$ of vector-indicator $\boldsymbol{s}_{ei}$ contains a number of elements belonging to CCG j that should also be taken into account when calculating the probability of the corresponding MSS state. Multiplying the α_{ei}^{*} coefficients by the probabilities that specific $s_{ei}(j)$ elements of each CCG j do not fail, one can obtain the probability of state i which should be the coefficient of the ith term of the u-function of the MSS calculated with respect to the CCF.

Thus, the u-function of an MSS can be obtained by applying the following Ω operator over the u-function of the MSS:

$$U(z) = \Omega(\widetilde{U}_e(z)) = \Omega[\sum_{i=1}^{k_e} \alpha_{ei}^{*} z^{g_{ei},(s_{ei}(1),\dots,s_{ei}(J))}]$$
$$= \sum_{i=1}^{k_e} \alpha_{ei}^{*} \prod_{j=1}^{J} R_{j,L_j}^{(s_{ei}(j))} z^{g_{ei}} \tag{4.54}$$

The numerical algorithm for the evaluation of the entire MSS u-function with respect to CCF is as follows:

1. Determine the reliabilities of the individual system elements p_i and $R_{j,L_j}^{(k)}$ $(0 \le k \le L_j)$ values for each CCG j $(1 \le j \le J)$ using (4.47) and (4.48).
2. Determine the u-functions of the individual MSS elements using definitions (4.51) and (4.52).
3. For a given MSS topology, obtain the entire system u-function $\widetilde{U}(z)$ by applying the composition operators (4.53) over the u-functions of the individual system elements (the ϕ functions should be chosen in accordance with the system type and connection between the elements).
4. Obtain the u-function of the MSS using the Ω operator (4.54) over $\widetilde{U}(z)$.

Example 4.13

Consider a series-parallel task processing MSS (with work sharing) containing two subsystems (components) connected in a series (Figure 4.14A).

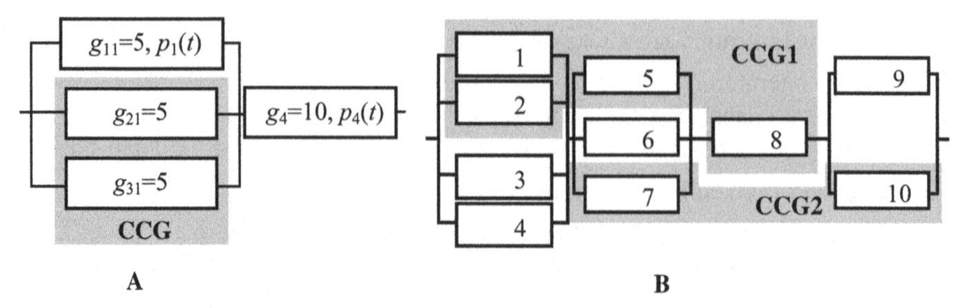

Figure 4.14. Examples of series-parallel MSS with CCF

The first component has three parallel elements with the same nominal performance rate: $g_{11} = g_{21} = g_{31} = 5$. The reliability of the first element is p_1. Two other elements of the first component compose a CCG, which is characterized by the probabilities $\tilde{P}_{11}$ and $\tilde{P}_{12}$. For this CCG, $R_{1,2}^{(1)} = \tilde{P}_{11}\tilde{P}_{12}$ and $R_{1,2}^{(2)} = \tilde{P}_{11}^2\tilde{P}_{12}$. The second component has a single element with a nominal performance rate of g_{41}=10 and the reliability p_4. All of the elements have the performance f=0 when they fail.

Following (4.51) and (4.52), we obtain the u-function for the first element as

$$u_1(z) = p_1 z^{5,(0)} + z^{0,(0)} - p_1 z^{0,(0)}$$

for elements belonging to the CCG as

$$u_2(z) = u_3(z) = z^{5,(1)} + z^{0,(0)} - z^{0,(1)}$$

and for element of the second component as

$$u_4(z) = p_4 z^{10,(0)} + (1-p_4) z^{0,(0)}$$

The u-function of the first component is obtained using the $\underset{+}{\otimes}$ operator:

$$U_1(z) = \underset{+}{\otimes}(u_1(z), u_2(z), u_3(z))$$

$$= (p_1 z^{5,(0)} + z^{0,(0)} - p_1 z^{0,(0)}) \underset{+}{\otimes} (z^{5,(1)} + z^{0,(0)} - z^{0,(1)}) \underset{+}{\otimes} (z^{5,(1)} + z^{0,(0)} - z^{0,(1)})$$

$$= p_1 z^{15,(2)} + (1-3p_1) z^{10,(2)} + 2p_1 z^{10,(1)} + (3p_1-2) z^{5,(2)}$$
$$+2(1-2p_1) z^{5,(1)} + p_1 z^{5,(0)} + (1-p_1) z^{0,(2)} + 2(p_1-1) z^{0,(1)} + (1-p_1) z^{0,(0)}$$

The u-function of the second component is equal to the u-function of its single element $U_2(z)=u_4(z)$.

To obtain the u-function $\tilde{U}(z)$ corresponding to the entire system we use the $\underset{\times}{\otimes}$ operator:

$$\tilde{U}(z) = \underset{\times}{\otimes}(U_1(z), U_2(z)) = p_1 p_4 z^{6,(2)} + p_4(1-3p_1) z^{5,(2)} + 2p_1 p_4 z^{5,(1)}$$

$$+p_4(3p_1-2)z^{3.3,(2)}+2p_4(1-2p_1)z^{3.3,(1)}+p_1p_4z^{3.3,(0)}+p_4(1-p_1)z^{0,(2)}$$

$$+p_4(p_1-1)z^{0,(1)}+(1-p_1p_4)z^{0,(0)}$$

Now, using operator Ω, we obtain the u-function for the entire system with respect to CCF:

$$U(z)=\Omega\,(\widetilde{U}(z))=q_1z^6+q_2z^5+q_3z^{3.3}+q_4z^0$$

where

$$q_1=R_{1,2}^{(2)}\,p_1p_4=\widetilde{P}_{11}^2\widetilde{P}_{12}p_1p_4$$

$$q_2=R_{1,2}^{(2)}\,p_4(1-3p_1)+2\,R_{1,2}^{(1)}\,p_1p_4=\widetilde{P}_{11}^2\widetilde{P}_{12}\,p_4(1-3p_1)+2\,\widetilde{P}_{11}\widetilde{P}_{12}p_1p_4$$

$$q_3=R_{1,2}^{(2)}\,p_4(3p_1-2)+2\,R_{1,2}^{(1)}\,p_4(1-2p_1)+p_1p_4$$

$$=\widetilde{P}_{11}^2\widetilde{P}_{12}p_4(3p_1-2)+2\,\widetilde{P}_{11}\widetilde{P}_{12}p_4(1-2p_1)+p_1p_4$$

$$q_4=R_{1,2}^{(2)}\,p_4(1-p_1)+2\,R_{1,2}^{(1)}\,p_4(p_1-1)+(1-p_1p_4)$$

$$=\widetilde{P}_{11}^2\widetilde{P}_{12}p_4(1-p_1)+2\,\widetilde{P}_{11}\widetilde{P}_{12}p_4(p_1-1)+(1-p_1p_4)$$

The system performance distribution is determined by the vectors

$$\boldsymbol{g}=\{6,5,3.3,0\},\ \boldsymbol{q}=\{q_1,q_2,q_3,q_4\}$$

Using the operators δ_w we can obtain the system reliability for any demand w:

$$R(w)=\delta_w(U(z))=\begin{cases}0, & w>6\\ q_1, & 5<w\le 6\\ q_1+q_2, & 3.3<w\le 5\\ q_1+q_2+q_3, & 0<w\le 3.3\\ q_1+q_2+q_3+q_4=1, & w=0\end{cases}$$

Example 4.14

The non-repairable series-parallel MSS (Figure 4.14B) consists of four components connected in a series. All of the MSS elements have Weibull cumulative hazard functions $H(t) = (\alpha t)^{\beta}$. Parameters of the elements are presented in Table 4.11. The performance of any element in a failed state is f=0.

Table 4.11. Parameters of MSS elements

No of element i	No of component	Nominal performance g_{i1}	Parameters of individual element cumulative hazard function $H(t)=(\alpha t)^{\beta}$ α	β	No of CCG
1	1	0.20	-	-	1
2	1	0.20	-	-	1
3	1	0.20	0.004	1.0	-
4	1	0.50	0.001	0.5	-
5	2	0.60	-	-	1
6	2	0.30	0.008	1.0	-
7	2	0.20	-	-	2
8	3	1.30	-	-	1
9	4	0.85	0.0012	1.0	-
10	4	0.25	-	-	2

There are two CCGs in the given MSS: C_1={1, 2, 5, 8}, C_2={7, 10}. The failure processes ϑ_{jk} in these CCGs that govern simultaneous failures of a specific set of k elements are characterized by the cumulative hazard functions $H_{jk}(t)$. The probability of the non-occurrence of the failure event governed by the process ϑ_{jk} in time interval [0, t] is $\widetilde{P}_{jk}(t) = \exp(-H_{jk}(t))$.

For CCG 1:

$$H_{11}(t)=(0.001t)^{0.8}, \quad H_{12}(t)=0.08H_{11}(t)$$

$$H_{13}(t)=0.02H_{11}(t), \quad H_{14}(t)=0.007H_{11}(t)$$

For CCG 2:

$$H_{21}(t)=0.003t, \quad H_{22}(t)=0.2H_{21}(t)$$

The structure presented is interpreted as flow transmission MSS with flow dispersion and task processing MSS with work sharing. The reliability functions $R(t,w)$ for both MSSs obtained using the numerical algorithm described above are presented in Figure 4.15. One can see that the task processing MSS has more different levels of PD than the flow transmission MSS. This is due to the nature of the operator $\otimes_{\min}$, which, as distinct from the operator $\otimes_{\times}$, reduces the diversity of the possible performance levels.

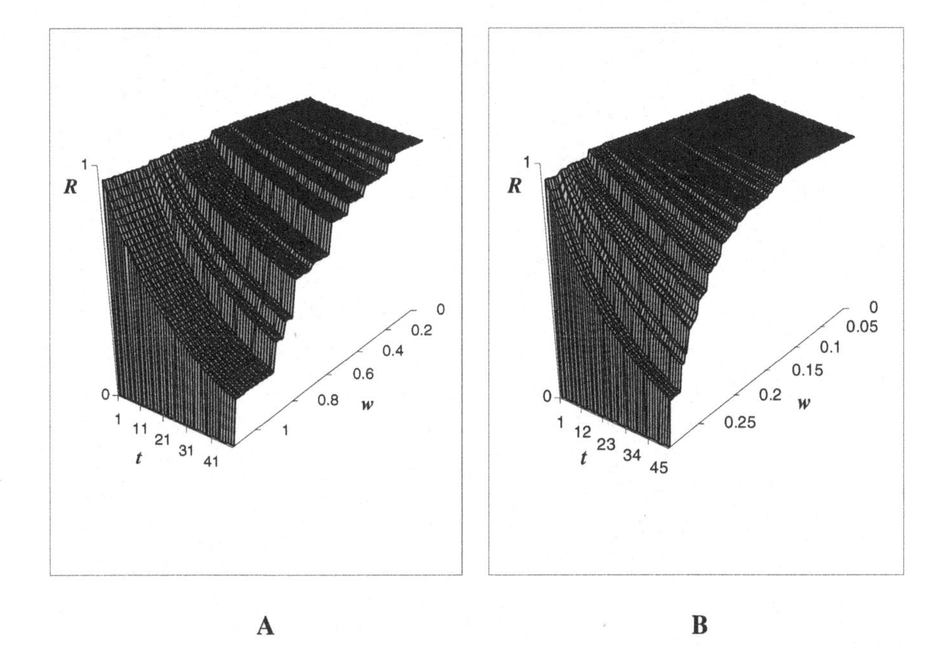

Figure 4.15. Reliability functions R (t, w) for MSSs with CCF (A: flow transmission system; B: task processing system)

To estimate the influence of CCF on MSS reliability we compare two systems of each type: an MSS without CCF in which elements belonging to the CCG j have their individual reliability functions equal to $\widetilde{P}_{j1}(t)$, and the same MSS with CCF. Since it is difficult visually to distinguish the differences between the three-dimensional representations of reliability functions for the MSSs with and without CCF, we present them for fixed values of t (Figure 4.16) as $R(w)$ and for fixed values of w (Figure 4.17) as $R(t)$. One can see the effect of CCF in decreasing MSS reliability. In addition, the expected MSS performances $\varepsilon(t)$ are presented for MSSs with and without CCF (Figure 4.17).

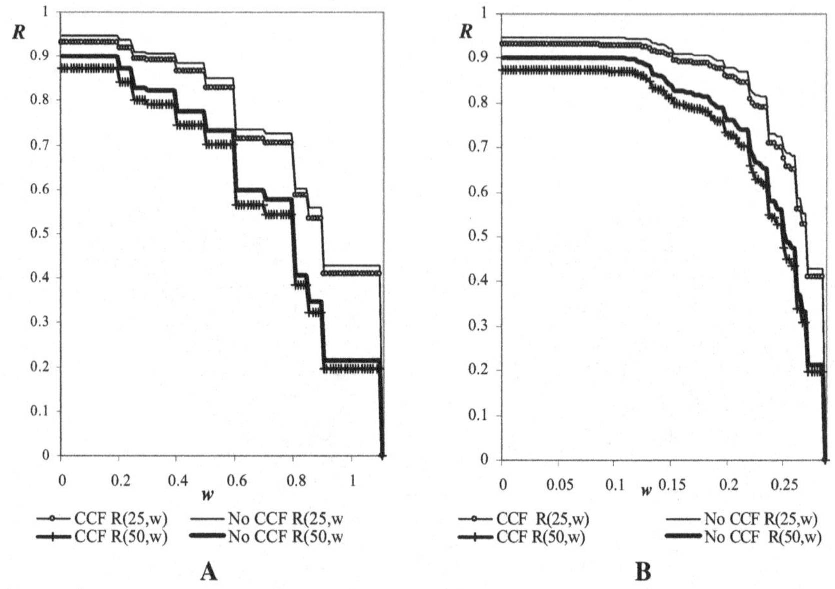

Figure 4.16. Reliability functions R (25, w) and R (50, w) for MSSs with and without CCF (A: flow transmission system; B: task processing system)

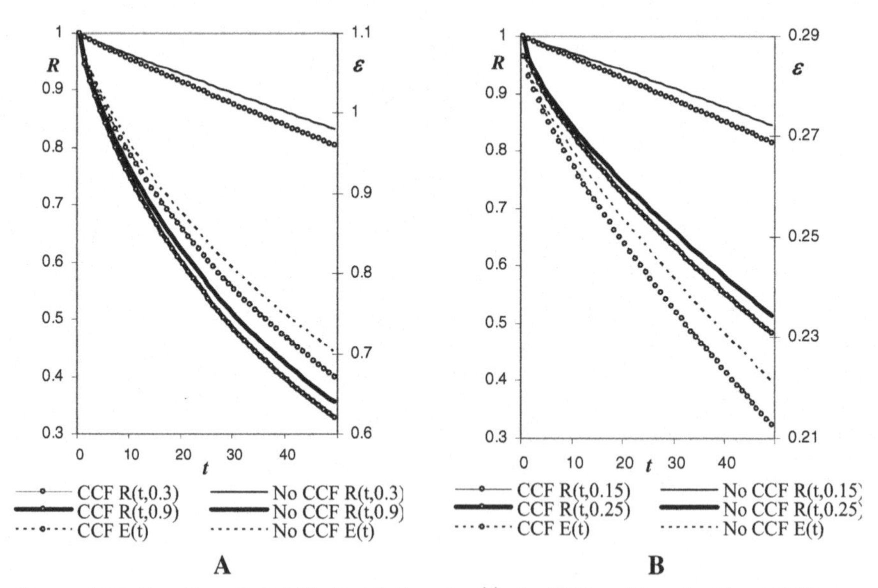

Figure 4.17. Functions R (t, 0.3), R (t, 0.9) and $\varepsilon(t)$ for MSSs with and without CCF (A: flow transmission system; B: task processing system)

4.4.2 Multi-state Systems with Total Common Cause Failures

In some cases CCFs lead to the total outage of all of the elements belonging to the corresponding CCG. Usually, such total failures occur when a group of elements share the same resource (energy source, space, protection, *etc.*) that has limited availability. Examples of such situations include an electrical supply failure that causes an outage of all production units supplied from the same source or the failure of a waterproof casing that causes water penetration into the hermetic compartment and destruction of all the equipment located there. The algorithm for incorporating the total CCF in reliability analysis of MSSs is simpler than the general algorithm considered in the previous section. This algorithm can be easily extended to MSS with multi-state elements [112].

Consider a subsystem consisting of several elements that compose a series-parallel structure. Assume that the elements are subject to a total CCF occurring with probability v. The total CCF leads to outage of all of the subsystem elements. The entire subsystem can have different performance rates, depending on the internal states of its elements. However, when the CCF occurs, the performance rate of the subsystem is f, which corresponds to its total failure.

The total or partial failures of subsystem elements and the entire subsystem failure due to common cause are independent events. Probabilities of all the states of the subsystem itself now should be treated as conditional probabilities, given the CCF does not occur. The only possible subsystem state, when the CCF occurs, is the state with the performance equal to f. If the u-function of a combination of elements composing the subsystem is $U_j(z)$, then the u-function of the subsystem which takes into account the CCF can be determined using the following operator ξ:

$$\xi(\mathrm{U}_\mathrm{j}(\mathrm{z})) = (1-v)U_j(z) + vz^f \tag{4.55}$$

One can model the subsystem with CCF as a series connection of the subsystem itself and an element representing the CCF, which has PD

$$\Pr(G = x_1) = 1-v, \ \Pr(G = x_2) = v \tag{4.56}$$

where x_1 corresponds to the state when CCF does not occur and x_2 corresponds to the state when CCF occurs. Such a model should reflect the fact that the subsystem performance rate will be changed to f with probability υ and will not be changed with probability $1-v$. In order to provide this property, one has to define the values of x_1 and x_2 such that for any G

$$\phi_{\text{ser}}(G, x_1) = G \quad \text{and} \quad \phi_{\text{ser}}(G, x_2) = f. \tag{4.57}$$

For any type of series-parallel systems described in Section 4.1, where f corresponds to the performance rate 0, x_1=∞ and x_2=0 meet the requirement (4.57).

Using the $\underset{\phi_{ser}}{\otimes}$ operator over $U_j(z)$ and the u-function representing the PD (4.56) one obtains

$$\xi(U_j(z)) = \underset{\phi_{ser}}{\otimes}(\mathrm{U_j(z)}, (1-v)\mathrm{z}^{\infty} + vz^0) = (1-v)U_j(z) + vz^0 \tag{4.58}$$

Replacing any CCG with the u-function $U_j(z)$ by an equivalent element with the u-function $\xi(U_j(z))$ one can use the reliability block diagram method for obtaining the reliability of series-parallel systems with total CCF.

Example 4.15

Consider the series-parallel flow transmission MSS with flow dispersion presented in Figure 4.18.

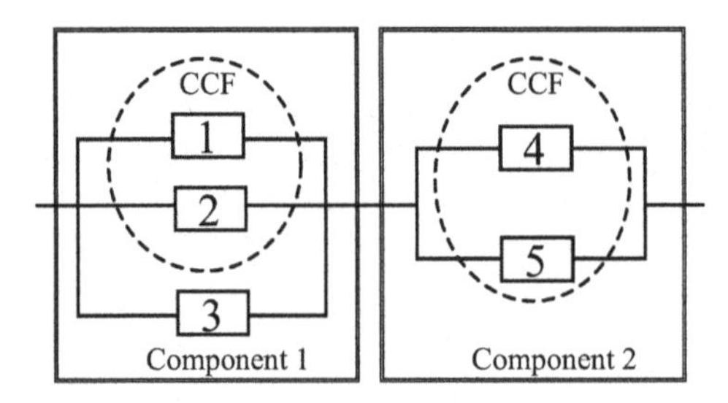

Figure 4.18. Series-parallel MSS with total CCF

The system consists of two components connected in a series. The first component contains three parallel elements. The first and second elements are subject to CCF, which has probability v_1 = 0.1. The second component contains two parallel elements that are subject to CCF with probability v_2 = 0.2. Each element j can have two states: total failure with performance rate zero and normal functioning with nominal performance g_{j1}. The availability p_{j1} and nominal performance of the elements are presented in Table 4.12. The system should meet the constant demand w = 2.

First, we determine the u-functions for the individual elements as follows:

$$u_1(z) = 0.9z^1+0.1z^0,\ u_2(z) = 0.8z^2+0.2z^0,\ u_3(z) = 0.8z^2+0.2z^0$$

$$u_4(z) = 0.9z^2+0.1z^0,\ u_5(z) = 0.8z^3+0.2z^0$$

Using the operator $\underset{+}{\otimes}$, we determine the u-functions for the subsystem consisting of two parallel elements, 1 and 2:

$$u_1(z)\underset{+}{\otimes} u_2(z) = (0.9z^1+0.1z^0)(0.8z^2+0.2z^0)=0.72z^3+0.08z^2+0.18z^1+0.02z^0$$

and for the subsystem consisting of two parallel elements 4 and 5:

$$u_4(z) \underset{+}{\otimes} u_5(z) = (0.9z^2+0.1z^0)(0.8z^3+0.2z^0) = 0.72z^5+0.08z^3+0.18z^2+0.02z^0$$

Table 4.12. Parameters of MSS elements

No of element j	Availability p_{j1}	Nominal performance rate g_{j1}
1	0.90	1.0
2	0.80	2.0
3	0.72	2.0
4	0.90	2.0
5	0.80	3.0

To incorporate the total CCF into *u*-functions of the subsystems, we use the operator ξ (4.58):

$$\xi(u_1(z) \underset{+}{\otimes} u_2(z)) = (1-v_1)(u_1(z) \underset{+}{\otimes} u_2(z)) + v_1 z^0 = 0.9(0.72z^3+0.08z^2$$

$$+ 0.18z^1+0.02z^0)+0.1z^0 = 0.648z^3+0.072z^2+0.162z^1+0.118z^0$$

$$\xi(u_4(z) \underset{+}{\otimes} u_5(z)) = (1-v_2)(u_4(z) \underset{+}{\otimes} u_5(z)) + v_2 z^0 = 0.8(0.72z^5+0.08z^3$$

$$+ 0.18z^2+0.02z^0)+0.2z^0 = 0.567z^5+0.064z^3+0.144z^2+0.216z^0$$

To obtain *u*-functions $U_1(z)$ for the entire first component, we consider it as a parallel connection of subsystem that has *u*-function $\xi(u_1(z) \underset{+}{\otimes} u_2(z))$ and the element 3 with *u*-function $u_3(z)$:

$$U_1(z) = \xi(u_1(z) \underset{+}{\otimes} u_2(z)) \underset{+}{\otimes} u_3(z)$$

$$= (0.648z^3+0.072z^2+0.162z^1+0.118z^0)(0.72z^2+0.28z^0)$$

$$= 0.4666z^5+0.0518z^4+0.298z^3+0.1051z^2+0.0454z^1+0.033z^0$$

The *u*-function of the second component, consisting of elements 4 and 5, is $U_2(z) = \xi(u_4(z) \underset{+}{\otimes} u_5(z))$. In order to obtain the *u*-function for the entire system consisting of two components connected in a series, we use the operator $\underset{\min}{\otimes}$ over *u*-functions $U_1(z)$ and $U_2(z)$:

$$U(z) = U_1(z) \underset{\min}{\otimes} U_2(z) = (0.4666z^5+0.0518z^4+0.298z^3+0.1051z^2$$

$$+0.0454z^1+0.033z^0) \underset{\min}{\otimes} (0.567z^5+0.064z^3+0.144z^2+0.216z^0)$$

$$= 0.269z^5+0.03z^4+0.224z^3+0.2z^2+0.035z^1+0.242z^0$$

This u-function represents the performance distribution of the entire MSS. Using the $\delta_2(U(z))$ operator we obtain the system availability as

$$A(2) = 0.269+0.03+0.224+0.2 = 0.723$$

4.4.3 Multi-state Systems with Nested Common Cause Groups

In the previous sections we assumed that the CCFs affecting different CCGs are independent. In many cases this model is not relevant because statistical dependence between the different CCFs exists. The typical examples of such a situation are systems with a multilevel protection. Such systems are used in many applications (nuclear, military, underwater, airspace systems, *etc.*) and are designed according to the so-called defence-in-depth methodology [113].

The multilevel protection means that a subsystem and its inner level protection are in turn protected by the protection of the outer level. This double-protected subsystem has its outer protection, and so forth. In such systems, the protected subsystems can be destroyed only if all of the levels of their protection are destroyed. Each level of protection can be destroyed only if all of the outer levels of protection are destroyed. This creates statistical dependence among the destruction events of the different protection levels (different CCFs). The systems with multilevel protection can be considered as systems with nested CCGs in which the CCF in any group can occur only if the CCFs in all CCGs containing this group have occurred.

In this section we consider series-parallel MSSs with nested CCGs and total CCFs and make the following assumptions:

- The elements belonging to any CCG compose a series-parallel structure (Figure 4.19A).
- Any CCG can belong to another CCG. For any pair of CCGs A and B $A\cap B\neq\emptyset$ means that $A\subseteq B$ or $B\subseteq A$, *i.e.* part of any CCG cannot belong to another CCG (Figure 4.19B).
- CCF in any group m cannot occur if this group belongs to another group and the CCF in the outer group has not occurred. If the CCFs in all of the outer CCGs that include the CCG m have occurred, the CCF in CCG m can occur with the probability v_m.
- Any element fails with probability 1 if CCFs in all of the CCGs that this element belongs to have occurred.
- The performance of any failed element is equal to f.
- The element failure caused by the CCFs and the transitions of this element in the space of states caused by its individual failures and repairs are independent events.

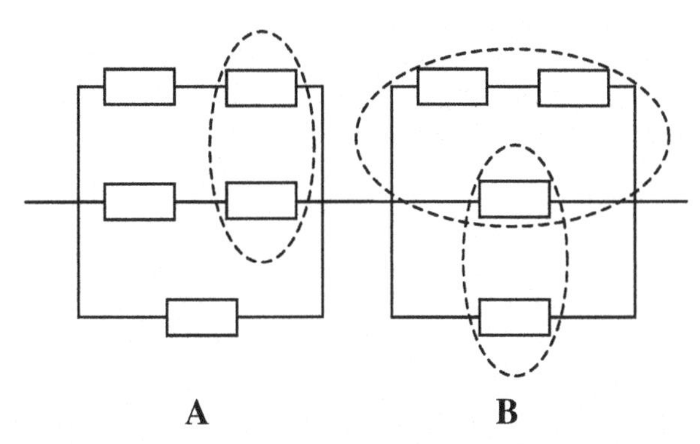

Figure 4.19. Impossible CCGs. (A: elements of CCG do not compose a series-parallel structure; B: two CCGs have common elements)

The probability of each state of an element (or subsystem) belonging to some CCG depends on the CC event. Therefore, each subsystem belonging to a CCG is characterized by two conditional performance distributions: the first corresponds to the case when the CCF in this group occurs and the second corresponds to the case when the CCF in the group does not occur. In order to represent the performance distributions of a subsystem m belonging to some CCG, we introduce the following double u-function (d-function) $d_m(z)=<U_m(z),\tilde{U}_m(z)>$, where $U_m(z)$ and $\tilde{U}_m(z)$ represent performance distributions for the first and second cases respectively.

If CCF in a group consisting of a single basic element occurs, then this element fails with probability 1 and has the performance rate f. Therefore, for a basic single element j that has a performance distribution represented by the u-function $u_j(z)$

$$d_j(z)=<z^f, u_j(z)> \tag{4.59}$$

It can easily be seen that any pair of elements with d-functions $d_j(z)$ and $d_i(z)$ belonging to the same CCG can be replaced by the equivalent element (Figure 4.20) with the d-function

$$\begin{aligned} d_j(z)\underset{\phi}{\otimes}d_i(z) &=< U_j(z),\tilde{U}_j(z)>\underset{\phi}{\otimes}< U_i(z),\tilde{U}_i(z)> \\ &=< U_j(z)\underset{\phi}{\otimes}U_i(z),\tilde{U}_j(z)\underset{\phi}{\otimes}\tilde{U}_i(z)> \end{aligned} \tag{4.60}$$

where ϕ should be substituted by ϕ_{ser} or ϕ_{par} in accordance with the type of connection between the elements.

Assume that the d-function of a series-parallel subsystem that constitutes CCG m obtained without respect to CCF in this group is $d_m(z)=<U_m(z),\tilde{U}_m(z)>$. Assume also that the group m belongs to an outer CCG h. If the CCF in group h occurs, then the CCF in group m can occur with probability v_m. If this CCF occurs, then the subsystem has its performance distribution represented by the u-function $U_m(z)$; if the CCF does not occur (with probability $1-v_m$), then the subsystem has its performance distribution represented by the *u*-function $\tilde{U}_m(z)$. Therefore, the conditional performance distribution of the group *m* given CCF in group *h* has occurred can be represented by the *u*-function

$$v_m U_m(z) + (1-v_m)\tilde{U}_m(z) \tag{4.61}$$

In the case when the CCF in CCG *h* have not occurred, CCF in the group *m* also cannot occur and its conditional performance distribution is represented by the *u*-function $\tilde{U}_m(z)$.

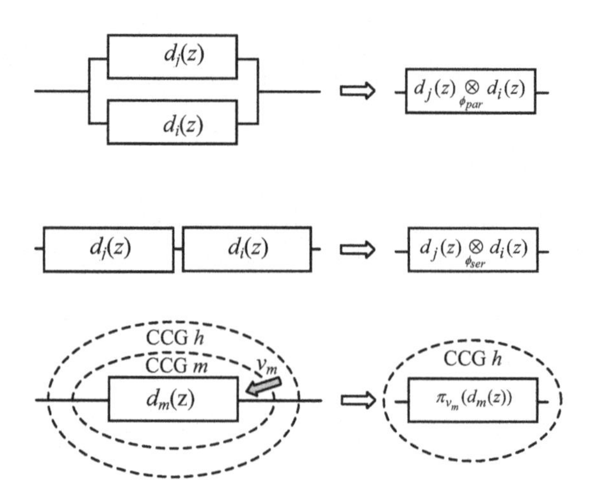

Figure 4.20. Basic equivalent transformations of system elements

These considerations allow one to incorporate the CCF that occurs in the CCG *m* with probability v_m into the *d*-function of this group by replacing the group with an equivalent element (Figure 4.20) with the *d*-function obtained by applying the following operator π_{v_m} over $d_m(z)$:

$$\begin{aligned}
&\pi_{v_m}(d_m(z)) = \pi_{v_m} < U_m(z), \tilde{U}_m(z) > \\
&=< v_m U_m(z) + (1-v_m)\tilde{U}_m(z), \tilde{U}_m(z) >
\end{aligned} \tag{4.62}$$

It can be seen that when $v_m = 1$ the operator π_{v_m} does not change the *d*-function. Indeed, the totally vulnerable protection (which is equivalent to absence of any protection) cannot affect the performance distribution of the subsystem it protects.

Consecutively applying the operators (4.60) and (4.62) and replacing the subsystems and the CCGs with equivalent elements, one can obtain the *d*-function representing the performance distribution of the entire system. The algorithm for obtaining the *d*-function is based on the assumption that any system element belongs to at least one CCG. In order to make this algorithm universal we can always assume that the entire system belongs to an outer CCG (is protected by an outer protection). If such protection does not exist, then the outer protection with vulnerability $v = 1$ can be added without changing the system performance distribution. The following recursive algorithm obtains the system *d*-function:

1. Obtain the *d*-functions of all of the system elements using Equation (4.59).
2. If the system contains a pair of elements connected in parallel or in a series and belonging to the same CCG, replace this pair with an equivalent element with the *d*-function obtained by the $\underset{\phi_{\text{par}}}{\otimes}$ or $\underset{\phi_{\text{ser}}}{\otimes}$ operator.
3. If the system contains a CCG consisting of a single element, replace this CCG with a single equivalent element with the *d*-function obtained using the π_{v_m} operator.
4. If the system contains more than one element or a CCG not replaced by a single element, return to step 2.
5. Determine the *d*-function of the entire series-parallel system as the *d*-function of the remaining single equivalent element $d(z)=<U(z),\tilde{U}(z)>$.

According to the definition of the *d*-function, the *u*-function $\tilde{U}(z)$ corresponds to the case when the CCFs in the system do not occur while the *u*-function $U(z)$ represents the entire system performance distribution in which all probabilities of the CCFs that can occur in the system are incorporated. The system reliability (or any other performance measure) can now be obtained by applying the corresponding operators over the *u*-function $U(z)$.

Example 4.16

Consider the system with multiple protection presented in Figure 4.21A. In this system, each CCG corresponds to a subsystem that has its own protection. Each CCG can contain other CCGs (protected subsystems). The CCF in any CCG corresponds to the destruction of the corresponding protection. If the protection of the CCG is destroyed, all unprotected elements in this CCG fail (the performance of a failed element is zero). The protection cannot be destroyed if an outer protection is not destroyed.

Assume that the performance distribution of each individual element j is represented by the *u*-function $u_j(z)$. The destruction probability v_m of each protection m is assumed to be known. The *d*-functions of the individual elements are

$$d_1(z) = <z^0, u_1(z)>, d_2(z) = <z^0, u_2(z)>, d_3(z) = <z^0, u_3(z)>$$

$$d_4(z) = <z^0, u_4(z)>, d_5(z) = <z^0, u_5(z)>$$

According to the recursive algorithm, in order to obtain the system's availability one has to perform the following steps:

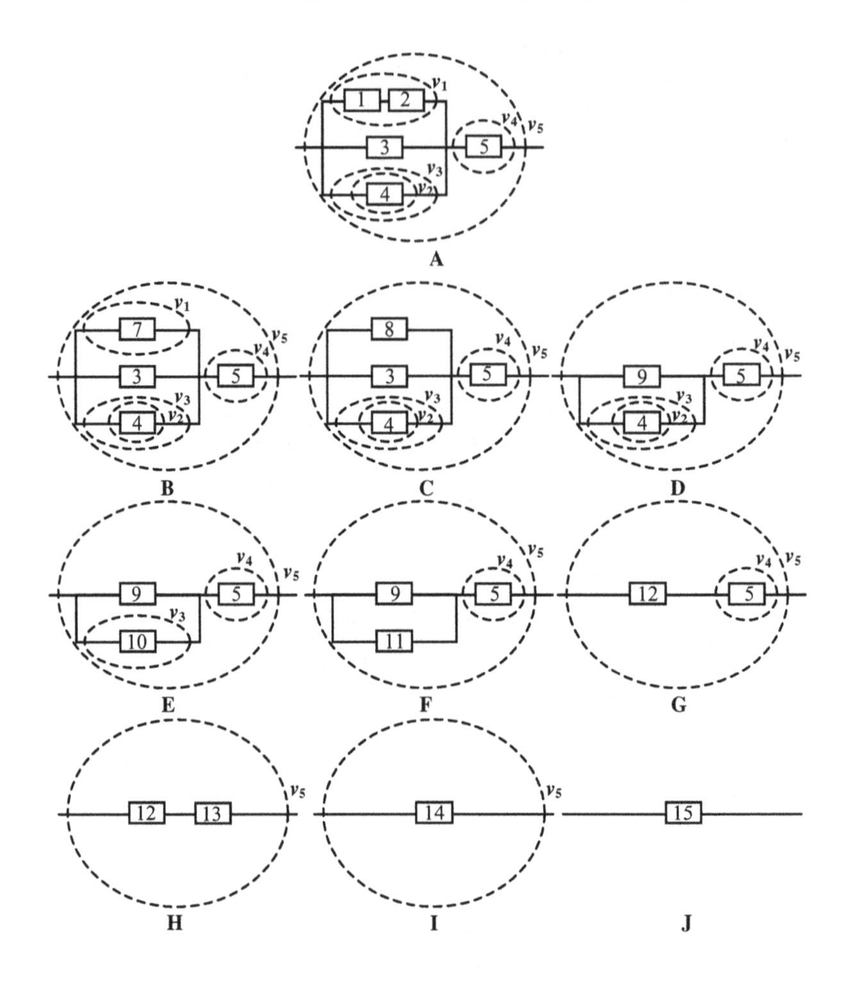

Figure 4.21. Example of recursive algorithm

Replace elements 1 and 2 connected in series by a single equivalent element 7 with the d-function

$$d_7(z) = d_1(z) \underset{\phi_{ser}}{\otimes} d_2(z) = \langle z^0, u_1(z)\rangle \underset{\phi_{ser}}{\otimes} \langle z^0, u_2(z)\rangle = \langle z^0, u_1(z) \underset{\phi_{ser}}{\otimes} u_2(z)\rangle$$

(see Figure 4.21B).

Replace element 7 with its protection by an equivalent element 8 with the d-function

$$d_8(z) = \pi_{v_1}(d_7(z)) = \pi_{v_1} \langle z^0, u_1(z) \underset{\phi_{ser}}{\otimes} u_2(z)\rangle$$

$$= \langle v_1 z^0 + (1-v_1)u_1(z) \underset{\phi_{ser}}{\otimes} u_2(z), u_1(z) \underset{\phi_{ser}}{\otimes} u_2(z)\rangle$$

(see Figure 4.21C).

Replace elements 8 and 3 connected in parallel by a single equivalent element 9 with the d-function (taking into account that $u(z) \underset{\phi_{par}}{\otimes} z^0 = u(z)$ for any $u(z)$)

$$d_9(z) = d_8(z) \underset{\phi_{par}}{\otimes} d_3(z)$$

$$= <v_1 z^0 + (1-v_1)u_1(z) \underset{\phi_{ser}}{\otimes} u_2(z), u_1(z) \underset{\phi_{ser}}{\otimes} u_2(z)> \underset{\phi_{par}}{\otimes} <z^0, u_3(z)>$$

$$= <v_1 z^0 + (1-v_1)u_1(z) \underset{\phi_{ser}}{\otimes} u_2(z), \quad (u_1(z) \underset{\phi_{ser}}{\otimes} u_2(z)) \underset{\phi_{par}}{\otimes} u_3(z)>$$

(see Figure 4.21D).

Replace element 4 with its inner protection by an equivalent element 10 with the d-function

$$d_{10}(z) = \pi_{v_2}(d_4(z)) = \pi_{v_2} <z^0, u_4(z)> = <v_2 z^0 + (1-v_2)u_4(z), u_4(z)>$$

see (Figure 4.21E).

Replace element 10 with its protection by an equivalent element 11 with the d-function

$$d_{11}(z) = \pi_{v_3}(d_{10}(z)) = \pi_{v_3} <v_2 z^0 + (1-v_2)u_4(z), u_4(z)>$$

$$= <v_3 v_2 z^0 + v_3(1-v_2)u_4(z) + (1-v_3)u_4(z), \quad u_4(z)>$$

$$= <v_2 v_3 z^0 + (1-v_3 v_2)u_4(z), u_4(z)>$$

(see Figure 4.21F).

Replace elements 9 and 11 connected in parallel by a single equivalent element 12 with the d-function

$$d_{12}(z) = d_9(z) \underset{\phi_{par}}{\otimes} d_{11}(z) = <v_1 z^0 + (1-v_1)u_1(z)$$

$$\underset{\phi_{ser}}{\otimes} u_2(z), (u_1(z) \underset{\phi_{ser}}{\otimes} u_2(z)) \underset{\phi_{par}}{\otimes} u_3(z)> \underset{\phi_{par}}{\otimes} <v_2 v_3 z^0 + (1-v_2 v_3)u_4(z), u_4(z)>$$

$$= <v_1 v_2 v_3 z^0 + (1-v_1)v_2 v_3 u_1(z) \underset{\phi_{ser}}{\otimes} u_2(z) + v_1(1-v_2 v_3)u_4(z)$$

$$+(1-v_1)(1-v_2 v_3)(u_1(z) \underset{\phi_{ser}}{\otimes} u_2(z)) \underset{\phi_{par}}{\otimes} u_4(z), (u_1(z) \underset{\phi_{ser}}{\otimes} u_2(z)) \underset{\phi_{par}}{\otimes} u_3(z) \underset{\phi_{par}}{\otimes} u_4(z)>$$

(see Figure 4.21G).

Replace element 5 with its protection by an equivalent element 13 with the d-function

$$d_{13}(z) = \pi_{v_4}(d_5(z)) = \pi_{v_4} <z^0, u_5(z)> = <v_4 z^0 + (1-v_4)u_5(z), u_5(z)>$$

(see Figure 4.21H).

Replace elements 12 and 13 connected in series by a single equivalent element 14 with the d-function

$$d_{14}(z) = d_{12}(z) \underset{\phi_{ser}}{\otimes} d_{13}(z) = <v_1v_2v_3z^0 + (1-v_1)v_2v_3u_1(z) \underset{\phi_{ser}}{\otimes} u_2(z)$$

$$+ v_1(1-v_2v_3)u_4(z) + (1-v_1)(1-v_2v_3)(u_1(z) \underset{\phi_{ser}}{\otimes} u_2(z)) \underset{\phi_{par}}{\otimes} u_4(z), (u_1(z) \underset{\phi_{ser}}{\otimes} u_2(z))$$

$$\underset{\phi_{par}}{\otimes} u_3(z) \underset{\phi_{par}}{\otimes} u_4(z)> \underset{\phi_{ser}}{\otimes} <v_4z^0 + (1-v_4)u_5(z), u_5(z)>$$

$$= <v_4z^0+v_1v_2v_3(1-v_4)z^0 + (1-v_1)v_2v_3(1-v_4)u_1(z) \underset{\phi_{ser}}{\otimes} u_2(z) \underset{\phi_{ser}}{\otimes} u_5(z)$$

$$+ v_1(1-v_2v_3)(1-v_4)u_4(z) \underset{\phi_{ser}}{\otimes} u_5(z)$$

$$+ (1-v_1)(1-v_2v_3)(1-v_4)((u_1(z) \underset{\phi_{ser}}{\otimes} u_2(z)) \underset{\phi_{par}}{\otimes} u_4(z))$$

$$\underset{\phi_{ser}}{\otimes} u_5(z), ((u_1(z) \underset{\phi_{ser}}{\otimes} u_2(z)) \underset{\phi_{par}}{\otimes} u_3(z) \underset{\phi_{par}}{\otimes} u_4(z)) \underset{\phi_{ser}}{\otimes} u_5(z)>$$

(see Figure 4.21I).

Finally, replace element 14 with its protection by an equivalent element 15 with the d-function

$$d_{15}(z) = \pi_{v_5}(d_{14}(z)) = \pi_{v_5} <v_4z^0 + v_1v_2v_3(1\text{-}v_4)z^0$$

$$+ (1-v_1)v_2v_3(1-v_4)u_1(z) \underset{\phi_{ser}}{\otimes} u_2(z) \underset{\phi_{ser}}{\otimes} u_5(z) + v_1(1-v_2v_3)(1-v_4)u_4(z) \underset{\phi_{ser}}{\otimes} u_5(z)$$

$$+ (1-v_1)(1-v_2v_3)(1-v_4)((u_1(z) \underset{\phi_{ser}}{\otimes} u_2(z)) \underset{\phi_{par}}{\otimes} u_4(z)) \underset{\phi_{ser}}{\otimes} u_5(z), ((u_1(z) \underset{\phi_{ser}}{\otimes} u_2(z))$$

$$\underset{\phi_{par}}{\otimes} u_3(z) \underset{\phi_{par}}{\otimes} u_4(z)) \underset{\phi_{ser}}{\otimes} u_5(z))>$$

$$= < v_4v_5z^0 + v_1v_2v_3(1-v_4)v_5z^0 + (1-v_1)v_2v_3(1-v_4)v_5u_1(z) \underset{\phi_{ser}}{\otimes} u_2(z) \underset{\phi_{ser}}{\otimes} u_5(z)$$

$$+ v_1(1-v_2v_3)(1-v_4)v_5u_4(z) \underset{\phi_{ser}}{\otimes} u_5(z)$$

$$+ (1-v_1)(1-v_2v_3)(1-v_4)v_5((u_1(z) \underset{\phi_{ser}}{\otimes} u_2(z)) \underset{\phi_{par}}{\otimes} u_4(z)) \underset{\phi_{par}}{\otimes} u_5(z)$$

$$+ (1-v_5)((u_1(z) \underset{\phi_{ser}}{\otimes} u_2(z)) \underset{\phi_{par}}{\otimes} u_3(z) \underset{\phi_{par}}{\otimes} u_4(z)) \underset{\phi_{par}}{\otimes} u_5(z), ((u_1(z) \underset{\phi_{ser}}{\otimes} u_2(z))$$

$$\underset{\phi_{par}}{\otimes} u_3(z) \underset{\phi_{par}}{\otimes} u_4(z)) \underset{\phi_{ser}}{\otimes} u_5(z) >$$

(see Figure 4.21J).

The entire system performance distribution is represented by the first u-function of $d_{15}(z)$

$$U(z) = v_4v_5z^0 + v_1v_2v_3(1-v_4)v_5z^0 + (1-v_1)v_2v_3(1-v_4)v_5u_1(z) \underset{\phi_{\text{ser}}}{\otimes} u_2(z) \underset{\phi_{\text{ser}}}{\otimes} u_5(z)$$

$$+ v_1(1-v_2v_3)(1-v_4)v_5u_4(z) \underset{\phi_{\text{ser}}}{\otimes} u_5(z)$$

$$+ (1-v_1)(1-v_2v_3)(1-v_4)v_5((u_1(z) \underset{\phi_{\text{ser}}}{\otimes} u_2(z)) \underset{\phi_{\text{par}}}{\otimes} u_4(z)) \underset{\phi_{\text{par}}}{\otimes} u_5(z)$$

$$+ (1-v_5)((u_1(z) \underset{\phi_{\text{ser}}}{\otimes} u_2(z)) \underset{\phi_{\text{par}}}{\otimes} u_3(z) \underset{\phi_{\text{par}}}{\otimes} u_4(z)) \otimes u_5(z)$$

Example 4.17

Consider a series-parallel MSS (power substation) that consists of three basic subsystems (Figure 4.22A):

1. blocks of commutation equipment (elements 1-5);
2. power transformers (elements 6-8);
3. output medium voltage line sections (elements 9-12).

All of the elements of this flow transmission system (with flow dispersion) are two-state units with nominal performance rates (the power that the elements can transform/transmit) g_{j1} and the availabilities p_{j1} presented in Table 4.13. The failed elements have performance zero.

Table 4.13. Parameters of elements of power substation

j	1	2	3	4	5	6	7	8	9	10	11	12
g_{j1}	2	6	6	3	5	5	4	5	4	3	4	5
p_{j1}	0.92	0.90	0.95	0.88	0.95	0.97	0.97	0.97	0.93	0.96	0.90	0.94

The d-function of two-state element j takes the form

$$d_j(z) = \langle z^0, p_{j1} z^{g_{j1}} + (1-p_{j1})z^0 \rangle$$

In order to increase the system survivability (the probability that the system meets demand w) in the case of an external attack, the system can be divided into four spatially separated groups represented by the following sets of elements: {1,2,3}, {4}, {6,7,9,10,11} and {5,8,12}. The probability of impact in the case of attack is $v_1 = 0.3$. Since the groups are separated, no more than one group can be affected by a single impact. Four subsystems belonging to the separated groups can be protected (located indoors within concrete constructions). These subsystems include elements 2 and 3, element 6, elements 9 and 10, elements 5, 8 and 12. The probability of protection destruction in the case of impact is $v_2 = 0.6$, while the probability of destruction of the unprotected elements in the case of impact is 1 (unprotected elements do not survive the impact).

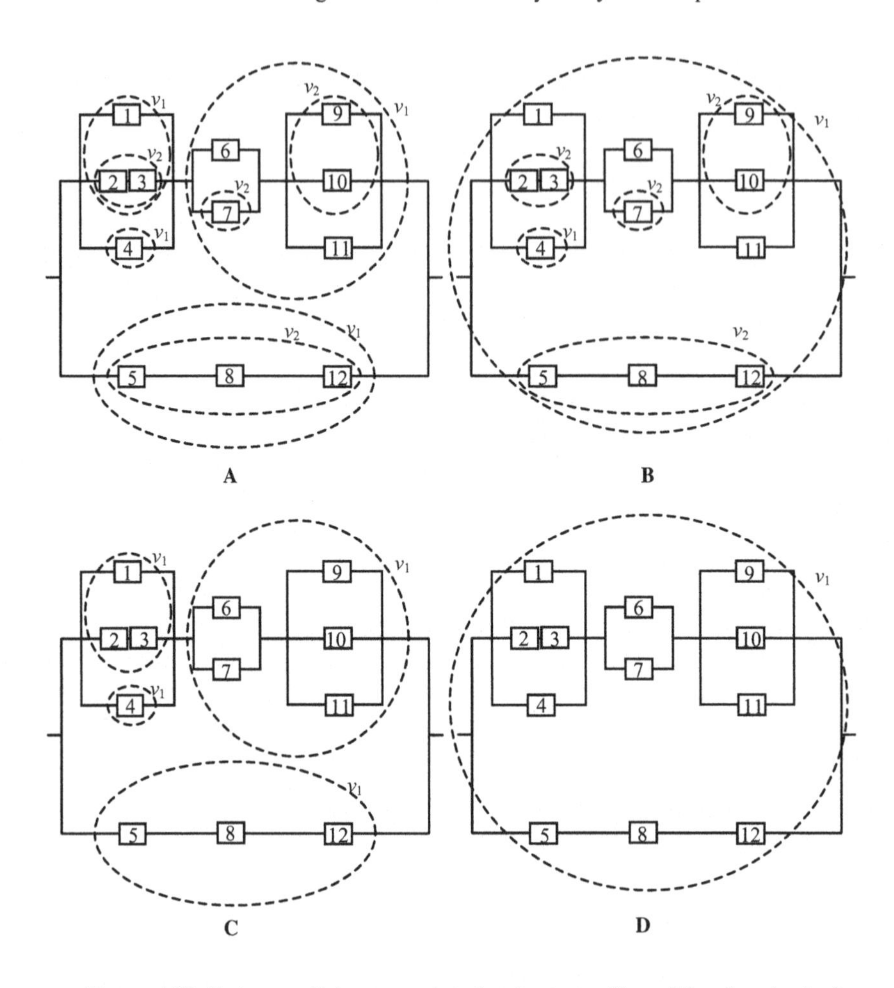

Figure 4.22. Series-parallel power substation (system with multilevel protection)

In order to evaluate the influence of each type of protection, four different configurations are compared:

A. Both separation and indoor allocation are applied (Figure 4.22A).

B. All of the elements are gathered in the same place (no separation). Indoor allocation is applied (Figure 4.22B).

C. The groups of elements are separated, but all of the elements are located outdoors (Figure 4.22C).

D. All of the elements are gathered in the same place and located outdoors (Figure 4.22D).

In Figure 4.23, one can see the system survivability (obtained using the method presented in this section) as a function of the demand for cases A, B, C, and D.

Observe that the protection of parts of the system is not effective when the system tolerates only a very small decrease of its performance below its maximal possible performance. In our case the indoor allocation of some system elements can increase the system survivability only when $w \le 9$ (compare curves B and D).

Indeed, in the case of impact, even if all of the elements located indoors survive, they cannot provide the system's performance greater than 9.

The separation is also effective only when the demand is considerably smaller than the maximal possible system performance. Moreover, the separation can decrease the system's survivability when the demand is close to its maximal performance. Indeed, by separating the system elements one creates additional vulnerable CCGs, which contribute to an additional overall system exposure to the impact. When the demand is relatively small, the separation increases the system's survivability because the smaller parts can be destroyed by a single impact. In our case, the separation is effective for $w \le 5$. When $w > 5$ the separation decreases the system's survivability (compare curves C and D).

The total survivability improvement achieved by separation and protection of its elements for $w \le 5$ is greater than 23%.

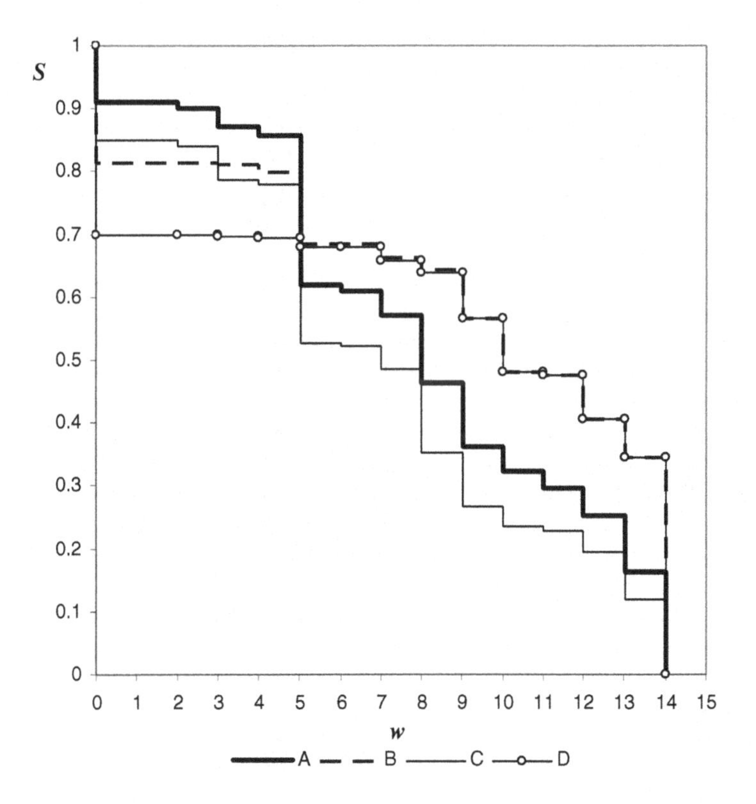

Figure 4.23. Survivability of power substation as a function of demand

4.5 Importance Analysis in Multi-state Systems

Information about the importance of the elements that constitute a system with respect to its safety, reliability, availability, and performance, is of great practical aid to system designers and managers. Indeed, the identification of which elements most influence the overall system performance allows one to trace technical bottlenecks and provides guidelines for effective actions of system improvement. In this sense, importance measures (IMs) are used to quantify the contribution of individual elements to the system's performance measures (*e.g.* reliability, availability, mean performance, expected performance deficiency).

IMs were first introduced by Birnbaum [114]. The Birnbaum importance measure gives the contributions to the system's reliability due to the reliability of the various system elements. Elements for which a variation in reliability results in the largest variation of the entire system's reliability have the highest importance. Fussell and Vesely later proposed a measure based on the cut-sets importance [115]. According to the Fussell-Vesely measure, the importance of an element depends on the number and the order of the cut-sets in which it appears. Other concepts of importance measures have been proposed and used based on different views of the elements' influence on the system's performance. Structural IMs account for the topographic importance of the logic position of the element in the system [116, 117]. Criticality IMs consider the conditional probability of the failure of an element, given that the system has failed [118, 119]. Joint IMs account for the introduction of the elements' interactions in their contribution to the system's reliability [120, 121].

IMs are being widely used in risk-informed applications of the nuclear industry to characterize the importance of basic events, *i.e.* element failures, human errors, common cause failures, etc., with respect to the risk associated to the system [122-125]. In this framework, the risk importance measures are based on two other IMs: the performance reduction worth and the performance achievement worth [122]. The former is a measure of the 'worth' of the basic failure event in achieving the present level of system performance and, when applied to elements, it highlights the importance of maintaining the current level of element reliability (with respect to the basic failure event). The latter, the performance achievement worth, is associated to the variation of the system's performance consequent to an improvement of the element reliability.

In a general context, the IMs reflect the changes in distribution of the performance of the entire system caused by constraints imposed n the performance of one of its elements. Once the system PD is determined, one can focus on specific system performance measures, *e.g.* system availability, for the definition of the relevant measures of element importance.

4.5.1 Reliability Importance of Two-state Elements in Multi-state Systems

Consider a system consisting of two-state elements. Each element j has performance g_{j1} in the state of perfect functioning and performance g_{i0} in the state of total failure, which corresponds to its u-function $u_j(z) = p_{j1}z^{g_{j1}} + (1-p_{j1})z^{g_{j0}}$.

Let O be a system output performance measure ($O \equiv A$ for availability or reliability; $O \equiv \varepsilon$ for mean system performance, $O \equiv \Delta^-$ for expected performance deficiency). The system performance measure (PM) O can be expressed for the given demand distribution as a function of parameters of system elements

$$O(p_{11}, g_{11}, g_{10}, \ldots, p_{j11}, g_{j1}, g_{j0}, \ldots, p_{n1}, g_{n1}, g_{n0}) \tag{4.63}$$

In order to obtain this index, one has to determine the u-functions of individual elements $u_j(z)$ for $1 \le i \le n$, to obtain the u-function of the PMs of interest (see Section 3.3) using the corresponding operators and to calculate the derivatives of these u-functions at $z = 1$.

Let O_{j0} and O_{j1} be the system PM when element j is fixed in its faulty and functioning state respectively, while the remainder of the elements are free to randomly change their states. The PMs O_{j0} and O_{j1} according to their definition are

$$O_{j0} = O(p_{11}, g_{11}, g_{10}, \ldots, 0, g_{j1}, g_{j0}, \ldots, p_{n1}, g_{n1}, g_{n0}) \tag{4.64}$$

$$O_{j1} = O(p_{11}, g_{11}, g_{10}, \ldots, 1, g_{j1}, g_{j0}, \ldots, p_{n1}, g_{n1}, g_{n0}) \tag{4.65}$$

O_{j0} corresponds to the system PM when the element j is in the state of total failure with probability $p_{j0} = 1$ (which can be represented by the u-function $u_j^-(z) = z^{g_{j0}}$). O_{j1} corresponds to the system PM when the element j is in the state of perfect functioning with probability $p_{j1} = 1$ (which can be represented by the u-function $u_j^+(z) = z^{g_{j1}}$). Therefore, O_{j0} and O_{j1} can be obtained by substituting $u_j(z)$ by $u_j^-(z)$ and $u_j^+(z)$ respectively before using the procedure of system PM determination.

The system output performance measure O can be expressed as

$$O = O_{j0}p_{j0} + O_{j1}p_{j1} = O_{j0}(1-p_{j1}) + O_{j1}p_j \tag{4.66}$$

Definitions of four of the most frequently used IMs with reference to PM O and element j are as follows

The *performance reduction worth* is the ratio of the actual system PM to the valueofthePMwhenelementjisconsideredas always failed:

$$I_{Oj} = O/O_{j0} \tag{4.67}$$

This index measures the potential damage to the system's performance caused by the total unavailability of element j.

The *performance achievement worth* is the ratio of the system PM obtained when element j is always in the operable state to the actual value of the system's PM (when all of the elements including element j are left free to change their states randomly in accordance with their PD):

$$I_O a_j = O_{j1}/O \tag{4.68}$$

This index measures the contribution of element j to enhancing the system's performance by considering the maximum improvement on the system's PM achievable by making the element fully available.

The *Fussell-Vesely measure* represents the relative PM reduction due to the total failure of element j:

$$I_O f_j = (O - O_{j0})/O = 1 - 1/I_O r_j \tag{4.69}$$

Similarly, one can define the relative PM achievement when element j is always in the operable state:

$$I_O v_j = (O_{j1} - O)/O = I_O a_j - 1 \tag{4.70}$$

The *Birnbaum importance measure* represents the variation of the system PM when element j switches from the condition of perfect functioning to the condition of total failure. It is a differential measure of the importance of element j, since it is equal to the rate at which the system PM changes with respect to changes in the reliability of element j:

$$I_O b_j = \partial O / \partial p_{j1} = \partial(p_{j1} O_{j1} + (1 - p_{j1}) O_{j0}) / \partial p_{j1} = O_{j1} - O_{j0} \tag{4.71}$$

Note that for the Fussel-Vesely and Birnbaum IMs, depending on the system's PM, an improvement in the system's performance can correspond either to an increase of the considered PM (*e.g.* the availability or mean performance) or to a decrease (*e.g.* the expected performance deficiency). In the latter case, the absolute values of Iv_j, If_j and Ib_j are taken as the importance values.

The IMs for each MSS element depend strongly on that element's place in the system, its nominal performance level, and the system's demand. The notion of element relevancy is closely connected to the element's importance. The element is relevant if some changes in its state that take place without changes in the states of the reminder of the elements cause changes in the PM of the entire system. According to this definition, if the element j is irrelevant then $O_{j0} = O_{j1} = O$. Therefore, for the irrelevant element

$$I_O r_j = I_O a_j = 1 \tag{4.72}$$

while

$$I_Q f_j = I_O v_j = I_O b_j = 0 \tag{4.73}$$

Example 4.18

Consider a system consisting of n elements with total failures connected in a series described in Example 4.1. For any element j $g_{j0} = 0$. The reliability measures of this system are presented in Table 4.1. The corresponding analytically obtained IMs are presented in Tables 4.14. - 4.18. In these tables $\pi = \sum_{i=1}^{n} p_{i1}$.

The element with the minimal availability has the greatest impact on MSS availability ("a chain fails at its weakest link"). The importance indices associated with the system's availability do not depend on the elements' performance rates or on demand. IMs associated with the system's mean performance and performance deficiency also do not depend on the performance rate of the individual element j; however, the performance rate g_{j1} can influence these indices if it affects the entire system performance $\hat{g}$.

Table 4.14. Performance reduction worth IMs for series MSS

w	$I_A r_j$	$I_{\Delta^-} r_j$	$I_\varepsilon r_j$
$w > \hat{g}$	not defined	$1-\hat{g}\pi / w$	not defined
$0 \le w \le \hat{g}$	not defined	$1-\pi$	

Table 4.15. Performance achievement worth IMs for series MSS

w	$I_A a_j$	$I_{\Delta^-} a_j$	$I_\varepsilon a_j$
$w > \hat{g}$	not defined	$\frac{wp_{j1} - \hat{g}\pi}{p_{j1}(w-\hat{g}\pi)}$	$1/p_{j1}$
$0 \le w \le \hat{g}$	$1/p_{j1}$	$\frac{p_{j1} - \pi}{p_{j1}(1-\pi)}$	

Table 4.16. Relative performance reduction IMs for series MSS

w	$I_A f_j$	$I_{\Delta^-} f_j$	$I_\varepsilon f_j$
$w > \hat{g}$	not defined	$\frac{\hat{g}\pi}{w-\hat{g}\pi}$	1
$0 \le w \le \hat{g}$	1	$\frac{\pi}{1-\pi}$	

Table 4.17. Relative performance achievement IMs for series MSS

w	$I_A v_j$	$I_{\Delta} v_j$	$I_{\varepsilon} v_j$
$w > \hat{g}$	not defined	$\frac{(1-p_{j1})\hat{g}\pi}{p_{j1}(w-\hat{g}\pi)}$	$1/p_{j1}-1$
$0 \le w \le \hat{g}$	$1/p_{j1}-1$	$\frac{(1-p_{j1})\pi}{p_{j1}(1-\pi)}$	

Table 4.18. Birnbaum importance IMs for series MSS

w	$I_A b_j$	$I_{\Delta} b_j$	$I_{\varepsilon} b_j$
$w > \hat{g}$	0	$\hat{g}\pi / p_{j1}$	$\hat{g}\pi / p_{j1}$
$0 \le w \le \hat{g}$	π / p_{j1}	$w\pi / p_{j1}$	

Example 4.19

Consider a task processing system without work sharing presented in Example 4.2. The system consists of two elements with total failures ($g_{10} = g_{20} = 0$) connected in parallel. The analytically obtained system reliability measures are presented in Table 4.3. The importance measures can also be obtained analytically. The measures $I_O r_j$, $I_O a_j$ and $I_O b_j$ are presented in Tables 4.19-4.21.

Table 4.19. Performance reduction worth IMs for parallel MSS

w	$I_A r_1$	$I_{\Delta} r_1$	$I_{\varepsilon} r_1$
$w>g_{21}$	not defined	$\frac{w-p_{11}g_{11}-p_{21}g_{21}+p_{11}p_{21}g_{11}}{w-p_{21}g_{21}}$	$1+\frac{p_{11}(1-p_{21})g_{11}}{p_{21}g_{21}}$
$g_{11}<w\le g_{21}$	1	$\frac{w-p_{11}g_{11}}{w}$	
$0<w\le g_{11}$	$1-p_{11}+p_{11}/p_{21}$	$1-p_{11}$	
	$I_A r_2$	$I_{\Delta} r_2$	$I_{\varepsilon} r_2$
$w>g_{21}$	not defined	$\frac{w-p_{11}g_{11}-p_{21}g_{21}+p_{11}p_{21}g_{11}}{w-p_{11}g_{11}}$	$1-p_{21}+\frac{p_{21}g_{11}}{p_{11}g_{11}}$
$g_{11}<w\le g_{21}$	not defined	$1-p_{21}$	
$0<w\le g_{11}$	$1-p_{21}+p_{21}/p_{11}$	$1-p_{21}$	

Table 4.20. Performance achievement worth IMs for parallel MSS

w	$I_A a_1$	$I_{\Delta^-} a_1$	$I_\varepsilon a_1$
$w>g_{21}$	not defined	$\frac{w-g_{11}-p_{21}(g_{21}-g_{11})}{w-p_{11}g_{11}-p_{21}g_{21}+p_{11}p_{21}g_{11}}$	
$g_{11}<w\le g_{21}$	1	$\frac{w-g_{11}}{w-p_{11}g_{11}}$	$\frac{(1-p_{21})g_{11}+p_{21}g_{21}}{(1-p_{21})p_{11}g_{11}+p_{21}g_{21}}$
$0<w\le g_{11}$	$1/(p_{11}+p_{21}-p_{11}p_{21})$	0	
	$I_A a_2$	$I_{\Delta^-} a_2$	$I_\varepsilon a_2$
$w>g_{21}$	not defined	$\frac{w-g_{21}}{w-p_{11}g_{11}-p_{21}g_{21}+p_{11}p_{21}g_{11}}$	
$g_{11}<w\le g_{21}$	$1/p_{21}$	0	$\frac{g_{21}}{(1-p_{21})p_{11}g_{11}+p_{21}g_{21}}$
$0<w\le g_{11}$	$1/(p_{11}+p_{21}-p_{11}p_{21})$	0	

Table 4.21. Birnbaum IMs for parallel MSS

w	$I_A b_1$	$I_{\Delta^-} b_1$	$I_\varepsilon b_1$
$w>g_{21}$	0	$(1-p_{21})g_{11}$	
$g_{11}<w\le g_{21}$	0	$(1-p_{21})g_{11}$	$g_{11}(1-p_{21})$
$0<w\le g_{11}$	$1-p_{21}$	$(1-p_{21})w$	
	$I_A b_2$	$I_{\Delta^-} b_2$	$I_\varepsilon b_2$
$w>g_{21}$	0	$g_{21}-p_{11}g_{11}$	
$g_{11}<w\le g_{21}$	1	$w-p_{11}g_{11}$	$g_{21}-p_{11}g_{11}$
$0<w\le g_{11}$	$1-p_{11}$	$(1-p_{11})w$	

Example 4.20

Consider the series-parallel system from Example 4.3 (Figure 4.1A). The IMs $I_A b_j$ of elements 1, 5, and 7 as functions of system demand w are presented in Figure 4.24A and B for the system interpreted as a flow transmission MSS with flow dispersion and task processing MSS without work sharing. Observe that $I_A b_j(w)$ are step nonmonotonic functions.

One can see that the values of w exist for which the importance of some elements is equal to zero. This means that these elements are irrelevant (have no influence on the system's entire availability). For example, in the case of the task processing system, the subsystem consisting of elements from 1 to 6 cannot have a performance that is greater then 1.154. Therefore, when $1.154<w\le 3$, the system satisfies the demand only when element 7 is available. In this case, the entire system availability is equal to the availability of element 7, which is reflected by the element's importance index: $I_A b_7(w) = 1$. The remainder of the elements are irrelevant for demands greater than 1.154: $I_A b_j(w) = 0$ for $1\le j\le 6$. Note that, although for the task processing system element 7 has the greatest importance, the importance of this element for the flow transmission system can be lower than the importance of some other elements at certain intervals of demand variance. For

example, for $3<w<5$ the importance of element 1 is greater than the importance of element 7.

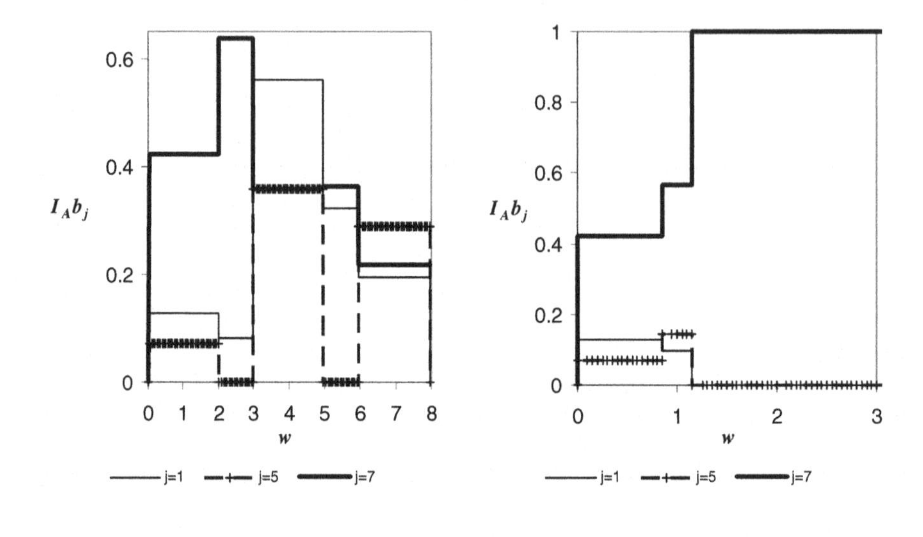

A **B**

Figure 4.24. IM $I_A b_j(w)$ of system elements in flow transmission MSS (A) and task processing MSS (B)

Unlike the IM associated with the system availability $I_A b_j$, the IM associated with the system mean performance $I_\varepsilon b_j$ for element 7 is the greatest for both types of system. The values of $I_\varepsilon b_j$ for $j = 1, \ldots, 7$ are presented in Table 4.22.

Table 4.22. The IMs $I_\varepsilon b_j$ for elements of series-parallel system

No of element	1	2	3	4	5	6	7
Flow transmission MSS	2.170	1.361	1.210	1.555	1.440	2.441	3.000
Task processing MSS	0.139	0.032	0.028	0.036	0.103	0.156	2.375

The IMs $I_{\Delta^-} b_j$ as functions of system demand w are presented in Figure 4.25 for j = 1, 5 and 7. Observe that $I_{\Delta^-} b_j(w)$ are piecewise linear functions. The demand intervals when the function $I_{\Delta^-} b_j(w)$ is constant always correspond to the irrelevancy of system element j.

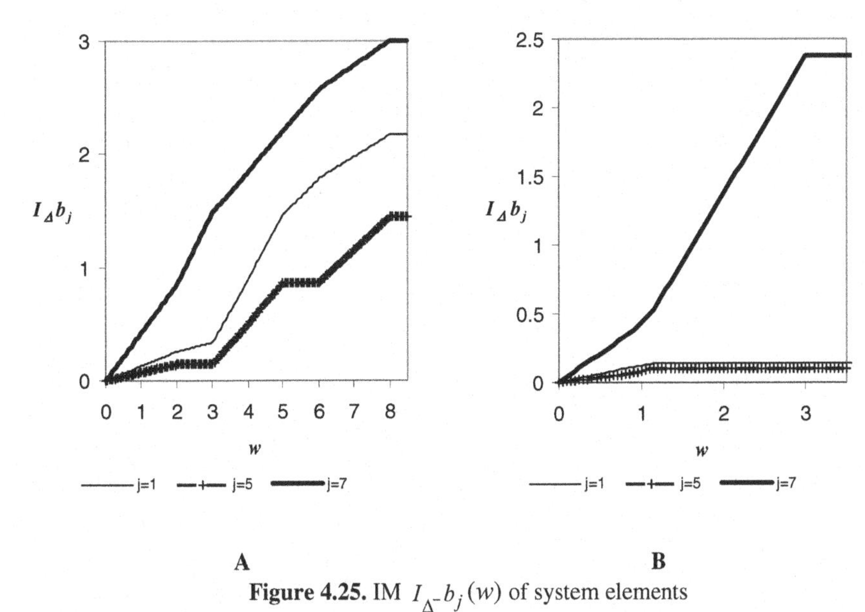

Figure 4.25. IM $I_{\Delta}\text{-}b_j(w)$ of system elements in flow transmission MSS (A) and task processing MSS (B)

4.5.2 Importance of Common Cause Group Reliability

In systems that contain CCGs with total CCF, the reliabilities of the groups (the probabilities that the groups do not fail) affect the reliability of the entire system. If a system consists of nonidentical elements and has a complex structure with nested CCGs, reliabilities of different groups play different roles in providing for the system's reliability. The evaluation of the relative influence of the group's reliability on the reliability of the entire system provides useful information about the importance of these groups.

For example, in systems with complex multilevel protection, the protection survivability (the ability to tolerate destructive external impacts) can depend on the type and location of the protection. The importance of each protection depends not only on its survivability but also on characteristics of the subsystem it protects.

Importance evaluation is an essential point in tracing bottlenecks in protected systems and in identifying the most important protections. The protection survivability importance analysis can also help the analyst to find the irrelevant protections, *i.e.* protections that have no impact on the entire system's reliability. Elimination of irrelevant protections simplifies the system and reduces its cost. In the complex multi-state systems with multilevel protection, finding the irrelevant protections is not a trivial task.

In order to evaluate the CCG reliability importance we use the MSS model with nested CCGs. The algorithm presented in Section 4.4.3 allows one to evaluate the system's performance measures O as a function of the probabilities of total CCFs in its CCGs.

Assume that the system has M CCGs. For the given system structure and the fixed performance distributions of the system elements, the system PM O is a function of the CCF probabilities in these CCGs: $O(v_1,\dots,v_m,\dots,v_M)$. Since the reliability of CCG m s_m is defined as the probability of non-occurrence of CCF in this group ($s_m = 1-v_m$) we can express the system PM as a function of CCG reliabilities $O(s_1,\dots,s_m,\dots,s_M)$ and define in accordance with (4.64) and (4.65):

$$O_{m0} = O(s_1, \dots, 0, \dots, s_M) \text{ and } O_{m1} = O(s_1, \dots, 1, \dots, s_M) \quad (4.74)$$

where O_{m0} corresponds to the system PM when the failure in CCG m has occurred (in accordance with Equation (4.62), this can be represented by the d-function $<U_m(z),\tilde{U}_m(z)>$ of this CCG) and O_{m1} corresponds to the system PM when the failure in CCG m has not occurred (which can be represented by the d-function $<\tilde{U}_m(z),\tilde{U}_m(z)>$). Therefore, O_{m0} and O_{m1} can be obtained by substituting $d_m(z)$ by $<U_m(z),\tilde{U}_m(z)>$ and $<\tilde{U}_m(z),\tilde{U}_m(z)>$ respectively in the procedure of determining the system's PM. The corresponding IMs can be obtained using Equation (4.67)-(4.71).

Example 4.21

Consider the simplest binary systems with multiple protections. In order to evaluate the protections' survivability importance we use the Birnbaum IM I_Ab_m.

The system consists of identical binary elements with availability a. The d-function of each element can be represented as:

$$d_m(z)=< z^0, az^1+(1-a)z^0 >$$

where performance 1 corresponds to its normal state and performance 0 corresponds to failure. The entire system succeeds (survives) if its performance is $G = 1$. Consider the following cases.

Case 1: n-level (concentric) protection of a single element (Figure 4.26A). The system's availability and the survivability importance of mth protection are respectively:

$$A = a[1-\prod_{i=1}^{n}(1-s_i)] \text{ and } I_Ab_m = a\frac{\prod_{i=1}^{n}(1-s_i)}{1-s_m}$$

This means that the protection with the greatest survivability has the greatest importance. The increase of protection survivability lowers the importance of the rest of the protections.

Case 2: n identical protected elements connected in a series (Figure 4.26B). The system's availability and the importance of mth protection are respectively

$$A = a^n \prod_{i=1}^{n} s_i \text{ and } I_A b_m = a^n \frac{\prod_{i=1}^{n} s_i}{s_m}$$

This means that the protection with the lowest survivability has the greatest importance. The increase of protection survivability increases the importance of the remainder of the protections. It can be easily seen that the absence of protection in at least one of the elements makes all of the protections irrelevant (if for any i $s_i = 0$ then $A = 0$ and $I_A b_j = 0$ for all of $j \neq i$). This means that the protection of the elements connected in a series has no sense if at least one element remains unprotected (see protection 1 in Figure 4.27).

Case 3: n identical protected elements connected in parallel (Figure 4.26C). The system's availability and the importance of the mth protection are respectively

$$A = 1 - \prod_{i=1}^{n} (1 - as_i) \text{ and } I_A b_m = a \frac{\prod_{i=1}^{n} (1 - as_i)}{1 - as_m}$$

As in the case of a single element with multiple protections, the protection with the greatest survivability has the greatest importance and the increase of protection survivability lowers the importance of the remainder of the protections.

While in complex systems composed of different multi-state elements, the relations between the elements' survivability and importance are more complicated, the general dependencies are the same as in the cases considered.

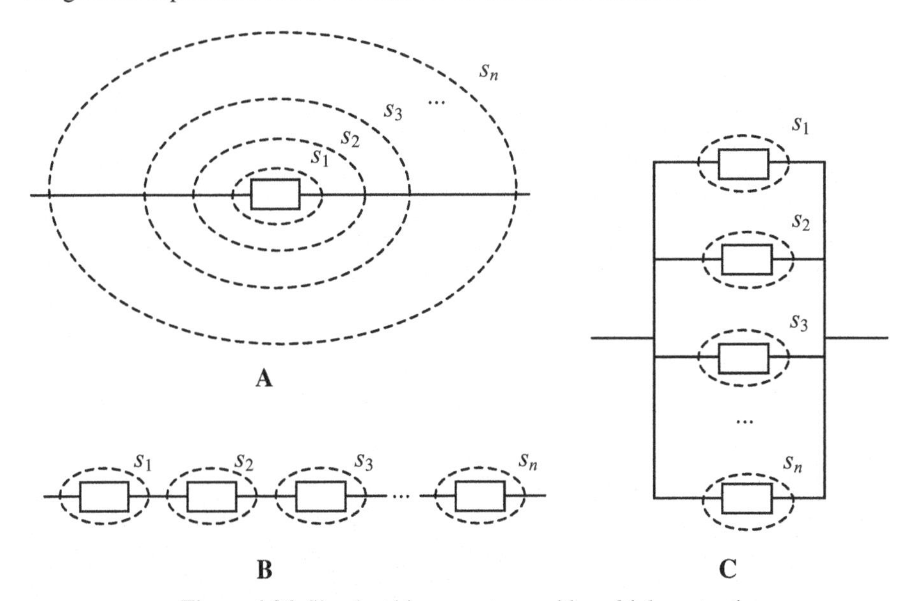

Figure 4.26. Simplest binary systems with multiple protections

Example 4.22

Consider the multi-state flow transmission series-parallel system (with flow dispersion) presented in Figure 4.27. The system consists of seven elements (with performance distributions as presented in Table 4.23) and six protection groups. The survivability of any protection is 0.8. The survivability importance of the protections as functions of demand w are presented in Figure 4.28.

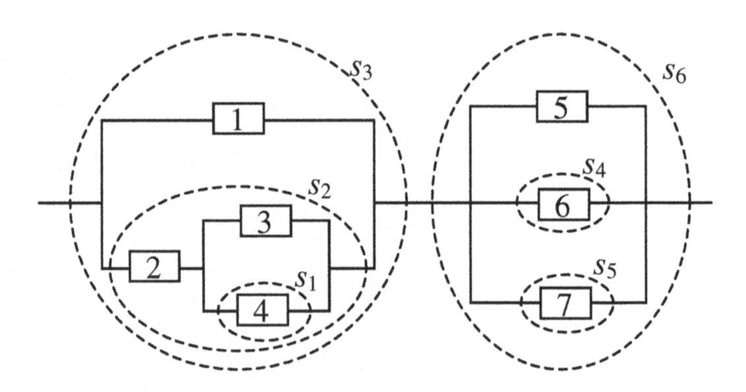

Figure 4.27. Structure of series-parallel MSS with multiple protections

Table 4.23. Performance distributions of multi-state elements

State (h)	No of element (j)													
	1		2		3		4		5		6		7	
	p_{jh}	g_{jh}	p_{jh}	g_{jh}	p_{jh}	g_{jh}	g_{jh}	g_{jh}	p_{jh}	g_{jh}	p_{jh}	g_{jh}	p_{jh}	g_{jh}
0	0.05	0	0.10	0	0.10	0	0.10	0	0.10	0	0.05	0	0.25	0
1	0.05	3	0.05	2	0.10	1	0.30	3	0.20	2	0.95	5	0.75	6
2	0.15	5	0.85	8	0.80	4	0.60	4	0.70	4	-	-	-	-
3	0.75	7	-	-	-	-	-	-	-	-	-	-	-	-

First observe that protection 1 is irrelevant for any w ($I_A b_1(w) = 0$). Indeed, when protection 2 is not destroyed, protection 1 does not affect the system's survivability. When protection 2 is destroyed then element 2 is always destroyed and the subsystem consisting of elements 2, 3 and 4 has a performance rate of 0 independent of the state of protection 1.

Some protections can be irrelevant only for certain intervals of w. For example, protection 2 affects the system's survivability only when protection 3 is destroyed. In this case, element 1 is always destroyed, which prevents the system from having a performance rate greater than 8. Therefore $I_A b_2(w) = 0$ for $w>8$.

Protection 4 affects the system's survivability only when protection 6 is destroyed. In this case, element 5 is always destroyed. If element 7 is in a normal state, then the performance rate of the subsystem remaining after the destruction of protection 6 (elements 6 and 7) is not less than 6. If element 7 does not perform its task, then the performance of the subsystem is no greater than 5 (maximal performance of element 6). This does not depend on the state of protection 4. Therefore, $I_A b_4(w) = 0$ for $5<w\leq 6$.

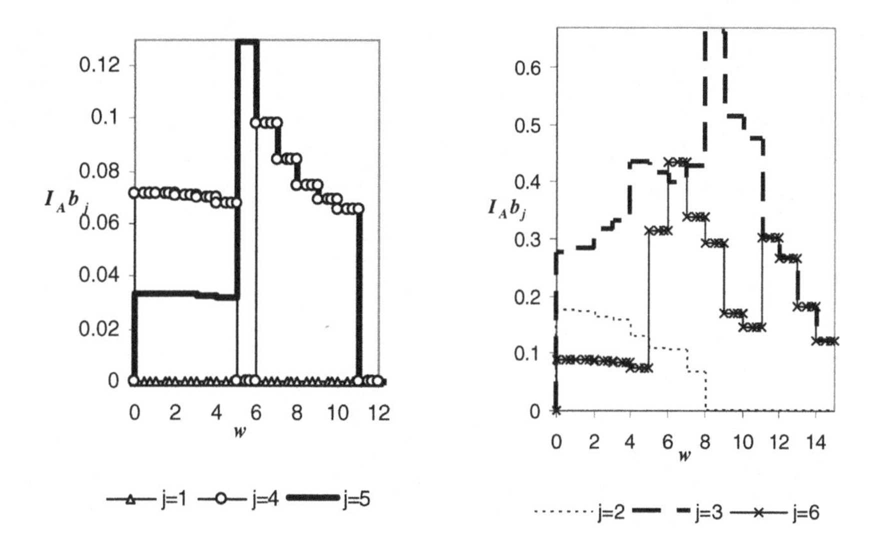

Figure 4.28. Survivability importance of protections as functions of demand

For $w>11$ $I_A b_3(w) = I_A b_6(w)$. Indeed, the system can provide a performance greater than 11 only if both protections 3 and 6 survive. It is the same for protections 4 and 5: when protection 6 is destroyed, the system can provide a performance greater than 6 only if both protections 4 and 5 survive. Therefore, for $w>6$ $I_A b_4(w) = I_A b_5(w)$.

Note also that the greater the availability of the two-state element, the greater the importance of its individual protection. For example, when $w\leq 5$ both elements 6 and 7 can meet the demand, but $I_A b_4(w)>I_A b_5(w)$.

In general, the outer-level protections are more important than the inner-level ones, since they protect more elements. In our case, protections 3 and 6 have the greatest importance for any w.

In order to estimate the effect of survivability of protections on their importance, consider Figure 4.29 representing the functions $I_A b_j(s_m)$ for different j and m when the system should meet the demand $w = 5$. Observe that although the relations among the different protections in complex MSSs are much more complicated than in the simple binary systems considered above, the general tendencies are the same. Observe, for example, that the mutual influence of the protections in pairs 2 and 3, 4 and 6, 5 and 7 resembles the mutual influence of protections in Case 1 of Example 4.21, since these pairs of protections are partly concentric (both protect the same subsystems). The greater the survivability of one of the protections in the pair the lower the importance of the other one. When the outer protection becomes invulnerable, the inner protection becomes irrelevant ($I_A b_2 = 0$ when $s_3 = 1$ and $I_A b_4 = I_A b_5 = 0$ when $s_6 = 1$).

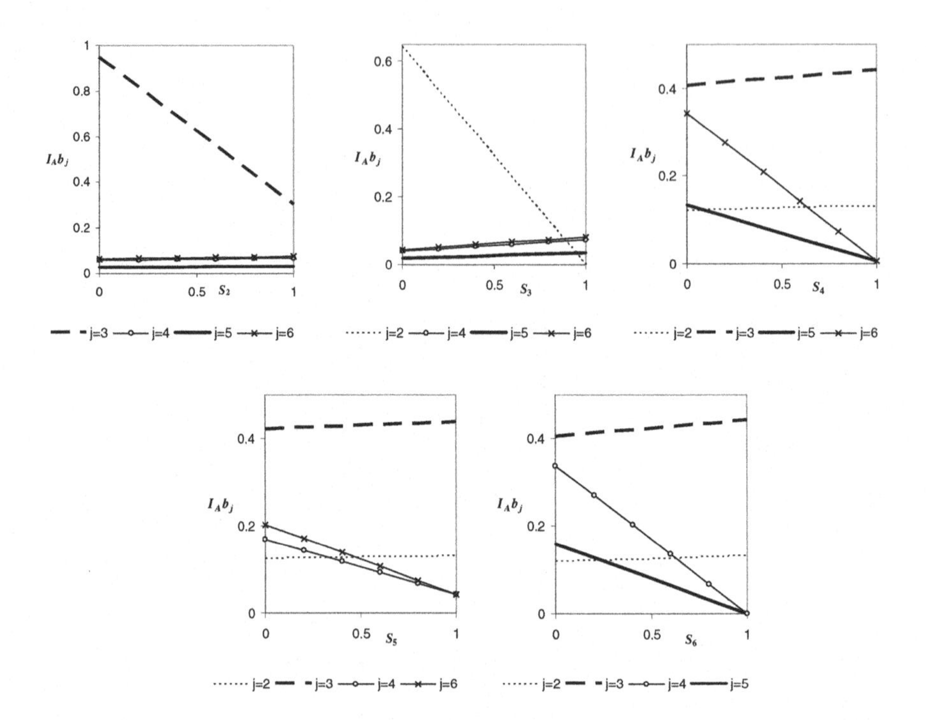

Figure 4.29. Survivability importance of protections as functions of protection survivability

The mutual influence of protections in pairs 2 and 4, 3 and 4, 2 and 6, and 3 and 6 resembles the mutual influence of the protections in Case 2 of Example 4.21, since these pairs of protections protect subsystems connected in the series. In this case the greater the survivability of one of the protections in the pair, the greater the importance of another one.

The mutual influence of protections 4 and 5 resembles the mutual influence of protections in Case 3 of Example 4.21, since this pair of protections protects parallel elements. In this case, the greater the survivability of one of the protection in the pair the lower the importance of another one.

4.5.3 Reliability Importance of Multi-state Elements in Multi-state Systems

Early progress towards the extension of IMs to the case of MSSs can be found in [126, 127], where the measures related to the occupancy of a given state by an element have been proposed. These measures characterize the importance of a given element being in a certain state or moving to the neighbouring state with respect to the system's performance. The IM of a given element is, therefore, represented by a vector of values, one for each state of the element. Such representation may be difficult for the practical reliability analyst to interpret. In the following sections we consider integrated IMs based on element performance restriction.

4.5.3.1 Extension of Importance Measures to Multi-state Elements

Assume that the states of each element j are ordered in such a manner that $g_{j0} \leq g_{j1} \leq .... \leq g_{jk_j-1}$. One can introduce a performance threshold α and divide this set into two ordered subsets corresponding respectively to the element performance above and below the level α. Let element j be constrained to a performance rate not greater than α, while the remainder of the elements of the MSS are not constrained: we denote by $O_j^{\leq\alpha|\mathrm{M}}$ the system PM obtained in this situation. Similarly, we denote by $O_j^{>\alpha|\mathrm{M}}$ the system PM resulting from the dual situation in which element j is constrained to performances above α. The MSS performance measures so introduced rely on a restriction of the achievable performance of the MSS elements. Different modelling assumptions in the enforcement of this restriction will lead to different performance values. The letter M in the definitions of $O_j^{\leq\alpha|\mathrm{M}}$ and $O_j^{>\alpha|\mathrm{M}}$ is used to code the modelling approach to the restriction of element behaviour. Substituting the measures $O_j^{\leq\alpha|\mathrm{M}}$ and $O_j^{>\alpha|\mathrm{M}}$ to the binary equivalents O_{j0} and O_{j1}, we can define importance measures for multi-state elements:

performance reduction worth

$$I_O r_j^{\alpha|\mathrm{M}} = O / O_j^{\leq\alpha|\mathrm{M}} \tag{4.75}$$

performance achievement worth

$$I_O a_j^{\alpha|\mathrm{M}} = O_j^{>\alpha|\mathrm{M}} / O \tag{4.76}$$

relative performance reduction (Fussell-Vesely)

$$I_O f_j^{\alpha|\mathrm{M}} = (O - O_j^{\leq\alpha|\mathrm{M}}) / O \tag{4.77}$$

relative performance achievement

$$I_O v_j^{\alpha|\mathrm{M}} = (O_j^{>\alpha|\mathrm{M}} - O) / O \tag{4.78}$$

Birnbaum importance

$$I_O b_j^{\alpha|\mathrm{M}} = O_j^{>\alpha|\mathrm{M}} - O_j^{\leq\alpha|\mathrm{M}} \tag{4.79}$$

This latter IM extends the concept of the IM introduced in [126]. Combining the different definitions of importance measures with different types of the system PM and different model assumptions M relative to the types of element restriction, one can obtain many different importance measures for MSS, each one bearing specific physical information. The choice of the proper IM to use depends on the system's mission and the type of improvement actions that one is aiming at in the system design or operation.

In the following section we consider two models of element performance restriction and discuss their application with respect to the importance measures $I_O f_j^{\alpha|\mathrm{M}}$ and $I_O b_j^{\alpha|\mathrm{M}}$.

4.5.3.2 State-space Reachability Restriction Approach

Let O_{jh} be the PM of the MSS when element j is in a fixed state h while the rest of the elements evolve stochastically among their corresponding states with performance distributions $\{g_{ih}, p_{ih}\}$, $1 \leq i \leq n$, $i \neq j$, $0 \leq h < k_i$. Using pivotal decomposition, we obtain the overall expected system performance

$$O = \sum_{h=0}^{k_j-1} p_{jh} O_{jh} \tag{4.80}$$

We denote by $h_{j\alpha}$ the state in the ordered set of states of element j whose performance $g_{jh_{j\alpha}}$ is equal to or immediately below α, *i.e.* $g_{jh_{j\alpha}} \leq \alpha < g_{jh_{j\alpha}+1}$. The conditional probability $\breve{p}_{jh}$ that element j is in a state h characterized by a performance g_{jh} not greater than a prespecified level threshold α $(h \leq h_{j\alpha})$ can be obtained as

$$\breve{p}_{jh} = \Pr\{G_j = g_{jh} \mid G_j \leq \alpha\} = p_{jh} / \Pr\{G_j \leq \alpha\}$$
$$= p_{jh} / \sum_{m=0}^{h_{j\alpha}} p_{jm} = p_{jh} / p_j^{\leq\alpha} \tag{4.81}$$

Similarly, the conditional probability $\hat{p}_{jh}$ of element j being in a state h when it is known that $h > h_{j\alpha}$ is

$$\hat{p}_{jh} = \Pr\{G_j = g_{jh} \mid G_j > \alpha\} = p_{jh} / \Pr\{G_j > \alpha\}$$
$$= p_{jh} / \sum_{m=h_{j\alpha}+1}^{k_j-1} p_{jm} = p_{jh} / p_j^{>\alpha} \tag{4.82}$$

In Equation (4.81) and (4.82), $p_j^{\le\alpha}$ and $p_j^{>\alpha}$ are probabilities that element j is in states with performance not greater than α and greater than α respectively.

The state-space reachability restriction model (coded with the letter s: M ≡ s) is based on the restrictive condition on the states reachable by element j. In this model we define as $O_j^{\le\alpha|s}$ the system PM obtained when element j is forced to visit only states with performance not greater than α:

$$O_j^{\le\alpha|s} = \sum_{m=0}^{h_{j\alpha}} \breve{p}_{jm} O_{jm} = \sum_{m=0}^{h_{j\alpha}} p_{jm} O_{jm} / p_j^{\le\alpha} \tag{4.83}$$

Similarly, we define as $O_j^{>\alpha|s}$ the system performance measure obtained under the condition that the element j stays in states with performance greater than α:

$$O_j^{>\alpha|s} = \sum_{m=h_{j\alpha}+1}^{k_j-1} \hat{p}_{jm} O_{jm} = \sum_{m=h_{j\alpha}+1}^{k_j-1} p_{jm} O_{jm} / p_j^{>\alpha} \tag{4.84}$$

According to these definitions

$$O = O_j^{\le\alpha|s} p_j^{\le\alpha} + O_j^{>\alpha|s} p_j^{>\alpha} \tag{4.85}$$

Using the definition of the performance measures O, $O_j^{\le\alpha|s}$ and $O_j^{>\alpha|s}$ we can specify the IMs. For example, the Birnbaum importance takes the form

$$I_O b_j^{\alpha|s} = O_j^{>\alpha|s} - O_j^{\le\alpha|s} \tag{4.86}$$

From (4.86) and since $p_j^{\le\alpha} + p_j^{>\alpha} = 1$

$$O = O_j^{\le\alpha|s} + I_O b_j^{\alpha|s} p_j^{>\alpha} = O_j^{>\alpha|s} - I_O b_j^{\alpha|s} p_j^{\le\alpha} \tag{4.87}$$

And thus

$$I_O b_j^{\alpha|s} = (O - O_j^{\le\alpha|s}) / p_j^{>\alpha} = (O_j^{>\alpha|s} - O) / p_j^{\le\alpha} \tag{4.88}$$

From Equation (4.66) and (4.71) we can see that for two-state elements:

$$O = O_{j0} + I_O b_j p_{j1} = O_{j1} - I_O b_j p_{j0} \tag{4.89}$$

Comparing (4.87) and (4.89) we see that the $I_O b_j^{\alpha|s}$ measure for MSSs is really an extension of the definition of the Birnbaum importance for two-state elements, for which $k_j = 2$ and $\alpha = 0$. As such, it measures the rate of improvement of the system PM deriving from improvements on the probability $p_j^{>\alpha}$ of element j occupying states characterized by performance higher than α.

The Fussel-Vesely importance measure (relative performance reduction) takes the form

$$I_O f_j^{\alpha|s} = 1 - O_j^{\leq\alpha|s} / O \tag{4.90}$$

It can be easily seen that

$$I_O b_j^{\alpha|s} = I_O f_j^{\alpha|s} O / p_j^{>\alpha} \tag{4.91}$$

The element IMs based on the state-space reachability restriction approach quantify the effect on the system performance of element j remaining confined in the dual subspaces of states corresponding to performances greater or not greater than α.

4.5.3.3 Performance Level Limitation Approach

We consider again a threshold α on the performance of element j. However, we assume that the space of reachable states of element j is not restricted, *i.e.* element j can visit any of its states independently on whether the associated performance is below or above α and it can do so with the original state probability distribution. Limitations, however, are imposed on the performance rate of element j: we consider a deteriorated version of the element that is not capable of providing a performance greater than α, in spite of the possibility of reaching any state, and an enhanced version of element j that provides performances always not less than α, even when residing in states below $h_{j\alpha}$. The limitation on the performance is such that, when in states $h > h_{j\alpha}$, the deteriorated element j is not capable of providing the design performance corresponding to its state; in these cases it is assumed that it provides the performance α. On the other hand, when the enhanced element is working in states $h < h_{j\alpha}$, it is assumed that it provides the performance α. We code this modelling approach by the letter w: $\mathrm{M} \equiv w$.

The output performance measures $O_j^{\leq\alpha|w}$ and $O_j^{>\alpha|w}$ in this model take the form

$$O_j^{\le\alpha|w} = \sum_{m=0}^{h_{j\alpha}} p_{jm} O_{jm} + O_j^{\alpha} \sum_{m=h_{j\alpha}+1}^{k_j-1} p_{jm}$$

$$= \sum_{m=0}^{h_{j\alpha}} p_{jm} O_{jm} + O_j^{\alpha} p_j^{>\alpha} \tag{4.92}$$

and

$$O_j^{>\alpha|w} = O_j^{\alpha} \sum_{m=0}^{h_{j\alpha}} p_{jm} + \sum_{m=h_{j\alpha}+1}^{k_j-1} p_{jm} O_{jm}$$

$$= O_j^{\alpha} p_j^{\le\alpha} + \sum_{m=h_{j\alpha}+1}^{k_j-1} p_{jm} O_{jm} \tag{4.93}$$

where O_j^{α} is the system PM when element j remains fixed operating with performance α while the remainder of the system elements visit their states in accordance with their performance distributions. It can be seen that

$$O_j^{\le\alpha|w} + O_j^{>\alpha|w} = O + O_j^{\alpha} \tag{4.94}$$

In this case, the Birnbaum importance takes the form

$$I_O b_j^{\alpha|w} = O_j^{>\alpha|w} - O_j^{\le\alpha|w}$$

$$= O_j^{\alpha} \sum_{m=0}^{h_{j\alpha}} p_{jm} + \sum_{m=h_{j\alpha}+1}^{k_j-1} p_{jm} O_{jm} - \sum_{m=0}^{h_{j\alpha}} p_{jm} O_{jm} - O_j^{\alpha} \sum_{m=h_{j\alpha}+1}^{k_j-1} p_{jm}$$

$$= \sum_{m=0}^{h_{j\alpha}} p_{jm}(O_j^{\alpha} - O_{jm}) + \sum_{m=h_{j\alpha}+1}^{k_j-1} p_{jm}(O_{jm} - O_j^{\alpha}) \tag{4.95}$$

$$= \sum_{m=0}^{k_j-1} p_{jm} \,|\, O_j^{\alpha} - O_{jm} \,|$$

Hence, in the performance level limitation model the Birnbaum IM is equal to the expected value of the absolute deviation of the system PM from its value when element j has performance α.

The Fussel-Vesely IM (relative performance reduction) takes the form

$$I_O f_j^{\alpha|w} = 1 - O_j^{\le\alpha|w} / O = (O_j^{>\alpha|w} - O_j^{\alpha}) / O \tag{4.96}$$

Birnbaum and Fussel-Vesely IMs are related as follows:

$$I_O b_j^{\alpha|w} = I_O f_j^{\alpha|w} \cdot O + (O_j^{\alpha} - O_j^{\leq\alpha|w}) \tag{4.97}$$

The element IMs based on a limitation of the achievable performance level give information on which level α of element performance is the most beneficial from the point of view of the entire system PM.

Observe that according to the definitions (4.83), (4.84) and (4.92), (4.93)

$$O_j^{\leq\alpha|w} = O_j^{\leq\alpha|s} p_j^{\leq\alpha} + O_j^{\alpha} p_j^{>\alpha} \tag{4.98}$$

and

$$O_j^{>\alpha|w} = O_j^{>\alpha|s} p_j^{>\alpha} + O_j^{\alpha} p_j^{\leq\alpha} \tag{4.99}$$

From these equations one can obtain relations between Birnbaum and Fussel-Vesely IMs as defined according to the two approaches

$$I_O b_j^{\alpha|w} = I_O b_j^{\alpha|s} p_j^{>\alpha} + (O_j^{\alpha} - O_j^{\leq\alpha|s})(2p_j^{\leq\alpha} - 1) \tag{4.100}$$

and

$$I_O f_j^{\alpha|w} = I_O f_j^{\alpha|s} - (O_j^{\alpha} - O_j^{\leq\alpha|s}) p_j^{>\alpha} / O \tag{4.101}$$

4.5.3.4 Evaluating System Performance Measures

In order to evaluate the system PM O when all of its elements are not restricted, one has to apply the reliability block diagram technique over u-functions of the individual elements representing their performance distributions in the form:

$$u_j(z) = \sum_{h=0}^{k_j-1} p_{jh} z^{g_{jh}} \tag{4.102}$$

In order to obtain the IMs in accordance with the state-space reachability restriction approach, one has to modify the u-function of element j as follows:

$$u_j(z) = \sum_{h=0}^{h_{j\alpha}} (p_{jh} / p_j^{\leq\alpha}) z^{g_{jh}} \tag{4.103}$$

when evaluating $O_j^{\leq\alpha|s}$ and

$$u_j(z) = \sum_{h=h_{j\alpha}+1}^{k_j-1} (p_{jh} / p_j^{>\alpha}) z^{g_{jh}} \quad (4.104)$$

when evaluating $O_j^{>\alpha|s}$ and then apply the reliability block diagram technique.

In order to obtain the PMs in accordance with the performance level limitation approach one has to modify the u-function of element j as follows:

$$u_j(z) = \sum_{h=0}^{h_{j\alpha}} p_{jh} z^{g_{jh}} + p_j^{>\alpha} z^{\alpha} \quad (4.105)$$

when evaluating $O_j^{\leq\alpha|w}$ and

$$u_j(z) = p_j^{\leq\alpha} z^{\alpha} + \sum_{h=h_{j\alpha}+1}^{k_j-1} p_{jh} z^{g_{jh}} \quad (4.106)$$

when evaluating $O_j^{>\alpha|w}$ and then apply the reliability block diagram technique.

Note that the PM O_j^{α} can also be easily obtained by using the u-function of element j in the form $u_j(z) = z^{\alpha}$.

Example 4.23

Consider the series-parallel flow transmission system (with flow dispersion) presented in Figure 4.30 with elements having performance distributions given in Table 4.24.

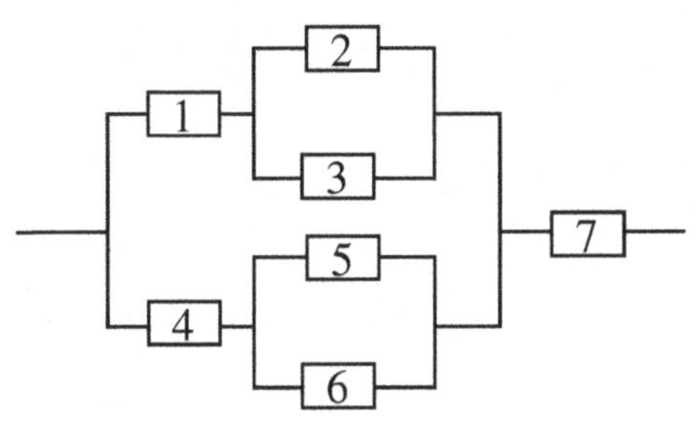

Figure 4.30. Structure of series-parallel MSS with multi-state elements

Elements 2, 3, 5, and 6 are identical. However, the pairs of elements 2, 3 and 5, 6 have different influences on the system's entire performance, since they are connected in a series with different elements (1 and 4 respectively). Therefore, while we expect elements 2 and 3 have the same importance (as well as elements 5 and 6), the importance of element 2 (or 3) differs from the importance of element 5 (or 6). The demand w is assumed to be constant in time, but different values of the constant will be considered.

Table 4.24. Performance distributions of multi-state elements

State (h)	No of element (j)													
	1		2		3		4		5		6		7	
	p_{jh}	g_{jh}	p_{jh}	g_{jh}	p_{jh}	g_{jh}	g_{jh}	g_{jh}	p_{jh}	g_{jh}	p_{jh}	g_{jh}	p_{jh}	g_{jh}
0	0.10	0	0.10	0	0.10	0	0.20	0	0.10	0	0.10	0	0.15	0
1	0.05	1	0.05	2	0.05	2	0.10	2	0.05	2	0.05	2	0.15	6
2	0.15	3	0.85	4	0.85	4	0.45	6	0.85	4	0.85	4	0.05	10
3	0.35	5	-	-	-	-	0.25	8	-	-	-	-	0.45	14
4	0.35	7	-	-	-	-	-	-	-	-	-	-	0.20	18

In this example we perform the importance analysis based on the Fussel-Vesely IM (relative performance reduction). In Figure 4.31 the $I_A f_j^{2|s}(w)$ and $I_A f_j^{2|w}(w)$ measures are presented for elements 1, 2 (identical to 3) and 4 and 5 (identical to 6) for different time-constant system demands w. The first measure shows how critical it is for the MSS availability that the element visits only states with performance below or equal to $\alpha = 2$. The second measure shows how critical for the MSS availability it is to limit the performance of the element below the threshold value $\alpha = 2$.

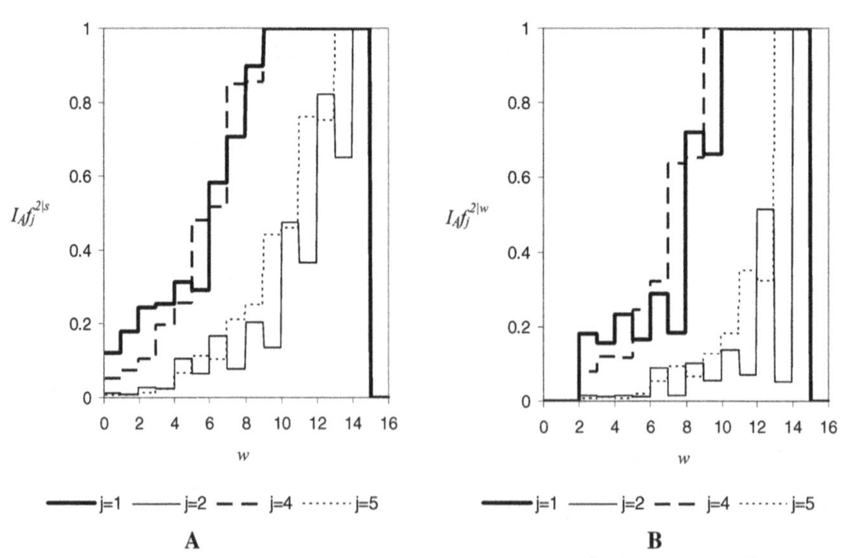

Figure 4.31. Behaviour of the elements' IMs. A: $I_A f_j^{2|s}(w)$; B: $I_A f_j^{2|w}(w)$

The functions $I_A f_j^{2|s}(w)$ and $I_A f_j^{2|w}(w)$ differ significantly. While $I_A f_j^{2|w}(w) = 0$ for $w \le 2$, since $O_j^{\le\alpha|w} = O$ for these demands, $I_A f_j^{2|s}(w) > 0$, since the reduction of the state-space for obtaining $O_j^{\le\alpha|s}$ changes the probabilities of being in the states with $g_{jh} \le 2$, and, therefore, $O_j^{\le\alpha|s} \ne O$.

Recall also that from the definitions, $I_A f_j^{2|s} = 1$ or $I_A f_j^{2|w} = 1$ means that, when the element j has a performance restricted below α, the entire system fails. The importance measure $I_A f_j^{2|s}$ for elements $j = 1$ and $j = 4$ becomes 1 when $w = 9$. Indeed, the greatest performance of the subsystem of unrestricted elements 4, 5, and 6 is 8 while the greatest performance of the subsystem of elements 1, 2, and 3 is 1 when element 1 is allowed to visit only states with a performance not greater than $\alpha = 2$ (*i.e.* $g_{10} = 0$ or $g_{11} = 1$). Therefore, the MSS cannot have a performance greater than 8+1 = 9. Similarly, the greatest performance of the subsystem of unrestricted elements 1, 2, and 3 is 7, while the greatest performance of the subsystem of elements 4, 5, and 6 is 2 when element 4 is allowed to visit only states with a performance not greater than $\alpha = 2$ ($g_{40} = 0$ or $g_{41} = 2$). Therefore, in this case the MSS cannot have a performance greater than 7+2 = 9.

On the contrary, the importance $I_A f_j^{2|w}$ for elements $j = 1$ and $j = 4$ becomes 1 for different values of w. When the performance of element 1 is restricted by $\alpha = 2$, the MSS cannot have a performance greater than 8+2 = 10; when the performance of element 4 is restricted by $\alpha = 2$, the MSS cannot have a performance greater than 7+2 = 9. Therefore, $I_A f_1^{2|w} = 1$ for $w > 10$ while $I_A f_4^{2|w} = 1$ for $w > 9$.

Figure 4.31 also shows that an element which is the most important with respect to a value of the demand w can be less important for a different value. This is a typical situation in MSSs. For example, when $5 < w < 6$ element 4 is the most important one among elements 1-6 when their ability to perform above $\alpha = 2$ is considered, while for $w \le 5$ it becomes less important than element 1.

The $I_\Delta f_j^{2|s}(w)$ and $I_\Delta f_j^{2|w}(w)$ functions are presented in Figure 4.32. Analogously to $I_A f_j^{2|w}(w)$, the function $I_\Delta f_j^{2|w}(w)$ is equal to zero when $w \le 2$, since in this case $\Delta_j^{\le 2|w} = \Delta$. For increasing demand values, the difference between $\Delta_j^{\le\alpha|w}$ and Δ (system performance deficiency when element j is not constrained) increases from zero to a constant level. Therefore, the ratio $I_\Delta f_j^{2|w}(w) = (\Delta_j^{\le 2|w} - \Delta)/\Delta$ first increases and then begins to decrease.

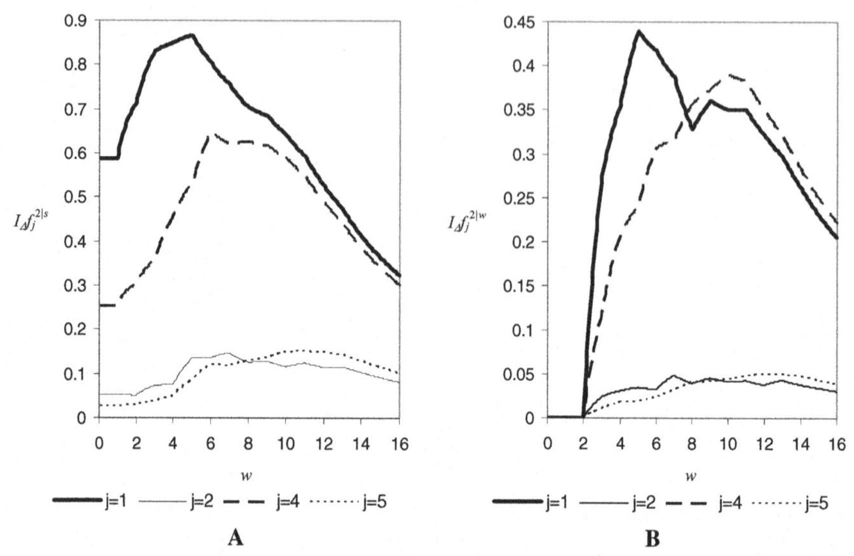

Figure 4.32. Behaviour of the elements' IMs. A: $I_{\Delta}f_j^{2|s}(w)$; B: $I_{\Delta}f_j^{2|w}(w)$

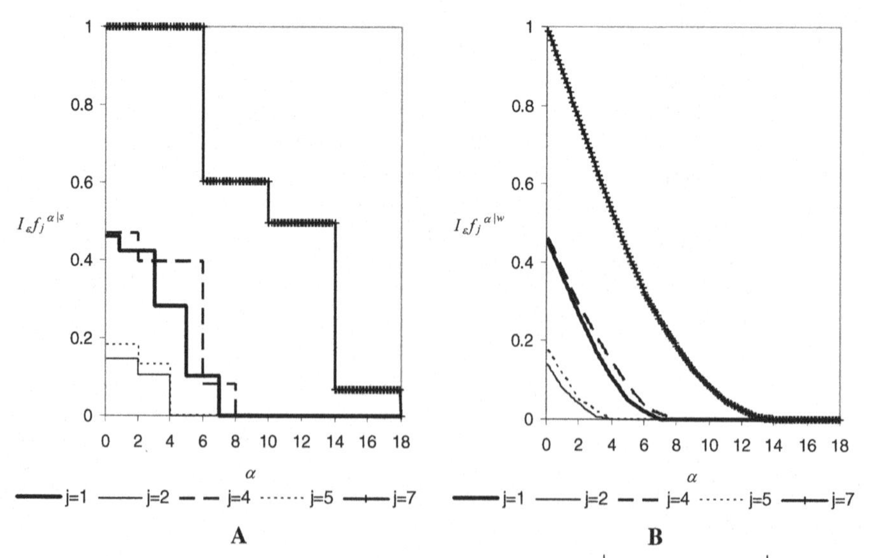

Figure 4.33. Behaviour of the elements' IMs. A: $I_{\varepsilon}f_j^{\alpha|s}(\alpha)$; B: $I_{\varepsilon}f_j^{\alpha|w}(\alpha)$

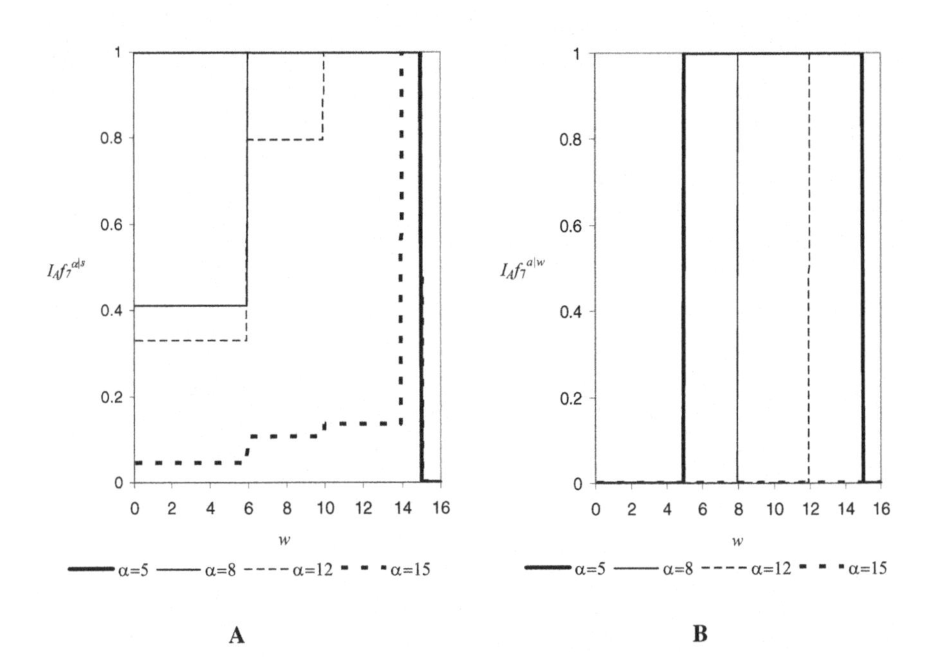

Figure 4.34. Behaviour of the elements' IMs. A: $I_A f_7^{\alpha|s}(w)$; B: $I_A f_7^{\alpha|w}(w)$

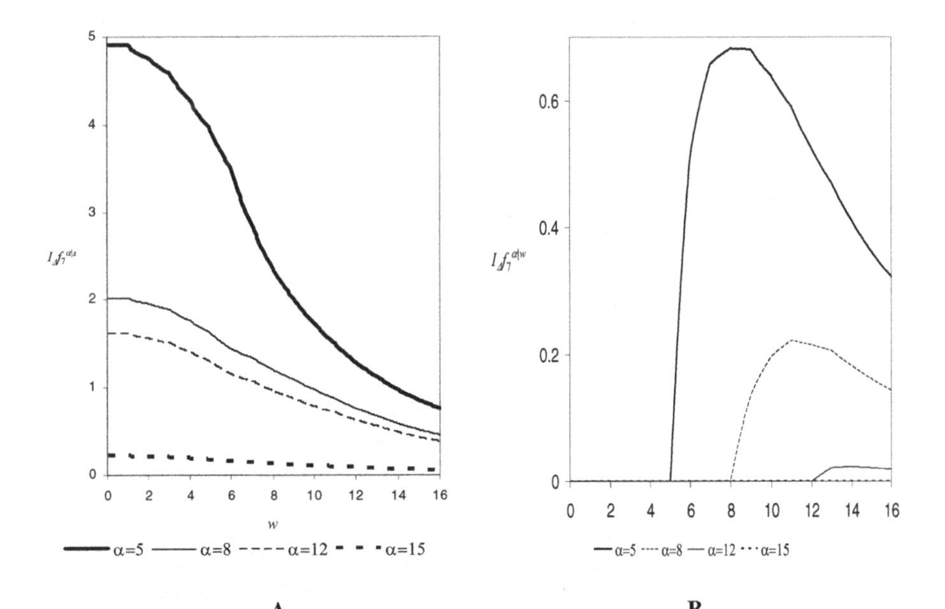

Figure 4.35. Behaviour of the elements' IMs. A: $I_\Delta f_7^{\alpha|s}(w)$; B: $I_\Delta f_7^{\alpha|w}(w)$

A similar behaviour is shown by $I_{\Delta}f_j^{2|s}(w)$. It can be seen that values of demand w exist for which the increase of the element performance above the threshold α causes the greatest relative reduction of the system performance deficiency (maxima of the curves $I_{\Delta}f_j^{2|w}(w)$ and $I_{\Delta}f_j^{2|s}(w)$ in Figure 4.32). It is also confirmed that the relative importance of different elements depends on the value of the demand (for example, element 2 is more important than element 5 for $w<8$ and less important for $w>8$).

The mean system performance ε does not depend on the demand. Figure 4.33 reports the indices $I_{\varepsilon}f_j^{\alpha|s}$ and $I_{\varepsilon}f_j^{\alpha|w}$ as functions of α. Note that while $I_{\varepsilon}f_j^{\alpha|w}(\alpha)$ are continuous functions, $I_{\varepsilon}f_j^{\alpha|s}(\alpha)$ are stepwise functions since $I_{\varepsilon}f_j^{\alpha|s}(\alpha_1)=I_{\varepsilon}f_j^{\alpha|s}(\alpha_2)$ for any α_1 and α_2 such that $g_{jh_{j\alpha}} \le \alpha_1 < \alpha_2 < g_{jh_{j\alpha}+1}$. Both functions are decreasing, which means that the higher levels of performance threshold α cause a less relative increase of the system's performance.

Note that both $I_{\varepsilon}f_j^{\alpha|s}$ and $I_{\varepsilon}f_j^{\alpha|w}$ take a value of zero (*i.e.* $O_j^{\le\alpha|w}=O_j^{\le\alpha|s}=O$) when the α level is above or equal to the maximum performance achievable by element j, g_{jk_j}.

Improvement of the performance of a certain element above a given threshold α may be achieved, either by increasing the probability of residing in states with performances larger than α (as indicated by the $I_O f_j^{\alpha|s}$ measures) or by increasing the performances of some states (as indicated by $I_O f_j^{\alpha|w}$ measures). Consider, for example, element 7, whose IMs for different threshold values α as functions of the demand w are given in Figures 4.34 and 4.35. Observe that $I_A f_7^{\alpha|w}(w)=0$ when $w\le\alpha$ and $I_A f_7^{\alpha|w}(w)=1$ when $w>\alpha$, since the logic of the system is such that its performance is not affected by limitations on the performance of element 7 if its threshold α is set to a value greater than the demand w, whereas the system fails completely if element 7 has a performance below the system's demand. Also, the $I_A f_7^{\alpha|s}(w)$ function does not depend on α, when α varies within the performance intervals 1-6, 6-10, 10-14, 14-18. The jumps in the step-functions $I_A f_7^{\alpha|s}(w)$ occur at values $w=g_{7h}$ and correspond to the restrictions to state h with $w>g_{7h}$. Functions $I_{\Delta}f_7^{\alpha|s}(w)$ and $I_{\Delta}f_7^{\alpha|w}(w)$ are continuous. When α increases, the relative reduction of the system's performance deficiency becomes smaller (because a smaller number of states are subject to restriction). Note that the demand w for which the greatest relative reduction of system performance deficiency is achieved (maximum of the function $I_{\Delta}f_7^{\alpha|w}(w)$) increases with the increase of α.

4.6 Universal Generating Function in Analysis of Continuum-state Systems

Some systems and elements exhibit continuous performance variation (for example, when their performance degrades due to gradual failures). In these cases, one can discern a continuum of different states. The structure functions $\phi(G_1,..., G_n)$ representing such continuous-state systems are mappings $[g_{1\min}, g_{1\max}] \times [g_{2\min}, g_{2\max}] \times ... \times [g_{n\min}, g_{n\max}] \rightarrow [g_{\min}, g_{\max}]$, where $[g_{j\min}, g_{j\max}]$ is the closed interval of performance variation of element j and $[g_{\min}, g_{\max}]$ is the closed interval of performance variation of the entire system. Such functions were introduced in [128-130] and are called the continuum structure functions.

The stochastic behaviour of continuous-state systems and elements may be specified through the complemented distribution functions [131]:

$$C_j(x) = \Pr\{G_j \geq x\}, \quad C(x) = \Pr\{\phi(G_1,..., G_n) \geq x\} \tag{4.107}$$

An example of such a function (cumulated curve) is presented in Figure 4.36.

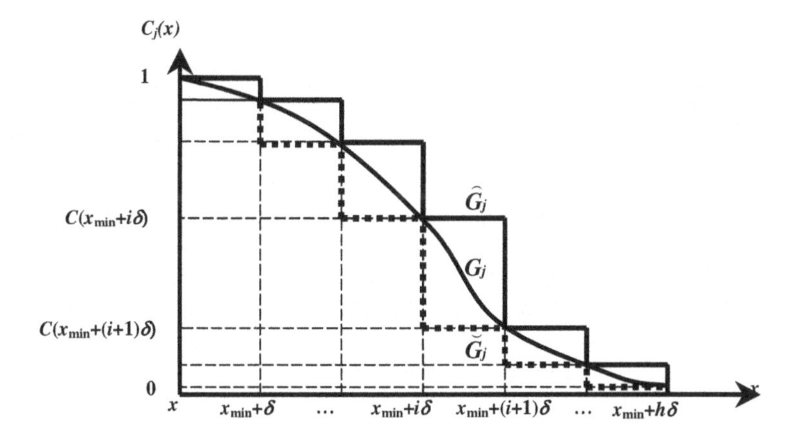

Figure 4.36. Complemented distribution functions for continuous and discrete variables

The method for estimating the boundary points for performance measures of continuum-state systems suggested by Lisnianski [132] uses the approximation of continuous performance distributions by discrete performance distributions. This method is based on the assumptions that the continuum structure functions are monotonic, *i.e.*

$$\phi(G_1,..., G_n) \leq \phi(\hat{G}_1,..., \hat{G}_n) \text{ if } G_j \leq \hat{G}_j \text{ for } 1 \leq j \leq n \tag{4.108}$$

or

$$\phi(G_1,...,G_n) \geq \phi(\hat{G}_1,...,\hat{G}_n) \text{ if } G_j \leq \hat{G}_j \text{ for } 1 \leq j \leq n \tag{4.109}$$

and that the functions $C_j(x)$ for all of the elements are continuous (with possible jumps at the end points). These assumptions are relevant for many types of practical system.

In order to obtain the discrete approximation of the continuous performance distribution of element j, we divide the interval $[g_{j\min}, g_{j\max}]$ into h equal subintervals. The length of each subinterval is

$$\partial_j = \frac{g_{j\max} - g_{j\min}}{h} \tag{4.110}$$

In order to obtain the lower and upper bound approximations of distribution of performance G_j, we introduce discrete random variables $\breve{G}_j$ and $\hat{G}_j$ such that

$$\Pr\{\breve{G}_j \geq g_{j\min} + i\partial_j\} = \Pr\{\hat{G}_j \geq g_{j\min} + i\partial_j\}$$
$$= \Pr\{G_j \geq g_{j\min} + i\partial_j\}, \ \ 0 \leq i \leq h \tag{4.111}$$

and

$$\Pr\{\breve{G}_j \geq g_{j\min} + i\partial_j + x\} = \Pr\{G_j \geq g_{j\min} + (i+1)\partial_j\}$$
$$< \Pr\{G_j \geq g_{j\min} + i\partial_j + x\}, \ \ 0 \leq i < h, \ \ 0 \leq x < \partial_j \tag{4.112}$$

$$\Pr\{\hat{G}_j \geq g_{j\min} + i\partial_j + x\} = \Pr\{G_j \geq g_{j\min} + i\partial_j\}$$
$$> \Pr\{G_j \geq g_{j\min} + i\partial_j + x\}, \ \ 0 \leq i < h, \ \ 0 \leq x < \partial_j \tag{4.113}$$

The complemented distribution functions of $\breve{G}_j$ and $\hat{G}_j$ are presented in Figure 4.36. Since for any variable X with a complemented distribution function $C(x)$ $\Pr\{x_1 \leq X < x_2\} = C(x_1) - C(x_2)$, we can obtain that for $\hat{G}_j$

$$\Pr\{\hat{G}_j = g_{j\min}\} = 0$$
$$\Pr\{\hat{G}_j = g_{j\min} + i\partial_j\}$$
$$= C_j(g_{j\min} + (i-1)\partial_j) - C_j(g_{j\min} + i\partial_j), \ \ 1 \leq i < h$$
$$\Pr\{\hat{G}_j = g_{j\min} + h\partial_j\} = \Pr\{\hat{G}_j = g_{j\max}\}$$
$$= C_j(g_{j\min} + (h-1)\partial_j) = C_j(g_{j\max} - \partial_j)$$
$$\Pr\{\hat{G}_j = g_{j\min} + i\partial_j + x\} = 0, \ \ 0 \leq i < h, \ \ 0 < x < \partial_j \tag{4.114}$$

and for $\breve{G}_j$

$$\begin{aligned}
&\Pr\{\breve{G}_j = g_{j\min} + i\partial_j\} \\
&= C_j(g_{j\min} + i\partial_j) - C_j(g_{j\min} + (i+1)\partial_j), \ \ 0 \le i < h \\
&\Pr\{\breve{G}_j = g_{j\min} + h\partial_j\} = \Pr\{\breve{G}_j = g_{j\max}\} = C_j(g_{j\max}) \\
&\Pr\{\breve{G}_j = g_{j\min} + i\partial_j + x\} = 0, \ \ 0 \le i < h, \ \ 0 < x < \partial_j
\end{aligned} \tag{4.115}$$

These expressions define the p.m.f. of the discrete variables $\breve{G}_j$ and $\widehat{G}_j$.

Observe that the inequalities (4.112) and (4.113) guarantee that for any j $E(\breve{G}_j) \le E(G_j)$ and $E(\widehat{G}_j) \ge E(G_j)$. Therefore, for any increasing monotonic function *f*:

$$E(f(\breve{G}_1,...,\breve{G}_n)) \le E(f(G_1,...,G_n)) \le E(f(\widehat{G}_1,...,\widehat{G}_n)) \tag{4.116}$$

and for any decreasing monotonic function *v*:

$$E(v(\widehat{G}_1,...,\widehat{G}_n)) \le E(v(G_1,...,G_n)) \le E(v(\breve{G}_1,...,\breve{G}_n)) \tag{4.117}$$

Since the system performance measures are defined as expected values of functions of performances of individual elements (see Section 3.3), the upper and lower bounds for these measures can be obtained by replacing the continuous-state elements with multi-state elements having discrete performances distributed as defined by Equations (4.114) and (4.115).

Having the complemented distribution functions $C_j(x)$ of system elements, one can determine the *u*-functions of the corresponding multi-state elements with discrete performance as

$$\begin{aligned}
\widehat{u}_j(z) = &\sum_{i=1}^{h-1} \{C_j[g_{j\min} + (i-1)\partial_j] - C_j(g_{j\min} + i\partial_j)\} z^{g_{j\min} + i\partial_j} \\
&+ C_j(g_{j\max} - \partial) z^{g_{j\max}}
\end{aligned} \tag{4.118}$$

$$\begin{aligned}
\breve{u}_j(z) = &\sum_{i=0}^{h-1} \{C_j(g_{j\min} + i\partial_j) - C_j[g_{j\min} + (i+1)\partial_j]\} z^{g_{j\min} + i\partial_j} \\
&+ C_j(g_{j\max}) z^{g_{j\max}}
\end{aligned} \tag{4.119}$$

Applying the reliability block diagram technique over *u*-functions $\widehat{u}_j(z)$ one obtains the *u*-function $\widehat{U}(z)$ representing the p.m.f. of the entire system consisting of elements with discrete performance distributions with p.m.f. (4.114). Applying this technique over u-functions $\breve{u}_j(z)$ one obtains the u-function $\breve{U}(z)$ representing the p.m.f. of the entire system consisting of elements with discrete performance distributions with p.m.f. (4.115). Having the u-functions $\widehat{U}(z)$ and $\breve{U}(z)$ one can obtain the boundary points for the system performance measures as described in Section 3.3.

Example 4.24

Consider the series-parallel continuum-state system presented in Figure 4.37. Each element of the system can be either available or totally unavailable due to a catastrophic failure. If the element is available, then its performance rate varies continuously depending on the state of the element's operating environment. The performance rate of the unavailable element is zero.

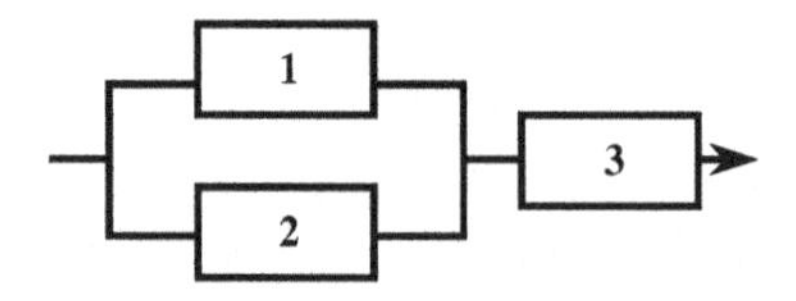

Figure 4.37. Structure of continuum-state system

Observe that if the element's availability is a_j and its complemented distribution function given that the element is available is $C_j^*(x)$, then the performance distribution of this element is defined by the complemented distribution function $C_j(x)$ that takes the form

$$C_j(x) = \begin{cases} 1, & x \leq 0 \\ a_2, & 0 < x \leq g_{\min} \\ a_2 C_j^*(x), & x > g_{\min} \end{cases}$$

The first element has the availability a_1 = 0.8 and exponentially distributed performance with mean $\mu_1 = 40$ and $g_{1\min}$ = 0 (the probability that $G_1 > g_{1\max}$ = 1000 is neglected). The second element has availability a_2 = 0.7 and uniformly distributed performance with $g_{2\min}$ = 30 and $g_{2\max}$ = 60. The third element has availability a_3 = 0.95 and normally distributed performance with mean $\mu_3 = 70$ and standard deviation $\sigma_3 = 10$ (the probabilities that $G_3 < g_{3\min}$ = 0 and that $G_3 > g_{3\max}$ = 1000 are neglected). In the state of failure, all the elements have performance zero. The system fails if its performance is less that the constant demand w = 20.

The complemented distribution functions of the element performances taking into account the element availabilities are

$$C_1(x) = \begin{cases} 1, & x \le 0 \\ a_1 e^{-x/\mu_1}, & x > 0 \end{cases}$$

$$C_2(x) = \begin{cases} 1, & x \le 0 \\ a_2, & 0 < x \le g_{\min} \\ a_2(g_{\max} - x)/(g_{\max} - g_{\min}), & g_{\min} \le x \le g_{\max} \\ 0, & x > g_{\max} \end{cases}$$

$$C_3(x) = \begin{cases} 1, & x \le 0 \\ a_3 (1 - \frac{1}{\sigma_3 \sqrt{2\pi}} \int_{-\infty}^{x} e^{-(t-\mu_3)^2/(2\sigma_3^2)} dt), & x > 0 \end{cases}$$

Considering the complemented distribution functions in the interval [0, 1000] and assigning $\partial = 1$, Lisnianski [132] has obtained for the system interpreted as a flow transmission MSS with flow dispersion $\hat{\varepsilon} = \hat{U}'(1) = 47.21$ and $\breve{\Delta}^- = 3.07$ when the element performance distributions are represented by u-functions $\hat{u}_j(z)$, and $\breve{\varepsilon} = \breve{U}'(1) = 46.23$ and $\hat{\Delta}^- = 3.19$ when the element performance distributions are represented by u-functions $\breve{u}_j(z)$. Using these boundary points, one can estimate the performance measures with maximal relative errors

$$100 \times (47.21 - 46.23)/46.23 = 2.1\%$$

for mean performance and

$$100 \times (3.19 - 3.07)/3.07 = 3.9\%$$

for expected performance deficiency.

For the system interpreted as a task processing MSS without work sharing, $\hat{\varepsilon} = \hat{U}'(1) = 24.23$ and $\breve{\Delta}^- = 3.32$ when the element performance distributions are represented by u-functions $\hat{u}_j(z)$, and $\breve{\varepsilon} = \breve{U}'(1) = 23.83$ and $\hat{\Delta}^- = 3.45$ when the element performance distributions are represented by u-functions $\breve{u}_j(z)$. This gives the estimations of the performance measures with maximal relative errors

$$100\times(24.23-23.83)/46.23=1.7\%$$

for mean performance and

$$100\times(3.45-3.32)/3.32=3.9\%$$

for expected performance deficiency.

The upper and lower boundary points for mean performance and expected unsupplied demand are presented in Figure 4.38 as functions of step ∂ for both types of systems. The decrease of step ∂ provides improvement in the accuracy of boundary points estimation. However, it considerably increases the computational burden, since the number of terms in the u-functions $\breve{u}_j(z)$ and $\hat{u}_j(z)$ is proportional to $1/\partial$.

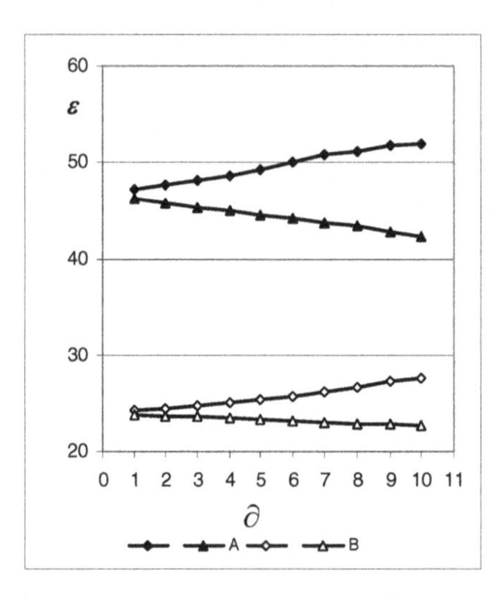

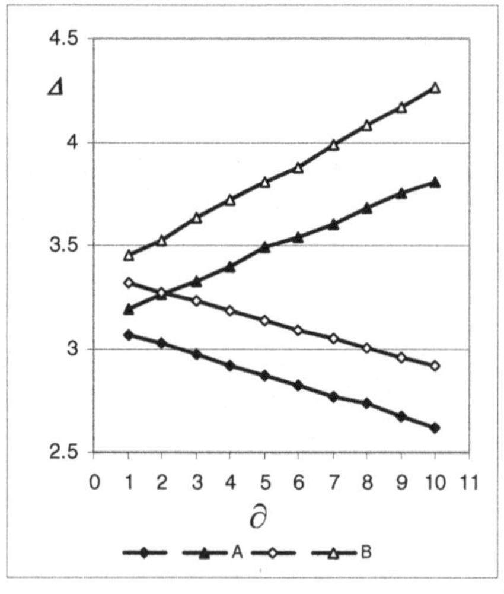

Figure 4.38. Boundary points for expected performance deficiency and mean performance (A: flow transmission system; B: task processing system)

5. Universal Generating Function in Optimization of Series-Parallel Multi-state Systems

Reliability improvement is of critical importance in various kinds of systems; however, any effort for this type of improvement usually requires resources that are limited by technical and/or economical constraints. Two approaches can be distinguished in the reliability optimization problem. The first one is aimed at achieving the greatest possible reliability subject to different constraints (according to Gnedenko and Ushakov [133], this problem is named the direct reliability optimization problem), and the second one focuses on minimizing the resources needed for providing a required reliability level (according to [133], this problem is named the inverse reliability optimization problem).

There are four general methods to improve system reliability:

- a provision of redundancy;
- an optimal adjustment of the system's parameters, an optimal arrangement of the existing elements or the assignment of interchangeable elements;
- an enhancement of the reliability (availability) and/or performance of the system's elements;
- a combination of the above-mentioned methods.

Applied to an MSS, these methods affect two basic system properties: its configuration (structure function) and the performance distribution of its elements.

The UGF method that allows for system performance distribution, and thereby its performance measures to be evaluated based on a fast procedure, opens new possibilities for solving MSS reliability optimization problems. Based on the UGF technique, the MSS reliability can be obtained as a function of the system structure and the performance distributions of its elements. Therefore, numerous optimization problems can be formulated in which the optimal composition of all or part of the factors influencing the entire MSS reliability has to be found subject to different constraints.

5.1 Structure Optimization Problems

In binary systems, providing redundancy means to incorporate several identical parallel elements into a certain functional subsystem (component). The redundancy optimization problem for an MSS, which may consist of elements with different performance distributions, is a problem of system structure optimization. Indeed, when solving practical system design problems, the reliability engineer deals with a variety of products existing on the market. Each product is characterized by its performance distribution, price, *etc.* To find the optimal system structure, one should choose the appropriate versions from a list of available products for each type of equipment, as well as the number of parallel elements of these versions.

In the simplest MSS structure optimization problem, each component can contain only identical elements. This optimization problem is relevant in cases where contracting or maintenance considerations prevent purchasing or using different elements of the same type. In some cases, such a limitation can be undesirable or even unacceptable for two reasons:

- Allowing different versions of the elements to be allocated in the same system component, one obtains a solution providing the desired reliability level at a lower cost than what appears in the solution containing identical parallel elements.
- In practice, when a system needs to be modernized according to new demand levels or new reliability requirements, the designer often has to include additional elements in the existing system. Some system components can contain versions of elements that are unavailable. In this case, some elements with the same functionality but with different parameters should make up the components. Therefore, in the general case, the MSS structure optimization problem should be solved without placing a limitation on the diversity of the versions of the elements.

The above-mentioned problem of optimal single-stage MSS expansion to enhance its reliability and/or performance is an important extension of the structure optimization problem. In this case, one has to decide which elements should be added to the existing system and to which component they should be added. There is a similar problem with the optimal single-stage replacement of the MSS elements. Here, one has to decide which elements should be replaced by new ones having better characteristics.

During the MSS's operation time, the demand and reliability requirements can change. To provide a desired level of MSS performance, management should develop a multistage expansion plan. For the problem of optimal multistage MSS expansion, it is important to answer not only the question of what must be included into the system, but also the question of when it must be included into the system.

5.1.1 Optimal Structure of Systems with Identical Elements in Each Component

5.1.1.1 Problem Formulation

An MSS consists of N components. Each component is a subsystem that can consist of parallel elements with the same functionality. The interaction between the system components is given by a reliability block diagram. Different versions of elements may be chosen for any given system component, but each component can contain only elements of the same version. For each component i there are B_i element versions available in the market. A PD $\boldsymbol{g}_i(b)$, $\boldsymbol{p}_i(b)$ and cost $c_i(b)$ can be specified for each version b of element of type i.

The structure of system component i is defined by the number of the element version chosen for this component b_i ($1 \le b_i \le B_i$) and by the number of parallel elements n_i ($1 \le n_i \le n_{\max i}$), where $n_{\max i}$ is the maximum allowed number of parallel elements of type i. The vectors $\boldsymbol{b} = (b_1, \ldots, b_N)$ and $\boldsymbol{n} = (n_1, \ldots, n_N)$ define the entire system structure.

For given $\boldsymbol{b}$ and $\boldsymbol{n}$ the total cost of the system can be calculated as

$$C = \sum_{i=1}^{N} n_i c_i(b_i) \quad (5.1)$$

To take into account price discounting, the element cost can be considered as a function of the number of elements purchased. In this case

$$C = \sum_{i=1}^{N} n_i c_i(n_i, b_i) \quad (5.2)$$

The problem of MSS structure optimization is formulated in [134] as finding the minimal cost system configuration $\boldsymbol{b}$, $\boldsymbol{n}$ that provides the required level O^* of the system performance measure O:

$$C(\boldsymbol{b}, \boldsymbol{n}) \to \min \text{ subject to } f(O, O^*) = 1 \quad (5.3)$$

where $f(O,O^*)$ is a function representing the desired relation between O and O^*. For example, if the system should provide a given level of availability, $O \equiv A$ and $f(A, A^*)=1$ ($A \ge A^*$).

5.1.1.2 Implementing the Genetic Algorithm

In order to represent the system structure in the GA one has to use $2N$-length integer strings: $\boldsymbol{a} = (a_1,\ldots,a_i,\ldots,a_{2N})$, where for each $i \le N$ $b_i = a_j$ and for $i>N$ $n_{i-N} = a_i$. An arbitrary integer string cannot represent a feasible solution because each b_i should vary within limits $1 \le b_i \le B_i$ and each n_i within limits $1 \le n_i \le n_{\max i}$. To provide solution feasibility, a decoding procedure should first transform each string $\boldsymbol{a}^*$ to a string $\boldsymbol{a}$ in which

$$a_i = \mathrm{mod}_{B_i}(a_i^*) + 1 \text{ for } i \le N \text{ and } a_i = \mathrm{mod}_{n_{\max i}}(a_i^*) + 1 \text{ for } i > N. \quad (5.4)$$

The solution decoding procedure, based on the UGF technique, performs the following steps:

1. Determines n_i and b_i for each system component from the string $\boldsymbol{a}$.

2. Determines u-functions $u_{ib_i}(z)$ of each version of elements according to their PD $\boldsymbol{g}_i(b_i)$, $\boldsymbol{p}_i(b_i)$.

3. Determines u-functions of each component i ($1 \le i \le N$) by assigning $U_1(z) = u_{1b_1}(z)$ and applying the recursive equation

$$U_j(z) = U_{j-1}(z) \underset{\phi_{par}}{\otimes} u_{ib_i}(z) \text{ for } j = 2,\dots,n_i \tag{5.5}$$

4. Determines the u-function of the entire MSS $U(z)$ by applying the corresponding composition operators using the reliability block diagram method (described in Section 4.1.3).

5. Having the system u-function, determines its performance measure as described in Section 3.3.

6. Determines the total system cost using Equation (5.2).

7. Determines the solution's fitness as a function of the MSS cost and performance measure as

$$M - C(\boldsymbol{a}) - \pi(1 + |O - O^*|)(1 - f(O, O^*)) \tag{5.6}$$

where π is a penalty coefficient and M is a constant value. The fitness function is penalized when $f(O, O^*) = 0$. The solution with minimal cost and with $f(O,O^*) = 1$ provides the maximal possible value of the fitness function.

Example 5.1

Consider a power station coal transportation system consisting of five basic components connected in series (Figure 3.3):

1. subsystem of primary feeders;
2. subsystem of primary conveyors;
3. subsystem of stackers-reclaimers;
4. subsystem of secondary feeders;
5. subsystem of secondary conveyors.

The system belongs to the type of flow transmission MSS with flow dispersion, since its main characteristic is the transmission capacity and parallel elements can transmit the coal simultaneously. The system should meet a variable demand W. Its acceptability function is defined as $F(G,W) = 1(G \ge W)$. The system should have availability not less than $A^* = 0.99$ for the given demand distribution $\boldsymbol{w}$, $\boldsymbol{q}$ presented in Table 5.1.

Table 5.1. Demand distribution

$\boldsymbol{w}$	1.00	0.80	0.50	0.20
$\boldsymbol{q}$	0.48	0.09	0.14	0.29

Each system element is an element with total failure (which means that it can have only two states: functioning with the nominal capacity and total failure, corresponding to a capacity of zero). For each type of equipment, a list of products available on the market exists. Each version of equipment is characterized by its nominal capacity g, availability p, and cost c. The list of available products is presented in Table 5.2. The maximal number of elements in each component should not exceed six.

Table 5.2. Parameters of available MSS elements

No. of version of MSS element	Component 1 Primary feeders			Component 2 Primary conveyors			Component 3 Stackers-reclaimers			Component 4 Secondary feeders			Component 5 Secondary conveyors		
	g	p	c	g	p	c	g	p	c	g	p	c	g	p	c
1	1.20	0.980	0.590	1.00	0.995	0.205	1.00	0.971	7.525	1.15	0.977	0.180	1.28	0.984	0.986
2	1.00	0.977	0.535	0.92	0.996	0.189	0.60	0.973	4.720	1.00	0.978	0.160	1.00	0.983	0.825
3	0.85	0.982	0.470	0.53	0.997	0.091	0.40	0.971	3.590	0.91	0.978	0.150	0.60	0.987	0.490
4	0.85	0.978	0.420	0.28	0.997	0.056	0.20	0.976	2.420	0.72	0.983	0.121	0.51	0.981	0.475
5	0.48	0.983	0.400	0.21	0.998	0.042				0.72	0.981	0.102			
6	0.31	0.920	0.180							0.72	0.971	0.096			
7	0.26	0.984	0.220							0.55	0.983	0.071			
8										0.25	0.982	0.049			
9										0.25	0.97	0.044			

The total number of components in the problem considered is N = 5. Each integer string containing 10 integer numbers can represent a possible solution. In order to illustrate the string decoding process performed by the GA, consider, for example, the string $\boldsymbol{a}^*$ = (9, 10, 4, 1, 3, 0, 7, 6, 1, 2). From Table 5.2 we have B_1 = 7, B_2 = 5, B_3 = 4, B_4 = 9, B_5 = 4; $n_{\max i}$ = 6 for any component. After transforming the string according to Equation (5.4) we obtain $\boldsymbol{a}$ = (3, 1, 1, 2, 4, 1, 2, 1, 2, 3). This string corresponds to one primary feeder of version 3, two primary conveyors of version 1, one stacker of version 1, two secondary feeders of version 2, and three secondary conveyors of version 4.

According to step 2 of the decoding procedure, the u-functions of the chosen elements are determined as $u_{13}(z) = 0.018z^0+0.982z^{0.85}$ (for the primary feeder), $u_{21}(z) = 0.005z^0+0.995z^{1.00}$ (for the primary conveyors), $u_{31}(z) = 0.029z^0+0.971z^{1.00}$ (for the stacker), $u_{42}(z) = 0.022z^0+0.978z^{1.00}$ (for the secondary feeders) and $u_{54}(z) = 0.019z^0+0.981z^{0.51}$ (for the secondary conveyor).

According to step 3, we determine the u-functions of the five system components using the composition operator $\underset{+}{\otimes}$:

$$U_1(z) = u_{13}(z) = 0.018z^0+0.982z^{0.85}$$

$$U_2(z) = u_{21}(z) \underset{+}{\otimes} u_{21}(z) = (0.005z^0+0.995z^1)^2$$

$$U_3(z) = u_{31}(z) = 0.029z^0+0.971z^{1.00}$$

$$U_4(z) = u_{42}(z) \underset{+}{\otimes} u_{42}(z) = (0.022z^0+0.978z^{1.00})^2$$

$$U_5(z) = u_{54}(z) \underset{+}{\otimes} u_{54}(z) \underset{+}{\otimes} u_{54}(z) = (0.019z^0+0.981z^{0.51})^3$$

The flow transmission system consisting of five components connected in series can be reduced to a binary system (see Section 4.1.4) and its availability can be obtained according to Equation (4.16) as

$$A = \sum_{m=1}^{4} q_m [\delta_{w_m}(U_1(z))\delta_{w_m}(U_2(z))\delta_{w_m}(U_3(z))\delta_{w_m}(U_4(z))\delta_{w_m}(U_5(z))]$$

The values $\delta_{w_m}(U_j(z))$ for each system component and each demand level are presented in Table 5.3.

Table 5.3. Availability of system components for different demand levels

w	$\delta_w(U_1(z))$	$\delta_w(U_2(z))$	$\delta_w(U_3(z))$	$\delta_w(U_4(z))$	$\delta_w(U_5(z))$
1.0	0.000000	0.999975	0.971000	0.999516	0.998931
0.8	0.982000	0.999975	0.971000	0.999516	0.998931
0.5	0.982000	0.999975	0.971000	0.999516	0.999993
0.2	0.982000	0.999975	0.971000	0.999516	0.999993

The overall system availability for the given demand distribution is $A = 0.495$.

The total system cost is

$$C = 0.47+2\times0.205+7.525+2\times0.16+3\times0.475=10.15$$

Since the given system configuration provides an availability that is less than the desired level A^*, the solution fitness function is penalized. The value of the fitness function obtained in accordance with Equation (5.6) for $M = 50$ and $\pi = 25$ is

$$50-10.15-25\cdot(1+0.99-0.495) = 2.475$$

The best solution obtained by the GA provides the system's availability as $A = 0.992$. According to this solution, the system should consist of two primary feeders of version 2, two primary conveyors of version 3, three stackers of version 2, three secondary feeders of version 7, and three secondary conveyors of version 4. The total system cost is $C = 17.05$. Since in this solution $f(A, A^*) = 1$, its fitness is $50-17.05 = 32.95$.

5.1.2 Optimal Structure of Systems with Different Elements in Each Component

5.1.2.1 Problem Formulation

The problem definition is similar to the one presented in Section 5.1.1.1. However, in this case different versions and the number of elements may be chosen for any given system component [135].

The structure of the system component i is defined by the numbers of the parallel elements of each version b chosen for this component: n_{ib} $(1 \le b \le B_i)$.

The vectors $\boldsymbol{n}_i = (n_{i1}, \ldots, n_{iB_i} : 1 \le i \le N)$ define the entire system structure. For a given set of vectors $\{\boldsymbol{n}_1, \boldsymbol{n}_2, \ldots, \boldsymbol{n}_N\}$ the total cost of the system can be calculated as

$$C(n_1, \ldots, n_N) = \sum_{i=1}^{N} \sum_{b=1}^{B_i} n_{ib} c_i(b) \tag{5.7}$$

Having the system structure defined by its components' reliability block diagram and by the set $\{\boldsymbol{n}_1, \boldsymbol{n}_2, \ldots, \boldsymbol{n}_N\}$, one can determine the entire MSS performance measure $O(\boldsymbol{w}, \boldsymbol{q}, \boldsymbol{n}_1, \boldsymbol{n}_2, \ldots, \boldsymbol{n}_N)$ for any given demand distribution $\boldsymbol{w}, \boldsymbol{q}$. The problem of MSS structure optimization is formulated as finding the minimal cost system configuration $\{\boldsymbol{n}_1, \boldsymbol{n}_2, \ldots, \boldsymbol{n}_N\}$ that provides the required level O^* of the system performance measure O:

$$C(\boldsymbol{n}_1, \boldsymbol{n}_2, \ldots, \boldsymbol{n}_N) \rightarrow \min \text{ subject to } f(O, O^*) = 1, \tag{5.8}$$

5.1.2.2 Implementing the Genetic Algorithm

The natural way of encoding the solutions of the problem (5.8) in the GA is by defining a B-length integer string, where B is the total number of versions available:

$$B = \sum_{i=1}^{N} B_i \tag{5.9}$$

Each solution is represented by string $\boldsymbol{a} = (a_1, \ldots, a_j, \ldots, a_B)$, where for each

$$j = \sum_{m=1}^{i-1} B_m + b \tag{5.10}$$

a_j denotes the number of parallel elements of type i and version b: $n_{ib} = a_j$. One can see that $\boldsymbol{a}$ is a concatenation of substrings representing the vectors $\boldsymbol{n}_1, \boldsymbol{n}_2, \ldots, \boldsymbol{n}_N$.

The solution decoding procedure, based on the UGF technique, performs the following steps:

1. Determines n_{ib} for each system component and each element version from the string $\boldsymbol{a}$.
2. Determines u-functions $u_{ib}(z)$ of each version of elements according to their PD $\boldsymbol{g}_i(b), \boldsymbol{p}_i(b)$.
3. Determines u-functions of subcomponents containing the identical elements by applying the composition operators $\otimes_{\phi_{\text{par}}}$ over n_{ib} identical u-functions $u_{ib}(z)$.
4. Determines u-functions of each component i ($1 \le i \le N$) by applying the composition operators $\otimes_{\phi_{\text{par}}}$ over the u-functions of all nonempty subcomponents belonging to this component.
5. Determines the u-function of the entire MSS $U(z)$ by applying the reliability block diagram method.
6. Determines the MSS performance measure as described in Section 3.3.

7. Determines the total system cost using Equation (5.7).
8. Determines the solution fitness as a function of the MSS cost and performance measure according to Equation (5.6).

Example 5.2

Consider the coal transportation system described in Example 5.1 and allow each system component to consist of different versions of the elements.

The total number of available versions in the problem considered is $B = 7+5+4+9+4 = 29$. Each string containing 29 integer numbers can represent a possible solution. In order to illustrate the string decoding process performed by the GA, consider the string

(0, 0, 0, 2, 0, 0, 1, 0, 0, 2, 0, 0, 1, 0, 0, 0, 0, 0, 0, 0, 0, 0, 3, 0, 0, 0, 1, 0, 0)

This string corresponds to two primary feeders of version 4, one primary feeder of version 7, two primary conveyors of version 3, one stacker of version 1, three secondary feeders of version 7, and one secondary conveyor of version 2.

According to step 2 of the decoding procedure, the u-functions of the chosen elements are determined as $u_{14}(z) = 0.022z^0+0.978z^{0.85}$, $u_{17}(z) = 0.016z^0+0.984z^{0.26}$ (for the primary feeders), $u_{23}(z) = 0.003z^0+0.997z^{0.53}$ (for the primary conveyors), $u_{31}(z) = 0.029z^0+0.971z^{1.00}$ (for the stacker), $u_{47}(z) = 0.017z^0+0.983z^{0.55}$ (for the secondary feeders) and $u_{52}(z) = 0.016z^0+0.984z^{1.28}$ (for the secondary conveyor).

According to steps 3 and 4, we determine the u-functions of the five system components using the composition operator $\underset{+}{\otimes}$:

$$U_1(z) = u_{14}(z) \underset{+}{\otimes} u_{14}(z) \underset{+}{\otimes} u_{17}(z)$$

$$= (0.022z^0+0.978z^{0.85})^2(0.016z^0+0.984z^{0.26})$$

$$U_2(z) = u_{23}(z) \underset{+}{\otimes} u_{23}(z) = (0.003z^0+0.997z^{0.53})^2$$

$$U_3(z) = u_{31}(z) = 0.029z^0+0.971z^{1.00}$$

$$U_4(z) = u_{47}(z) \underset{+}{\otimes} u_{47}(z) \underset{+}{\otimes} u_{47}(z) = (0.017z^0+0.983z^{0.55})^3$$

$$U_5(z) = u_{52}(z) = 0.016z^0+0.984z^{1.28}$$

As in the previous example, the entire system availability can be obtained using Equation (4.16). We obtain $A = 0.95$ for the given system structure and demand distribution.

The total system cost, according to Equation (5.7), is

$$C = 2\times0.42+0.22+2\times0.091+7.525+3\times0.071+0.825 = 9.805$$

The fitness of the solution is estimated using Equation (5.6), where $M = 50$ and $\pi = 25$. For the desired value of system availability $A^* = 0.99$ the fitness takes the value

$$50-9.805-25(1+0.99-0.95) = 14.195$$

For the desired value of system availability $A^* = 0.95$, the fitness takes the value

50–9.805 = 40.195

The minimal cost solutions obtained by the GA for different desired availability levels A^* are presented in Table 5.4. This table presents the cost, calculated availability, and structure of the minimal cost solutions obtained by the GA. The structure of each system component i is represented by a string of the form $n_{i1}*b_{i1}$, ..., $n_{im}*b_{im}$, where n_{im} is the number of identical elements of version b_{im} belonging to this component.

Table 5.4. Optimal solution for MSS structure optimization problem

	$A^* = 0.95$	$A^* = 0.97$	$A^* = 0.99$
System availability A	0.950	0.970	0.992
System cost C	9.805	10.581	15.870
	System structure		
Primary feeders	1*7, 2*4	2*2	2*4, 1*6
Primary conveyors	2*3	6*5	2*3
Stackers-reclaimers	1*1	1*1	2*2, 1*3
Secondary feeders	3*7	6*9	3*7
Secondary conveyors	1*2	3*3	3*4

Consider, for example, the best solution obtained for $A^* = 0.99$. The minimal cost system configuration that provides the system availability $A = 0.992$ consists of two primary feeders of version 4, one primary feeder of version 6, two primary conveyors of version 3, two stackers of version 2, one stacker of version 3, three secondary feeders of version 7, and three secondary conveyors of version 4. The cost of this configuration is 15.870, which is 7% less than the cost of the optimal configuration with identical parallel elements obtained in Example 5.1.

5.1.3 Optimal Single-stage System Expansion

In practice, the designer often has to include additional elements in the existing system. It may be necessary, for example, to modernize a system according to new demand levels or new reliability requirements. The problem of minimal cost MSS expansion is very similar to the problem of system structure optimization [135]. The only difference is that each MSS component already contains some working elements. The initial structure of the MSS is defined as follows: each component of type i contains B'_i different subcomponents connected in parallel. Each subcomponent j in its turn contains n'_{ij} identical elements, which are also connected in parallel. Each element is characterized by its PD $\mathbf{g}'_i(j)$, $\mathbf{p}'_i(j)$. The entire initial system structure can, therefore, be defined by a set $\{\mathbf{g}'_i(j), \mathbf{p}'_i(j), n'_{ij}: 1 \le i \le N, 1 \le j \le B'_i\}$ and by a reliability block diagram representing the interconnection among the components.

The optimal MSS expansion problem formulation is the same as in Section 5.1.2 and the GA implementation is the same as in Section 5.1.2.2. The only difference is that for the availability evaluation one should take into account u-functions of both the existing elements and the new elements chosen from the list, while the cost of the existing elements should not be taken into account when the MSS expansion cost is calculated.

Example 5.3

Consider the same coal transportation system that was presented in Example 5.1. The initial structure of this MSS is given in Table 5.5. Each component contains a single subcomponent consisting of identical elements: $B'_i = 1$ for $1 \le i \le 5$. The existing structure can satisfy the demand presented in Table 5.1 with the availability $A(\boldsymbol{w},\boldsymbol{q}) = 0.506$. In order to increase the system's availability to the level of A^*, the additional elements should be included. These elements should be chosen from the list of available products (Table 5.2).

Table 5.5. Parameters of initial system structure

Component no.	Capacity	Availability	Number of parallel elements
1	0.75	0.988	2
2	0.28	0.997	3
3	0.66	0.972	2
4	0.54	0.983	2
5	0.66	0.981	2

The minimal cost MSS expansion solutions for different desired values of system availability A^* are presented in Table 5.6.

Table 5.6. Optimal solutions for MSS expansion problem

	$A^* = 0.95$	$A^* = 0.97$	$A^* = 0.99$
System availability	0.950	0.971	0.990
Expansion cost	0.630	3.244	4.358
	Added elements		
Primary feeders	–	1*6	1*6
Primary conveyors	2*5	1*5, 1*4	1*5
Stackers-reclaimers	–	1*4	1*3
Secondary feeders	1*7	1*7	1*7
Secondary conveyors	1*4	1*4	1*4

Consider, for example, the best solution obtained for $A^* = 0.99$, encoded by the string (0, 0, 0, 0, 0, 1, 0, 0, 0, 0, 0, 1, 0, 0, 1, 0, 0, 0, 0, 0, 0, 0, 1, 0, 0, 0, 0, 0, 1). The corresponding minimal cost system expansion plan that provides the system availability $A = 0.990$ presumes the addition of the primary feeder of version 6, the primary conveyor of version 5, the stacker of version 3, the secondary feeder of version 7 and the secondary conveyor of version 4. The cost of this expansion plan is 4.358.

5.1.4 Optimal Multistage System Expansion

In many cases, when the demand and/or reliability requirements increase with time, the design problem concerns multiple expansions (reinforcements) rather than the construction of complete new systems or their single-stage expansions. While in the problems of system structure optimization or single-stage expansion it is sufficient to ask what should be done and where, in designing the expansion plan an answer is also needed to the question of when.

Using the single-stage system expansion algorithm suggested in Section 5.1.3, one can consider the reinforcement planning problem as a sequence of expansions so that each expansion is a separate problem. However, this may not lead to the optimization of the whole plan, because, in general, partial optimal solutions cannot guarantee an overall optimal solution. Therefore, a more complex method should be developed to solve the system-expansion planning problem in several stages, rather than in a single stage [136].

This section addresses the multistage expansion problem for MSSs. The study period is divided into several stages. At each stage, the demand distribution is predicted. The additional elements chosen from the list of available products may be included into any system component at any stage to increase the total system availability. The objective is to minimize the sum of costs of investments over the study period while satisfying availability constraints at each stage.

5.1.4.1 Problem Formulation

As in the single-stage expansion problem, the initial system structure is defined by a reliability block diagram representing the interconnection among MSS components and by a set $\{\boldsymbol{g}'_i(j), \boldsymbol{p}'_i(j), n'_{ij} : 1 \le i \le N, 1 \le j \le B'_i\}$.

In order to provide the desired level of availability, the system should be expanded at different stages of the study period. Each stage y $(1 \le y \le Y)$ begins $\tau(y)$ years after the base stage (stage 0) and is characterized by its demand distribution $\boldsymbol{w}(y)$, $\boldsymbol{q}(y)$. The timing of stages can be chosen for practical reasons so that $\tau(y)$ must not be evenly spaced. Increasing the number of stages Y during the planning horizon increases the solution flexibility. Indeed, the addition of intermediate stages may allow investments to be postponed without violating the system availability constraints. On the other hand, the increase of number of stages expands drastically the search space and, therefore, slows the algorithm convergence. The recommended strategy is to obtain a number of solutions for different Y in a reasonable time and to choose the best one.

Different versions and numbers of elements may be chosen for expansion of any given system component. The versions can be chosen from the list of elements available in the market. The expansion of system component i at stage y is defined by the numbers $n_{ib}(y)$ of parallel elements of each version b $(1 \le b \le B_i)$ included into the component at this stage The set $\boldsymbol{n}_y = \{ n_{ib}(y) : 1 \le i \le N, 1 \le b \le B_i\}$ defines the entire system expansion at stage y.

For a given set $\boldsymbol{n}_y$ the total cost of the system expansion at stage y can be calculated in present values as

$$C(y) = \frac{1}{(1+\mathrm{IR})^{\tau(y)}} \sum_{i=1}^{N} \sum_{b=1}^{B_i} n_{ib}(y) c_i(b) \tag{5.11}$$

where IR is the interest rate. The total system expansion cost for the given expansion plan $\{\boldsymbol{n}_1,...,\boldsymbol{n}_Y\}$ is

$$C=\sum_{y=1}^{Y}\frac{1}{(1+\mathrm{IR})^{\tau(y)}}\sum_{i=1}^{N}\sum_{b=1}^{B_i}n_{ib}(y)c_i(b) \tag{5.12}$$

The system structure at stage y (after expansion) is defined by a set of elements composing the initial structure $\boldsymbol{n}'=\{n'_{ij}: 1\leq i\leq N, 1\leq j\leq B'_i\}$ and by a set of included elements $\boldsymbol{m}_y=\{m_{ib}(y): 1\leq i\leq N, 1\leq b\leq B_i\}$, where

$$m_{ib}(y)=\sum_{k=1}^{y}n_{ib}(k) \tag{5.13}$$

Having the system structure and demand distribution for each stage y, one can evaluate the MSS performance measures. The problem of system expansion optimization can be formulated as follows

Find the minimal cost system expansion plan that provides the required level O^* of the performance measure O_y at each stage y:

$$C(\boldsymbol{n}_1,...,\boldsymbol{n}_Y)\rightarrow \min$$

$$\text{subject to } f(O_y(\boldsymbol{n}_1,...,\boldsymbol{n}_Y, \boldsymbol{w}(y), \boldsymbol{q}(y)), O^*)=1 \text{ for } 1\leq y\leq Y \tag{5.14}$$

where $f(O_y,O^*)$ is a function representing the desired relation between O_y and O^*.

The problem of system structure optimization can be considered as a special case of this formulation. Indeed, if the set $\boldsymbol{n}'$ of the elements composing the system at stage 0 is empty and $Y=1$, then the problem is reduced to system structure optimization subject to reliability constraints. If $\boldsymbol{n}'$ is empty but $Y\neq 1$, then the problem is that of multistage system structure optimization, which includes both determination of the initial system structure and its expansion plan.

5.1.4.2 Implementing the Genetic Algorithm

The natural way to represent the expansion plan $\boldsymbol{n}_1,...,\boldsymbol{n}_Y$ is to use a string containing integer numbers corresponding to $n_{ib}(y)$ for each type i, version b_i and stage y. Such a string contains $Y\sum_{i=1}^{N}B_i$ elements, which can result in an enormous growth of the length of the string even for problems with a moderate number of available versions of elements and number of stages. Besides, to represent a reasonable solution, such a string should contain a large percentage of zeros, because only small number of elements should be included into the system at each stage. This redundancy causes an increase in the need for computational resources and lowers efficiency of the GA.

In order to reduce the redundancy of expansion plan representation, each inclusion of n elements of version b into the system at stage y is represented by the triplet $\{n, b, y\}$. The maximum number of possible inclusions into each component (diversity factor) K and the maximum number of identical elements in each inclusion n_{max} are defined as parameters which can be preliminarily specified for each problem. Thus, the first K triplets represent the total expansion plan for the first component, the second K triplets for the second component, and so on. The length of string representing a solution is in this case $3KN$, which can be much smaller than that in the first case.

Consider a string $\boldsymbol{a} = (a_1,...,a_{3KN})$ where each element a_j is generated in the range

$$0 \le a_j \le \max\{n_{\max}, Y, \max_{1\le i\le N} B_i\} \tag{5.15}$$

For each system component i, each of K triplets (a_x, a_{x+1}, a_{x+2}), where

$$x = 3K(i-1)+3(j-1) \text{ for } 1\le j\le K \tag{5.16}$$

determines the addition of $\text{mod}_{n_{\max}+1}(a_x)$ elements of version $\text{mod}_{B_i}(a_{x+1})+1$ at stage $\text{mod}_Y(a_{x+2})+1$. This transform provides solution feasibility by mapping each of the variables into a range of its possible values, since $n_{ib}(y)$ varies from 0 to n_{max}, b varies from 1 to B_i and y varies from 1 to Y.

For example, consider a problem with $K = 3$, $Y = 5$, $n_{max} = 4$, where $\max_{1\le i\le N}\{B_i\} = 9$. Substring (7, 3, 2, 0, 2, 6, 1, 7, 0) corresponding to a component for which eight different versions are available, represents addition of $\text{mod}_{4+1}7 = 2$ elements of version $\text{mod}_8 3+1 = 4$ at stage $\text{mod}_5 2+1 = 3$ and $\text{mod}_{4+1}1 = 1$ element of version $\text{mod}_8 7+1 = 8$ at stage $\text{mod}_5 0+1 = 1$. The second triplet (0, 2, 6) represents addition of zero elements and, therefore, should be discarded. The existence of such "dummy" triplets allows the number of additions to each component to vary from 0 to K, providing flexibility of solution representation.

Having the vector $\boldsymbol{a}$, one can determine the MSS structure in each stage y using (5.13) and obtain the system PD using the UGF technique. For this PD and for the given demand distribution $\boldsymbol{w}(y)$, $\boldsymbol{q}(y)$ one obtains the MSS availability A_y. The solution fitness for the estimated values of A_y and C calculated using (5.12) should be determined as

$$M - C(\boldsymbol{a}) - \pi \sum_{y=1}^{Y} [(1+|O^* - O_y(\boldsymbol{a})|)(1-f(O_y(\boldsymbol{a}), O^*)] \tag{5.17}$$

Example 5.4

In this example, we consider the same coal transportation system from Example 5.1. The initial system structure is presented in Table 5.5. Table 5.7 contains the boiler system demand distributions at five different stages and times from the present to the beginning of these future stages.

Because of demand increase, the availability index of the system becomes $A_1(\boldsymbol{w}(1), \boldsymbol{q}(1)) = 0.506$ at the first stage. To provide a desired availability level at all the stages, the system should be expanded. The characteristics of products available in the market for each type of equipment are presented in Table 5.2.

Table 5.7. Demand distributions

Stage no.	$\tau(y)$		Demand distribution			
1	0	w	1.00	0.80	0.50	0.20
		q	0.48	0.09	0.14	0.29
2	3	w	1.20	0.80	0.50	-
		q	0.43	0.32	0.25	-
3	6	w	1.40	1.20	0.80	0.50
		q	0.10	0.39	0.31	0.20
4	9	w	1.40	1.20	0.80	-
		q	0.41	0.24	0.35	-
5	12	w	1.60	1.40	1.00	-
		q	0.30	0.45	0.25	-

The best expansion plans obtained by the GA for different values of A^* are presented in Table 5.8. The interest rate for the problems considered is IR = 0.1. Expansion of each system component i at stage y is presented in the form $n*b$, where n is a number of identical elements of version b to be included into the component at a given stage. The table also contains system availability A_y obtained at each stage, total expansion costs in present values C, and costs of the system expansions at the first stage $C(1)$.

Table 5.8. Optimal MSS multistage expansion plans

	$A^* = 0.95$					$A^* = 0.97$					$A^* = 0.99$				
Expansion cost	$C = 4.127$, $C(1) = 0.645$					$C = 6.519$, $C(1) = 5.598$					$C = 7.859$, $C(1) = 5.552$				
Stage	1	2	3	4	5	1	2	3	4	5	1	2	3	4	5
System availability	0.951	0.951	0.963	0.952	0.960	0.972	0.970	0.987	0.975	0.970	0.995	0.991	0.990	0.996	0.994
Added elements															
Primary feeders		1*4				1*4				1*7	1*6	1*6	2*6		
Primary conveyors	2*5		1*5	1*5		3*5			2*5		1*3		2*4		
Stacker-reclaimers			1*4		1*2	1*4		1*2			1*2			1*3	
Secondary feeders	1*7	1*7				2*7					1*7	2*7			
Secondary conveyors	1*3			1*4		1*3				1*4	1*3		1*3		

One can compare expansion at the first stage of multistage plans with the single-stage expansion plans presented in Table 5.6. The demand distribution for the single-stage problem is the same as that in the first stage of the multistage problem. The solutions for the first stage of the multistage plan differ from ones that are obtained for single-stage expansion. The comparison shows that a single-stage expansion plan is less expensive than the first stage of a multistage plan, which should consider the effect of the first stage expansion on the further stages.

To demonstrate the advantages offered by incremental expansion, the solutions that satisfy reliability constraints for the final stage demand distribution but require all expansion investments at time zero were obtained (Table 5.9). One can see that incremental expansion provides considerably less expensive solutions.

Table 5.9. Optimal MSS single-stage expansion plans

	$A^* = 0.95$	$A^* = 0.97$	$A^* = 0.99$
System availability	0.951	0.970	0.990
Expansion cost	8.660	8.895	10.192
	Added elements		
Primary feeders	1*7	1*4	2*6, 1*7
Primary conveyors	4*5	1*3, 2*4	5*5
Stacker-reclaimers	2*3	2*3	1*2, 1*3
Secondary feeders	2*7	2*7	2*7
Secondary conveyors	2*4	2*4	2*4

5.1.5 Optimal Structure of Controllable Systems

The problem of structure optimization of controllable systems is similar to the problem considered in Section 5.1.2 [105]. Consider a system with fixed resource consumption that consists of an MPS and H RGSs. Each system component (RGS and MPS) can consist of parallel elements with the same functionality. The interaction between the system components is determined by the control rule that provides the maximal possible system performance by choosing the optimal set of working MPS units for any combination of available MPS and RGS units.

Different versions of the elements may be chosen for any given system component from the list of element versions available in the market.

The structure of any one out of H+1 system components is defined by the numbers of the parallel elements of each version chosen for this component. For a given set of chosen elements, the total cost of the system can be calculated using Equation (5.7) and the entire MSS performance measure for any given demand distribution $\boldsymbol{w}$, $\boldsymbol{q}$ can be obtained using the algorithms presented in Section 4.2. The problem of controllable MSS structure optimization is formulated as finding the minimal cost system configuration that provides the required level O^* of the system performance measure O (5.8).

Example 5.5

The MPS may have up to six parallel producing elements (chemical reactors) working in parallel. To perform their task, producing elements require three different resources:

- Power, generated by energy supply subsystem (group of converters).
- Computational resource, provided by control subsystem (group of controllers).
- Cooling water, provided by water supply subsystem (group of pumps).

Each of these RGSs can have up to five parallel elements. Both producing units and resource-generating units may be chosen from the list of products available in the market. Each producing unit is characterized by its availability, productivity, cost and amount of resources required for its work. The characteristics of available producing units are presented in Table 5.10. The resource-generating units are characterized by their availability, generating capacity (productivity) and cost. The characteristics of available resource-generating units are presented in Table 5.11. Each element of the system is considered to be a unit with total failures.

Table 5.10. Parameters of the available MPS units

Version	Cost	Nominal performance	Availability	Resources required w		
				Resource 1	Resource 2	Resource 3
1	9.9	30.0	0.970	2.8	2.0	1.8
2	8.1	25.0	0.954	0.2	0.5	1.2
3	7.9	25.0	0.960	1.3	2.0	0.1
4	4.2	13.0	0.988	2.0	1.5	0.1
5	4.0	13.0	0.974	0.8	1.0	2.0
6	3.0	10.0	0.991	1.0	0.6	0.7

Table 5.11. Parameters of the available RGS units

Type of resource	Version	Cost	Nominal performance	Availability
	1	0.590	1.8	0.980
1	2	0.535	1.0	0.977
	3	0.370	0.75	0.982
	4	0.320	0.75	0.978
	1	0.205	2.00	0.995
2	2	0.189	1.70	0.996
	3	0.091	0.70	0.997
	1	2.125	3.00	0.971
3	2	2.720	2.60	0.973
	3	1.590	2.40	0.971
	4	1.420	2.20	0.976

The demand for final product varies with time. The demand distribution is presented in Table 5.12.

Table 5.12. Demand distribution

w	65	48	25	8
q	0.6	0.1	0.1	0.2

Table 5.13 contains minimal cost solutions for different required levels of system availability. The structure of each subsystem is presented by the list of numbers of versions of the elements included in the subsystem. The estimated availability of the system and its total cost are also presented in the table for each solution.

Table 5.13. Parameters of the optimal solutions for system with different MPS elements

A^*	A	C	System structure			
			MPS	RGS 1	RGS 2	RGS 3
0.950	0.951	27.790	3,6,6,6,6	1,1,1,1,1	1,1,2,3	1,1
0.970	0.973	30.200	3,6,6,6,6,6	1,1,1,1	1,1,2,3	1,1
0.990	0.992	33.690	3,3,6,6,6,6	1,1,1,1	1,1,2,3	4,4
0.999	0.999	44.613	2,2,3,3,6,6	1,1,1	1,2,2	4,4,4

The solutions of the system structure optimization problem when the main producing subsystem can contain only identical elements are presented in Table

5.14 for comparison. Note that when the MPS is composed from elements of different types the same system availability can be achieved by much lower cost. Indeed, using elements with different availability and productivity provides much greater flexibility for optimizing the entire system performance in different states. Therefore, the algorithm for solving the problem for different MPS elements (Section 4.2.2), which requires much greater computational effort, usually yields better solutions than one for identical elements (Section 4.2.1).

Table 5.14. Parameters of the optimal solutions for system with identical MPS elements

A^*	A	C	System Structure			
			MPS	RGS 1	RGS 2	RGS 3
0.950	0.951	34.752	4,4,4,4,4,4	1,4,4,4,4	2,2,2	1,3,3,3,4
0.970	0.972	35.161	4,4,4,4,4,4	1,1,1,4	2,2,2,2	1,3,3,3,4
0.990	0.991	37.664	2,2,2,2	4,4	3,3,3,3	4,4,4
0.999	0.999	47.248	2,2,2,2,2	3,4	2,2	4,4,4,4

5.2 Structure Optimization in the Presence of Common Cause Failures

The reliability enhancement of systems that are the subject of CCFs presumes such an arrangement of system elements that can reduce or limit the influence of coupling factors affecting the system. In many cases, the CCFs occur due to intentional attacks, accidental failures, or errors. The ability of a system to tolerate such impacts is called survivability. This ability is becoming especially important when a system operates in battle conditions, or is affected by a corrosive medium or another hostile environment. The measure of the system's survivability is the probability that the system's acceptability function is equal to one in the presence of the CCFs mentioned.

A survivable system is one that is able to "complete its mission in a timely manner, even if significant portions are incapacitated by attack or accident" [137]. This definition presumes two important things:

- First, the impacts of both external factors (attacks) and internal causes (failures) affect system survivability. Therefore, it is important to take into account the influence of reliability (availability) of system elements on the entire system survivability.

- Second, a system can have different states corresponding to different combinations of failed or damaged elements composing the system. Each state can be characterized by a system performance rate, which is the quantitative measure of a system's ability to perform its task. Therefore, a system should be considered a multi-state one when its survivability is analyzed.

Common destructive factors usually cause total failures of all of the elements belonging to the same CCG. Therefore, adding more redundant parallel elements into the same group, while improving the system's reliability, is not effective from a vulnerability standpoint. The effective way of reducing the influence of coupling factors on the system's survivability is separating elements and protecting them against impacts.

5.2.1 Optimal Separation of System Elements

One of the ways to enhance system survivability is to separate elements with the same functionality (parallel elements). The separation can be performed by spatial dispersion, by encapsulating different elements into different protective casings, *etc.*

Parallel elements not separated from one another are considered to belong to the same CCG. All elements belonging to the same CCG are destroyed in the case of group destruction. The destructions of different CCGs are independent events. Obviously, separation has its price. Allocating all the parallel elements together (within a single CCG) is usually cheaper than separating them. The separation usually requires additional areas, constructions, communications *etc.*

Since system elements of the same functionality can have a different PD, the way in which elements are partitioned into CCGs strongly affects system survivability. In this section, we consider the problem of how to separate the elements of series-parallel MSSs in order to achieve a maximal possible level of system survivability by the limited cost.

5.2.1.1 Problem formulation

An MSS consists of N components connected according to a reliability block diagram. Each component contains different elements connected in parallel. The total number of mutually independent elements in each component i is E_i. Each element j is characterized by its PD $\mathbf{g}_j$, $\mathbf{p}_j$. In order to survive, the MSS should meet demand w. The probability that the demand is met is the system's survivability S.

The elements belonging to component i can be separated into B_i independent groups, where B_i can vary from 1 (all the elements are gathered together) to E_i (all the elements are separated one from another). It is assumed that all of the elements of the ith component belonging to the same CCG can be destroyed by a total CCF with the probability v_i, which characterizes the group's vulnerability. The failures of individual MSS elements and CCFs are considered independent events.

In the general case, the vulnerability of each CCG can be a function of the number of elements belonging to this group. It can also depend on the total number of groups within a component. These dependencies can be easily included into the model. For the sake of simplicity, we consider the case in which all CCGs within the component have the same vulnerability. In this case, the separation increases the composite probability of damage caused by the external impact, but it makes the implications of the impact much less severe. This assumption is relevant, for example, when the separated groups of elements encapsulated into identical protective casings can be destroyed by a corrosive medium. The increase of number of separated groups increases the overall system exposure to the medium (the overall area of contact surface of the casings).

The cost of each CCG $c_i(h)$ depends on both the types of element protected (number of component i) and the number of elements in the group h.

The elements' separation problem for each component i can be considered as a problem of partitioning a set of E_i items into a collection of E_i mutually disjoint subsets. This partition can be represented by the vector $\boldsymbol{x}_i = (x_{ij}: 1\le j\le E_i)$, $1\le x_{ij}\le E_i$, where x_{ij} is the number of the subset to which element j belongs. One can obtain the number of elements in each subset m of component i as

$$h_{im}= \sum_{j=1}^{E_i} 1(x_{ij} = m) \tag{5.18}$$

If the number of CCGs in component i is less than E_i, then for some subset m h_{im} must be equal to zero.

The cost of element separation for the component i can be determined as

$$C_i = \sum_{m=1}^{E_i} c_i(h_{im}) \tag{5.19}$$

where $c_i(0) = 0$ by definition. (Note that we include in the sum the maximal possible number of separated groups E_i. If $B_i<E_i$, the empty groups do not contribute to the cost C_i since $c_i(0) = 0$.)

Concatenation of vectors $\boldsymbol{x}_i$ for $1\le i\le N$ determines the separation of elements in the entire system. The total MSS separation cost can be determined as

$$C_{\text{tot}}(\boldsymbol{x}) = \sum_{i=1}^{N} \sum_{m=1}^{E_i} c_i(h_{im}) \tag{5.20}$$

where $\boldsymbol{x}=\{\boldsymbol{x}_1, \ldots, \boldsymbol{x}_N\}$.

The optimal separation problem is formulated in [112] as finding the vectors $\boldsymbol{x}_1, \ldots, \boldsymbol{x}_N$ that maximize MSS survivability S subject to cost limitation:

$$S(\boldsymbol{x}) \to \max \text{ subject to } C_{\text{tot}}(\boldsymbol{x}) \le C^* \tag{5.21}$$

In general, the problem (5.21) can be formulated for any other system performance measure $O(\boldsymbol{x})$ obtained in presence of CCFs.

5.2.1.2 Implementing the Genetic Algorithm

In the problem considered, element separation is determined by vector $\boldsymbol{x}$ that contains $E = \sum_{i=1}^{N} E_i$ values. The solutions are represented by integer strings $\boldsymbol{a} = (a_1, \ldots, a_E)$. For each $f= j + \sum_{k=1}^{i-1} E_k$ item a_f of the string corresponds to item x_{ij} of the vector $\boldsymbol{x}$ and determines the number of group, the jth element of ith component belongs to. Therefore, all the items a_f of the string $\boldsymbol{a}$, corresponding to component i $(1+\sum_{k=1}^{i-1} E_k \le f \le \sum_{k=1}^{i} E_k)$, should vary in the range $(1, E_i)$. Since the random solution generation procedure can produce strings with elements randomized within the same range, to provide solution feasibility one must use a transformation procedure that makes each string element belonging to the proper interval. This procedure determines the value of x_{ij} as $1+\text{mod}_{E_i}(a_f)$. The range of values produced by the random generation procedure should be $(1, \max_{i=1}^{N} E_i)$.

Consider, for example, an MSS with $N = 2$, $E_1 = 3$, $E_2 = 4$. The string of integers (4, 1, 2, 4, 3, 4, 2), generated in the range (1, 4), after applying the transformation procedure takes the form (2, 2, 3, 1, 4, 1, 3), which corresponds to the separation solution presented in Table 5.15.

Table 5.15. Example of element separation solution

Component 1		Component 2	
CCG	Elements	CCG	Elements
1	-	1	1,3
2	1,2	2	-
3	3	3	4
		4	2

The following procedure determines the fitness value for an arbitrary solution defined by integer string ***a***

1. Assign zero to the total cost C_{tot}.
2. For $1 \le i \le N$:

   2.1. Determine the number of the group for each element of component i:

$$x_{ij} = 1+\text{mod}_{E_i}(a_{j_0+j}),\ 1 \le j \le E_i, \text{ where } j_0 = \sum_{k=1}^{i-1} E_k \tag{5.22}$$

   2.2. For each group m, $1 \le m \le E_i$:

   2.2.1. Create a list of MSS elements belonging to the group. (Their numbers are j_0+j for all j such that $x_{ij} = m$.)

   2.2.2. If the list is not empty ($h_{im}>0$), determine the u-function of the parallel elements belonging to the group using individual u-functions of the elements and the composition operators $\otimes_{\varphi_{par}}$. In order to incorporate the group vulnerability into its u-function, apply the ξ operator (4.58) with $\upsilon = \upsilon_i$ over the u-function of the group.

   2.2.3. Add cost $c_i(h_{im})$ to the total cost C_{tot}.

   2.3. Determine the u-function of the entire component i using the composition operator $\otimes_{\varphi_{par}}$ over the u-functions of groups.

3. Determine the u-function of the entire MSS using the composition operators $\otimes_{\varphi_{par}}$ and $\otimes_{\varphi_{ser}}$ over the u-functions of system components.
4. Determine the system survivability index S for the given demand distribution.
5. In order to let the GA look for the solution with maximal survivability S and with total cost C_{tot} not greater than the required value C^*, evaluate the solution fitness as follows:

$$MS(\boldsymbol{a}) - \pi\,(C_{\mathrm{tot}}(\boldsymbol{a}) - C^*)1(C_{\mathrm{tot}}(\boldsymbol{a}) > C^*) \qquad (5.23)$$

Example 5.6

Consider the series-parallel oil transportation subsystem presented in Figure 5.1. The system belongs to the type of flow transmission MSS with flow dispersion and consists of four components with a total of 16 two-state elements. The nominal performance rates g (oil transmission capacity) and availability indices p of the elements are presented in Table 5.16.

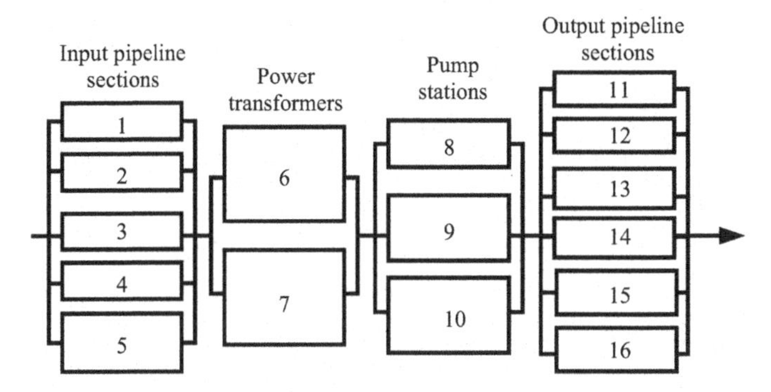

Figure 5.1. Structure of oil transportation subsystem

The system should meet constant demand $w = 5$ (tons per minute). One can see that, in order to enhance the system reliability, redundant elements are included in each of its components.

For each system component, the cost of gathering elements within a group is defined as a function of the number of elements in that group. This function is presented in Table 5.17. Since for each system component i $c_i(n+k)<c_i(n)+c_i(k)$ for arbitrary n and k, by minimizing the number of different groups in which MSS elements are distributed one can minimize the system cost.

Table 5.16. Parameters of the MSS elements

Element no.	1	2	3	4	5	6	7	8	9	10	11	12	13	14	15	16
g	1.2	1.4	1.6	1.8	2	5	5	2	2.5	3.5	1.1	1.1	1.3	1.3	1.4	1.4
p	0.97	0.95	0.94	0.93	0.98	0.98	0.98	0.99	0.97	0.98	0.98	0.98	0.99	0.99	0.98	0.98

Table 5.17. Elements grouping cost

No. of component	No. of elements per group					
	1	2	3	4	5	6
1	5.0	6.0	7.0	8.0	9.0	-
2	9.0	11.0	-	-	-	-
3	2.0	3.0	4.0	-	-	-
4	4.0	4.4	4.7	5.0	5.2	5.5

The optimal separation problem was solved for two different values of group vulnerability $v_i = 0.05$ and $v_i = 0.01$. (The same vulnerability was considered for groups of elements belonging to different MSS components $1 \le i \le N$.) The solutions obtained by the GA are presented in Tables 5.18 and 5.19. These tables contain the distribution of MSS elements between groups within each component (the elements belonging to the same group are put in square brackets) and corresponding MSS survivability and total separation cost.

Table 5.18. System separation solutions for $v = 0.05$

Condition	Separation	S	C_{tot}
Total separation	{[1][2][3][4][5]} {[6][7]} {[8][9][10]} {[11][12][13][14][15][16]}	0.8618	73.0
$S \rightarrow \max$	{[1,4] [2,3] [5]} {[6] [7]} {[8,9,10]} {[11,15] [12,16] [13,14]}	0.8771	52.2
$S \rightarrow \max$, C_{tot}<50	{[1,2,3,4,5]} {[6] [7]} {[8,9,10]} {[11,15] [12,16] [13,14]}	0.8552	44.2
$S \rightarrow \max$, C_{tot}<40	{[1,2,3,4,5]} {[6] [7]} {[8,9,10]} {[11,12,13,14,15,16]}	0.8255	36.5
$C_{tot} \rightarrow \min$	{[1,2,3,4,5]} {[6,7]} {[8,9,10]} {[11,12,13,14,15,16]}	0.7877	29.5

Table 5.19. System separation solutions for $v = 0.01$

Condition	Separation	S	C_{tot}
Total separation	{[1][4][2][3][5]} {[6] [7]} {[8][9][10]} {[11][12][13][14][15][16]}	0.9495	73.0
$S \rightarrow \max$	{[1,4] [2,3] [5]} {[6] [7]} {[8,9,10]} {[11] [12] [13,14] [15] [16]}	0.9509	59.4
$S \rightarrow \max$, C_{tot}<50	{[1,2,3,4,5]} {[6] [7]} {[8,9,10]} {[11,15] [12,16] [13,14]}	0.9453	44.2
$S \rightarrow \max$, C_{tot}<40	{[1,2,3,4,5]} {[6] [7]} {[8,9,10]} {[11,12,13,14,15,16]}	0.9379	36.5
$C_{tot} \rightarrow \min$	{[1,2,3,4,5]} {[6,7]} {[8,9,10]} {[11,12,13,14,15,16]}	0.9290	29.5

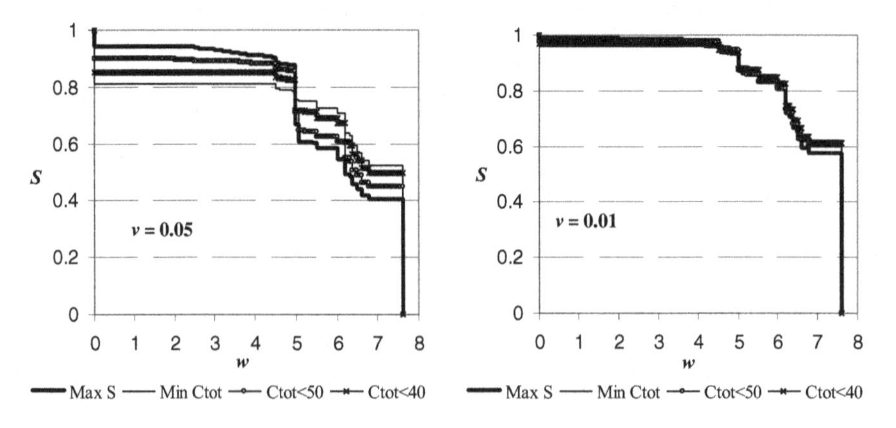

Figure 5.2. MSS survivability functions for different separation solutions

First, the maximal survivability solutions were obtained (no cost limitations). The maximal MSS survivability is achieved when not all the elements are separated from one another. One can see that total separation provides poorer solutions. It is natural that, with an increase in vulnerability, the optimal number of different CCGs decreases, since additional groups contribute to additional overall exposure of the MSS to the impact (despite the fact that the smaller parts can be destroyed by an impact).

The same separation was obtained for both vulnerability values when MSS cost was limited. The less the separation cost allowed, the less the number of separated groups in the optimal solution. If all the elements within each component are gathered in the same group (no separation), then MSS survivability is at its lowest.

In Figure 5.2 one can see the system survivability as a function of demand. For solutions which provide the greatest survivability S for $w = 5$, this index for $w>5$ is less than for the rest of the solutions. Indeed, the optimal separation provides the greatest probability of survival for combinations of elements that correspond to states in which the system performance is equal to five, while combinations corresponding to greater output performance rates are of lower importance.

5.2.2 Optimal System Structure in the Presence of Common Cause Failures

The generalization of the separation problem considered in the previous section is the problem of finding the structure of series-parallel MSSs (including the choice of system elements, their separation, and protection) in order to achieve a desired level of system survivability at a minimal cost [138]. This problem can be considered as a combination of the structure optimization (Section 5.1.2) and optimal separation (Section 5.2.1) problems.

5.2.2.1 Problem Formulation

An MSS consists of N components connected according to a reliability block diagram. Each component can contain a number of different but functionally equivalent mutually independent elements connected in parallel. Each element of type i and version b is characterized by its performance distribution $\boldsymbol{g}_i(b)$, $\boldsymbol{p}_i(b)$ and cost $c_i(b)$. The elements for each component can be chosen from the list of available elements that contains B_i versions of elements for each component i.

The structure of system component i is defined by the numbers of the parallel elements of each version n_{ib} $(1 \le b \le B_i)$ chosen for this component. The vectors $\boldsymbol{n}_i = (n_{i1}, \ldots, n_{iB_i})$ $(1 \le i \le N)$ define the entire system structure. The total cost of the elements chosen for each component i can be calculated as

$$C_i^{\mathrm{el}} = \sum_{b=1}^{B_i} n_{ib} c_i(b) \tag{5.24}$$

The total number of chosen elements in each component i is

$$E_i = \sum_{b=1}^{B_i} n_{ib} \tag{5.25}$$

The elements belonging to component i can be separated into e independent groups, where e can vary from 1 to E_i. For each group m within component i, different levels of protection can be chosen $\gamma_{im} \in \{1,\dots,\Gamma_i\}$. It is assumed that all the elements of component i belonging to the same protection group m can be destroyed by a total CCF with probability $v_i(\gamma_{im})$, which depends on the chosen protection level γ_{im} and characterizes the group vulnerability. The failures of individual elements and CCFs are considered independent events.

The cost σ_m of each CCG m depends on the types of element protected (number of components i), on the number of elements belonging to the group and on the chosen protection level γ_{im}.

The elements' separation for each component i can be represented by the vector $\boldsymbol{x}_i = (x_{ij}: 1 \le j \le E_i)$, $1 \le x_{ij} \le E_i$, where x_{ij} is the number of the group to which element j belongs. The number of elements h_{im} in each separated group m of component i can be obtained using Equation (5.18). Vectors $\boldsymbol{\gamma}_i = (\gamma_{im}: 1 \le m \le E_i)$ determine the protection levels of each group within component i.

The cost of element separation and protection for component i can be determined as

$$C_i^{\text{sep}} = \sum_{m=1}^{E_i} \sigma_i(h_{im}, \gamma_{im}) \tag{5.26}$$

where $\sigma_i(0, \gamma_{im}) = 0$ by definition.

Concatenation of vectors $\boldsymbol{n}_i$, $\boldsymbol{x}_i$ and $\boldsymbol{\gamma}_i$ for $1 \le i \le N$ determines the structure of the entire system, including element separation and protection. The total MSS cost can be determined as

$$C_{\text{tot}}(\boldsymbol{n}, \boldsymbol{x}, \boldsymbol{\gamma}) = \sum_{i=1}^{N} [C_i^{\text{sep}} + C_i^{\text{el}}] = \sum_{i=1}^{N} [\sum_{m=1}^{E_i} \sigma_i(h_{im}, \gamma_{im}) + \sum_{b=1}^{B_i} n_{ib} c_i(b)] \tag{5.27}$$

where $\boldsymbol{n} = \{\boldsymbol{n}_1, \dots, \boldsymbol{n}_N\}$, $\boldsymbol{x} = \{\boldsymbol{x}_1, \dots, \boldsymbol{x}_N\}$ and $\boldsymbol{\gamma} = \{\boldsymbol{\gamma}_1, \dots, \boldsymbol{\gamma}_N\}$.

The optimization problem is formulated as follows. Find vectors $\boldsymbol{n}_i$, $\boldsymbol{x}_i$ and $\boldsymbol{\gamma}_i$ for $1 \le i \le N$ that provide the desired MSS survivability S^* with the minimal cost:

$$C_{\text{tot}}(\boldsymbol{n}, \boldsymbol{x}, \boldsymbol{\gamma}) \to \min \text{ subject to } S(\boldsymbol{n}, \boldsymbol{x}, \boldsymbol{\gamma}) \ge S^* \tag{5.28}$$

5.2.2.2 Implementing the Genetic Algorithm

Consider substring $\boldsymbol{a}_m = (a_{m1}, \dots, a_{mF})$, corresponding to ith system component. Let all the elements of the substring belong to the range $(0, B_i + \Gamma_i)$. We use the following rules to encode the structure of the ith MSS component using the substring.

1. If $a_{ij} \le B_i$, a_{ij} determines the version of element included in the component: $b = a_{ij}$ ($a_{ij} = 0$ means that no element is included).
2. If $a_{ij} > B_i$, a_{ij} constitutes a separator and determines the protection level of the group of elements: $\gamma = a_{ij} - B_i$.
3. The adjacent string elements located between the beginning of the string and separator or between two separators are considered to belong to the same group with the protection level determined by the right separator.
4. The last element of each substring a_{iF} is always considered to be a separator. The corresponding protection level is determined as $\gamma = \mathrm{mod}_{\Gamma_i}(a_{iF})+1$.

In order to determine an arbitrary structure of component i that contains up to E_i elements, one has to use substring $\boldsymbol{a}_i$ of length $F = 2E_i$ (for the case when all the elements are separated). The string of this length, corresponding to exactly E_i elements gathered within a single protection group, contains one separator (a_{iF}) and E_i-1 zeros.

Concatenating substrings $\boldsymbol{a}_i$ for $i = 1, \ldots, N$, one obtains the string $\boldsymbol{a}$ which determines the structure of the entire MSS. In order to allow all the string elements distributed within the same range to represent feasible solutions, we determine this range as $(0, \max_{1 \le i \le N} B_i + \max_{1 \le i \le N} \Gamma_i)$. When the string is decoded, we transform each string element a^*_{ij}, corresponding to ith component in the following way:

$$a_{ij} = \mathrm{mod}_{B_i+\Gamma_i+1}(a^*_{ij}) \tag{5.29}$$

and apply rules 1-4 to the obtained value of a_{ij}. The transform (5.29) ensures that each a_{ij} belongs to the range $(0, B_i+\Gamma_i)$.

Example 5.7

Consider, an MSS with $N = 2$, $B_1 = 3$, $\Gamma_1 = 2$, $B_2 = 4$, $\Gamma_2 = 4$. Let $E_1 = E_2 = 5$ ($F = 10$). The string of integers generated in the range (0, 8)

(4, 8, 1, 0, 5, 6, 3, 0, 1, 0, 3, 0, 8, 7, 1, 5, 4, 2, 1, 6)

after transformation takes the form

(4, 2, 1, 0, 5, 0, 3, 0, 1, 0, 3, 0, 8, 7, 1, 5, 4, 2, 1, 6)

After decoding the string using rules 1-4, we obtain

($\underline{1}$, 2, 1, 0, $\underline{2}$, 0, 3, 0, 1, $\underline{1}$, 3, 0, $\underline{4}$, $\underline{3}$, 1, $\underline{1}$, 4, 2, 1, $\underline{3}$)

(separators are underlined and their values are replaced with corresponding values of protection levels), which corresponds to the structure presented in Table 5.20. One can see that the string determines two empty groups for which protection level values have no meaning and should not be considered.

Table 5.20. Example of MSS structure encoded by integer string

No. of component	No. of group	Versions of elements belonging to the group	Protection level
	1	-	1
1	2	2, 1	2
	3	3, 1	1
	1	3	4
2	2	-	3
	3	1	1
	4	4, 2, 1	3

The following procedure determines the fitness value for an arbitrary solution defined by integer string $\boldsymbol{a}$:

1. Assign 0 to the total cost C_{tot}.
2. For $1 \le i \le N$:
   2.1. Decode substring $\boldsymbol{a}_i$, containing elements $a_{(i-1)F+1}$ to a_{iF}, of string $\boldsymbol{a}$ to obtain versions of elements belonging to different groups and corresponding protection levels.
   2.2. For each nonempty group m:
      2.2.1. Determine cost of elements it contains in accordance with their versions and define the u-functions of these elements in accordance with their performance distributions.
      2.2.2. Determine the u-function for the entire group using composition operator $\otimes_{\varphi_{\text{par}}}$.
      2.2.3. Determine group vulnerability $v_i(\gamma_{im})$ as a function of chosen protection level and apply the operator ξ (4.58) in order to obtain the u-function for the protected group.
   2.3. Obtain the u-function of the component using $\otimes_{\varphi_{\text{par}}}$ operator over the u-functions of the protected groups.
   2.4. Determine the cost of component in accordance with (5.24) and (5.26) and add this value to C_{tot}.
3. Determine the u-function of the entire MSS $U(z)$ by applying the reliability block diagram method.
4. Determine the MSS survivability for the system u-function and given demand distribution.
5. Determine the solution fitness as a function of the MSS cost and survivability as

$$M - C_{\text{tot}}(\boldsymbol{a}) - \pi(1+S^* - S)\times 1(S < S^*) \tag{5.30}$$

Example 5.8

A series-parallel MSS (power substation) consists of four components: power transformers; capacitor banks; output medium voltage line sections; blocks of commutation equipment.

Each component should be built from elements of certain functionality (all the elements are elements with total failure). The cost, capacity and availability of elements that can be included in each component are presented in Table 5.21.

Table 5.21. Parameters of available elements

No. of component	No. of version	Capacity	Availability	Cost
1	1	1.2	0.97	3.1
	2	1.6	0.92	4.2
	3	1.8	0.94	4.7
	4	2.0	0.93	5.0
	5	5.0	0.86	11.0
	6	5.0	0.91	14.5
2	1	1.8	0.98	3.1
	2	3.6	0.98	6.0
	3	5.4	0.96	8.8
3	1	1.4	0.9	6.6
	2	1.6	0.93	7.0
	3	1.8	0.91	7.9
	4	2.0	0.95	9.4
4	1	1.4	0.86	2.6
	2	2.6	0.91	6.0
	3	3.8	0.93	7.9
	4	5.0	0.85	9.4

Within each component, the elements can be separated in an arbitrary way, but each new protection group requires additional investment. For the sake of simplicity in this example, the cost of separation and/or protection for each protection group does not depend on the number of elements it contains, but does depend strictly on the level of protection provided for this group. The descriptions and costs of different protection levels available for each component and the vulnerabilities corresponding to these protection levels are presented in Table 5.22. The system should meet constant demand $w = 5$.

Table 5.22. Characteristics of available protection levels

No. of component	Type of component	Protection level	Protection description	Vulnerability	Cost
1	Transformers	1	Outdoor location	0.35	0.1
		2	Indoor location	0.15	4.1
		3	Underground	0.05	15.7
2	Capacitors	1	Outdoor location	0.01	1.0
3	Lines	1	Overhead	0.60	1.0
		2	Overhead insulated	0.35	5.5
		3	Underground	0.15	17.0
4	Commutation blocks	1	Outdoor location	0.10	1.1
		2	Indoor location	0.03	4.2

The solutions of the optimization problem for four different values of desired system survivability $S^* \in \{0.85, 0.90, 0.95, 0.99\}$ are presented in Figure 5.3. In this figure each system element is marked with its version number and each protection group is encased into a rectangle (dashed for protection level 1, regular for protection level 2 and double lined for protection level 3). In Figure 5.4 the system survivability functions $S(w)$ for the four structures obtained are presented.

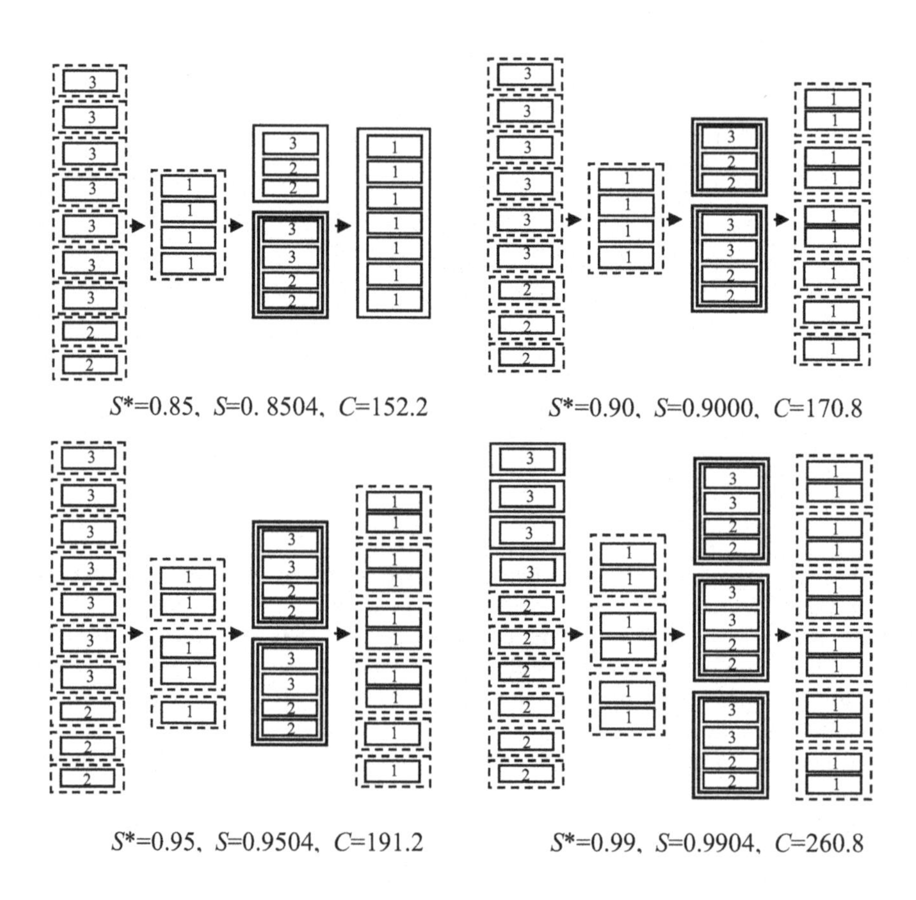

Figure 5.3. Lowest cost MSS structures obtained by the GA

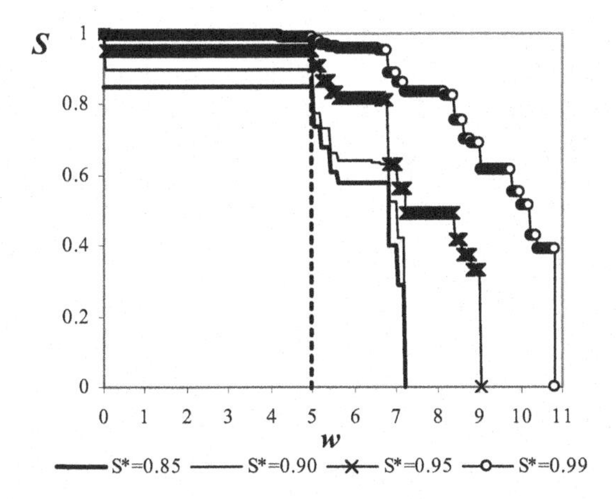

Figure 5.4. MSS survivability functions $S(w)$ for four obtained structures

5.2.3 Optimal Multilevel Protection

In systems consisting of nonidentical elements and having complex multilevel protection, different protections play different roles in providing for the system's survivability. Subject to investment cost limitations, one usually has to find a minimal cost configuration of protections that provides desired system survivability [139].

5.2.3.1 Problem Formulation

Assume that H different protections can be installed in a given series-parallel system and for each protection h its cost c_h and vulnerability v_h are known. Define a binary vector $\boldsymbol{a}$ in which $a(h) = 1$ ($1 \le h \le H$) means that protection h is installed and $a(h) = 0$ means that the protection is not installed. Note that in the model considered the protection decisions are binary (either protect or not) and the parameters of each protection are fixed.

For the given system structure and the given parameters of system elements and protections the system's survivability depends only on the vector $\boldsymbol{a}$. The total protection cost is

$$C_{\text{tot}}(\boldsymbol{a}) = \sum_{h=1}^{H} c_h a(h) \tag{5.31}$$

The optimal protection problem is formulated as finding the minimal cost set of protections that provides the required level of the system survivability:

$$C_{\text{tot}}(\boldsymbol{a}) \to \min \text{ subject to } S(\boldsymbol{a}) \ge S^* \tag{5.32}$$

where S^* is a desired level of system survivability.

5.2.3.2 Implementing the Genetic Algorithm

In the problem considered, the protections chosen are determined by binary vectors $\boldsymbol{a}$ that contain H elements. Any binary vector $\boldsymbol{a}$ can represent a solution in the GA. For each given vector $\boldsymbol{a}$ the GA solution decoding procedure determines system protection cost using Equation (5.31) and evaluates the system survivability $S(\boldsymbol{a})$ using the algorithm described in Section 4.4.3. In order to let the GA look for the solution with minimal total cost and with $S(\boldsymbol{a})$ not less than the required value S^*, the solution quality (fitness) is evaluated as follows:

$$M - C(\boldsymbol{a}) - \pi(1 + S^* - S(a)) \times 1(S(a) < S^*) \tag{5.33}$$

Example 5.9

Consider a series-parallel system consisting of 12 multi-state elements (Figure 5.5). The performance distributions of the elements are presented in Table 5.23. Each element can have individual protection. In addition, different groups of elements can be protected. The list of possible protections is presented in Table 5.24. This list includes the set of protected elements, the expected survivability and the cost of each protection.

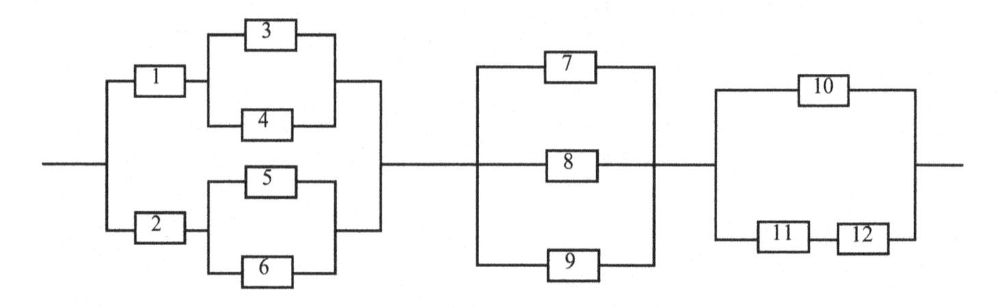

Figure 5.5. Structure of series-parallel system

All of the protections can be chosen independently except protections 20, 21, 22 and 27, 28. Only one out of three protections 21, 22 and 23 can be chosen in order to protect two out of three elements from the set {7, 8, 9}. Similarly, only one out of two protections 27 and 28 can be chosen in order to protect the first or last pair of subsystems out of three subsystems connected in series. This constraint can be easily taken into account in the GA solution decoding procedure by using the following rules:

if $a(20) = 1$ assign $a(21) = a(22) = 0$

then if $a(21) = 1$ assign $a(22) = 0$

and if $a(27) = 1$ assign $a(28) = 0$.

Table 5.23. Performance distributions of available elements

No. of element (j)	1		2		3		4		5		6	
State (k)	p_{jk}	g_{jk}	p_{jk}	g_{jk}	p_{jk}	g_{jk}	p_{jk}	g_{jk}	p_{jk}	g_{jk}	p_{jk}	g_{jk}
1	0.75	7	0.75	7	0.75	4	0.75	4	0.75	4	0.75	4
2	0.15	5	0.15	5	0.15	2	0.15	2	0.15	2	0.15	2
3	0.05	3	0.05	3	0.1	0	0.1	0	0.1	0	0.1	0
4	0.05	0	0.05	0	-	-	-	-	-	-	-	-
No. of element (j)	7		8		9		10		11		12	
State (k)	p_{jk}	g_{jk}	p_{jk}	g_{jk}	p_{jk}	g_{jk}	p_{jk}	g_{jk}	p_{jk}	g_{jk}	p_{jk}	g_{jk}
1	0.85	5	0.85	5	0.10	6	0.80	8	0.95	8	0.85	10
2	0.05	3	0.05	3	0.70	4	0.15	5	0.05	0	0.10	7
3	0.10	0	0.10	0	0.15	2	0.05	0	-	-	0.05	0
4	-	-	-	-	0.05	0	-	-	-	-	-	-

Table 5.24. Characteristics of possible protections

No. of protection	Set of protected elements	Protection vulnerability	Protection cost	No. of protection	Set of protected elements	Protection vulnerability	Protection cost
1	1	0.05	1.5	16	5, 6	0.15	2.2
2	2	0.05	1.5	17	1, 3, 4	0.20	3.8
3	3	0.10	1.0	18	2, 5, 6	0.20	3.8
4	4	0.10	1.0	19	1-6	0.40	7.2
5	5	0.10	1.0	20	7, 8	0.30	3.0
6	6	0.10	1.0	21	8, 9	0.30	3.0
7	7	0.15	0.8	22	7, 9	0.30	3.0
8	8	0.15	0.8	23	7-9	0.15	4.2
9	9	0.15	0.8	24	11, 12	0.35	3.0
10	10	0.05	2.2	25	11, 12	0.10	5.0
11	11	0.15	1.3	26	10-12	0.25	4.3
12	12	0.15	1.5	27	1-9	0.35	5.2
13	3, 4	0.40	1.5	28	7-12	0.35	5.2
14	3, 4	0.15	2.2	29	1-12	0.30	7.0

Table 5.25. Solutions obtained for $w = 5$

S^*	S	C_{tot}	Chosen protections
0.85	0.862	16.8	1, 2, 4, 5, 6, 7, 8, 10, 29
0.90	0.901	20.1	1-10, 15, 29
0.95	0.950	42.7	1-12, 17, 18, 23, 26, 27, 29
Max possible	0.953	79.7	All except 21, 22, 28

Table 5.26. Solutions obtained for $w = 6$

S^*	S	C_{tot}	Chosen protections
0.80	0.807	17.2	10, 11, 12, 27, 29
0.85	0.851	22.6	1-5, 7-12, 16, 29
0.90	0.901	35.1	1-12, 23, 26, 27, 29
Max possible	0.913	79.7	All except 21, 22, 28

Table 5.27. Solutions obtained for $w = 7$

S^*	S	C_{tot}	Chosen protections
0.75	0.752	14.4	10, 27, 29
0.80	0.813	21.4	1-12, 29
0.85	0.857	28.8	1-10, 25, 27, 29
Max possible	0.882	79.7	All except 21, 22, 28

The solutions obtained for different w and S^* are presented in Tables 5.25, 5.26, and 5.27, and include the system survivability obtained, the total protection cost and the list of chosen protections. The maximal possible system survivability (achieved when all the protections except 21, 22 and 28 are chosen) is given for each demand level for comparison. The system performance distributions for the solutions obtained are given in Figure 5.6. The optimal protection configurations for $S^* = 0.85$ are presented in Figure 5.7.

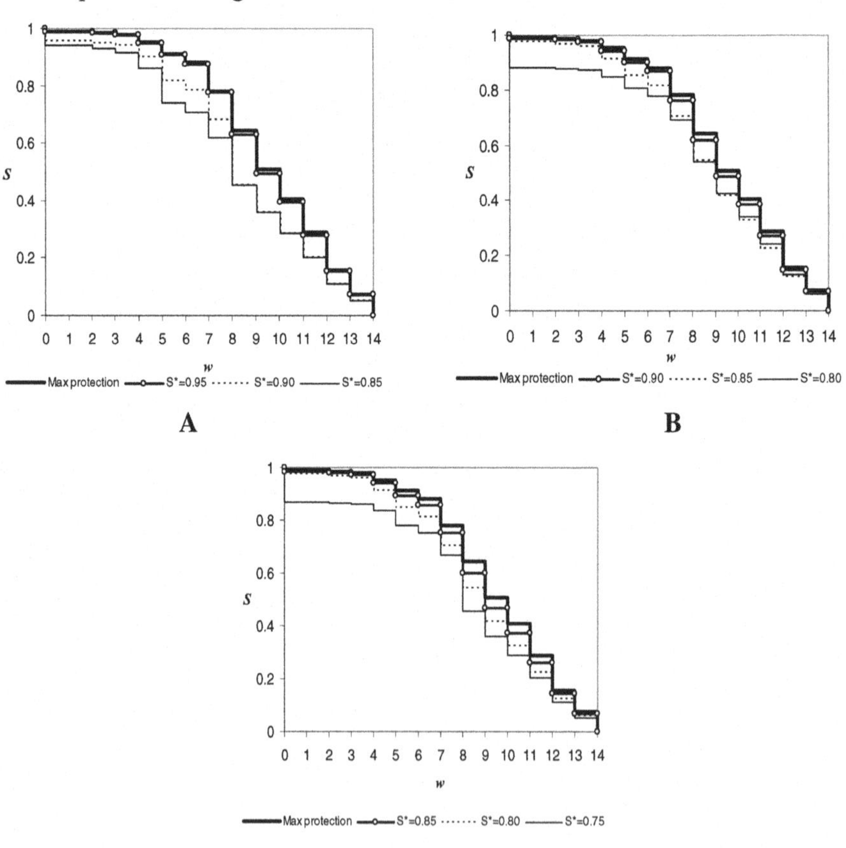

Figure 5.6. System performance distributions for solutions obtained for $w = 5$ (A), $w = 6$ (B) and $w = 7$ (C)

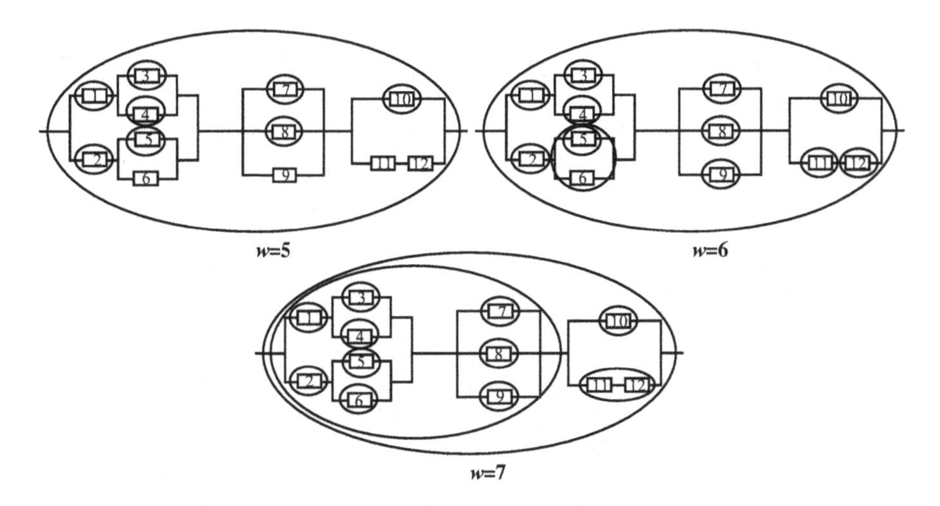

Figure 5.7. Configurations of protections obtained for $S^* = 0.85$.

Observe that the greater the survivability level achieved, the greater is the cost of further survivability improvement. The cost-survivability curves are presented in Figure 5.8. Each point of these curves corresponds to an optimal or near-optimal solution obtained by the GA.

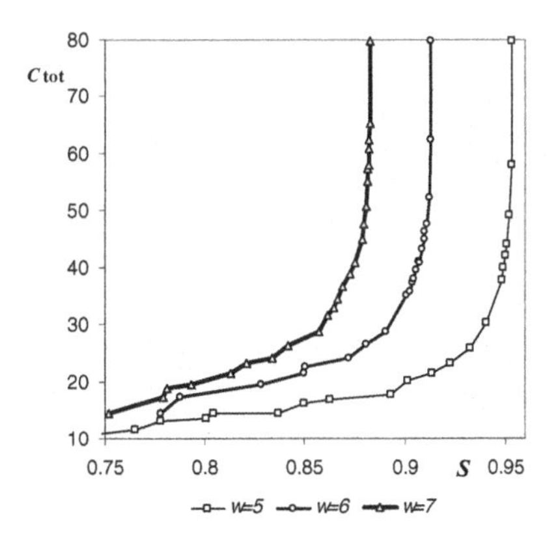

Figure 5.8. Survivability-cost curves for the optimal solutions

5.2.4 Optimal Structure of System with Multilevel Protection

When systems aimed at performing their task in a hostile environment are developed, the designer has to make a decision concerning the system's structure. The proper combination of system structure and system protection allows for its greatest survivability.

In the generalized structure optimization problem one has to:

- find the optimal system structure by choosing the appropriate product (version of a system element) from the list of available products for each type of equipment;
- allocate the chosen elements among the different protection groups and define the hierarchy of the multilevel protection in the system;
- choose the method of protection for each protection group.

5.2.4.1 Problem Formulation

Consider a series-parallel system consisting of N components connected in series. Each component contains elements connected in parallel. Different versions and numbers of elements may be chosen for any given system component. Elements are characterized by performance distributions and costs, according to their versions. The states of MSS elements are mutually statistically independent.

For each component i, there exists a list of B_i different versions of available elements. A performance distribution $\boldsymbol{g}_i(b)$, $\boldsymbol{p}_i(b)$ ($1\le b\le B_i$) and cost $c_i(b)$ can be specified for each version b of element of type i. Let E_i be the maximal possible number of elements in component i. The structure of this component can be defined by a vector containing numbers of versions of elements chosen for the component $\boldsymbol{b}_i$ = (b_{ij}: $1\le j\le E_i$), where $0\le b_{ij}\le B_i$. Including a dummy version zero corresponding to the absence of elements allows one to represent a different number of elements chosen for component i by vectors $\boldsymbol{b}_i$ of the same length E_i. The total cost of elements chosen for the component i is

$$C_i^{\mathrm{el}} = \sum_{j=1}^{E_i} c_i(b_{ij}) \tag{5.34}$$

The chosen E_i elements can be separated into E_i independent protection groups (some of these groups can be empty and some can contain several elements). The partitioning among the groups can be represented by vector $\boldsymbol{r}^1_i = (r^1_{ij}$: $1\le j\le E_i$), where r^1_{ij} is the number of the first-level group to which element j belongs. The groups are protected by first-level protection. For each group m within component i different methods of protection γ^1_{im} can be chosen $\gamma^1_{im}\in\{0,\ldots,\Gamma^1_i\}$, where $\gamma^1_{im}=0$ corresponds to absence of protection. The vector $(\gamma^1_{i1},\ldots,\gamma^1_{iE_i})$ defines the choice of the first-level protection methods in component i.

The E_i protected first-level groups can be further separated into E_i second-level groups and protected using methods $\gamma^2_{im}\in\{0,\ldots,\Gamma^2_i\}$. These second-level protection groups can be further separated and protected by the third-level protection and so forth up to the L_i protection level.

The vectors $\boldsymbol{r}^e_i$ = (r^e_{ij}: $1\le j\le E_i$) and $\boldsymbol{\gamma}^e_i$ = (γ^e_{im}: $1\le j\le E_i$) determine the separation and protection of each level e. Having the vector $\boldsymbol{r}^e_i$ one can obtain the number of nonempty inner level groups belonging to each group m of level e: x^e_{im}.

Each e-level protection in component i that uses protection method γ can be destroyed with probability $v^e_i(\gamma)$. Therefore, the vulnerability of protection of

e-level protection group m is $v^e_i(\gamma^e_{im})$. Any unprotected group has vulnerability $v^e_i(0) = 1$.

The cost of protection of the e-level group m depends on the types of element protected (number of component i), on the number of nonempty inner groups belonging to the given group x^e_{im} and on the chosen protection method γ^e_{im}. This cost can be expressed as $c^e_i(x^e_{im}, \gamma^e_{im})$, where $c^e_i(0, \gamma^e_{im}) = 0$ by definition.

Assume that component i has L_i protection levels. Each level has E_i protection groups (some of these groups can be empty). The total cost of protections in the component i is

$$C_i^{\text{prot}} = \sum_{e=1}^{L_i} \sum_{m=1}^{E_i} c^e{}_i(x^e{}_{im}, \gamma^e{}_{im}) \tag{5.35}$$

In a similar way one can define the cost protection of components connected in series $C_{\text{comp}}^{\text{prot}}$. Considering each component as equivalent element that can belong to any protection group one can define the partition of components among e-level groups using vector $\boldsymbol{r}^e_{\text{comp}}$ and the methods of protection for these groups using vector $\boldsymbol{\gamma}^e_{\text{comp}}$ (the number of protection groups on this level can not be greater than N). The vectors $\boldsymbol{r}^e_{\text{comp}}$ determine number of components in each group m: $x^e_{\text{comp},m}$.

Having the vectors $\boldsymbol{r}^e_i$ and $\boldsymbol{\gamma}^e_i$ for $1 \le e \le L_i$ and $1 \le i \le N$, the vectors $\boldsymbol{r}^e_{\text{comp}}$ and $\boldsymbol{\gamma}^e_{\text{comp}}$ for $1 \le e \le L_{\text{comp}}$ and the vectors $\boldsymbol{b}_i$ for $1 \le i \le N$ one can determine the structure of the entire system. The total MSS cost can be determined as

$$C_{tot}(\beta, \rho, \eta) = \sum_{i=1}^{N} (C_i^{\text{el}} + C_i^{\text{prot}}) + C_{\text{comp}}^{\text{prot}} = \sum_{i=1}^{N} [\sum_{j=1}^{E_i} c_i(b_{ij})$$
$$+ \sum_{e=1}^{L_i} \sum_{m=1}^{E_i} c^e{}_i(x^e{}_{im}, \gamma^e{}_{im})] + \sum_{e=1}^{L_{\text{comp}}} \sum_{m=1}^{N} c^e{}_{\text{comp}}(x^e{}_{\text{comp},m}, \gamma^e{}_{\text{comp},m})] \tag{5.36}$$

where

$$\beta = \{\boldsymbol{b}_1, \ldots, \boldsymbol{b}_N\}$$
$$\rho = \{\boldsymbol{r}_1^1, \ldots, \boldsymbol{r}_1^{L_1}, \ldots, \boldsymbol{r}_N^1, \ldots, \boldsymbol{r}_N^{L_N}, \boldsymbol{r}_{\text{comp}}^1, \ldots, \boldsymbol{r}_{\text{comp}}^{L_{\text{comp}}}\}$$
$$\eta = \{\boldsymbol{\gamma}_1^1, \ldots, \boldsymbol{\gamma}_1^{L_1}, \ldots, \boldsymbol{\gamma}_N^1, \ldots, \boldsymbol{\gamma}_N^{L_N}, \boldsymbol{\gamma}_{\text{comp}}^1, \ldots, \boldsymbol{\gamma}_{\text{comp}}^{L_{\text{comp}}}\} \tag{5.37}$$

The problem of structure optimization is formulated in [140] as finding the sets β, ρ, η that provide the desired level of MSS survivability S^* with the minimal cost:

$$C_{\text{tot}}(\beta, \rho, \eta) \to \min \text{ subject to } S(\beta, \rho, \eta) \ge S^* \tag{5.38}$$

5.2.4.2 Implementing the Genetic Algorithm

As described in the previous section, three things determine the structure of an MSS:

- the list of version numbers of elements chosen for each component (determined by the set β),
- partition of elements and lower level protection groups between protection groups of each level within each component and partition of components and lower level groups between protection groups consisting of serially connected components (determined by the set ρ),
- methods of protection for each protection group (determined by the set η).

These three sets can be combined within an integer string $\boldsymbol{a} = (\boldsymbol{a}_1, \ldots, \boldsymbol{a}_N, \boldsymbol{a}_{\text{comp}})$, where $\boldsymbol{a}_i$ ($1 \le i \le N$) is a concatenation of the vectors $\boldsymbol{b}_i$, $\boldsymbol{r}^e{}_i$ and $\boldsymbol{\gamma}^e{}_i$ (for $1 \le i \le L_i$) and $\boldsymbol{a}_{\text{comp}}$ is a concatenation of the vectors $\boldsymbol{r}^e{}_{\text{comp}}$ and $\boldsymbol{\gamma}^e{}_{\text{comp}}$ (for $1 \le e \le L_{\text{comp}}$). Each one of the vectors $\boldsymbol{b}_i$, $\boldsymbol{r}^e{}_i$ and $\boldsymbol{\gamma}^e{}_i$ has E_i elements, while each one of vectors $\boldsymbol{r}^e{}_{\text{comp}}$ and $\boldsymbol{\gamma}^e{}_{\text{comp}}$ has N elements. The total length of the string representing the solution is

$$\sum_{i=1}^{N}(E_i + 2L_i E_i) + 2NL_{\text{comp}} \tag{5.39}$$

Note that all of the string values make sense only in the case when each component has the maximal possible number of elements and all of these elements are separated from one another (all of the protection groups are not empty). Otherwise, some of values of the string should be ignored by the decoding procedure.

In the feasible solution the values of substring $\boldsymbol{b}_i$ should be distributed in the range $(0, B_i)$, the values of substrings $\boldsymbol{r}^e{}_i$ and $\boldsymbol{r}^e{}_{\text{comp}}$ should be distributed in the range $(1, E_i)$ and $(1, N)$ respectively, and the values of substrings $\boldsymbol{\gamma}^e{}_m$ and $\boldsymbol{\gamma}^e{}_{\text{comp}}$ should be distributed in the range $(0, \Gamma^e{}_i)$ and $(0, \Gamma^e{}_{\text{comp}})$ respectively.

In order to allow all the string elements distributed within the same range to represent feasible solutions, we determine this range as

$$(0, \max\{N, \max_{1 \le i \le N}\{B_i, E_i, \max_{1 \le e \le L_i} \Gamma_i^e\}, \max_{1 \le e \le L_{\text{comp}}} \Gamma_{\text{comp}}^e\}) \tag{5.40}$$

When the string is decoded, we transform each string element a corresponding to each substring $\boldsymbol{b}_i$, $\boldsymbol{r}^e{}_i$, $\boldsymbol{r}^e{}_{\text{comp}}$, $\boldsymbol{\gamma}^e{}_m$ and $\boldsymbol{\gamma}^e{}_{\text{comp}}$ in the following way:

$$b_{ij} = \text{mod}_{B_i+1}(a), \quad r^e{}_{ij} = \text{mod}_{E_i}(a) + 1, \quad r^e{}_{\text{comp},j} = \text{mod}_N(a) + 1$$

$$\gamma^e{}_{im} = \text{mod}_{\Gamma^e{}_i+1}(a), \quad \gamma^e{}_{\text{comp},m} = \text{mod}_{\Gamma^e{}_{\text{comp}}+1}(a) \tag{5.41}$$

The unification of the distribution range of all the string elements simplifies the string generation procedure, as well as mutation and crossover operators.

Example 5.10

Consider a series-parallel MSS with $N = 2$, $B_1 = B_2 = 3$, $E_1 = E_2 = 3$, $L_1 = 2$, $L_2 = 1$, $L_{\text{comp}} = 2$, $\Gamma_1^1 = 3$, $\Gamma_1^2 = 2$, $\Gamma_2^1 = 3$, $\Gamma_{\text{comp}}^1 = \Gamma_{comp}^2 = 2$. In this example we will use the notation PGm_i^e to designate protection group m of level e within component i.

According to (5.40) the solution encoding string should consist of integer numbers distributed in the range (0, 3). Consider, for example, the following string $\boldsymbol{a}$ obtained after transformation (5.41):

(1,3,1, 3,3,2, 2,3,2, 1,2,2, 1,1,2, 2,0,2, 1,1,1, 1,2,3, 1,2, 0,2, 1,1, 2,2).

In the first part of this string ($\boldsymbol{a}_1$) the substrings (1,3,1), (3,3,2), (2,3,2), (1,2,2) and (1,1,2) represent $\boldsymbol{b}_1$, $\boldsymbol{r}^1{}_1$, $\boldsymbol{\gamma}^1{}_1$, $\boldsymbol{r}^2{}_1$ and $\boldsymbol{\gamma}^2{}_1$ respectively. The versions of elements chosen to fill positions 1, 2 and 3 of the first component are, according to $\boldsymbol{b}_1$, 1, 3 and 1 respectively. According to $\boldsymbol{r}^1{}_1$, elements located at positions 1 and 2 belong to PG$3^1{}_1$, and element located in position 3 belongs to PG$2^1{}_1$ (PG$1^1{}_1$ is empty). According to $\boldsymbol{\gamma}^1{}_1$, the PG$2^1{}_1$ has protection method 3 and the PG$3^1{}_1$ has protection method 2 (the first number of $\boldsymbol{\gamma}^1{}_1$ is ignored because PG$1^1{}_1$ is empty). Substring $\boldsymbol{r}^2{}_1$ defines the distribution of first-level protection groups among the groups of the second level. According to $\boldsymbol{r}^2{}_1$, PG$2^1{}_1$ and PG$3^1{}_1$ belong to PG$2^2{}_1$ (first element of $\boldsymbol{r}^2{}_1$ is ignored because PG$1^1{}_1$ is empty). PG$1^2{}_1$ and PG$3^2{}_1$ remain empty. The protection method for PG$2^2{}_1$ according to second element of $\boldsymbol{\gamma}^2{}_1$ is 1 (the first and third elements of $\boldsymbol{\gamma}^2{}_1$ are ignored).

In the second part of the string ($\boldsymbol{a}_2$) substrings (2,0,2), (1,1,1) and (1,2,3) represent $\boldsymbol{b}_2$, $\boldsymbol{r}^1{}_2$ and $\boldsymbol{\gamma}^1{}_2$ respectively. Two elements of version 2 are chosen to fill positions 1 and 3 of the second component according to $\boldsymbol{b}_2$ (0 corresponds to absence of any element). According to $\boldsymbol{r}^1{}_2$, elements located at positions 1 and 3 belong to PG$1^1{}_2$ (PG$2^1{}_2$ and PG$3^1{}_2$ remain empty). According to $\boldsymbol{\gamma}^1{}_2$, the PG$1^1{}_2$ has protection method 1 (the second and third numbers of $\boldsymbol{\gamma}^1{}_2$ are ignored because PG$2^1{}_2$ and PG$3^1{}_2$ are empty).

In the last part of string ($\boldsymbol{a}_{\text{comp}}$), substrings (1, 2), (0, 2), (1, 1) and (2, 2) represent $\boldsymbol{r}^1{}_{\text{comp}}$, $\boldsymbol{\gamma}^1{}_{\text{comp}}$, $\boldsymbol{r}^2{}_{\text{comp}}$ and $\boldsymbol{\gamma}^2{}_{\text{comp}}$ respectively. According to $\boldsymbol{r}^1{}_{\text{comp}}$, component 1 belongs to PG$1^1{}_{\text{comp}}$ and component 2 belongs to PG$2^1{}_{\text{comp}}$. According to $\boldsymbol{\gamma}^1{}_{\text{comp}}$, the PG$1^1{}_{\text{comp}}$ has no protection (protection method 0) and the PG$2^1{}_{\text{comp}}$ has protection method 2. According to $\boldsymbol{r}^2{}_{\text{comp}}$, both PG$1^1{}_{\text{comp}}$ and PG$2^1{}_{\text{comp}}$ belong to PG$1^2{}_{\text{comp}}$. According to $\boldsymbol{\gamma}^2{}_{\text{comp}}$, this protection group has protection method 2 (the second number of $\boldsymbol{\gamma}^2{}_{\text{comp}}$ is ignored because PG$2^2{}_{\text{comp}}$ is empty).

One can see the structure of the system encoded by the given string in Figure 5.9. In this figure, each protection denoted by an ellipse is numbered according to the protection method chosen.

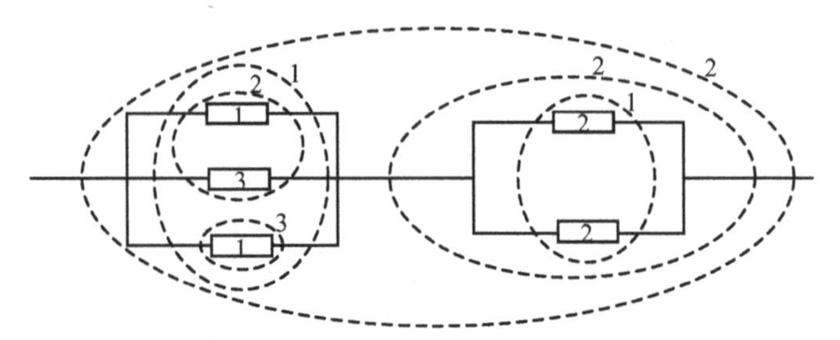

Figure 5.9. MSS structure encoded by the integer string

The following procedure determines the fitness value for an arbitrary solution defined by integer string $\boldsymbol{a}$:

1. Assign 1 to the number of component i. Assign 0 to the total cost C_{tot}.

2. Decode substring $\boldsymbol{a}_i$ and obtain versions of elements belonging to the component i, structure of protection groups and corresponding protection methods at each protection level. Determine cost and performance distributions of elements in accordance with their versions and define d-functions of these elements.

3. Calculate the cost of elements and protections in the component i using Equation (5.34) and (5.35) and add this cost to C_{tot}.

4. For each protection level e (from $e = 1$ to $e = L_i$), obtain d-functions of protected groups and replace them by equivalent elements using operators (4.60) and (4.62) over the corresponding d-functions.

5. Obtain the d-function of the component i using operator $\otimes_{\varphi_{\text{par}}}$ over the d-functions of nonempty protection groups of L_i level.

6. Increment i and if $i \leq N$ return to step 2.

7. Decode substring $\boldsymbol{a}_{\text{comp}}$ and obtain structure of protection groups and corresponding protection methods at each protection level.

8. Calculate the cost of protections of serially connected components and add this cost to C_{tot}.

9. For each protection level e (from $e = 1$ to $e = L_{\text{comp}}$), obtain the d-functions of the protection groups and replace them by equivalent elements using operators (4.60) and (4.62).

10. Obtain the d-function of the entire MSS using operator $\otimes_{\varphi_{\text{ser}}}$ over the d-functions of the nonempty protection groups of the L_{comp} level.

11. Evaluate the system survivability S for the given demand distribution using the d-function corresponding to the entire MSS.

12. So that the GA will search for the solution with minimal total cost and with survivability not less than the required value S^*, evaluate the solution fitness using Equation (5.30).

Example 5.11

Consider the same series-parallel multi-state power substation system that was presented in Example 5.8. The parameters of this system are $N = 4$, $B_1 = 6$, $B_2 = 3$, $B_3 = B_4 = 4$; $E_i = 10$ for $e = 1, \ldots, 4$; $L_1 = L_2 = L_3 = L_4 = L_{comp} = 2$; $\Gamma^e{}_1 = 3$, $\Gamma^e{}_2 = 1$, $\Gamma^e{}_3 = 3$, $\Gamma^e{}_4 = 2$ for $e = 1,2$; $\Gamma^1{}_{comp} = 3$, $\Gamma^2{}_{comp} = 2$.

The elements within each component can have two level protections (different types of protection shield and casing). The entire components can be allocated within protecting constructions and distributed among different protected sites. While the cost and vulnerability of the shields and the casings do not depend on the number of protected elements, the number of protected components strictly affects the protection cost and vulnerability. The parameters of the available protections within the components are presented in Table 5.28. The parameters of protections of the groups of the entire components are presented in Table 5.29.

The structure optimization problem was solved for four different values of desired system survivability $S^* \in \{0.85, 0.90, 0.95, 0.99\}$. One can see the solutions obtained in Figures 5.10-5.13, where the system elements are marked with their version numbers and each protection group is encased in an ellipse numbered in accordance with the level and the chosen method of protection (the marks have the form level/method). Ellipses corresponding to the different protection levels have different types of line (solid lines represent the protections for elements within the components and dashed lines represent the protections of the entire components). Observe that in the optimal solutions some protections do not appear (for example, the protection of level 1 in components 2 and 3 in Figure 5.10). This corresponds to protection method 0 chosen by the GA for the given protection level.

Table 5.28. Characteristics of protections available within components

No. of component	Protection level e	Protection method γ	Protection vulnerability v	Protection cost c
1	1	1	0.4	0.2
		2	0.3	2.1
		3	0.1	10.7
	2	1	0.35	0.1
		2	0.15	4.1
		3	0.05	15.7
2	1	1	0.1	1.2
	2	1	0.01	1.0
3	1	1	0.5	2.0
		2	0.37	4.5
		3	0.13	12.0
	2	1	0.60	1.0
		2	0.35	5.5
		3	0.15	17.0
4	1	1	0.2	1.5
		2	0.05	4.7
	2	1	0.10	1.1
		2	0.03	4.2

Table 5.29. Characteristics of protections available for entire components

Protection level e	Protection method γ	No. of protected components	Protection vulnerability v	Protection cost c
1	1	1	0.42	2.0
		2	0.44	2.2
		3	0.50	2.9
		4	0.55	3.5
	2	1	0.23	4.1
		2	0.25	4.6
		3	0.30	6.4
		4	0.35	7.3
	3	1	0.17	5.4
		2	0.19	6.0
		3	0.25	7.7
		4	0.30	8.5
2	1	1	0.38	4.0
		2	0.39	4.2
		3	0.40	4.3
		4	0.41	4.4
	2	1	0.33	8.1
		2	0.35	8.6
		3	0.38	9.4
		4	0.39	10.3

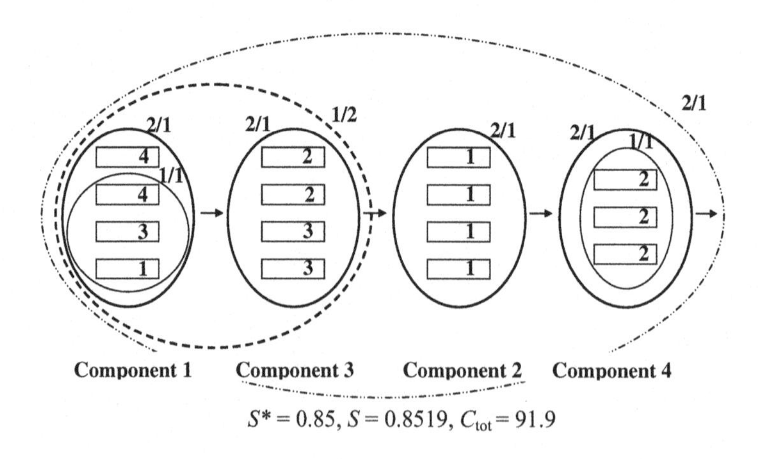

Figure 5.10. Lowest cost MSS with multilevel protection for $S^* = 0.85$

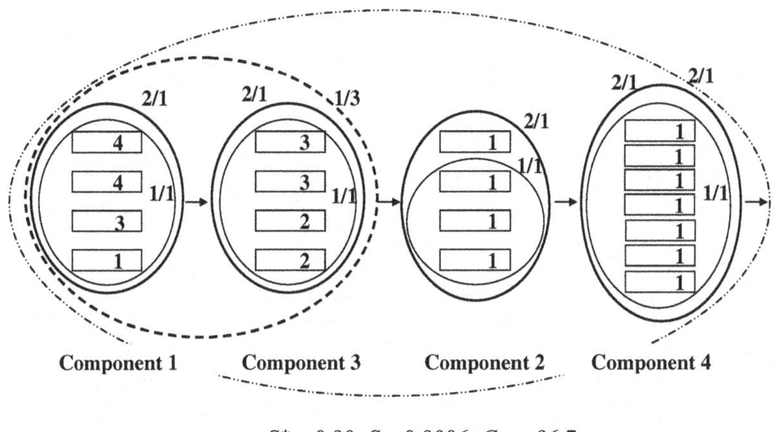

$S^{*} = 0.90$, $S = 0.9006$, $C_{tot} = 96.7$

Figure 5.11. Lowest cost MSS with multilevel protection for $S^{*} = 0.90$

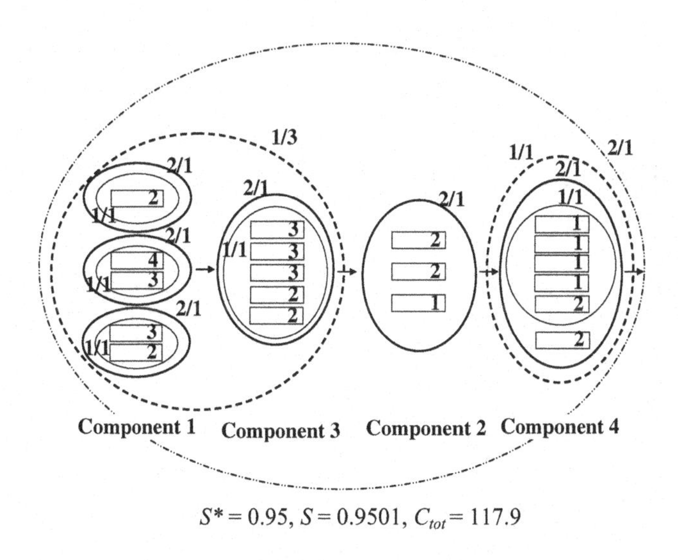

$S^{*} = 0.95$, $S = 0.9501$, $C_{tot} = 117.9$

Figure 5.12. Lowest cost MSS with multilevel protection for $S^{*} = 0.95$

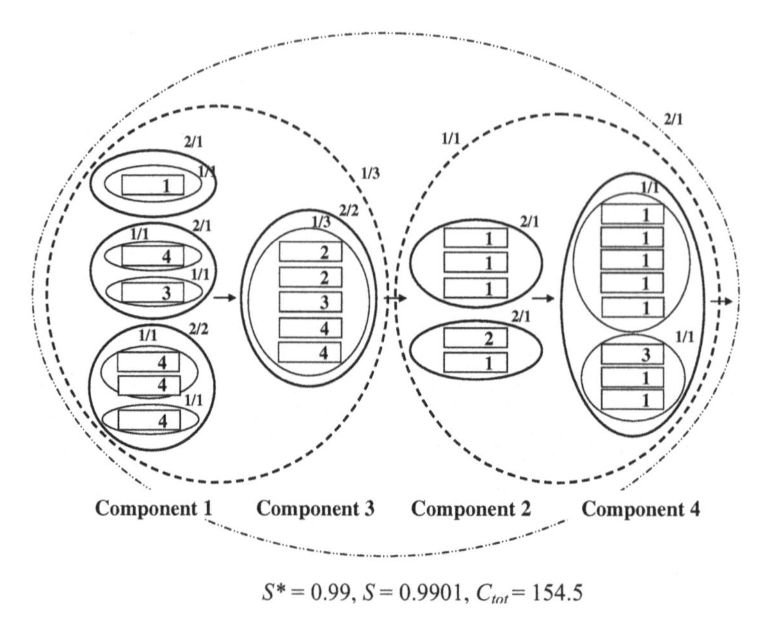

$S^* = 0.99$, $S = 0.9901$, $C_{tot} = 154.5$

Figure 5.13. Lowest cost MSS with multilevel protection for $S^* = 0.99$

5.3 Optimal Reliability Enhancement of Multi-state System Elements

The optimal element reliability enhancement problems belong to the class of reliability allocation problems [141]. Enhancing the reliability of the system's elements by performing special actions leads to the improvement of the system's overall performance; however, it increases the system's cost. In MSSs, where the roles of different elements for improving the system's performance depend on both their performance distribution and their place in the system, the optimal distribution of the limited maintenance resources is a complicated combinatorial problem.

The optimal element reliability enhancement presumes knowledge of dependencies between the amount of resources applied to improve the reliability of the system's elements and the extent of the increase of their reliability indices. The optimization problem lies in the optimal distribution of the limited resources among the MSS elements in order to achieve the greatest possible system performance (or in order to provide the desired level of performance by the minimal amount of resources).

An example of such a problem is a reliability growth test allocation problem. In this problem, models are used that predict the reliability of the system elements as a function of testing time. The testing time for each system element should be determined in order to maximize the entire system's reliability when total testing resources are limited.

Another example is optimizing the maintenance policy. Having estimations as to the influence of different maintenance actions on the elements' reliability, one can evaluate their influence on the entire complex MSS containing elements with different performance rates and reliabilities. An optimal maintenance policy can be developed that would answer the following questions: "Which elements should be the focus of the maintenance activity?" and "What should be the intensity of this activity?".

Since both the maintenance activity and the incorporation of redundancy improve MSS reliability, the question arises as to what is more effective. In other words, should the designer prefer a structure with more redundant elements and invest less in maintenance or *vice versa*? The optimal compromise should minimize the MSS cost while providing its desired reliability. The joint maintenance and redundancy optimization problem is to find this optimal compromise while taking into account the differences in the reliability and the performance rates of the elements that compose the MSS.

Finally, the most general optimization problem is optimal multistage modernization of MSSs subject to the reliability and the performance requirements. In order to solve this problem, one should develop a minimal-cost modernization plan that includes maintenance, modernization of the elements, and system expansion actions. The objective is to provide the desired reliability level while meeting the increasing demand during the entire operation time of the MSS.

5.3.1 Optimization of Multi-state System Reliability Growth Testing

Reliability growth testing (RGT) is an important part of "chamber-type" reliability testing programmes that are aimed at developing corrective actions. Analysis of hardware-experienced RGT and field usage shows that RGT combined with environmental stress screening and reliability qualification testing is the most effective method of improving system reliability when system design is new and unproven. RGT begins in the early stages of the design and development process. It involves conducting tests in which actual usage stresses are simulated. All failures observed are analyzed to identify their causes and to make corresponding changes in the design aimed at preventing or minimizing future occurrences of failures of the observed type.

RGT is widely adopted as a systematic method of planning developmental testing and measuring the effectiveness of reliability growth actions. It is effectively implemented in numerous fields from complex electronic systems [142], nuclear technology [143] and computer hardware design [144] to automotive and cellular telephone industries [145], defence [146] and software development [147]. Different models of reliability growth during RGT were developed for tracking the progress of reliability improvement, measuring the effectiveness of RGT, and anticipating the rate of further improvements [147-149].

For complex and large systems, it is usually impossible or prohibitively expensive to conduct RGT for an entire system. In these cases, RGT is conducted at the level of system elements or subsystems. As demonstrated by Rajgopal and Mazumdar [150], conducting reliability qualification testing at the subsystem level is, in many cases, less expensive (for example, when the system assembly cost is very high). The benefit from RGT of different elements can be different depending on the maturity of their design and on their importance in the system. In conditions of limited development times and budgets it is important to allocate testing resources in a manner that maximizes resultant benefits. The problem of optimal allocation of testing resources for subsystem-level RGT was formulated by Coit [151]. He developed a method for optimally allocating testing resources between different elements arranged in a series configuration in a binary system. In this section, we consider an extension of Coit's method to MSSs.

5.3.1.1 Problem Formulation

Following [151], we adopt the Crow/AMSAA reliability model in which failure intensity $h(\tau)$ for each system element is expressed as a function of development test time τ as follows:

$$h(\tau)=\lambda\beta\tau^{\beta-1} \tag{5.42}$$

where λ and β are model parameters ($\lambda>0$, $0<\beta<1$). This model is based on the assumption that the failures during time τ occur as a non-homogeneous Poisson process with decreasing failure intensity. After the completion of RGT at time τ, subsequent failures occur in accordance with a homogeneous Poisson process at a constant rate of $h(\tau)$. It is also assumed that failure inter-arrival times are exponentially distributed after the completion of RGT. Therefore, the mean inter-arrival time (MTTF) can be expressed as

$$\text{MTTF} = h^{-1}(\tau) = \frac{\tau^{1-\beta}}{\lambda\beta} \tag{5.43}$$

For non-repairable elements, their reliability r after performing RGT during time τ can be expressed by the following function of time t:

$$r(t) = \exp(-h(\tau)t) = \exp(-\lambda\beta\tau^{\beta-1}t) \tag{5.44}$$

while for repairable elements their steady-state availability A is estimated as

$$A=\text{MTTF}/(\text{MTTF+MTTR}) \tag{5.45}$$

If some initial testing has already been performed during time τ_0 before the testing resource allocation, the function $h(\tau)$ should be replaced by $h(\tau+\tau_0)$.

Consider a series-parallel system consisting of N two-state elements. Each element j is characterized by its nominal performance rate g_{j1}, parameters of reliability growth model λ_j and β_j and mean time to repair MTTR (for repairable systems). The MSS is supposed to meet a variable demand with distribution $\boldsymbol{w}$, $\boldsymbol{q}$.

For a given RGT time τ_j one can evaluate the expected reliability (or availability) of element j after completion of development testing. Therefore, vector $\boldsymbol{\tau} = (\tau_j: 1 \le j \le N)$ defines the entire MSS reliability $R(\boldsymbol{\tau}, \boldsymbol{w}, \boldsymbol{q}, t)$ (or availability $A(\boldsymbol{\tau}, \boldsymbol{w}, \boldsymbol{q})$) achieved due to RGT. If for each system element j a resource c_j needed for its testing per unit time is given, then the total resource for the RGT C is determined as

$$C(\boldsymbol{\tau}) = \sum_{j=1}^{N} c_j \tau_j \tag{5.46}$$

For the total resource C^* available for RGT, the following two optimization problems can be formulated [152]:

1. Find testing time distribution $\boldsymbol{\tau}$ which provides maximal MSS reliability during specified operation time T subject to the resource constraint (for non-repairable MSS):

$$R(\boldsymbol{\tau}, \boldsymbol{w}, \boldsymbol{q}, T) \to \max \text{ subject to } C(\boldsymbol{\tau}) \le C^* \tag{5.47}$$

2. Find testing time distribution $\boldsymbol{\tau}$ which provides maximal MSS availability $A(\boldsymbol{\tau}, \boldsymbol{w}, \boldsymbol{q})$ subject to the resource constraint (for repairable MSS):

$$A(\boldsymbol{\tau}, \boldsymbol{w}, \boldsymbol{q}) \to \max \text{ subject to } C(\boldsymbol{\tau}) \le C^* \tag{5.48}$$

5.3.1.2 Implementing the Genetic Algorithm

The natural representation of testing times distribution is by an N-length integer string in which the value in the jth position corresponds to the testing time (in hours) of the jth element of the MSS. One can see that arbitrary integer strings cannot guarantee feasibility of solution because of constraint violation. To provide solution feasibility, the string $\boldsymbol{a}^* = (a^*_1,\dots,a^*_N)$ should be normalized in the following way:

$$a_j = C^* a_j^* / \sum_{k=1}^{N} c_k a_k^* \tag{5.49}$$

for each $1 \le j \le N$. The values of τ_j for each MSS element can be obtained now from an arbitrary integer string: $\tau_j = \lfloor a_j \rfloor$. Note that the range of initial distribution of integer numbers a^*_j does not affect solution feasibility; rather it defines the measure of "recognition" provided by optimization. Indeed, the greater the range, the smaller the variation of time distribution that can be encoded by string $\boldsymbol{a}^*$.

The following procedure determines the fitness value for an arbitrary solution defined by integer string $\boldsymbol{a}^*$:

1. Normalize string $\boldsymbol{a}^*$ using (5.49) and assign values of testing time τ to each MSS element.

2. Determine reliability $r_j(T)$ of each element $1 \leq j \leq N$ at time T using (5.44) or availability A_j using (5.43) or (5.45).

3. Define for each element j its u-function in the form

$$u_j(z)=r_j(T)\,z^{g_{j1}}+(1-r_j(T))z^0 \text{ (for non-repairable systems) or}$$

$$u_j(z)=A_j\,z^{g_{j1}}+(1-A_j)z^0 \text{ (for repairable systems).}$$

4. Determine the entire system u-function using the reliability block diagram method.

5. Obtain the MSS reliability at time T, $R(\boldsymbol{\tau}, \boldsymbol{w}, \boldsymbol{q}, T)$, (for non-repairable systems) or steady-state availability $A(\boldsymbol{\tau}, \boldsymbol{w}, \boldsymbol{q})$ (for repairable systems) for the given demand distribution $\boldsymbol{w}, \boldsymbol{q}$ using the system u-function obtained.

6. Evaluate the solution fitness as $R(\boldsymbol{\tau}, \boldsymbol{w}, \boldsymbol{q}, T)$ or $A(\boldsymbol{\tau}, \boldsymbol{w}, \boldsymbol{q})$.

Example 5.12

Consider the series-parallel MSS consisting of four components connected in series in Figure 5.14. The system contains 11 elements with different performance rates and reliability-related parameters.

First, assume that all the system elements are repairable. The element's nominal performance rate, RGT model parameters λ and β, initial testing times τ_0, and MTTR (in hours) are presented in Table 5.30. The allowable cumulative test time is t_{test} =30,000 h. (Since the test time is the only limited resource, c_j is set to 1 for each element and $C^* = t_{test}$.) The demand distribution is presented in Table 5.31.

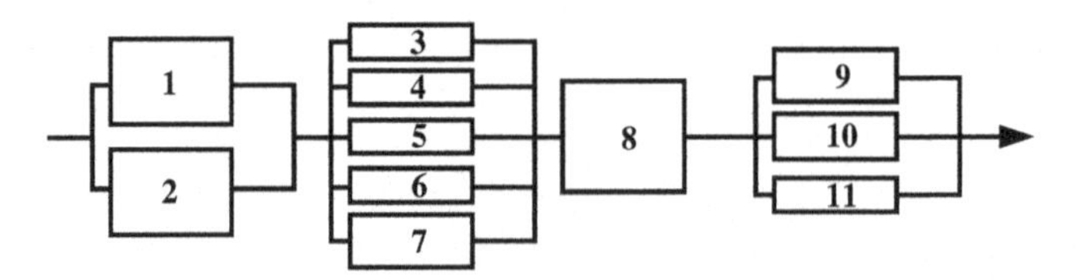

Figure 5.14. Reliability block diagram of series-parallel MSS

Table 5.30. Parameters of repairable MSS elements

No. of element	g	λ	β	τ_0	c	MTTR
1	0.8	0.02	0.57	125	1	2
2	1.0	0.04	0.51	125	1	2
3	0.2	0.05	0.58	1	1	8
4	0.2	0.05	0.58	1	1	8
5	0.2	0.10	0.40	250	1	8
6	0.2	0.10	0.40	250	1	8
7	0.3	0.06	0.65	100	1	12
8	1.2	0.01	0.52	1	1	24
9	0.4	0.02	0.60	150	1	15
10	0.4	0.02	0.60	150	1	15
11	0.4	0.02	0.60	150	1	15

Table 5.31. Demand distribution

w	1.0	0.8	0.6	0.4	0.2
q	0.10	0.55	0.20	0.05	0.10

Table 5.32. RGT time distribution for repairable MSS

No. of element	τ	MTTF		A	
		No RGT	RGT	No RGT	RGT
1	0	699	699	0.9971	0.9971
2	522	522	1169	0.9962	0.9983
3	2984	34	993	0.8117	0.9920
4	2984	34	993	0.8117	0.9920
5	1634	687	2307	0.9885	0.9965
6	1634	687	2307	0.9885	0.9965
7	6183	129	547	0.9146	0.9785
8	6786	192	13279	0.8890	0.9982
9	2424	618	1928	0.9763	0.9923
10	2424	618	1928	0.9763	0.9923
11	2425	618	1928	0.9763	0.9923

The system availability before conducting RGT for the given demand distribution was $A(\boldsymbol{w}, \boldsymbol{q}) = 0.813$. The best solution obtained by the GA yields $A(\boldsymbol{w}, \boldsymbol{q}) = 0.991$. The RGT time distribution for this solution is presented in Table 5.32.

This table also contains the MTTF and availability estimates obtained for each element of the MSS before and after conducting RGT. One can also compare the system availability functions $A(w)$ obtained from system PD before and after conducting RGT that are presented in Figure 5.15.

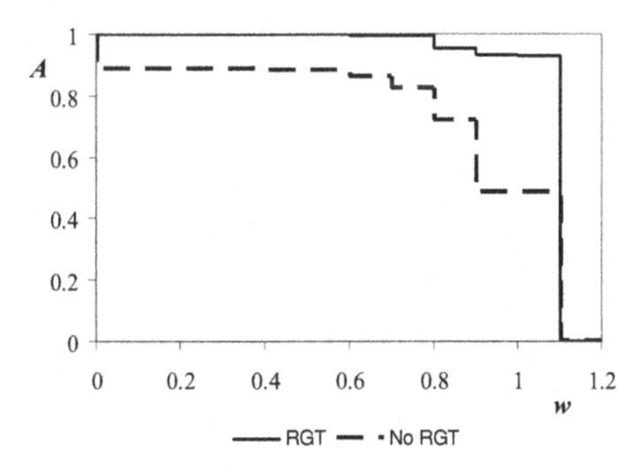

Figure 5.15. Availability function for repairable MSS

Now, assume that a system with the same structure consists of non-repairable elements. The nominal performance and reliability parameters of these elements are presented in Table 5.33, as well as the testing cost per time unit. The maximal allowable total cost of RGT is $C^* = 10{,}000$. The system operation time during which its reliability index $R(T)$ should be maximized is $T = 8760$ h (1 year).

Table 5.33. Parameters of non-repairable MSS elements

No. of element	g	λ	β	τ_0	c
1	1.0	0.003	0.44	25	0.20
2	1.0	0.005	0.41	25	0.28
3	0.4	0.008	0.38	0	0.28
4	0.4	0.008	0.38	0	0.28
5	0.4	0.007	0.40	10	0.26
6	0.4	0.007	0.40	10	0.26
7	0.6	0.009	0.35	20	0.40
8	1.3	0.001	0.32	0	0.19
9	0.6	0.005	0.40	20	0.30
10	0.5	0.008	0.28	10	0.30
11	0.4	0.010	0.24	10	0.36

The best solution obtained for demand distribution presented in Table 5.31 yields $R(\boldsymbol{w}, \boldsymbol{q}, T) = 0.9548$. One can see the variation of reliability index $R(\boldsymbol{w}, \boldsymbol{q}, t)$ of the entire MSS for $0 \le t \le T$ in Figure 5.16. The best RGT time distribution obtained is presented in Table 5.34, along with testing cost, MTTF, and reliability at time T of each element.

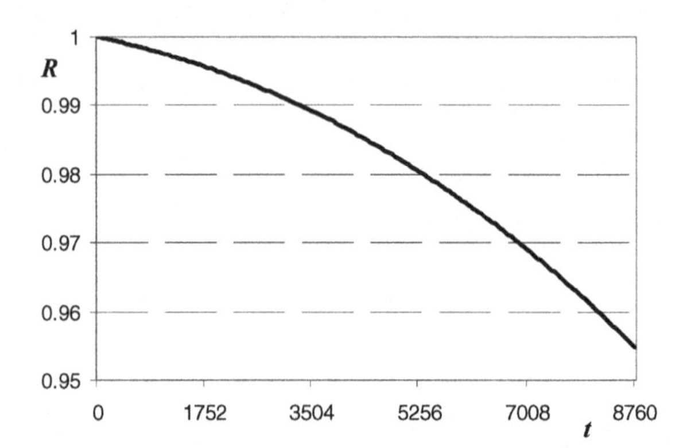

Figure 5.16. Reliability function for non-repairable MSS

Table 5.34. RGT time distribution for non-repairable MSS

No. of element	τ	Testing cost $c\tau$	MTTF	$r(T)$
1	5667	1133.40	96023	0.9128
2	4143	1160.04	66681	0.8769
3	2081	582.68	37537	0.7919
4	2081	582.68	37537	0.7919
5	2191	569.66	36175	0.7849
6	2192	569.92	36185	0.7850
7	2080	832.00	45828	0.8260
8	5558	1056.02	1100005	0.9921
9	6139	1841.70	93902	0.9109
10	2813	843.90	136227	0.9377
11	2300	828.00	150015	0.9433

5.3.2 Optimization of Cyclic Replacement of Multi-state System Elements

When a system consists of elements having hazard rates that increase with time, a perfect preventive maintenance is aimed at reducing the hazard rates by making the elements "as good as new", or by replacing the elements with new ones. Further on, we will refer to such perfect preventive maintenance as preventive replacement (PR). An alternative type of maintenance activity, corrective maintenance, is aimed at making the system operable by the minimal cost when a failure occurs. Such an activity, named minimal repair (MR), that enables the system element to continue its work but does not affect the hazard rate of the element may be much less expensive. PR of elements with high risk of failure reduces the chance of failure, but can cause significant expenses, especially in systems with a high replacement rate. Maintenance policies of compromising between PRs and MRs aim at achieving an optimal solution for problems with different criteria.

It is recognized that obtaining the element lifetime distributions is the bottleneck in the implementation of existing maintenance optimization approaches. The expected number of element failures in any time interval can be obtained either from mathematical models and experimental data or from expert opinion. The analytical expressions for this function have been covered extensively in [153] and used in many problems of maintenance optimization based on increasing failure rate models [154-156]. The failure rate function obtained from experts can be represented in a tabular form [157]. For MR, the duration of which is relatively small compared with the time between failures, the expected number of failures is equal to the expected number of repairs for any time interval. Thus, it is possible to obtain the renewal function of each element (expected number of its repairs in time interval $(0, t)$). This expected number of element failures/repairs $f(t_j)$ can be estimated for different time intervals $(0, t_j)$ between consecutive PRs.

In this section, we consider the determination of the optimal schedule of cyclic PRs for an MSS with a given series-parallel configuration and two-state elements. Each element of this system is characterized by its nominal performance and renewal function, obtained from mathematical models or elicited from expert opinion. The times and costs of two types of maintenance activity (PR and MR) are also available for each system element. The objective is to provide the desired system availability by the minimal sum of maintenance cost and penalty costs caused by system mission losses (performance deficiency).

The method presented presumes independence between replacement and repair activities for different system elements. Such an assumption is justified, for example, in complex distributed systems (power systems, computer networks, *etc.*) where the information about system element repairs and replacements may be inaccessible for the maintenance staff serving the given element. In the general case, the method, which assumes independence of maintenance actions in the system, gives the worst estimation of system availability.

Another important assumption is that repair and replacement times are much smaller than time between failures. In this case, the probability of replacement and repair events coincidences may be neglected.

In systems with cyclic variable demand (double-shift job shop production, power or water supply, *etc.*), the PR can be performed in periods of low demand even if repairs of some of the elements of the system are not finished. For example, in power generation systems some important elements may be replaced at night when the power demand is much lower than the nominal demand. In these cases, the replacement time may be neglected and all the maintenance actions may be considered as independent.

5.3.2.1 Problem Formulation

A system consists of N two-state elements connected according to a reliability block diagram. For each element j ($1 \leq j \leq N$) its nominal performance rate g_{j1}, expected preventive, and corrective maintenance times and costs are given, as well as a renewal function representing the expected number of elements failures/repairs in the time interval $(0, t)$.

For any replacement interval t_j for each element j, one can obtain the expected number of failures and repairs during the period between preventive replacement actions $f_j(t_j)$. The replacement interval may be alternatively defined by the number of preventive replacement actions x_j during the system operation time T: $t_j = (T - x_j\hat{\tau}_j)/(x_j+1)$ where $\hat{\tau}_j$ is PR time. Usually, $\hat{\tau}_j \ll T$ and the replacement interval can be estimated as $t_j = T/(x_j+1)$. The expected number of failures of element j during the system operation time is $(x_j+1)f_j(T/(x_j+1))$.

Under the assumptions formulated, the expected time that the jth system element is unavailable can be estimated by the following expression:

$$(x_j+1)f_j(T/(x_j+1))\tilde{\tau}_j + x_j\hat{\tau}_j \tag{5.50}$$

where $\tilde{\tau}_j$ is MR time.

Now one can define the availability of each element as

$$p_{j1} = \frac{T - (x_j+1)f_j(T/(x_j+1))\tilde{\tau}_j - x_j\hat{\tau}_j}{T} \tag{5.51}$$

the total expected maintenance time τ_{tot} during the system operation time as

$$\tau_{\text{tot}} = \sum_{j=1}^{N}[(x_j+1)f_j(T/(x_j+1))\tilde{\tau}_j + x_j\hat{\tau}_j] \tag{5.52}$$

and the expected maintenance cost C_{m} during the system operation time as

$$C_{\text{m}} = \sum_{j=1}^{N}[(x_j+1)f_j(T/(x_j+1))\tilde{c}_j + x_j\hat{c}_j] \tag{5.53}$$

where $\tilde{c}_j$ and $\hat{c}_j$ are the corrective and preventive maintenance costs respectively.

Having the PD of each system element j ($\boldsymbol{g}_j = \{0, g_{j1}\}$, $\boldsymbol{p}_j = \{(1-p_{j1}), p_{j1}\}$), one can obtain the entire system PD using the reliability block diagram method, and for the given demand distribution $\boldsymbol{w}$, $\boldsymbol{q}$ one can obtain the system performance measures: the availability $A(\boldsymbol{w}, \boldsymbol{q})$ and the expected performance deficiency $\Delta^{-}(\boldsymbol{w}, \boldsymbol{q})$.

The total unsupplied demand cost during the system operation time T can be estimated as

$$C_{\text{ud}} = Tc_{\text{u}}\Delta^{-}(\boldsymbol{w},\boldsymbol{q}) \tag{5.54}$$

where c_{u} is a cost of unsupplied demand per time unit.

Defining the system replacement policy by the vector $\boldsymbol{x} = (x_j\text{: } 1 \le j \le N)$ one can give two formulations of the problem of system maintenance optimization [158]:

1. Find the minimal maintenance cost system replacement policy $\boldsymbol{x}$ that provides the required MSS availability level A^* while the total maintenance time does not exceed preliminarily specified limitation τ^*:

$$C_{\text{m}}(\boldsymbol{x}) \to \min \quad \text{subject to} \quad A(\boldsymbol{x},\boldsymbol{w},\boldsymbol{q}) \ge A^*, \tau_{\text{tot}}(\boldsymbol{x}) \le \tau^* \tag{5.55}$$

2. Find the system replacement policy $\boldsymbol{x}$ that minimizes the total maintenance and unsupplied demand cost while the total maintenance time does not exceed the preliminary specified limitation τ^*:

$$C_{\text{tot}} = C_{\text{m}}(\boldsymbol{x}) + C_{\text{ud}}(\boldsymbol{x},\boldsymbol{w},\boldsymbol{q}) \to \min \text{ subject to } \tau_{\text{tot}}(\boldsymbol{x}) \le \tau^* \tag{5.56}$$

In the general case, one can use the following formulation:

$$C_{\text{tot}} = C_{\text{m}}(\boldsymbol{x}) + C_{\text{ud}}(\boldsymbol{x},\boldsymbol{w},\boldsymbol{q}) \to \min$$

$$\text{subject to } A(\boldsymbol{x},\boldsymbol{w},\boldsymbol{q}) \ge A^*, \tau_{\text{tot}}(\boldsymbol{x}) \le \tau^* \tag{5.57}$$

which can be reduced to (5.55) by defining $c_{\text{u}} = 0$ and to (5.56) by defining $A^* = 0$.

5.3.2.2 Implementing the Genetic Algorithm

Different elements can have different possible numbers of PR actions during the system operation time. The possible maintenance alternatives (number of PR actions) for each system element j can be ordered in vector $\boldsymbol{Y}_j = (y_{j1},\ldots,y_{jK})$, where y_{ji} is the number of preventive maintenance actions corresponding to alternative i for the system element j. The same number K of possible alternatives (length of vectors $\boldsymbol{Y}_j$) can be defined to each element. If, in practical problems, the number of alternatives differs for different elements, then some elements of shorter vectors $\boldsymbol{Y}_j$ can be duplicated to provide equality of the vector's length.

Each solution is represented by the integer string $\boldsymbol{a} = (a_j: 1 \le a_j \le K)$ that represents the number of maintenance alternative applied to element j. Hence, the vector $\boldsymbol{x}$ for the given solution, represented by the string $\boldsymbol{a}$ is $\boldsymbol{x} = (y_{1a_1}, \ldots, y_{na_n})$. For example, for a problem with $N = 5$, $K = 4$, $\boldsymbol{Y}_1 = \boldsymbol{Y}_2 = \boldsymbol{Y}_3 = (2, 3, 4, 5)$ and $\boldsymbol{Y}_4 = \boldsymbol{Y}_5 = (20, 45, 100, 100)$ string $\boldsymbol{a} = (1, 4, 4, 3, 2)$ represents a solution with $\boldsymbol{x} = (2, 5, 5, 100, 45)$. Any arbitrary integer string with elements belonging to the interval $(1, K)$ represents a feasible solution.

For each given string $\boldsymbol{a}$ the decoding procedure first obtains the vector $\boldsymbol{x}$ and estimates $f_j(t_j) = f_j(T/(x_j+1))$ for all the system elements $1 \le j \le N$, then calculates availability indices of each two-state system element using expression (5.51) and determines the entire system PD using the reliability block diagram method. It also determines τ_{tot} and C_{m} using expressions (5.52) and (5.53). After obtaining the entire system PD, the procedure evaluates $A(\boldsymbol{w}, \boldsymbol{q})$ and $\Delta^{-}(\boldsymbol{w}, \boldsymbol{q})$ and computes $C_{\text{ud}}(\boldsymbol{w}, \boldsymbol{q})$ using expression (5.54).

In order to let the GA look for the solution with minimal total cost, and with A that is not less than the required value A^*, and τ_{tot} not exceeding τ^*, the solution fitness is evaluated as follows:

$$\begin{aligned} &M - C_{\text{tot}}(\boldsymbol{a}) - \pi_1(1 + A^* - A(\boldsymbol{a})) \times 1(A(\boldsymbol{a}) < A^*) \\ &-\pi_2(1 + \tau_{\text{tot}}(\boldsymbol{a}) - \tau^*) \times 1(\tau^* < \tau_{\text{tot}}(\boldsymbol{a})) \end{aligned} \tag{5.58}$$

where π_1 and π_2 are penalty coefficients and M is a constant value.

Example 5.13

A series-parallel water desalination system consists of four components containing 14 elements of eight different types. The structure of the system (which belongs to the type of flow transmission MSS with flow dispersion) is presented in Figure 5.17. Each element is marked with its type number j. Table 5.35 contains parameters of each element, including its renewal function $f_j(t)$ (replacement period t in months is given for each x), estimated using expert judgments. Times are measured in months. The element nominal performance is measured as a percentage of a maximal system demand.

All the replacement times in the system are equal to 0.5 h (0.0007 month). The corrective maintenance includes a fault location search and tuning of the elements, so it takes much more time than preventive replacement, but repairs are much cheaper than replacements.

The demand distribution is presented in Table 5.36. The total operation time T is 120 months, and the cost of 1% of unsupplied demand during one month is $c_{\text{u}} =$ 10.

For the sake of simplicity, we use in this example the same vector of replacement frequency alternatives for all the elements. The possible number of replacements during the system operation time varies from 5 to 30 with step 5.

The first solutions obtained are for the first formulation of the problem in which unsupplied demand cost was not considered and the total maintenance time was not constrained: $\tau^* = \infty$. (Three different solutions are presented in Table 5.37.) One can see the total maintenance time and cost as functions of system availability index A in Figure 5.18 (each point of the graph corresponds to an optimal solution).

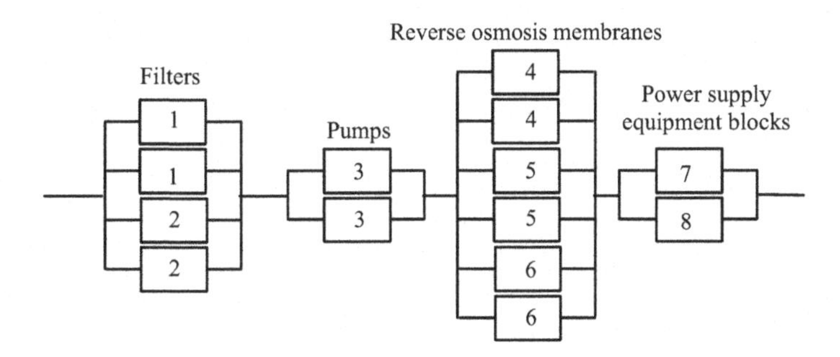

Figure 5.17. Series-parallel water desalination system

Table 5.35. Characteristics of the system elements

Element	g_{j1}	$\hat{c}_j$	$\tilde{c}_j$	$\tilde{\tau}_j$	Renewal function $f_j(t)$					
					t: 24	12	8	6	4.8	4
j					*x:* 5	10	15	20	25	30
1	0.40	3.01	0.019	0.002	25.00	10.00	5.00	2.0	1.00	0.50
2	0.30	2.21	0.049	0.004	26.00	9.00	2.00	0.6	0.20	0.05
3	0.60	2.85	0.023	0.008	20.00	4.00	1.00	0.3	0.08	0.01
4	0.15	2.08	0.017	0.005	36.00	14.00	9.00	6.0	4.00	3.00
5	0.15	1.91	0.029	0.003	55.00	15.00	7.00	4.0	0.32	0.30
6	0.25	0.95	0.031	0.009	31.00	9.50	5.60	4.0	2.70	2.00
7	1.00	5.27	0.050	0.002	13.00	3.20	1.40	0.8	0.50	0.10
8	0.70	4.41	0.072	0.005	5.00	2.00	1.00	0.4	0.10	0.01

Table 5.36. Demand distribution

w	1.00	0.80	0.50	0.20
q	0.60	0.25	0.05	0.10

Then the unsupplied demand cost was introduced and the problem was solved in its second formulation. The solutions corresponding to the minimal and maximal possible system availability (minimal and maximal maintenance cost) are presented in Table 5.37, as well as the optimal solution, which minimizes the total cost. One can see that the optimal maintenance solution allows about 50% total cost reduction to be achieved in comparison with minimal C_m and minimal C_{ud} solutions.

Table 5.37. Optimal maintenance solutions obtained

	C_{ud}	C_m	C_{tot}	τ_{tot}	A	x
			Formulation 1			
$A^* = 0.96$	0.0	263.1	263.1	9.2	0.9606	(5,5,5,5,5,5,5,5,10,10,10,10,5,5)
$A^* = 0.97$	0.0	296.6	296.6	7.7	0.9700	(5,5,5,5,10,10,5,5,10,10,25,25,5,5)
$A^* = 0.98$	0.0	384.4	384.4	5.85	0.9800	(5,5,5,5,15,15,10,10,10,25,25,25,5,5)
			Formulation 2			
Minimal C_m $\tau^* = \infty$	1029.5	249.1	1278.6	11.61	0.9490	(5,5,5,5,5,5,5,5,5,5,5,5,5,5)
Minimal C_{ud} $\tau^* = \infty$	156.4	1060.3	1216.7	2.47	0.9880	(30,30,30,30,30,30,30,30,25,25,30,30,30,30)
Minimal C_{tot} $\tau^* = \infty$	256.2	397.4	653.5	6.02	0.9800	(5,5,5,5,20,20,10,10,10,10,30,30,5,5)
Minimal C_{tot} $\tau^* = 3$	181.7	674.7	856.4	2.99	0.9877	(10,20,20,20,25,20,30,30,25,25,30,30,10,5)
			General formulation			
Minimal C_{tot} $\tau^* = 5.5$, $A^* = 0.985$	192.8	498.1	690.9	4.96	0.9850	(5,5,5,5,25,25,10,10,25,25,30,30,10,5)

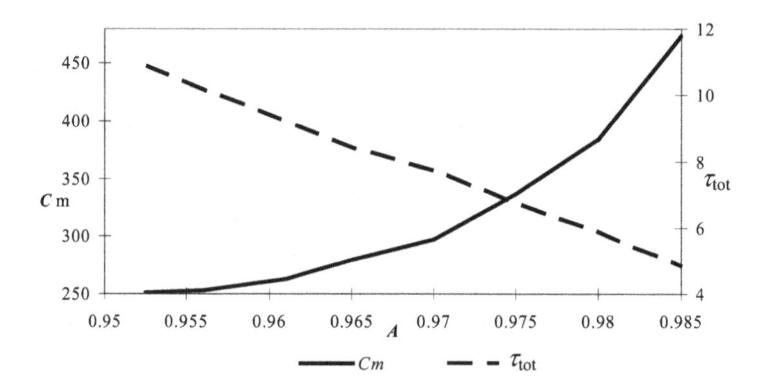

Figure 5.18. Total maintenance time and cost as functions of system availability

The influence of the total maintenance time constraints is illustrated in Figure 5.19, where the costs and the system availability index are presented as functions of τ_{tot}. Observe that reduction of allowable maintenance time causes the system availability and total cost to increase. The interesting exception is when maintenance time decreases from 5 to 4.5 months. In this case, the variations in maintenance policy cause additional expenses without system availability enhancement. The solution of the general formulation of the problem where τ^* = 5.5 and A^* = 0.985 is also presented in Table 5.37.

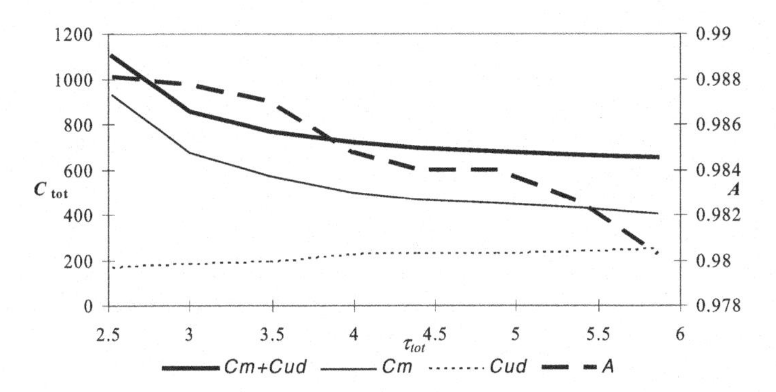

Figure 5.19. System cost and availability under maintenance time constraints

5.3.3 Joint Redundancy and Cyclic Replacement Optimization

Both introducing the performance redundancy (considered in Sections 5.1.1 and 5.1.2) and providing for preventive replacement of the elements (considered in Section 5.3.2) are aimed at improving the system's availability and are associated with certain expenses. Therefore, a trade-off exists between the investments into the system's redundancy and its preventive maintenance cost. The optimal availability design should consider both of these factors in order to reach a solution that provides for the system's desired reliability at minimal cost. Since these factors influence each other, the problem cannot be solved in two separate stages, *i.e.* by finding the minimal cost PR policy for the preliminarily determined optimal structure. In this section, we consider the problem of determining both the optimal structure for a series-parallel MSS and the optimal PR of the system's elements.

It is assumed that the PD, the time and cost of repair, and its replacement are available for each system element. The cost of the element's inclusion within the system is also given. In general, this cost is not equal to the cost of the element's replacement because the inclusion of the element can incur additional expenses, such as investments into the corresponding infrastructure (communication, foundation, *etc.*). The objective is to provide the desired system availability with a minimal sum of the costs of the system's structure, maintenance, and penalties caused by the unsupplied demand.

5.3.3.1 Problem Formulation

An MSS consists of N components. Each component is a subsystem that can consist of parallel two-state elements with the same functionality. Different versions and numbers of elements may be chosen for any given system component.

For each component i there are B_i element versions available in the market. Each element of version b and type i is characterized by the following parameters: nominal performance rate $g_{i1}(b)$, cost of including the element in the system $c_i(b)$,

PR and MR costs $\hat{c}_i(b)$ and $\tilde{c}_i(b)$, mean PR and MR times $\hat{\tau}_i(b)$ and $\tilde{\tau}_i(b)$, renewal function $f_{ib}(t)$.

The structure of system component i is defined by the numbers of parallel elements of each version n_{ib} ($1 \le b \le B_i$) chosen for this component. The vectors $\boldsymbol{n}_i = (n_{ib}: 1 \le b \le B_i)$ for $1 \le i \le N$ define the entire system structure. For a given set of vectors $\{\boldsymbol{n}_1, ..., \boldsymbol{n}_N\}$ the total cost of the system C can be calculated using Equation (5.7).

The PR policy is defined by assigning to each element of type i and version b a number of preventive replacement actions x_{ib} during the system operation time T. Having the set of vectors $\boldsymbol{x}_i = (x_{ib}: 1 \le b \le B_i)$ for $1 \le i \le N$, one can determine the availability of each MSS element in analogy with Equation (5.51) as

$$p_{i1}(b) = \frac{T - (x_{ib}+1) f_{ib}(T/(x_{ib}+1)) \tilde{\tau}_i(b) - x_{ib} \hat{\tau}_i(b)}{T} \tag{5.59}$$

the total expected MSS maintenance time τ_{tot} during the system operation time as

$$\begin{aligned} &\tau_{\text{tot}}(\boldsymbol{n}_1,...,\boldsymbol{n}_N,\boldsymbol{x}_1,...,\boldsymbol{x}_N) \\ &= \sum_{i=1}^{N} \sum_{b=1}^{B_i} n_{ib} [(x_{ib}+1) f_{ib}(T/(x_{ib}+1)) \tilde{\tau}_i(b) + x_{ib} \hat{\tau}_i(b)] \end{aligned} \tag{5.60}$$

and the expected maintenance cost C_{m} during the system operation time as

$$\begin{aligned} &C_{\text{m}}(\boldsymbol{n}_1,...,\boldsymbol{n}_N,\boldsymbol{x}_1,...,\boldsymbol{x}_N) \\ &= \sum_{i=1}^{N} \sum_{b=1}^{B_i} n_{ib} [(x_{ib}+1) f_{ib}(T/(x_{ib}+1)) \tilde{c}_i(b) + x_{ib} \hat{c}_i(b)] \end{aligned} \tag{5.61}$$

Having the PD of each system element of type i and version b ($\boldsymbol{g}_i(b) = \{0, g_{i1}b)\}$, $\boldsymbol{p}_i(b) = \{(1-p_{i1}(b)), p_{i1}(b)\}$), one can obtain the entire system PD using the reliability block diagram method and for the given demand distribution $\boldsymbol{w}$, $\boldsymbol{q}$ one can obtain the system availability $A(\boldsymbol{w}, \boldsymbol{q})$ and the expected performance deficiency $\Delta^-(\boldsymbol{w}, \boldsymbol{q})$. The total unsupplied demand cost during the system operation time T can be estimated using (5.54).

The general formulation of the system redundancy and maintenance optimization problem is as follows [159]: find the system configuration $\{\boldsymbol{n}_1, ..., \boldsymbol{n}_N\}$ and PR policy $\{\boldsymbol{x}_1, ..., \boldsymbol{x}_N\}$ that minimize the sum of costs of the system equipment, maintenance and unsupplied demand. The required MSS availability level should be provided while the total maintenance time should not exceed the preliminarily specified limitation:

$$\begin{aligned} C_{\text{tot}} &= C(\boldsymbol{n}_1,...,\boldsymbol{n}_N,\boldsymbol{x}_1,...,\boldsymbol{x}_N) + C_{\text{m}}(\boldsymbol{n}_1,...,\boldsymbol{n}_N,\boldsymbol{x}_1,...,\boldsymbol{x}_N) \\ &+ C_{\text{ud}}(\boldsymbol{n}_1,...,\boldsymbol{n}_N,\boldsymbol{x}_1,...,\boldsymbol{x}_N,\boldsymbol{w},\boldsymbol{q}) \to \min \end{aligned}$$

subject to (5.62)

$$A(\boldsymbol{n}_1,...,\boldsymbol{n}_N,\boldsymbol{x}_1,...,\boldsymbol{x}_N,\boldsymbol{w},\boldsymbol{q}) \geq A^*, \tau_{\text{tot}}(\boldsymbol{n}_1,...,\boldsymbol{n}_N,\boldsymbol{x}_1,...,\boldsymbol{x}_N) \leq \tau^*$$

5.3.3.2 Implementing the Genetic Algorithm

In analogy with Section 5.3.2.2, the possible PR alternatives for each MSS element of type i and version b are ordered in vectors $\boldsymbol{Y}_i(b)=\{y_{i1}(b),..., y_{iK}(b)\}$, where $y_{ij}(b)$ is thenumber of PR actions corresponding to alternative j; K is the total number of possible PR alternatives.

To represent the vectors $\boldsymbol{n}_1$, ..., $\boldsymbol{n}_N$ and $\boldsymbol{x}_1$, ..., $\boldsymbol{x}_N$ we use integer strings $\boldsymbol{a} = (a_1,...,a_B)$, where B is the total number of versions available (determined according to Equation (5.9)) and a_j corresponds to the element of type i and version b (where j is determined according to (5.10)).

Each vector element a_j determines both the number of parallel elements n_{ib} and the number of replacement frequency alternative m_j corresponding to all of the elements of this type and version. To provide this property, all the numbers a_j ($0\leq j\leq B$) are generated in the range $0\leq a_j<(n_{\max}+1)K$ (where $n_{\max}$ is the maximal allowable number of identical parallel elements) and are decoded in the following manner:

$$n_{ib} = \lfloor a_j / K \rfloor$$
$$m_j = 1 + \text{mod}_K a_j \tag{5.63}$$

For any given n_{ib} and m_j the corresponding a_j is composed as

$$a_j = n_{ib}K + m_j - 1 \tag{5.64}$$

Note that all $a_j<K$ correspond to solutions where elements of type i and version b are not included in the system because $n_{ib} = 0$.

Having m_j for the given element one can define the number of its replacements as $x_{ib} = y_{im_j}(b)$.

Example 5.14

Consider a problem in which $N = 3$, $B_1 = 3$, $B_2 = 2$, and $B_3 = 3$, $K = 3$, $n_{\max} = 5$ and $\boldsymbol{Y} = \{3, 10, 15\}$ for all the elements. Table 5.38 contains the parameters of solution ($\boldsymbol{n}_1$, $\boldsymbol{n}_2$, $\boldsymbol{n}_3$, $\boldsymbol{x}_1$, $\boldsymbol{x}_2$, $\boldsymbol{x}_3$) encoded by the string (11, 0, 6, 5, 14, 4, 9, 2).

For each given string $\boldsymbol{a}$ the decoding procedure first obtains the vectors $\boldsymbol{n}_1$, ..., $\boldsymbol{n}_N$ and $\boldsymbol{x}_1$, ..., $\boldsymbol{x}_N$ and estimates $f_{ib}(x_{ib})$ for all the system elements $1\leq i\leq N$, $1\leq b\leq B_i$, then calculates the availability of each system element using expression (5.59), determines the u-functions of the elements and obtains the entire system PD using the reliability block diagram method in accordance with the specified system structure. It also determines C, τ_{tot} and C_{m} using expressions (5.7), (5.60) and (6.61) respectively. After obtaining the entire system PD, the procedure evaluates $A(\boldsymbol{w},\boldsymbol{q})$, and $\Delta^{-}(\boldsymbol{w},\boldsymbol{q})$ and computes $C_{\text{ud}}(\boldsymbol{w},\boldsymbol{q})$ using expression (5.54). The total system cost

C_{tot} is determined as the sum of C, C_{m} and C_{ud}. In order to let the GA look for the solution with minimal total cost and with A which is not less than the required value A^* and τ_{tot} not exceeding τ^*, the solution fitness is evaluated using Equation (5.58).

Table 5.38. Example of solution string decoding

String element a_j	Type of element i	Version of element b	No. of parallel elements n_{ib}	No. of PR alternative m_j	No. of PR actions x_{ib}
11	1	1	3	3	15
0	1	2	0	0	-
6	1	3	2	1	3
5	2	1	1	3	15
14	2	2	4	3	15
4	3	1	1	2	10
9	3	2	3	1	3
2	3	3	0	3	-

Example 5.15

Consider the water desalination system described in Example 5.13. Each of four basic system components can consist of no more than seven parallel elements. The elements may be chosen from the list of available products. This list contains parameters of each element version (Table 5.39) and the corresponding $f(t)$ functions (Table 5.40), estimated using expert judgments.

Table 5.39. Parameters of the MSS elements

Type i	Description	Version b	$g_{i1}(b)$	$c_i(b)$	$\hat{c}_i(b)$	$\tilde{c}_i(b)$	τ_c
1	Filter	1	0.60	29.0	8.11	0.650	0.010
		2	0.40	20.0	8.01	0.810	0.016
		3	0.30	17.1	4.61	0.810	0.016
		4	0.30	15.9	3.23	0.490	0.004
		5	0.25	14.0	3.01	0.400	0.012
		6	0.20	12.3	2.91	0.400	0.012
		7	0.15	10.1	2.24	0.400	0.012
		8	0.10	9.40	1.90	0.450	0.022
2	Pump	1	0.65	8.8	6.85	0.430	0.024
		2	0.25	5.1	3.08	0.390	0.020
		3	0.15	3.7	1.71	0.490	0.012
3	Membrane	1	1.00	9.0	5.27	0.190	0.018
		2	0.70	7.1	3.41	0.190	0.018
		3	0.25	4.6	0.95	0.190	0.018
4	Power supply equipment	1	0.50	4.5	4.01	0.109	0.008
		2	0.35	2.9	2.21	0.121	0.012
		3	0.22	1.95	1.36	0.138	0.009

The demand distribution, the total operation time, the replacement times, the cost of unsupplied demand, and the possible number of replacements are the same as in Example 5.13.

Table 5.40. Renewal functions of MSS elements

Element			$f_{ib}(t)$					
Type	Version	t:	24	12	8	6	4.8	4
i	b	x:	5	10	15	20	25	30
1	1		33.0	8.0	4.1	2.0	1.0	0.7
	2		45.0	10.0	5.0	2.0	1.0	0.75
	3		36.0	8.0	2.0	0.6	0.2	0.15
	4		56.0	9.0	1.8	0.9	0.2	0.15
	5		43.0	6.0	3.0	0.9	0.2	0.16
	6		42.0	7.0	3.0	1.0	0.3	0.22
	7		16.0	4.0	0.6	0.1	0.03	0.02
	8		6.0	2.3	0.4	0.05	0.03	0.02
2	1		120.0	14.0	1.0	0.3	0.08	0.06
	2		36.0	4.0	0.9	0.2	0.06	0.04
	3		155.0	35.0	7.0	4.0	0.32	0.23
3	1		23.0	3.2	1.4	0.8	0.5	0.4
	2		15.0	4.0	1.0	0.4	0.1	0.01
	3		31.0	9.5	5.6	4.0	2.7	2.0
4	1		64.0	21.0	3.0	1.1	0.4	0.28
	2		58.0	19.0	2.2	0.8	0.3	0.24
	3		47.0	11.0	2.0	0.8	0.6	0.5

The problem is to find a minimal cost system structure and PR policy subject to the constraints of availability and total maintenance time.

Table 5.41 contains optimal solutions obtained for different A^* and τ^*.

Table 5.41. Examples of the obtained solutions

Constraints	C	C_{ud}	C_m	C_{tot}	τ_{tot}	A	Component			
							1	2	3	4
–	95.25	107.26	591.90	794.41	4.34	0.9706	3*4(15) 1*8(5)	4*2(15)	1*1(10)	1*2(15) 3*3(15)
A^*=0.980	108.55	40.00	652.86	801.41	8.09	0.9804	3*4(15) 1*8(15)	4*2(15)	3*2(5)	5*3(15)
A^*=0.990	109.10	28.44	683.67	821.21	8.18	0.9903	3*4(15) 1*8(15)	1*1(15) 2*2(15)	3*2(5)	6*3(15)
A^*=0.999	131.50	2.53	767.09	901.13	12.39	0.9990	4*4(15) 1*8(5)	5*2(10)	3*2(5)	6*3(15)
A^*=0.980 τ^*=3	95.20	75.93	648.33	819.47	2.78	0.9803	3*4(15) 1*8(15)	4*2(15)	1*1(15)	3*2(20)
A^*=0.990 τ^*=3	98.10	49.41	712.08	859.56	2.95	0.9909	3*4(15) 1*8(20)	4*2(20)	1*1(15)	4*2(15)
A^*=0.999 τ^*=3	124.30	2.86	918.94	1050.41	3.00	0.9990	4*4(15) 1*8(15)	5*2(20)	2*2(25)	4*2(15)
A^*=0.980 τ^*=2	101.70	57.49	725.20	884.40	1.96	0.9868	4*4(15)	4*2(20)	1*1(15)	3*2(20)
A^*=0.990 τ^*=2	98.10	41.05	772.63	911.79	1.98	0.9917	3*4(15) 1*8(20)	4*2(20)	1*1(20)	4*2(20)
A^*=0.999 τ^*=2	124.30	2.69	970.25	1097.79	1.99	0.9991	4*4(15) 1*8(15)	5*2(20)	2*2(25)	2*2(25) 3*3(20)

The solutions are presented in the form $n*b(x)$. The b corresponds to the version chosen for the given element from the list (Tables 5.39 and 5.40), n corresponds to the number of such elements, and x corresponds to the number of replacements of each element during the system operation time. For example, the solution obtained for $A^* = 0.98$ and $\tau^* = 2$ corresponds to the system consisting of four filters of version 4, four pumps of version 2, one membrane block of version 1, and three blocks of power commutation units of version 2. The number of replacements during $T = 120$ months for the filters, the pumps, the membrane block and for the power commutation equipment are 15, 20, 15, and 20 respectively. The estimated availability of such system is 0.9868; the estimated repair time is 1.96 months. The total cost is 884.40.

Reducing the allowable total maintenance time causes a reduction in the number of units installed. This lowers system redundancy and increases unsupplied demand. Alternatively, the total maintenance time can be reduced by increasing the replacement frequencies. (Compare, for example, solutions for $A^* = 0.990$ and $A^* = 0.999$ that have equal C for $\tau^* = 2$ and $\tau^* = 3$.)

5.3.4 Optimal Multistage System Modernization

When the system demand and/or reliability requirements increase over time, multiple system modernization actions are usually performed. Besides the system's multistage expansion (considered in Section 5.1.4) the system modernization plan can include modification of the existing equipment and changes in the maintenance policy. Both types of action lead to changes in the performance distribution of the elements already presented in the system. The multistage modernization plan should consider all these possible alternatives.

5.3.4.1 Problem Formulation

The initial system structure is defined by a reliability block diagram representing the interconnection among N MSS components and by a set $\Phi(0) = \{\boldsymbol{g}'_i(j), \boldsymbol{p}'_i(j), n'_{ij}: 1 \le i \le N, 1 \le j \le B'_i\}$ (see Section 5.1.3). Each component i can contain up to E_i different elements connected in parallel. In order to allow number of elements in each component to vary we introduce "dummy" elements of version 0.

In order to provide the desired levels of the system performance O^*_y at different stages y of the study period, the system should be modernized. Each stage y ($1 \le y \le Y$) begins $\tau(y)$ years after the base stage (stage 0) and is characterized by its demand distribution $\boldsymbol{w}(y)$, $\boldsymbol{q}(y)$. At each stage, three different types of action may be undertaken: modification of some of the system's elements, addition of new elements, and removal of some of the system's elements. In order to manipulate the system structure in a uniform way, we will formulate all of these actions in terms of element replacements:

The modification of element j of component i or changing the procedure of the element maintenance leads to changes in its performance distribution. We can consider this action as replacement of the element of version b_{ij} with the element of version b^*_{ij}. Parameters of new versions of the element should be specified for each element modification action.

The addition of a new element into the system can be considered as the replacement of the element of the corresponding type i and version 0 with a new element with version b^*_{ij}. For each component i there are B_i element versions available in the market.

The removal of the element of version b_{ij} from the system can be formulated as its replacement with the element of version 0.

Now one can define each system modernization action x by vector

$$\{i_x, b_{i_x x}, b^*_{i_x x}, c_x\} \tag{5.65}$$

where i_x is the number of the component in which the element of version $b_{i_x x}$ should be replaced with the element of version $b^*_{i_x x}$ and c_x is the cost of the action. Modernization action is feasible if the system component i_x already contains elements of version $b_{i_x x}$. Unfeasible actions should be replaced with "dummy" ones which correspond to $x = 0$. Action $x = 0$ is associated with cost $c_0 = 0$ and implies that no replacements are performed.

The system modernization plan at stage y can be defined by the vector $\boldsymbol{x}_y = (x_{ys}: 1 \le h \le H)$ where H is the maximal number of possible actions at a single stage. The cost of plan $\boldsymbol{x}_y$ can be calculated in present values as

$$C_y = \frac{1}{(1+\mathrm{IR})^{\tau(y)}} \sum_{h=1}^{H} c_{x_{yh}} \tag{5.66}$$

where IR is the interest rate.

Each modernization plan $\boldsymbol{x}_y$ transforms the set of versions of elements presented in the system at stage $y-1$ into the set of versions of elements presented in the system at stage y: $(\Phi(y-1), \boldsymbol{x}_y) \to \Phi(y)$ for $1 \le y \le Y$, where $\Phi(0)$ is the initial structure of the system considered.

The total system modernization cost for the given modernization plan $\boldsymbol{x}_1, \ldots, \boldsymbol{x}_Y$ is

$$C = \sum_{y=1}^{Y} C_y = \sum_{y=1}^{Y} \frac{1}{(1+\mathrm{IR})^{\tau(y)}} \sum_{h=1}^{H} c_{x_{yh}} \tag{5.67}$$

Having the system structure $\Phi(y)$ and demand distribution $\boldsymbol{w}(y)$, $\boldsymbol{q}(y)$ for each stage y, one can evaluate the MSS performance measures. The problem of optimal system expansion can be formulated as follows [160]:

Find the minimal cost system modernization plan that provides the required level O^* of the performance measure O_y at each stage y:

$$C(\boldsymbol{x}_1, \ldots, \boldsymbol{x}_Y) \to \min$$

$$\text{subject to } f(O_y(\boldsymbol{x}_1, \ldots, \boldsymbol{x}_y, \boldsymbol{w}(y), \boldsymbol{q}(y)), O^*) = 1 \text{ for } 1 \le y \le Y. \tag{5.68}$$

where $f(O_y, O^*)$ is a function representing the desired relation between O_y and O^*.

5.3.4.2 Implementing the Genetic Algorithm

The natural way to represent the multistage modernization plan $\boldsymbol{x}_1, \ldots, \boldsymbol{x}_Y$ is to use a string $\boldsymbol{a}$ containing HY integer numbers corresponding to the numbers of the actions to be undertaken during the study period. In the string $\boldsymbol{a}$, each number a_m for $H(y-1)+1 \le m \le yH$ corresponds to the action number $m - H(y-1)$ undertaken at stage y. All of the numbers should be generated in the range $0 \le a_m \le L$, where L is the total number of available actions. Action with number 0 means "do nothing"; it has no cost and is used in order to allow the total number of actions to vary. One can see that the solution representation string $\boldsymbol{a}$ contains sequentially allocated vectors $\boldsymbol{x}_y$ of the modernization plan at each stage $y = 1,\ldots, Y$.

By having the vector $\boldsymbol{a}$, one can determine the MSS structure in each stage y by replacing the elements of version $b_{i_x x}$ with elements of version $b^*_{i_x x}$ in accordance with the modernization plan x_y. Using the u-functions of the elements composing the system at each stage y after the corresponding replacements, one obtains the entire system PD by applying the UGF technique. For this PD and for the given demand distribution $\boldsymbol{w}(y)$, $\boldsymbol{q}(y)$ one obtains the MSS performance measure O_y. The solution fitness for the estimated values of O_y and the total cost C calculated using (5.76) is determined according to Equation (5.17).

Example 5.16

Consider two different optimization problems applied to the coal transportation system from Example 5.1. In the first problem we have to determine the initial system structure as well as its modernization plan. In this case, the initial structure set is empty: $\Phi(0) = \varnothing$. In the second problem we have to find the optimal modernization plan for the system with the initial structure presented in Table 5.42. Table 5.43 contains the demand distributions at five different stages and times from the present to the beginning of these future stages.

Table 5.42. Initial coal transportation system structure

No. of system component	1	2	3	4	5
Versions of elements, present in the component	1, 1	2, 2, 2	3, 3	4, 4	5, 5

Table 5.43. Demand distributions

Stage no.	$\tau(y)$		Demand distribution			
1	0	*w*	1.0	0.8	0.5	0.2
		q	0.48	0.09	0.14	0.29
2	3	*w*	1.0	0.8	0.5	-
		q	0.63	0.12	0.25	-
3	6	*w*	1.0	0.9	0.8	-
		q	0.65	0.19	0.16	-
4	9	*w*	2.0	1.6	0.8	-
		q	0.41	0.24	0.35	-
5	12	*w*	2.0	1.7	1.2	0.9
		q	0.52	0.15	0.13	0.20

The demand distributions have the same maximal load at stages 1-3 and 4-5. Within these periods, variations of the load demand are caused by intensification of boiler use. At stage 4, the installation of a new boiler is planned, which should be supplied by the same coal supply system.

Because of increased demand, the availability index of the system becomes $A_1(\boldsymbol{w}(1), \boldsymbol{q}(1)) = 0.506$ at the first stage. To provide a desired availability level $A^* = 0.95$ at all of the stages, the system should be modernized. The characteristics of the available modernization actions are presented in Table 5.44 in the form (5.74). The action costs c_x comprise both the investment and operational costs.

Table 5.44. List of available actions

x	i_x	$b_{i_x x}$	$b^*_{i_x x}$	c_x	Description
1	1	0	6	18.0	Addition of new feeder of version 6
2	1	0	7	42.0	Addition of new feeder of version 7
3	1	1	8	5.0	Electric drive system modification aimed at providing feeder speed increase
4	1	1	9	2.0	Installation of engines monitoring system in feeder of version 1
5	1	6	10	1.6	Installation of engines monitoring system in feeder of version 6
6	2	0	11	9.1	Addition of new conveyor of version 11
7	2	0	12	5.6	Addition of new conveyor of version 12
8	2	0	13	4.2	Addition of new conveyor of version 13
9	2	2	14	0.8	Changes in mechanical equipment to provide load increase
10	2	14	15	0.3	Installation of engines monitoring system in conveyor of version 14
11	2	14	16	0.5	Reduction of flat-belt condition inspection (replacement) period in conveyor of version 14
12	2	13	17	0.4	Installation of engines monitoring system in conveyor of version 13
13	3	0	3	567.0	Addition of new stacker of version 3
14	3	0	18	359.0	Addition of new stacker of version 18
15	3	3	19	6.0	Installation of hydraulics monitoring system in stacker of version 3
16	4	0	20	7.1	Addition of new feeder of version 20
17	4	4	21	2.0	Reduction of engines inspection period in feeder of version 4
18	4	20	22	1.8	Reduction of engines inspection period in feeder of version 20
19	5	0	23	47.5	Addition of new conveyor, version 23
20	5	5	24	1.1	Installation of engines monitoring system in conveyor of version 5
21	5	5	25	1.4	Reduction of flat-belt inspection (replacement) period
22	5	25	26	1.1	Action 20 performed after action 21
23	5	24	26	1.4	Action 21 performed after action 20

Each action leads to the appearance of an element of a certain version either purchased in the market or obtained by modification of some existing element or its maintenance procedure (equipment removal actions are not considered in this example). The list of parameters of different versions of elements is presented in Table 5.45. The table also contains the origins of the versions (P corresponds to versions already present in the system, N corresponds to new elements available on the market, and M corresponds to versions obtained by the modification of existing versions or their maintenance procedures).

Table 5.45. Parameters of available MSS elements

No. of version	g	p	Component	Origin
1	0.75	0.998	1	P
2	0.28	0.997	2	P
3	0.66	0.972	3	P
4	0.54	0.983	4	P
5	0.66	0.981	5	P
6	0.31	0.920	1	N
7	0.85	0.978	1	N
8	0.93	0.970	1	M
9	0.75	0.995	1	M
10	0.31	0.985	1	M
11	0.53	0.997	2	N
12	0.28	0.997	2	N
13	0.21	0.988	2	N
14	0.33	0.980	2	M
15	0.33	0.991	2	M
16	0.33	0.992	3	M
17	0.21	0.999	3	M
18	0.40	0.971	3	N
19	0.66	0.980	3	M
20	0.55	0.983	4	N
21	0.54	0.998	4	M
22	0.55	0.995	4	M
23	0.51	0.981	5	N
24	0.66	0.990	5	M
25	0.66	0.988	5	M
26	0.66	0.997	5	M

The best modernization plans obtained by the suggested algorithm for the first and the second problems are presented in Tables 5.46 and 5.47 respectively. The tables contain the list of actions to be undertaken for each stage, the structure of each of the system's components, and the system's availability index (if there are a few identical actions or elements, they are presented in the form $n*x$, where n corresponds to the number of identical actions of type x). For the interest rate

R = 0.1, the total modernization costs for problems 1 and 2 are in the present values C = 1783.23 and C = 542.55 respectively.

Table 5.46. Optimal modernization plan for problem the with $\Phi(0) = \varnothing$

Stage No.	List of actions	A	Structure of system components				
			1	2	3	4	5
1	1, 2*2, 2*6, 13, 14, 3*16, 3*19	0.958	6, 2*7	2*11	3, 18	3*20	3*23
2	15	0.951	6, 2*7	2*11	18, 19	3*20	3*23
3	2, 6, 18	0.950	6, 3*7	3*11	18, 19	2*20, 22	3*23
4	2*6, 2*13, 2*16, 2*19	0.952	6, 3*7	5*11	2*3, 18, 19	4*20, 22	5*23
5	5, 2*15, 18	0.950	10, 3*7	5*11	18, 3*19	3*20, 2*22	5*23

Table 5.47. Optimal modernization plan for problem the with $\Phi(0) \neq \varnothing$

Stage No	List of actions	A	Structure of system components				
			1	2	3	4	5
1	6, 16, 2*21, 2*22	0.950	2*1	3*2, 11	2*3	2*4, 20	2*26
2	4, 2*15	0.951	1, 9	3*2, 11	2*19	2*4, 20	2*26
3	2	0.951	1, 7, 9	3*2, 11	2*19	2*4, 20	2*26
4	2, 2*6, 13, 14, 2*16, 3*19	0.955	1, 2*7, 9	3*2, 3*11	3, 18, 2*19	2*4, 3*20	2*26, 3*23
5	2*9, 15	0.951	1, 2*7, 9	2, 3*11, 2*14	18, 3*19	2*4, 3*20	2*26, 3*23

5.3.5 Optimal Imperfect Maintenance

In Sections 5.3.2 and 5.3.3 we considered the preventive maintenance actions (PMAs) that return the element to its initial condition by replacing it by a new one or making it "as good as new". However, some actions (such as cleaning, adjustment, *etc.*), while improving the condition of system elements, cannot return them to the initial condition. Such actions are named imperfect preventive maintenance and are modelled using the age reduction concept [161].

The evolution of the reliability of system elements is a function of element age on a system's operating life. Element aging is strongly affected by maintenance activities performed on the system. Surveillance and maintenance reduce this age. In the case when the element becomes "as good as new", its age is reduced to zero. In the cases when the preventive maintenance (surveillance) does not affect the state of the element but ensures that the element is in an operating condition, the element remains "as bad as old" and its age does not change. All actions that do not zero element age can be considered to be imperfect PMAs. When an element of the system fails, corrective maintenance in the form of MR is performed. The MR returns the element to the operating condition without affecting its failure rate.

In this section, we consider the problem of determining a minimal cost plan of PMAs during MSS operation time T, which provides the required level of system reliability. The algorithm answers the questions of in what sequence, where (to which element), and what kind of available PMA should be applied to keep the system reliability R on the required level during a specified time.

5.3.5.1 Element Age Reduction Model

According to the age reduction concept the PMA reduces the effective age of the two-state element that it has immediately before it enters maintenance. The proportional age setback model used [156] assumes that the effective age τ_j of element j which undergoes PMAs at chronological times $(t_{j1},\dots,t_{jY_j})$ is

$$\tau_j(t) = \tau_j^+(t_{jy}) + (t - t_{jy}) \ \text{ for } \ t_{jy}< t < t_{jy+1} \quad (0\le y\le Y_j)$$

$$\tau_j^+(t_{jy}) = \varepsilon_{jy}\cdot \tau_j(t_{jy}) = \varepsilon_{jy} \cdot[\tau_j^+(t_{jy-1}) + (t_{jy} - t_{jy-1})] \tag{5.69}$$

where $\tau_j^+(t_{jy})$ is the age of the element immediately after the yth PMA, ε_{jy} is the age reduction coefficient associated with this PMA, which ranges in the interval [0, 1], and $\tau_j(0) = t_{j0} = 0$ by definition.

The two extreme effects of PMA on the state of the element correspond to the cases when $\varepsilon_{jy} = 1$, or $\varepsilon_{jy} = 0$. In the first case, the model simply reduces to "as bad as old", which assumes that PMA does not affect the age of the element. In the second case the model reduces to "as good as new", which means that the element's age is restored to zero (replacement). All the PMAs with $0<\varepsilon_{jy}<1$ lead to a partial improvement in the state of the element.

The induced hazard function of the MSS element j can be expressed as

$$h^*_j(t) = h_j(\tau_j(t))+h_{j0} \tag{5.70}$$

where $h_j(t)$ is the hazard function of the element defined for the case when it does not undergo PMAs and h_{j0} is the term corresponding to the initial age of the element, which can differ from zero at the beginning of MSS operation time.

The reliability of element j in the interval between PMAs y and y+1 is

$$r_j(t) = \exp[-\int_{t_{jy}}^{t} h_j^*(x)\ \mathrm{d}x] = \exp[-\int_{\tau_j^+(t_{jy})}^{\tau_j(t)} h_j(x)\ \mathrm{d}x]$$

$$= \exp[H_j(\tau_j^+(t_{jy})) - H_j(\tau_j(t))],\ t_{jy} \le t \le t_{jy+1} \tag{5.71}$$

where $H_j(\tau)$ is the accumulated hazard function for element j. One can see that immediately after PMA y, when $t = t_{jy}$, the reliability of element $r_j(t_{jy}) = 1$.

MRs are performed if MSS elements fail between PMAs. The cost associated with MRs depends on the failure rates of the elements. According to [162], the expected minimal repair cost of element j in interval $[0, t]$ is

$$c_{MRj}(t) = \tilde{c}_j \int_0^t h_j(x)\ dx \tag{5.72}$$

where $\tilde{c}_j$ is the cost of a single MR of this element. When the element undergoes PMAs at times $t_{j1},\ldots,t_{jY_j}$, the total MR cost is

$$C_{MRj} = \tilde{c}_j \sum_{y=0}^{Y_j} \int_{\tau_j^+(t_{jy})}^{\tau_j(t_{jy+1})} h_j(x)\ dx$$
$$= \tilde{c}_j \sum_{y=0}^{Y_j} [H(\tau_j(t_{jy+1})) - H(\tau_j^+(t_{jy}))] \tag{5.73}$$

where $t_{j0} = 0$ and $t_{jY_j+1} = T$ by definition, $\tau_j(t_{jy+1}) = \tau_j^+(t_{jy}) + (t_{jy+1} - t_{jy})$.

5.3.5.2 Problem Formulation

Consider a system consisting of N two-state elements. Each element j is characterized by its nominal performance rate g_{j1}, hazard function $h_j(t)$ and associated MR cost $\tilde{c}_j$. The MSS is supposed to meet a demand w.

PMAs and MRs can be realized on each MSS element. PMAs modify element reliability, whereas minimal repairs do not affect it. The effectiveness of each PMA is defined by the age reduction coefficient ε ranging from 0 ("as good as new") to 1 ("as bad as old"). The time in which the element is not available due to a PMAs or MR activity is negligible if compared with the time elapsed between the consecutive activities. A list of possible PMAs is available for a given MSS. In this list, each PMA x is associated with the cost of its implementation $c(x)$, the number of affected element $e(x)$ and the age reduction coefficient $\varepsilon(x)$.

System operation time T is divided into M intervals, each with duration d_m ($1 \le m \le M$). PMAs can be performed at the end of each interval. They are performed if the MSS reliability $R(t,w)$ becomes lower than the desired level R^*. It should be mentioned that d_m of different intervals must not necessarily be equal to each other and can be chosen for practical reasons.

The sequence of PMAs performed to maintain MSS reliability can be defined by a vector $\boldsymbol{x}$ of numbers of these actions as they appear on the PMA list. Each time PMA is necessary to improve system reliability, the action to be performed is defined by the next number from this vector. Note that the chosen PMA x_i may be insufficient to increase MSS reliability to the desired level. In this case, the next action x_{i+1} should be performed at the same time, and so on. One can see that the total number Y of PMAs cannot be predefined and depends on the composition of vector $\boldsymbol{x}$.

For a given vector $\boldsymbol{x}$ the total number Y_j and chronological times of PMAs are determined for each system element j ($1\le j\le n$). If $Y_j = 0$, which corresponds to the case in which $e(x_i) \neq j$ for all $x_i\in\boldsymbol{x}$, the minimal repair cost $C_{\mathrm{MR}j}$ given by Equation (5.82) is defined by $t_{j0} = 0$ and $t_{jY_j+1} = t_{j1} = T$.

The vector $\boldsymbol{x}$ defines both the total cost of PMAs as

$$C_{\mathrm{PM}}(\mathbf{x}) = \sum_{i=1}^{N} c(x_i) \tag{5.74}$$

and the total cost of MRs as

$$C_{\mathrm{MR}}(\boldsymbol{x}) = \sum_{j=1}^{N} \tilde{c}_j \sum_{y=0}^{Y_j} [H(\tau_j(t_{jy+1})) - H(\tau_j^{+}(t_{jy}))] \tag{5.75}$$

The optimal imperfect maintenance problem is formulated as follows [163]: find the optimal sequence $\boldsymbol{x}$ of PMAs chosen from the list of available actions which minimizes the total maintenance cost while providing the desired MSS reliability

$$C_{\mathrm{tot}}(\boldsymbol{x}) = C_{\mathrm{PM}}(\boldsymbol{x}) + C_{\mathrm{MR}}(\boldsymbol{x}) \rightarrow \min$$

$$\text{subject to } R(\boldsymbol{x}, t, w) \ge R^*,\ 0\le t\le T \tag{5.76}$$

5.3.5.3 Implementing the Genetic Algorithm

The natural representation of a vector of PMA numbers $\boldsymbol{x}$ corresponding to problem formulation (5.85) is by an integer string $\boldsymbol{a}$ containing numbers generated in the range $(1, L)$, where L is the total number of possible types of PMA (the length of the PMA list). Since the number of PMAs performed can vary from solution to solution, a redundant number of positions Y^* should be provided in each string. In this case, after decoding the solution represented by a string $\boldsymbol{a} = (a_1,\ldots,a_{Y^*})$, only the first Y numbers will define the PMA plan. The elements of the string from a_{Y+1} to a_{Y^*} do not affect the solution but can affect its offspring by participating in crossover and mutation procedures.

The following procedure determines the fitness value for an arbitrary solution defined by integer string $\boldsymbol{a}$:

1. Define for all the MSS elements ($1\le j\le N$) effective ages $\tau_j = -d_1$. Define $H_j(\tau_j^{+}) = 0$. Assign 0 to chronological time t, the interval number m, the number of PMAs performed y, and total maintenance cost C_{tot}.
2. Increment the interval number m by 1. Increment chronological time t and ages τ_j of all the system elements ($1\le j\le N$) by d_m.

3. Calculate $H_j(\tau_j)$ and reliability $r_j(\tau_j) = \exp[H_j(\tau_j^+)-H_j(\tau_j)]$ for all MSS elements ($1 \le j \le N$).

4. Using the reliability block diagram method, define the MSS PD and calculate system reliability $R(t,w)$ for the given demand w.

5. If $R(t,w)<R^*$ increment y by 1 and define the PMA to be performed at time t as $x = a_y$. Add the PMA cost $c(x)$ to C_{tot}.

Determine the cost of MRs for element $e(x)$ in the interval between the previous and current PMAs as $\tilde{c}_{e(x)}\,[H_{e(x)}(\tau_{e(x)})-H_{e(x)}(\tau_{e(x)}^+)]$ and add this value to C_{tot}.

Modify the age $\tau_{e(x)}$ of element $e(x)$ by multiplying it by the age reduction coefficient $\varepsilon(x)$. Calculate the new value of $H_{e(x)}(\tau_{e(x)}^+)$ for the modified age $\tau_{e(x)}$ and assign $H_{e(x)}(\tau_{e(x)}) = H_{e(x)}(\tau_{e(x)}^+)$.

Assign the value of 1 to the reliability of element $e(x)$ and return to step 4.

6. If $R(t, w) \ge R^*$ and $t < T$, return to step 2.

7. If $R(t, w) \ge R^*$ and $t \ge T$, evaluate the costs of MRs during the last interval for all the elements ($1 \le j \le N$) as $\tilde{c}_j\,[H_j(\tau_j)-H_j(\tau_j^+)]$ and add these costs to C_{tot}. Determine the solution fitness as $M-C_{tot}(\boldsymbol{a})$.

Example 5.17

Consider the series-parallel oil refinery subsystem consisting of four components connected in series in Figure 5.20. The system belongs to the type of flow transmission MSS with flow dispersion and contains 11 two-state elements with different performance rates and reliability functions. The reliability of each element is defined by a Weibull hazard function

$$h_j(t)=\lambda^{\gamma}\gamma[\tau(t)]^{\gamma-1}+h_0$$

which is widely adopted to fit repairable equipment. The accumulated hazard function takes the form

$$H(t)=[\lambda\tau(t)]^{\gamma}+h_0\tau(t)$$

The nominal performance rate g_{j1}, intensity function scale parameter λ_j, shape parameter γ_j and hazard constant h_{j0} are presented in Table 5.48 for each element j. This table also contains the cost of MR for each element.

A set of possible PMAs is defined for the given MSS in Table 5.49. Each action is characterized by its cost, the number of the element affected and the age reduction coefficient ε. PMA can be replacement ($\varepsilon = 0$), surveillance ($\varepsilon = 1$) and imperfect PMA with a partial improvement effect ($0<\varepsilon<1$).

MSS operation time is 25 years. The times for possible PMAs are evenly spaced intervals of $d_m = 1.5$ months (0.125 years). The problem is to generate a

PMA sequence which provides the system work during its operation time with a performance rate not less than w and reliability not less than R^*.

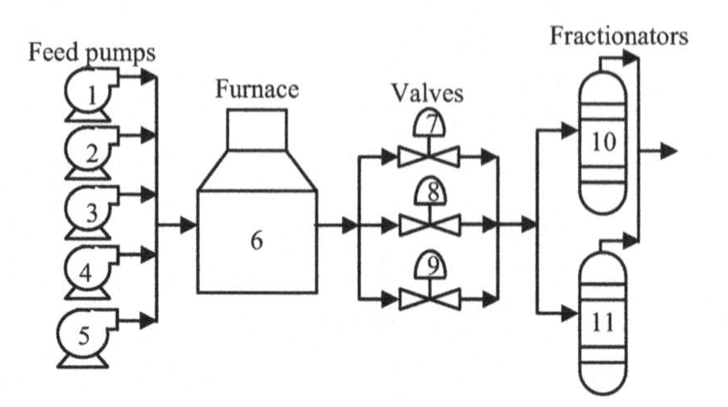

Figure 5.20. Oil refinery subsystem

The length of the integer string representing solutions was chosen to be $Y^* = 25$. To see the influence of parameters w and R^* on the optimal solution, four optimal PMA sequences were obtained for different compositions of these parameters: $w = 0.8$, $w = 1.0$ and $R^* = 0.90$, $R^* = 0.95$. The sequences obtained are presented in Tables 5.50-5.53. The tables contain the total cost of the maintenance plan, the list of the PMAs performed and includes the time from the beginning of the MSS life in which the action should be performed, the number of the action (as appears in Table 5.49), the number of the element affected by the action, and the entire MSS reliability immediately after performing the action.

Table 5.48. Parameters of MSS elements

No. of element	g_{j1}	λ_j	γ_j	h_{j0}	$\tilde{c}_j$
1	0.4	0.050	1.8	0.0001	0.9
2	0.4	0.050	1.8	0.0001	0.9
3	0.4	0.050	1.8	0.0	0.9
4	0.4	0.070	1.2	0.0003	0.8
5	0.6	0.010	1.5	0.0	0.5
6	1.3	0.010	1.8	0.00007	2.4
7	0.6	0.020	1.8	0.0	1.3
8	0.5	0.008	2.0	0.0001	0.4
9	0.4	0.020	2.1	0.0	0.7
10	1.0	0.034	1.6	0.0	1.2
11	1.0	0.008	1.9	0.0004	1.9

Table 5.49. Parameters of PMAs

No. of PMA	No. of element	Age reduction ε	PMA cost c	No. of PMA	No. of element	Age reduction ε	PMA cost c
1	1	1.00	2.2	16	6	0.00	19.0
2	1	0.56	2.9	17	7	0.75	4.3
3	1	0.00	4.1	18	7	0.00	6.5
4	2	1.00	2.2	19	8	0.80	5.0
5	2	0.56	2.9	20	8	0.00	6.2
6	2	0.00	4.1	21	9	1.00	3.0
7	3	1.00	2.2	22	9	0.65	3.8
8	3	0.56	2.9	23	9	0.00	5.4
9	3	0.00	4.1	24	10	1.00	8.5
10	4	0.76	3.7	25	10	0.70	10.5
11	4	0.00	5.5	26	10	0.00	14.0
12	5	1.00	7.3	27	11	1.00	8.5
13	5	0.60	9.0	28	11	0.56	12.0
14	5	0.00	14.2	29	11	0.00	14.0
15	6	0.56	15.3				

Table 5.50. The best PMA sequence obtained for $w = 0.8$, $R^* = 0.9$

C_{tot}	t (years)	No. of PMA	Element affected	$R(t, w)$
	14.250	6	2	0.949
	17.875	8	3	0.923
34.824	19.500	15	6	0.948
	21.750	21	9	0.932
	23.000	2	1	0.947

Table 5.51. The best PMA sequence obtained for $w = 1.0$, $R^* = 0.9$

C_{tot}	t (years)	No. of PMA	Element affected	$R(t, w)$
	10.625	18	7	0.956
	13.625	3	1	0.939
	16.000	15	6	0.934
51.301	17.625	6	2	0.925
	19.000	8	3	0.930
	20.500	10	4	0.913
	21.250	18	7	0.956
	24.375	7	3	0.915

Table 5.52. The best PMA sequence obtained for $w = 0.8$, $R^* = 0.95$

C_{tot}	t (years)	No. of PMA	Element affected	$R(t, w)$
	11.750	8	3	0.969
	13.500	18	7	0.955
	14.000	6	2	0.963
63.669	15.875	3	1	0.953
	16.500	16	6	0.988
	21.625	9	3	0.962
	23.125	21	9	0.963
	24.500	1	1	0.955

Table 5.53. The best PMA sequence obtained for $w = 1.0$, $R^* = 0.95$

C_{tot}	t (years)	No. of PMA	Element affected	$R(t, w)$
	7.750	18	7	0.982
	10.750	3	1	0.963
	11.875	10	4	0.959
	12.625	18	7	0.964
	14.000	16	6	0.978
82.625	16.125	6	2	0.969
	17.625	9	3	0.965
	18.875	3	1	0.955
	19.375	27	11	0.963
	20.375	18	7	0.983
	23.125	10	4	0.965
	24.250	17	7	0.958
	24.750	4	2	0.956

The functions $R(t, 1.0)$ and $R(t, 0.8)$ for the optimal PMA sequences obtained are presented in Figure 5.21. One can see that the timing of PMAs is defined by the behaviour of the entire MSS reliability function. Indeed, the PMA are accomplished when the reliability curve approaches the level R^*.

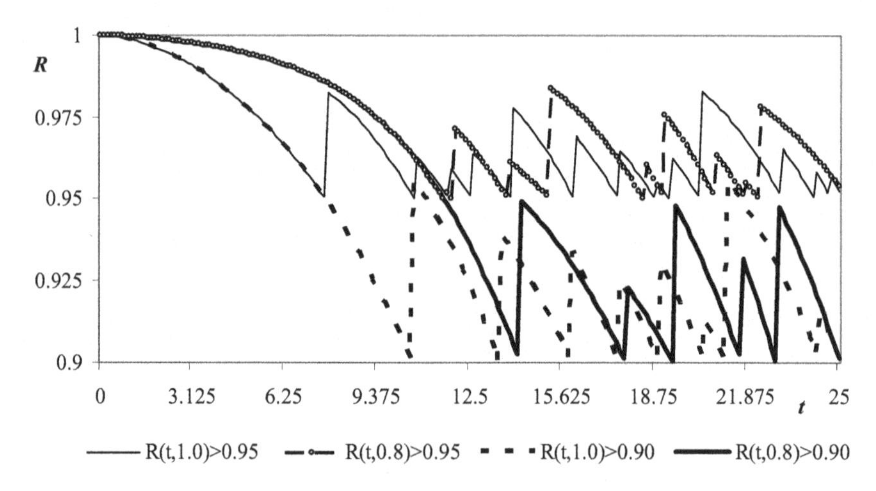

Figure 5.21. MSS reliability function $R(t, w)$ corresponding to PMA plans obtained

6. Universal Generating Function in Analysis and Optimization of Special Types of Multi-state System

6.1 Multi-state Systems with Bridge Structure

The bridge structure (Figure 6.1) is an example of a complex system for which the u-function cannot be evaluated by decomposing it into series and parallel subsystems. Each of the five bridge components can in turn be a complex composition of the elements. After obtaining the equivalent u-functions of these components one should apply the general composition operator in the form (1.20) over all five u-functions of the components in order to obtain the u-function of the entire bridge. The choice of the structure function in this composition operator depends on the type of system.

By having the u-function of the entire bridge system, one can use it either directly for evaluating the system performance measures (as shown in Section 3.3) or use it as a u-function of an equivalent element that replaces the bridge structure for evaluating the u-function of a higher level system when applying the block diagram method.

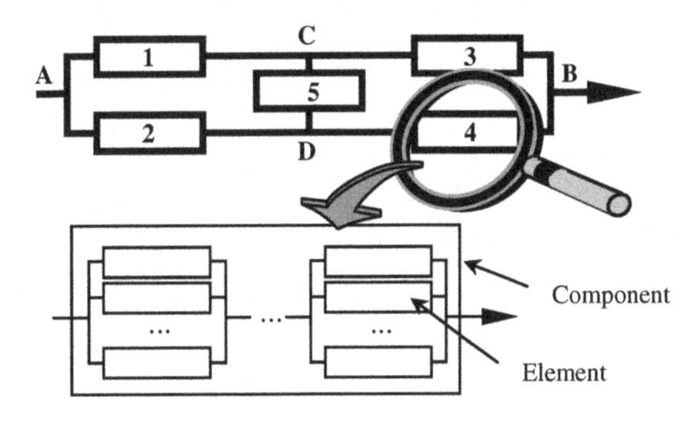

Figure 6.1. Bridge structure

6.1.1 *u*-function of Bridge Systems

6.1.1.1 Flow Transmission Multi-state Systems

In order to evaluate the performance of the flow transmission MSS with the flow dispersion, consider the flows through the bridge structure presented in Figure 6.1. First, there are two parallel flows through components 1, 3 and 2, 4. To determine the capacities of each of the parallel substructures composed from components connected in series, the function ϕ_{ser} (4.2) should be used. The function ϕ_{par} (4.9) should be used afterwards to obtain the total capacity of the two parallel substructures. Therefore, the structure function of the bridge, which does not contain diagonal component, is

$$\phi(G_1, G_2, G_3, G_4) = \phi_{par}(\phi_{ser}(G_1, G_3), \phi_{ser}(G_2, G_4)) \tag{6.1}$$

and its total capacity for the flow transmission MSS with the flow dispersion is equal to $\min\{G_1, G_3\}+\min\{G_2, G_4\}$.

The surplus of the transferred product on one of end nodes of component 5 can be expressed as

$$s = \max\{(G_1-G_3),(G_2-G_4), 0\} \tag{6.2}$$

and the deficit of the transferred product on one of the end nodes of component 5 can be expressed as

$$d = \max\{(G_3-G_1), (G_4-G_2), 0\} \tag{6.3}$$

The necessary condition for the existence of the flow through component 5 is the simultaneous existence of a surplus on one end node and a deficit on the other end: $s \neq 0$, $d \neq 0$. This condition can be expressed as $(G_1-G_3)(G_2-G_4)<0$.

If the condition is met, the flow through the component 5 will transfer the amount of the product which cannot exceed the capacity of the component G_5 and the amount of the surplus product s. The deficit d on the second end of component 5 is the amount of the product that can be transferred by the component that follows the diagonal (component 3 or 4). Therefore, the flow through the diagonal component is also limited by d. Thus, the maximal flow through the diagonal component is $\min\{s, d, G_5\}$.

Now we can determine the total capacity of the bridge structure when the capacities of its five components are given:

$$\begin{aligned}&\phi_{br}(G_1, G_2, G_3, G_4, G_5) = \min\{G_1, G_3\}+\min\{G_2, G_4\}\\&+ \min\{|G_1-G_3|, |G_2-G_4|, G_5\} \times 1((G_1-G_3)(G_2-G_4)<0)\end{aligned} \tag{6.4}$$

Now consider the performance of the flow transmission MSS without flow dispersion. In such a system a single path between points A and B providing the greatest flow should be chosen. There exist four possible pathsconsisting of groups of components (1, 3), (2, 4), (1, 5, 4) and (2, 5, 3) connected in a series. The transmission capacity of each path is equal to the minimum transmission capacity of the elements belonging to this path. Therefore, the structure function of the entire bridge takes the form

$$\phi_{br}(G_1, G_2, G_3, G_4, G_5) = \max\{\min\{G_1,G_3\} \min\{G_2,G_4\}, \min\{G_1,G_5,G_4\}, \min\{G_2,G_5,G_3\}\} \quad (6.5)$$

Note that the four parallel subsystems (paths) are not statistically independent, since some of them contain the same elements. Therefore, the bridge u-function cannot be obtained by system decomposition as for the series-parallel systems. Instead, one has to evaluate the structure function (6.5) for each combination of states of the five independent components.

6.1.1.2 Task Processing Multi-state Systems

In these types of system a task is executed consecutively by components connected in series. No stage of work execution can start until the previous stage is entirely completed. Therefore, the total processing time of the group of elements connected in series is equal to the sum of the processing times of the individual elements.

First, consider a system without work sharing in which the parallel components act in a competitive manner. There are four alternative sequences of task execution (paths) in a bridge structure. These paths consist of groups of components (1, 3), (2, 4), (1, 5, 4) and (2, 5, 3). The total task can be completed by the path with a minimal total processing time

$$T = \min\{t_1+t_3, t_2+t_4, t_1+t_5+t_4, t_2+t_5+t_3\} \quad (6.6)$$

where t_j and $G_j = 1/t_j$ are respectively the processing time and the processing speed of element j.

The entire bridge performance defined in terms of its processing speed can be determined as

$$G = 1/T = \phi_{br}(G_1, G_2, G_3, G_4, G_5) = \max\{\phi_{ser}(G_1,G_3), \phi_{ser}(G_2,G_4), \phi_{ser}(G_1,G_4,G_5), \phi_{ser}(G_2,G_3,G_5))\} \quad (6.7)$$

where ϕ_{ser} is defined in Equation (4.5).

Now consider a system with work sharing for which the same three assumptions that were made for the parallel system with work sharing (Section 4.1.2) are made. There are two stages of work performing in the bridge structure. The first stage is performed by components 1 and 2 and the second stage is performed by components 3 and 4. The fifth component is necessary to transfer work between nodes C and D. Following these assumptions, the decision about work sharing can be made in the nodes of bridge A, C or D only when the entire amount of work is available in this node. This means that components 3 or 4 cannot start task processing before both the components 1 and 2 have completed their tasks and all of the work has been gathered at node C or D.

There are two ways to complete the first stage of processing in the bridge structure, depending on the node in which the completed work is gathered. To complete it in node C, the amount of work $(1-\alpha)x$ should be performed by component 1 with processing speed G_1 and the amount of work αx should be performed by component 2 with processing speed G_2 and then transferred from node D to node C with speed G_5 (α is the work sharing coefficient). The time the work performed by component 1 appears at node C is $t_1 = (1-\alpha)x/G_1$. The time the work performed by component 2 and transferred by component 5 appears at node C is t_2+t_5, where $t_2 = \alpha x/G_2$ and $t_5 = \alpha x/G_5$. The total time of the first stage of processing is $T_{1C} = \max\{t_1, t_2+t_5\}$. It can be easily seen that T_C is minimized when the α is chosen that provides equality $t_1 = t_2+t_5$. The work sharing coefficient obtained from this equality is $\alpha = G_2G_5/(G_1G_2+G_1G_5+G_2G_5)$ and the minimal processing time is

$$T_{1C} = x(G_2+G_5)/(G_1G_2+G_1G_5+G_2G_5) \tag{6.8}$$

To complete the first stage of processing in node D, the amount of work $(1-\beta)x$ should be performed by component 2 with processing speed G_2 and the amount of work βx should be performed by component 1 with processing speed G_1 and then transferred from node C to node D with speed G_5. The minimal possible processing time can be obtained in the same manner as T_{1C}. This time is

$$T_{1D} = x(G_1+G_5)/(G_1G_2+G_1G_5+G_2G_5) \tag{6.9}$$

If the first stage of processing is completed in the node C, then the amount of work $(1-\gamma)x$ should be performed by component 3 in the second stage of processing, which takes time $t_3 = (1-\gamma)x/G_3$. The rest of the work γx should be first transferred to node D by component 5 and then performed by component 4. This will take time $t_5+t_4 = \gamma x/G_5+\gamma x/G_4$. Using the optimal work sharing (when $t_3 = t_4+t_5$) with $\gamma = G_4G_5/(G_3G_4+G_3G_5+G_4G_5)$ we obtain the minimal time of the second stage of processing:

$$T_{2C} = x(G_4+G_5)/(G_3G_4+G_3G_5+G_4G_5) \tag{6.10}$$

Using the same technique we can obtain the minimal processing time when the second stage of processing starts from node D:

$$T_{2D} = x(G_3+G_5)/(G_3G_4+G_3G_5+G_4G_5) \tag{6.11}$$

Assuming that the optimal way of work performing can be chosen in node A, we obtain the total bridge processing time T as

$$T = \min\{T_{1C}+T_{2C}, T_{1D}+T_{2D}\} \tag{6.12}$$

where

$$T_{1C}+T_{2C} = x[(G_2+G_5)/\sigma+(G_4+G_5)/\pi]$$

$$T_{1D}+T_{2D} = x[(G_1+G_5)/\sigma+(G_3+G_5)/\pi]$$

$$\sigma = G_1G_2+G_1G_5+G_2G_5$$

$$\pi = G_3G_4+G_3G_5+G_4G_5$$

The condition $T_{1C}+T_{2C} \le T_{1D}+T_{2D}$ is satisfied when $(G_2-G_1)\pi \le (G_3-G_4)\sigma$.

The expressions obtained can be used to estimate the processing speed of the entire bridge:

$$G = 1/T = \phi_{br}(G_1, G_2, G_3, G_4, G_5)=\sigma\pi/[(f+G_5)\sigma+(e+G_5)\pi] \quad (6.13)$$

where

$$f = G_4,\ e = G_2 \ \text{ if } \ (G_2-G_1)\pi \le (G_3-G_4)\sigma$$

$$f = G_3,\ e = G_1 \ \text{ if } \ (G_2-G_1)\pi > (G_3-G_4)\sigma$$

6.1.1.3 Simplification Technique

Note that in the special case when one of the bridge elements is in a state of total failure, the bridge structure degrades to a series-parallel one. All five possible configurations of this degraded bridge are presented in Figure 6.2.

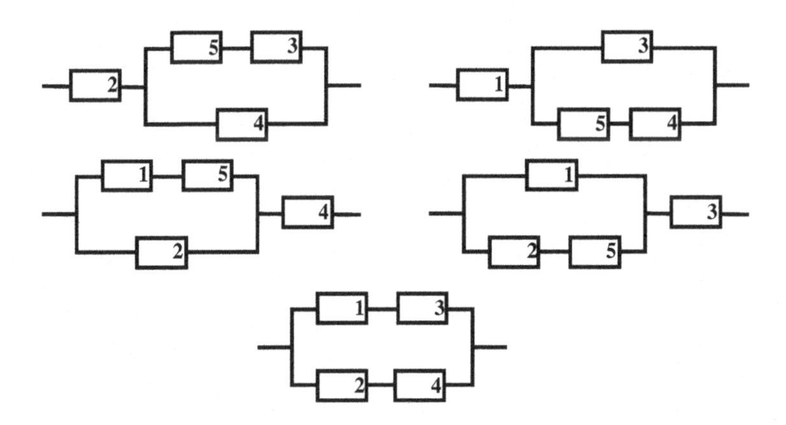

Figure 6.2. Degraded bridge structures in the case of single-element total failure

There is no need to use Equations (6.4), (6.5), (6.7) or (6.13) in order to evaluate the structure function of the bridge when one of the random values $G_1,\ldots,G_5$ is equal to zero. A simpler way to evaluate it is by using the reliability block diagram technique.

The following simplification rules can be used when more than one element is in a state of total failure:

1. If $G_1 = G_2 = 0$ or $G_3 = G_4 = 0$ the total bridge performance is equal to zero.

2. If $G_1 = G_4 = G_5 = 0$ or $G_2 = G_3 = G_5 = 0$ the total bridge performance is equal to zero.
3. If any two out of three random values composing groups $\{G_1, G_3, G_5\}$ or $\{G_2, G_4, G_5\}$ are equal to zero, the third value can also be zeroed. In this case, the bridge is reduced to two components connected in a series (2, 4 and 1, 3 respectively).

Example 6.1

Consider a bridge consisting of five elements with performance distributions presented in Table 6.1. The elements can have up to three states.

Table 6.1. Performance distributions for bridge elements

No. of element	Component performance distribution					
	State 0		State 1		State 2	
	g	p	g	p	g	p
1	0	0.10	6	0.60	8	0.30
2	0	0.05	7	0.95	-	-
3	0	0.10	4	0.10	6	0.80
4	0	0.05	6	0.20	9	0.75
5	0	0.15	2	0.85	-	-

The reliability and the performance deficiency for this bridge structure as functions of system demand are presented in Figure 6.3 for the structure interpreted as a system of four different types (numbered according to Table 4.4). The values of the expected performances obtained for these four different systems are ε = 8.23, ε = 6.33, ε = 3.87 and ε = 3.55.

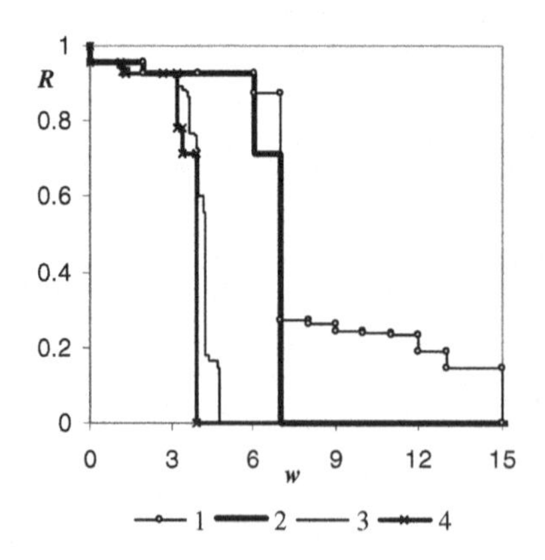

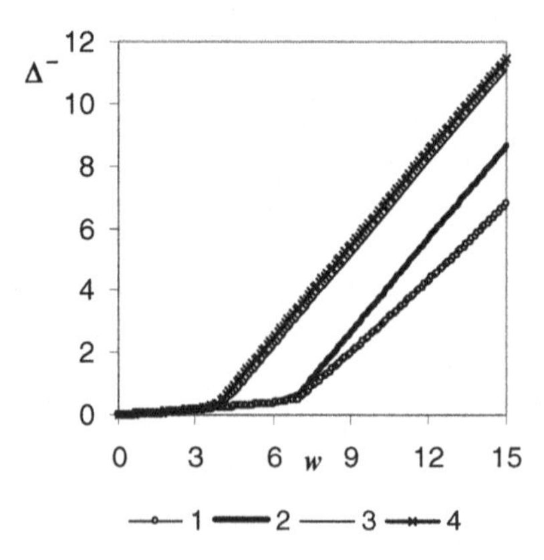

Figure 6.3. Reliability and performance deficiency for different types of bridge MSS

6.1.2 Structure Optimization of Bridge Systems

By having the technique for determining the u-function of bridge subsystems, one can apply it for solving structure optimization problems for systems with complex topology consisting of series-parallel and bridge subsystems. The formulation of the generalized structure optimization problem is similar to that presented in Section 5.1.2.1:

An MSS consists of N components connected in series, in parallel, or composing a bridge (according to a given reliability block diagram). Each component is a subsystem that can consist of parallel elements with the same functionality. For each component i, different versions of elements may be chosen from the list of element versions available in the market. The optimal solution corresponds to the minimal cost system configuration that provides the desired level of the given system performance measure.

When the system reliability is optimized and no constraints are imposed on the system configuration, the solutions for the structure optimization problem for the bridge system always produce a degraded series-parallel system. This happens because, from the reliability point of view, building a system with bridge topology is not justified. Indeed, when no allocation constraints are imposed on the system, the series-parallel solution is more reliable and less expensive than one based on the bridge architecture. One can see that for an arbitrary bridge structure (Figure 6.1) the less expensive and more reliable solution can be obtained by uniting elements of components 1 and 2 in component 1 and elements of components 3 and 4 in component 3 and by removing diagonal component 5. The existence of the bridge systems is justified only when some constraints are imposed on the allocation of the system elements or when the elements belonging to the same component are subject to CCFs.

6.1.2.1 Constrained Structure Optimization Problems

Connecting the system components in a bridge topology is widely used in design practice. There can be many different reasons for a system to take the bridge form. For example:

- the bridge configuration of the system is determined by factors not related to its reliability;
- in order to provide the redundancy on the component level the system should have parallel functionally equivalent components;
- the system contains parallel functionally equivalent but incompatible components;
- the number of elements that can be allocated within each component is limited.

In order to take into account such constraints, when the optimal system configuration is determined one has to modify the objective function in a way that penalizes the constraint violation. The methodology of solving the structure optimization problem for the systems containing series-parallel and bridge structures presumes using the optimization problem definition and GA implementation technique presented in Sections 5.1.2.1 and 5.1.2.2 and including the corresponding penalties to the solution fitness.

The solution decoding procedure, based on the UGF technique, performs the following steps:

1. Determines number of chosen elements n_{ib} for each system component and each element version from the string $\boldsymbol{a}$.

2. Determines u-functions $u_{ib}(z)$ of each version of elements according to their PD $\boldsymbol{g}_i(b)$, $\boldsymbol{p}_i(b)$.

3. Determines u-functions of each component i ($1\leq i\leq N$) by applying the composition operator $\otimes_{\varphi_{\text{par}}}$ over u-functions of the elements belonging to this component.

4. Determines the u-function of the entire MSS $U(z)$ by applying the corresponding composition operators using the reliability block diagram method and composition operators $\otimes_{\varphi_{\text{par}}}$, $\otimes_{\varphi_{\text{ser}}}$ and $\otimes_{\varphi_{\text{br}}}$.

5. Having the u-function of the entire system and its components, determines their performance measures as described in Section 3.3.

6. Determines the total system cost using Equation (5.2).

7. Determines the solution's fitness as a function of the MSS cost and performance measure as

$$M-C(\boldsymbol{a})-\pi_1(1+|O-O^*|)(1-f(O,O^*))-\pi_2\eta \tag{6.14}$$

where π_1 and π_2 are penalty coefficients, M is a constant value and η is a measure of constraint violation. For example, if it is important to provide the expected performance ε_j of each bridge component j ($1\leq j\leq 5$) at a level not less than ε_j*, then η takes the form

$$\eta=\sum_{j=1}^{5}\max(\varepsilon_j^*-\varepsilon_j,0) \tag{6.15}$$

If no more than I_j elements can be allocated in each bridge component j, then η takes the form

$$\eta=\sum_{j=1}^{5}\max\Big(\sum_{b=1}^{B_j}n_{jb}-I_j,0\Big) \tag{6.16}$$

If a pair of parallel components j and i should provide an identical nominal performance (the components consist of two state elements with nominal performances $g_{j1}(b)$ and $g_{i1}(b)$, then η takes the form

$$\eta=\left|\sum_{b=1}^{B_j}n_{jb}g_{j1}(b)-\sum_{b=1}^{B_i}n_{ib}g_{i1}(b)\right| \tag{6.17}$$

Example 6.2

Consider a power station coal transportation system that receives the coal carried by sea [164]. There may be up to two separate piers where a ship with coal can be berthed. Each pier should be provided with a separate coal transportation line. The lines can be connected by an intermediate conveyor in order to balance their load.

This flow transmission system (with flow dispersion) contains six basic components:

1, 2. Coal unloading terminals, including a number of travelling rail cranes for the ship discharge with adjoining primary conveyors.

3, 4. Secondary conveyors that transport the coal to the stacker-reclaimer.

5. An intermediate conveyor that can be used for load balancing between the lines.

6. The stacker-reclaimer that transfers the coal to the boiler feeders.

Each element of the system is considered to be a two-state unit. Table 6.2 shows availability, nominal capacity, and unit cost for equipment available in the market. The system demand is $w = 1$.

First consider an optimization problem in which the number of cranes at any pier cannot exceed three because of allocation constraints. The penalty (6.16) with $I_1 = I_2 = 3$ and $I_3 = I_4 = I_5 = \infty$ is incorporated into the solution fitness function.

Table 6.2. Characteristics of available system elements

Component	Description	No. of version	*g*	*p*	*c*
1,2	Crane with primary conveyor	1	0.80	0.930	0.750
		2	0.60	0.920	0.590
		3	0.60	0.917	0.535
		4	0.40	0.923	0.400
		5	0.40	0.860	0.330
		6	0.25	0.920	0.230
3,4	Secondary conveyor	1	0.70	0.991	0.325
		2	0.70	0.983	0.240
		3	0.30	0.995	0.190
		4	0.25	0.936	0.150
5	Intermediate conveyor	1	0.70	0.971	0.1885
		2	0.60	0.993	0.1520
		3	0.40	0.981	0.1085
		4	0.20	0.993	0.1020
		5	0.10	0.990	0.0653
6	Stacker-reclaimer	1	1.30	0.981	1.115
		2	0.60	0.970	0.540
		3	0.30	0.990	0.320

The results obtained for different values of required availability A^* are presented in Figure 6.4 (each element is marked by its version number). One can see the modifications of the optimal structure of the system corresponding to different levels of the availability provided. The simple series-parallel system appears to be optimal when providing relatively little availability. In this case, the single-pier system satisfies the availability requirement (note that in this case there are no constraints on the interchangeability of the piers, on the number of piers, or on the number of ships to be discharged simultaneously).

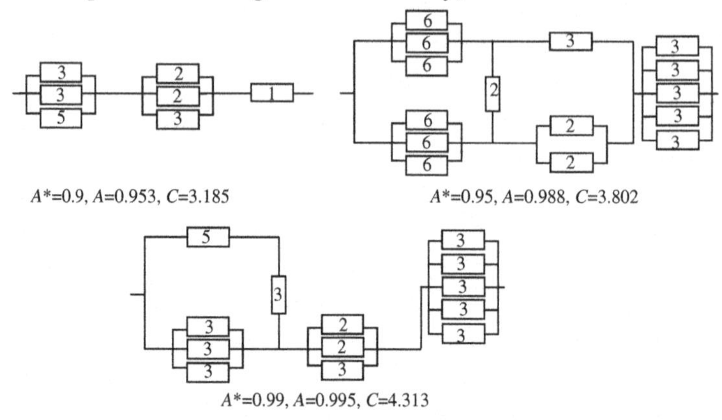

Figure 6.4. Optimal structures for a problem with allocation constraints

The bridge system becomes the best solution as A^* grows and, finally, for $A^* = 0.99$ the system returns to the series-parallel configuration. In this case the necessary redundancy of the ship discharge facilities is provided by the second pier connected with the single coal transportation line by the intermediate conveyor.

Now consider the problem in which the equality of the total installed capacities of components 1 and 2 is required to make the piers interchangeable (symmetry constraint). To meet this requirement, the additional penalty (6.17) with $j = 1$ and $i = 2$ is added to the solution fitness function.

The results obtained for different desired values of the system availability A^* are presented in Figure 6.5. One can see that the load-balancing diagonal element (intermediate conveyor) appears only in the solutions with relatively high availability.

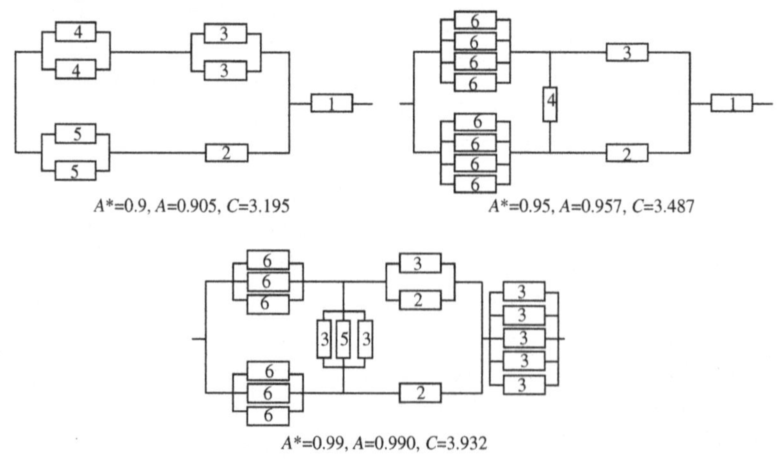

Figure 6.5. Optimal structures for a problem with symmetry constraints

For the next example consider the existing obsolete coal transportation system consisting of elements with low availability. This system can supply the boiler with availability A = 0.532. The parameters of the components of the existing system are presented in Table 6.3. The problem is to achieve a desired availability level A* by including additional elements from Table 6.2 into the system.

Table 6.3. Structure of the obsolete system

Component	No. of parallel elements	Parameters of elements	
		g	p
1	2	0.30	0.918
2	2	0.30	0.918
3	1	0.60	0.907
4	1	0.60	0.903
5	0	-	-
6	1	1.00	0.911

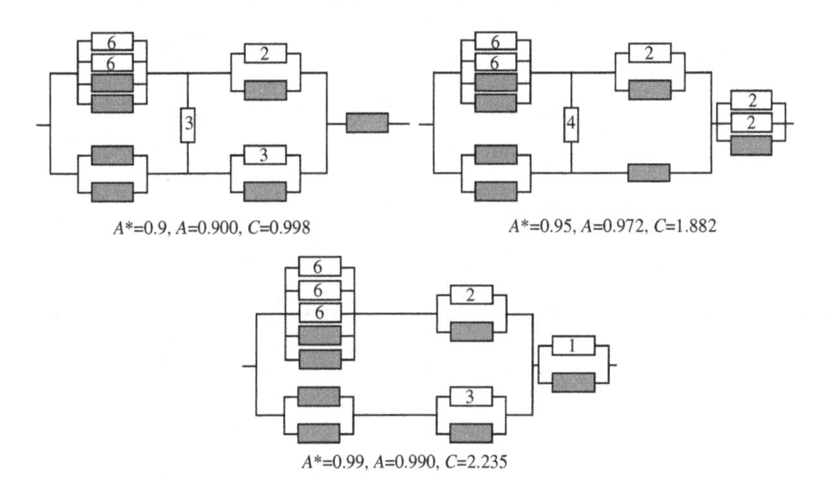

Figure 6.6. Optimal structures for the system extension problem

The minimal cost solutions of the system extension problem in which the cost of additional elements alone is considered obtained for this type of problem are presented in Figure 6.6 (the elements belonging to the initial system are depicted by grey rectangles).

Example 6.3

An alarm data-processing system has processing units of different types and data transmission/conversion facilities (see Figure 6.7). The system is composed of the following subsystems:

- primary data-processing subsystem containing components 1 and 2;
- alarm-type recognition subsystem containing components 3 and 4;
- data transmission subsystem (component 6);
- remote output data processing subsystem containing components 7 and 8;

- decision-support subsystem containing components 9 and 10.

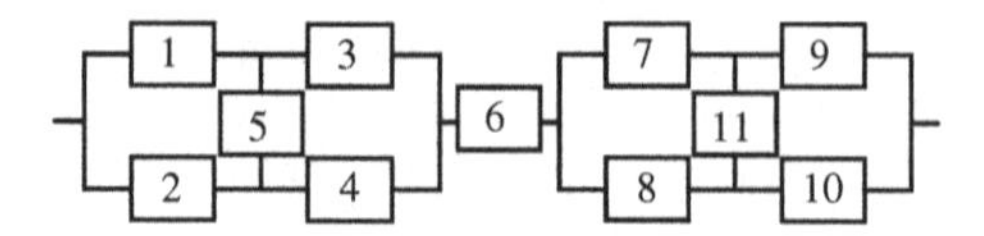

Figure 6.7. Structure of the alarm data-processing system

Table 6.4. Characteristics of available system elements

Component	Description	No. of version	g	p	c
1,2		1	8.0	0.830	0.750
		2	6.0	0.820	0.590
		3	6.0	0.807	0.535
		4	4.0	0.860	0.400
		5	4.0	0.825	0.330
		6	2.5	0.820	0.230
3,4		1	7.0	0.821	0.325
		2	7.0	0.803	0.240
		3	3.0	0.915	0.190
		4	2.5	0.806	0.150
5	Data transmission (conversion) units	1	20.0	0.871	0.1885
		2	18.0	0.893	0.1520
		3	14.0	0.881	0.1085
		4	12.0	0.893	0.1020
		5	11.0	0.890	0.0653
6	Data transmission line	1	11.3	0.881	1.115
		2	6.6	0.870	0.540
		3	4.3	0.890	0.320
7,8	Output data processing unit	1	3.2	0.801	0.750
		2	2.8	0.827	0.590
		3	2.8	0.801	0.535
9,10	Decision-support unit	1	5.7	0.891	0.325
		2	5.7	0.813	0.240
		3	5.3	0.925	0.220
		4	4.25	0.836	0.150
11	Data transmission (conversion) units	1	70.0	0.971	0.031
		2	60.0	0.993	0.025
		3	40.0	0.981	0.020

Pairs of components {1, 3}, {2, 4}, {7, 9} and {8, 10} have compatible data exchange protocols, whereas data transmission between pairs of components {1, 4}, {2, 3} and {7, 10}, {8, 9} requires its conversion, which can be performed by components 5 and 11 respectively. The set of available versions of two-state elements for each system component is presented in Table 6.4. The system performance (processing speed) should be no less than w = 1. The structure optimization problems were solved for two types of this task processing system: with and without work sharing [165].

First, the solutions of the unconstrained optimization problem were obtained for values of desired system reliability R^* = 0.95 and R^* = 0.99. The parameters of solutions obtained are presented in Table 6.5 and the optimal system structures are presented in Figures 6.8 and 6.9 (structures A and C). One can see that the simple series-parallel system appears to be optimal in all of the cases.

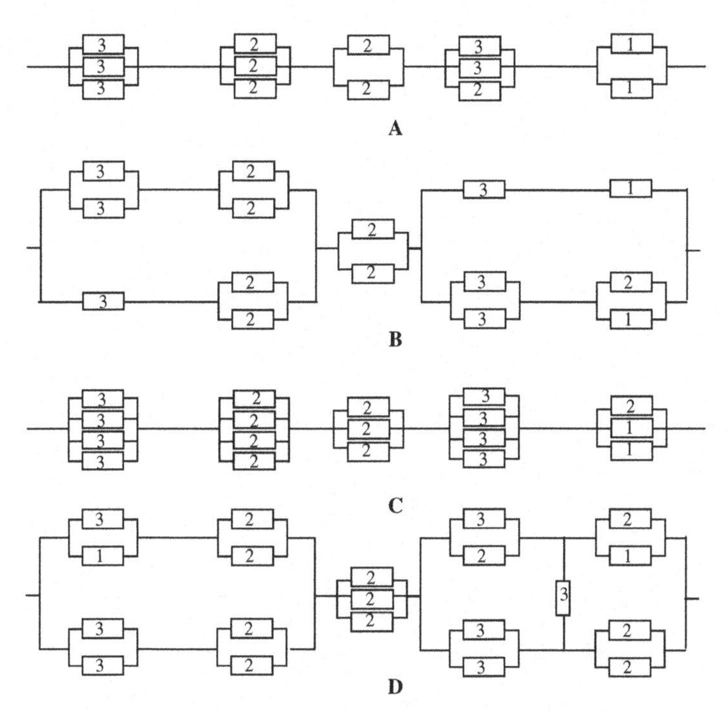

Figure 6.8. Optimal structures of system without work sharing.
A: $R^* = 0.95$, no constraints; B: $R^* = 0.95$, constraints;
C: $R^* = 0.99$, no constraints; D: $R^* = 0.99$, constraints

To force the GA to obtain the solution based on bridge structures, allocation constraints were imposed that forbid allocation of more than two parallel elements in all of the components except 5, 6 and 11. The parameters of the solutions obtained are also presented in Table 6.5 and the optimal system structures are presented in Figures 6.8 and 6.9 (structures B and D). The modification of the structure of the system without work sharing for different levels of required reliability is apparent. The transmission/conversion units appear unnecessary when providing $R>0.95$, but one such unit is included in the system providing $R>0.99$. For the system with work sharing, the use of a bridge diagonal element is more justified because its contribution to the total performance increases.

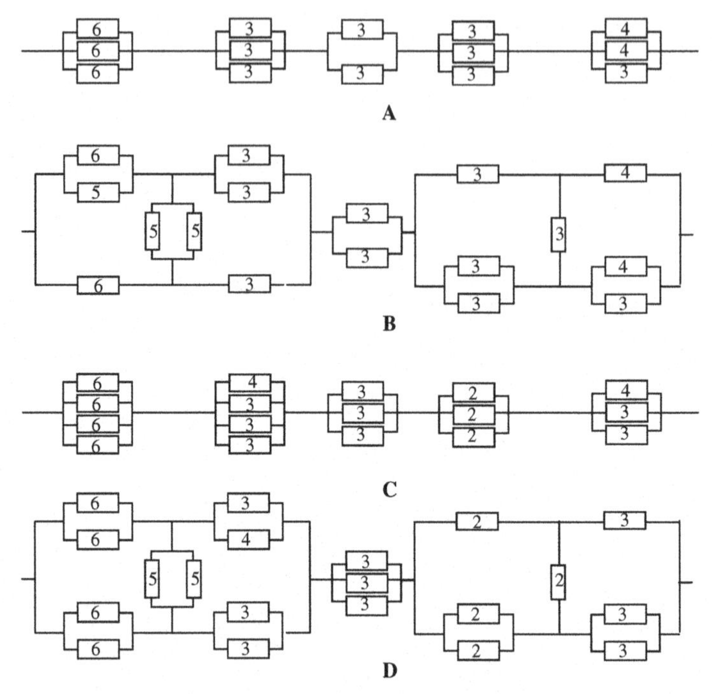

Figure 6.9. Optimal structures of system with work sharing.
A: R^* = 0.95, no constraints; B: R^* = 0.95, constraints;
C: R^* = 0.99, no constraints; D: R^* = 0.99, constraints

The work sharing allows the system to perform faster. Therefore, less expensive solutions were obtained for this type of system than for the system without work sharing.

Table 6.5. Parameters of the obtained solutions

R^*	0.99		0.95	
Constraints	No	Yes	No	Yes
	System without work sharing			
C	7.750	8.195	5.715	6.140
R	0.9911	0.9901	0.9505	0.9503
ε	0.997	1.032	0.960	0.960
	System with work sharing			
C	4.960	5.1856	4.02500	4.2756
R	0.9901	0.9901	0.9511	0.9518
ε	1.801	1.744	1.425	1.445

The systems without work sharing have few different possible levels of performance because the operator $\otimes_{\max}$ used in this case provides the same performance level for many different system states. In the solutions presented, only the system obtained for R^* = 0.99 with allocation constraints (Figure 6.8D) has different performance levels. The remainder have the single possible nonzero performance level because their components contain elements with the same nominal performance rates. The system performance distributions for systems without work sharing are presented in Figure 6.10A.

In contrast, the systems with work sharing have many possible performance values depending on their states, because the failure of each element affects the ability of the corresponding component to participate in the work sharing. The performance distributions for systems with work sharing are presented in Figure 6.10B.

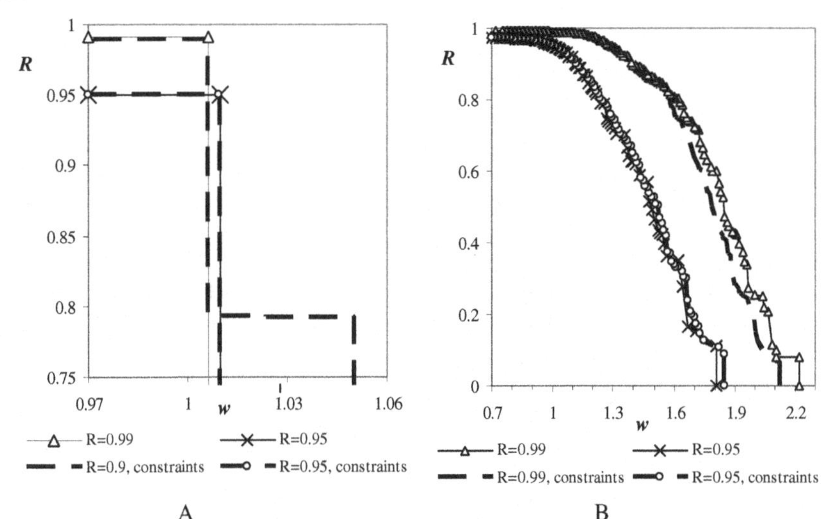

Figure 6.10. System reliability as a function of demand.
A: system without work sharing; B: system with work sharing

6.1.2.2 Structure Optimization in the Presence of Common Cause Failures

When the system components are subject to CCFs, the separation of elements among different parallel components can improve the overall system survivability (see Section 5.2.1). Such separation can justify the appearance of a bridge structure.

Consider, for example, an MSS containing two components A and B connected in series (Figure 6.11A). The components consist of M and L different elements respectively. All elements belonging to the same component are of the same functionality and are connected in parallel. The components A and B are subject to total CCF (they can be destroyed by hostile environments with the probabilities v_A and v_B respectively). The component destruction means that all of its elements are damaged and cannot perform their task. The system survivability is defined as the probability that a given demand w is met. This probability is affected by both the failures of the elements and the vulnerability of the components.

To enhance the system's survivability its components can be separated into two independent subcomponents. Let $\{1, \ldots, M\}$ and $\{1, \ldots, L\}$ be sets of numbers of elements belonging to components A and B respectively. The elements' separation problem can be considered as a problem of partitioning these sets inttwo mutually disjoint subsets. The partition can be represented by the binary vectors $\boldsymbol{x}_A = \{x_{Aj}: 1\le j\le M\}$ and $\boldsymbol{x}_B = \{x_{Bj}: 1\le j\le L\}$, where $x_{Aj}, x_{Bj}\in\{0,1\}$ and two elements i and j belong to the same subset if and only if $x_{Ai} = x_{Aj}$ or $x_{Bi} = x_{Bj}$ for components A and B respectively.

Actually, the separation leads to the appearance of two independent parallel subsystems containing components connected in series: 1, 3 and 2, 4 (see Figure 6.11C). In these systems, components 1 and 2 have the same vulnerability as the basic component A, and components 3 and 4 have the same vulnerability as the basic component B. Usually, the separation disconnects component 1 from component 4 and component 2 from component 3 (for example, when components 1 and 3 are spatially separated from components 2 and 4). To provide a connection between these components, a diagonal component 5 can be included that delivers an output of component 1 to the input of component 4 or an output of component 2 to the input of component 3. The diagonal component can also consist of different multi-state elements. This component can also be characterized by its vulnerability v_5.

The problem of bridge structure optimization is to find the optimal separation $\boldsymbol{x}_A$, $\boldsymbol{x}_B$ of elements from components A and B which provides the maximal system survivability for the given demand w when the structure of the diagonal component is given:

$$S(\boldsymbol{x}_A, \boldsymbol{x}_B, v_A, v_B, v_5, w) \rightarrow \max \tag{6.18}$$

The separation solution can be represented in the GA by the binary string $\boldsymbol{a}$, which is a concatenation of the binary vectors $\boldsymbol{x}_A$ and $\boldsymbol{x}_B$. The following procedure determines the fitness value for an arbitrary solution defined by the string $\boldsymbol{a}$.

1. According to $\boldsymbol{a}$, determine lists of elements belonging to components 1, 2, 3 and 4.
2. For $1\le i\le 5$, determine the u-function of the entire component i using the composition operator $\otimes_{\varphi_{\text{par}}}$ over the u-functions of elements belonging to this component (the list of elements belonging to the diagonal component is given).
3. In order to incorporate the component vulnerability into its u-function, apply the ξ operator (4.58) with $v = v_i$ over the u-function of the component.
4. Determine the u-function of the entire bridge system using the composition operator $\otimes_{\varphi_{\text{br}}}$ over the u-functions of its components.
5. Determine the system survivability index S for the given demand w and evaluate the solution fitness as $S(\boldsymbol{a})$.

Example 6.4

Consider a system initially consisting of two components connected in series (see Figure 6.11A). The first component consists of 12 two-state elements and the second consists of eight two-state elements. All of the elements within each component are connected in parallel. The sets of elements belonging to the components consist of pairs of identical elements (six pairs for the first component and four pairs for the second one). This provides the possibility of symmetric separation. Parameters of elements of each pair are presented in Table 6.6.

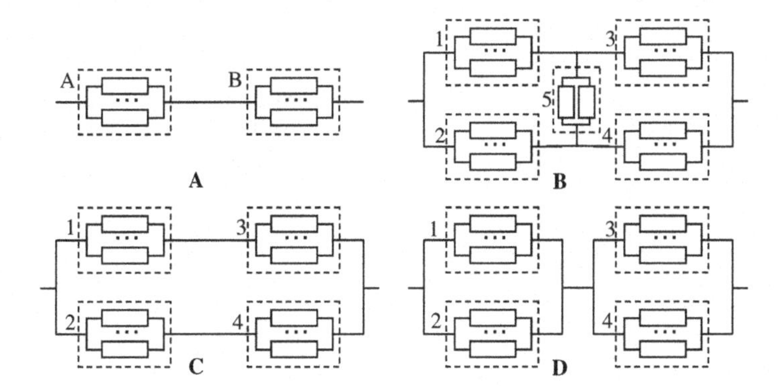

Figure 6.11. Special cases of bridge structure

It is assumed that the elements belonging to each component are subject to total CCFs (external impact) and the probability of the CCF (component vulnerability) is the same for each component.

Table 6.6. Parameters of MSS elements

No. of component	No. of element	G	p
1, 2	1	8.0	0.830
	2	6.0	0.820
	3	6.0	0.807
	4	4.0	0.860
	5	4.0	0.825
	6	2.0	0.820
3, 4	1	15.0	0.821
	2	10.0	0.803
	3	10.0	0.815
	4	5.0	0.806
5	1	6.0	0.871
	2	4.0	0.893

To enhance the system's survivability, each component can be divided into two subcomponents and a diagonal component consisting of two parallel elements can be included (Figure 6.11B). The parameters of the element belonging to diagonal component 5 are also presented in Table 6.6.

The system is interpreted first as a flow transmission one with flow dispersion and then as a task processing one without work sharing. The optimal solutions were obtained for both types of system for the same parameters of elements [166]. The system demands are $w = 25$ and $w = 4$ for the flow transmission system and the task processing system respectively. Solutions were obtained for three different values of component vulnerability: $v = 0$ (pure reliability optimization problem), $v = 0.05$ and $v = 0.5$.

The bridge structures providing the maximal survivability for the flow transmission system are presented in Table 6.7 and for the task processing system in Table 6.8. These solutions are compared with solutions having no component separation (Figure 6.11A) and with solutions having symmetric separation. The lists of elements belonging to each component, as well as the corresponding system survivability index $S(w)$ and mean performance ε, are presented for each solution.

Table 6.7. Solutions for flow transmission system (vulnerability variation)

Solution description	System structure		$v = 0$		$v = 0.05$		$v = 0.5$	
	No. of component	Elements included	S	ε	S	ε	S	ε
Optimal for $v = 0.05$	1	114466	0.9968	48.268	0.9333	43.957	0.2837	12.713
	2	223355						
	3	112						
	4	23344						
Symmetric	1	123456	0.9968	48.266	0.9244	43.969	0.2565	12.722
	2	123456						
	3	1234						
	4	1234						
Optimal for $v = 0$ (no separation)	1	112233445566	0.9974	48.518	0.9001	43.787	0.2493	12.129
	2	-						
	3	11223344						
	4	-						
Optimal for $v = 0.5$	1	23345566	0.9960	48.185	0.9319	43.870	0.2866	12.652
	2	1124						
	3	113						
	4	22344						

Table 6.8. Solutions for task processing system (vulnerability variation)

Solution description	System structure		$v = 0$		$v = 0.05$		$v = 0.5$	
	No. of component	Elements included	S	ε	S	ε	S	ε
Optimal for $v = 0.05$ and for $v = 0.5$	1	223346	0.9990	4.434	0.9841	4.381	0.4255	2.047
	2	114556						
	3	1144						
	4	2233						
Symmetric	1	123456	0.9983	5.135	0.9820	5.058	0.4221	2.325
	2	123456						
	3	1234						
	4	1234						
Optimal for $v = 0$ (no separation)	1	112233445566	0.9990	5.168	0.9016	4.664	0.2497	1.292
	2	-						
	3	11223344						
	4	-						

One can see that, when the probability of CCFs in the components is neglected ($v = 0$), the best solution is one without elements separation. The optimal solutions for different values of component vulnerability can differ (as in case of the flow transmission system, where the optimal solution for $v = 0.05$ is not optimal for $v = 0.5$ and *vice versa*). For both types of system the optimal solutions are not symmetric.

The solution that provides the maximal system survivability for a given demand w does not necessarily provide the greatest system mean performance. Indeed, the system resources are distributed in such a way that maximizes only the probability of demand w satisfaction, while the rest of the performance levels can be provided with probabilities lower than those obtained by an alternative solution. In Figure 6.12 one can see $S(w)$ functions for three different solutions for both types of system when $v = 0.05$. While the probability $\Pr\{G \geq w\}$ for the optimal solutions is maximal, the probabilities $\Pr\{G \geq w'\}$ for $w' > w$ is often greater for the symmetric solution and the solution without separation. Note that the greatest mean performance ε is achieved for the symmetric solutions for both types of system.

It is interesting that the optimal solution for the task processing system when $v = 0.05$ cannot even provide a processing speed $G > 4.44$, while the rest of the solutions provide a processing speed $G = 5.22$ with a probability close to 0.85. Indeed, the fastest elements in the optimal solution are located in components 1 and 3. Therefore, the path including only these two elements with nominal processing speeds $g_{11} = 8$ and $g_{31} = 15$ does not exist in this solution, since the two elements are not directly connected, though they are through the diagonal element. In the remainder of the solutions such a path exists. Creating the fastest path by exchanging elements between components 3 and 4 in the optimal solution improves the system's average processing speed (it grows from $\varepsilon = 4.381$ to $\varepsilon = 4.968$); however, this drastically decreases the system's survivability for the given demand $w = 4$ (from $S(4) = 0.984$ to $S(4) = 0.848$).

In order to estimate the effect of diagonal element parameters on the optimal separation, compare the optimal solution obtained above (Figure 6.11B) with two extreme cases. In the first case (Figure 6.11C) no diagonal component is available: $\Pr\{G_5 = 0\} = 1$. In the second case (Figure 6.11D), no capacity limitations are imposed on the fully reliable diagonal component: $\Pr\{G_5 = \infty\} = 1$. The solutions obtained for both types of system are presented in Tables 6.9 and 6.10.

Table 6.9. Solutions for flow transmission system (different diagonal elements)

Solution description	System structure		Case B		Case C		Case D	
	No. of component	Elements included	S	ε	S	ε	S	ε
Optimal for case B	1	114466	0.9333	43.957	0.9038	42.306	0.9432	44.505
	2	223355						
	3	112						
	4	23344						
Optimal for cases C and D	1	1124	0.9328	43.911	0.9053	42.202	0.9445	44.500
	2	23345566						
	3	112						
	4	23344						

One can see that the optimal separation solutions coincide in cases B and C for the flow transmission system and in cases C and D for the task processing system. Obviously, the highest system survivability is achieved for case D and the lowest for case C.

Table 6.10. Solutions for task processing system (different diagonal elements)

Solution description	System structure		Case B		Case C		Case D	
	No. of component	Elements included	S	ε	S	ε	S	ε
Optimal for cases B and C	1	223346	0.9841	4.381	0.9841	4.370	0.9862	5.026
	2	114556						
	3	1144						
	4	2233						
Optimal for case D	1	1234455	0.9820	5.057	0.9820	5.045	0.9914	5.113
	2	12366						
	3	123						
	4	12344						

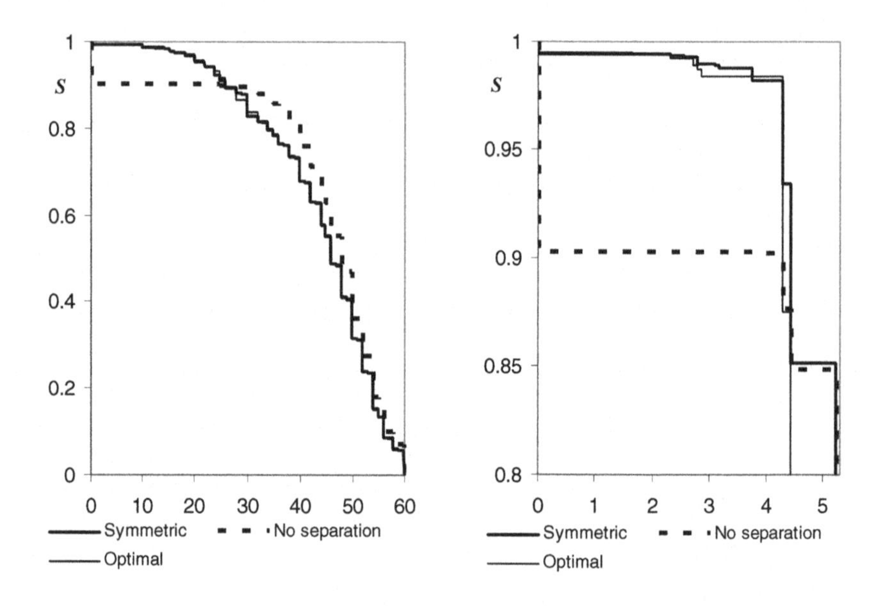

Figure 6.12. System survivability as function of demand for different element separation solutions in flow transmission system (A) and in task processing system (B)

6.2 Multi-state Systems with Two Failure Modes

Systems with two failure modes consist of statistically independent devices (elements) that are all to operate in the same two modes (the operation commands in each mode arrive at all the elements simultaneously). Each element can fail in either of two modes. A typical example of systems with two failure modes are switching systems that not only can fail to close when commanded to close but can also fail to open when commanded to open. Two different types of switching system can be distinguished:

- flow transmission systems, in which the main characteristic of each switching device is the flow controlled by this device (for example, fluid flow valve);
- task processing systems, in which the main characteristic of each switching device is the switching time of this device (for example, electronic diode).

The study of the systems with two failure modes started as early as in the 1950th [167-169] and still attracts interest of researchers [170-173]. In 1984, Barlow and Heidtman [14] suggested using a generating function method for computing k-out-of-n reliability of systems with two failure modes.

The aforementioned studies consider only reliability characteristics of elements composing the system. In many practical cases, some measures of element (system) performance must be taken into account. For example, fluid-transmitting capacity is the performance of fluid valves and of switching systems that consist of such valves, while operating time is the performance of electronic switches and of switching systems that consist of such switches. Each system element can be characterized in each mode by its nominal performance and the element fails if it is unable to provide its nominal performance. A system can have different levels of output performance depending on its structure and on the combination of elements available at any given moment. Therefore, the system is multi-state.

The system is considered to be in an operational state if its performance rate in open mode G_o satisfies condition $F_o(G_o, w_o)=1$ and its performance rate in closed mode G_c satisfies condition $F_c(G_c, w_c)=1$, where w_o and w_c are the required levels of system performance in the open and closed modes respectively and F_o and F_c are the acceptability functions in open and closed modes respectively.

Since the failures in open and closed modes, which have probabilities $Q_o = \Pr\{F_o(G_o, w_o)=0\}$ and $Q_c = \Pr\{F_c(G_c, w_c)=0\}$ respectively, are mutually exclusive events and the probabilities of both modes are equal to 0.5 (each command to close is followed by a command to open and *vice versa*), the entire system availability can be defined as

$$A = 1-0.5\,(Q_o + Q_c) \tag{6.19}$$

In order to characterize the expected performance of MSSs with two failure modes, one has to evaluate this index for both its modes: ε_o and ε_c.

Having u-functions $U_o(z)$ and $U_c(z)$ representing the system performance distributions in open and closed modes, one can obtain the probabilities Q_o and Q_c and the expected performance rates ε_o and ε_c using the technique presented in Section 3.3 over $U_o(z)$ and $U_c(z)$ respectively.

Usually, the switching systems consist of elements with total failures. Each element j has nominal performance rates g_{jo} and g_{jc}, performance rates in fault state $\tilde{g}_{jo}$ and $\tilde{g}_{jc}$, and probabilities of normal functioning p_{jo} and p_{jc} in open and closed modes respectively. The individual *u*-functions of the system element can be defined as

$$u_{jo}(z) = p_{jo} z^{g_{jo}} + (1 - p_{jo}) z^{\tilde{g}_{jo}}$$
$$u_{jc}(z) = p_{jc} z^{g_{jc}} + (1 - p_{jc}) z^{\tilde{g}_{jc}} \quad (6.20)$$

In order to obtain the system *u*-functions $U_o(z)$ and $U_c(z)$, one has to determine the parameters of the *u*-functions of individual system elements (6.20) and define the structure functions used in the composition operators for both modes. Having these *u*-functions one can easily evaluate the system availability using Equation (6.19) and operators (3.8) and (3.12):

$$A(w_o, w_c) = 1 - 0.5[1 - E(U_o(z) \underset{F_o}{\otimes} z^{w_o}) + 1 - E(U_c(z) \underset{F_c}{\otimes} z^{w_c})]$$
$$= 0.5[E(U_o(z) \underset{F_o}{\otimes} z^{w_o}) + E(U_c(z) \underset{F_c}{\otimes} z^{w_c})] \quad (6.21)$$

In the following sections we consider two typical switching systems.

6.2.1 Flow Transmission Multi-state System

In this model the performance of the switching element (flow valve) is defined as its transmitting capacity. To determine the *u*-function of an individual element *j* in the closed mode, note that in the operational state, which has the probability p_{jc}, the element should transmit a nominal flow f_j ($g_{jc} = f_j$) and in the failure state it fails to transmit any flow ($\tilde{g}_{jc} = 0$). Therefore, according to (6.20), the *u*-function of the element takes the form

$$u_{jc}(z) = p_{jc} z^{f_j} + (1 - p_{jc}) z^0 \quad (6.22)$$

In the open mode the element has to prevent the flow transmission through the system. If it succeeds in doing this (with probability p_{jo}), then the flow is zero ($g_{jo} = 0$), and if it fails to do so the flow is equal to its nominal value in the closed mode ($\tilde{g}_{jo} = f_j$). The *u*-function of the element in the open mode takes the form

$$u_{jo}(z) = p_{jo} z^0 + (1 - p_{jo}) z^{f_j} \quad (6.23)$$

The structure functions for subsystems of elements connected in a series, in parallel or composing a bridge structure for the flow transmission MSS with flow dispersion are defined by Equations (4.2), (4.9) and (6.4) respectively. Using the reliability block diagram method one can obtain the u-function of the arbitrary system by consecutively applying the corresponding composition operators.

Note that the u-function of a subsystem containing n identical parallel elements ($p_{jc} = p_c$, $p_{jo} = p_o$, $f_j = f$ for any j) can be obtained by applying the operator $\otimes_{+}(u(z),...,u(z))$ over n functions $u(z)$ of an individual element represented by (6.22) or (6.23). The u-function of this subsystem takes the form

$$U_c(z) = \sum_{k=0}^{n} \frac{n!}{k!(n-k)!} p_c^k (1-p_c)^{n-k} z^{kf} \tag{6.24}$$

for the closed mode and

$$U_o(z) = \sum_{k=0}^{n} \frac{n!}{k!(n-k)!} p_o^{n-k} (1-p_o)^k z^{kf} \tag{6.25}$$

for the open mode. The u-function of a subsystem containing n identical elements connected in a series can be obtained by applying operator $\otimes_{\min}$ over n functions $u(z)$ of an individual element. The u-function of this subsystem takes the form

$$U_c(z) = p_c^n z^f + (1-p_c^n) z^0 \tag{6.26}$$

for the closed mode and

$$U_o(z) = [1-(1-p_o)^n] z^0 + (1-p_o)^n z^f \tag{6.27}$$

for the open mode.

To determine the system's reliability one has to define its acceptability function. For the flow transmission system, it is natural to require that in its closed mode the amount of flow should not be lower than the demand w_c, while in the open mode it should not exceed a value of w_o. Therefore, the conditions of the system's success are

$$F_c(G_c,w_c) = 1(G_c \ge w_c) \text{ and } F_o(G_o,w_o) = 1(G_o \le w_o) \tag{6.28}$$

6.2.2 Task Processing Multi-state Systems

In this type of switching system the task of each element is to connect (or disconnect) a circuit. Since the task processing in each mode is associated with a single switching action, the performance of element j is defined not as its processing speed but as its operation time.

To determine the u-function of an individual element with total failures (for example, an electronic diode) in closed and open modes, note that the element j operates in times $g_{jc} = t_{jc}$ and $g_{jo} = t_{jo}$ with the probabilities p_{jc} and p_{jo} respectively. If the element fails to operate, then its operation time is equal to infinity ($\tilde{g}_{jo} = \tilde{g}_{jc} = \infty$). Therefore, according to (6.20), the u-functions of the element for the two modes take the form

$$u_{jo}(z) = p_{jo} z^{t_{jo}} + (1 - p_{jo}) z^{\infty}$$
$$u_{jc}(z) = p_{jc} z^{t_{jc}} + (1 - p_{jc}) z^{\infty} \tag{6.29}$$

If several elements are connected in parallel within a subsystem, then the subsystem disconnection is completed only when all the elements including the slowest one are opened. Therefore, the operation time of n elements in the open mode is equal to the greatest of the operation times of the elements. The structure function for the open mode takes the form

$$\phi_{\text{par}}(G_1,...,G_n) = \max\{G_1,...,G_n\} \tag{6.30}$$

For n elements connected in series, the first disconnected element disconnects the subsystem in the open mode. Therefore, the structure function takes the form

$$\phi_{\text{ser}}(G_1,...,G_n) = \min\{G_1,...,G_n\} \tag{6.31}$$

If n elements are connected in parallel within a subsystem, then the first connected element makes the subsystem connected. Therefore, the operation time of the group of elements in closed mode is equal to the least of the operation times of the elements. The structure function for the closed mode takes the form

$$\phi_{\text{par}}(G_1,...,G_n) = \min\{G_1,...,G_n\} \tag{6.32}$$

For n elements connected in series, all of the elements, including the slowest one, should be connected to make the subsystem connected in the closed mode. Therefore, the structure function takes the form:

$$\phi_{\text{ser}}(G_1,...,G_n) = \max\{G_1,...,G_n\} \tag{6.33}$$

Combining the two operators one can obtain a u-function representing the performance distribution of an arbitrary series-parallel system in both modes. Note that the u-function of a subsystem containing n identical parallel elements is

$$U_c(z) = [1 - (1 - p_c)^n] z^{t_c} + (1 - p_c)^n z^{\infty} \tag{6.34}$$

for the closed mode and

$$U_o(z) = p_o^n z^{t_o} + (1 - p_o^n) z^\infty \tag{6.35}$$

for the open mode. The u-function of a subsystem containing n identical elements connected in series takes the form

$$U_c(z) = p_c^n z^{t_c} + (1 - p_c^n) z^\infty \tag{6.36}$$

for the closed mode and

$$U_o(z) = [1 - (1 - p_o)^n] z^{t_o} + (1 - p_o)^n z^\infty \tag{6.37}$$

for the open mode.

In order to evaluate the operation time of a bridge structure, notice that there are four possible parallel ways to connect input and output of the bridge (see Figure 6.1): through groups of elements {1, 3} or {2, 4} or {1, 5, 4} or {2, 5, 3} connected in series. Therefore, the entire bridge operation time can be obtained as

$$\begin{aligned}&\phi_{br}(G_1, G_2, G_3, G_4, G_5)\\ &= \phi_{par}(\phi_{ser}(G_1, G_3), \phi_{ser}(G_2, G_4), \phi_{ser}(G_1, G_5, G_4), \phi_{ser}(G_2, G_5, G_3))\end{aligned} \tag{6.38}$$

For the open mode this expression takes the form

$$\begin{aligned}&\phi_{br}(G_1, G_2, G_3, G_4, G_5)\\ &= \max(\min(G_1, G_3), \min(G_2, G_4), \min(G_1, G_5, G_4), \min(G_2, G_5, G_3))\end{aligned} \tag{6.39}$$

and for the closed mode it takes the form

$$\begin{aligned}&\phi_{br}(G_1, G_2, G_3, G_4, G_5)\\ &= \min(\max(G_1, G_3), \max(G_2, G_4), \max(G_1, G_5, G_4), \max(G_2, G_5, G_3))\end{aligned} \tag{6.40}$$

For a system in which operation time is the crucial factor, it is natural to require that in its closed and open modes the operation times should not exceed the values w_c and w_o respectively. The system's acceptability functions are

$$F_c(G_c, w_c) = 1(G_c \le w_c) \text{ and } F_o(G_o, w_o) = (G_o \le w_o) \tag{6.41}$$

Having these acceptability functions one can easily evaluate the system's availability using Equation (6.21).

Since, in the worst case, the operation time of the entire system is equal to infinity, determining the expected operation time makes no sense. A more natural way of evaluating expected performance is by using the conditional expected operation time (expected operation time given the system manages to operate). In this case, Equation (3.7) with the acceptability functions $F_o(G_o) = 1(G_o<\infty)$ and $F_c(G_c) = 1(G_c<\infty)$ should be used.

Example 6.5

Consider a switching series-parallel subsystem consisting of three elements 1-3 connected as depicted in Figure 6.13. The elements are characterized by their availability and performance level (transmitting capacity *f*) in open and closed modes. The parameters of the system elements are presented in Table 6.11 (first three rows). The *u*-functions of the individual elements according to (6.22) and (6.23) are

$$u_{1o}(z) = 0.87z^0+0.13z^{1.5};\ u_{1c}(z) = 0.89z^{1.5}+0.11z^0$$

$$u_{2o}(z) = 0.78z^0+0.22z^{3.5};\ u_{2c}(z) = 0.82z^{3.5}+0.18z^0$$

$$u_{3o}(z) = 0.82z^0+0.18z^{2.5};\ u_{3c}(z) = 0.91z^{2.5}+0.09z^0$$

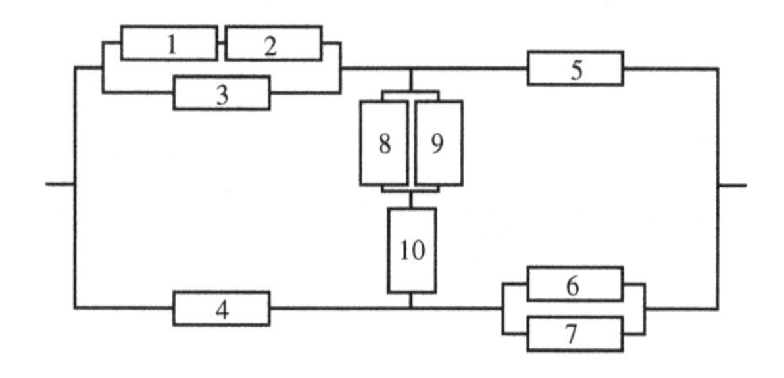

Figure 6.13. Reliability block diagram of MSS with two failure modes

In order to determine the system performance distribution in the open and closed modes we have to obtain the *u*-function of the entire system using composition operators over the *u*-functions of individual elements.

$$U_o(z) = [u_{1o}(z) \underset{\min}{\otimes} u_{2o}(z)] \underset{+}{\otimes} u_{3o}(z)$$

$$= [(0.87z^0+0.13z^{1.5}) \underset{\min}{\otimes} (0.78z^0+0.22z^{3.5})] \underset{+}{\otimes} (0.82z^0+0.18z^{2.5})$$

$$= (0.9714z^0+0.0286z^{1.5}) \underset{+}{\otimes} (0.82z^0+0.18z^{2.5})$$

$$= 0.7965z^0+0.0234z^{1.5}+0.1749z^{2.5}+0.0051z^4$$

$$U_c(z) = [u_{1c}(z) \underset{\min}{\otimes} u_{2c}(z)] \underset{+}{\otimes} u_{3c}(z)\ U_o(z)$$

$$= [(0.89z^{1.5}+0.11z^0) \underset{\min}{\otimes} (0.82z^{3.5}+0.18z^0)] \underset{+}{\otimes} (0.91z^{2.5}+0.09z^0)$$

$$= (0.7298z^{1.5}+0.2702z^0) \underset{+}{\otimes} (0.91z^{2.5}+0.09z^0)$$

$$= 0.6641z^4+0.0657z^{1.5}+0.2459z^{2.5}+0.0243z^0$$

Having the system u-functions for the open and closed modes one can determine the expected flows through the system in these modes by applying the operator (3.11):

$$\varepsilon_o = 0.7965\times0+0.0234\times1.5+0.1749\times2.5+0.0051\times4 = 0.4927$$

$$\varepsilon_c = 0.6641\times4+0.0657\times1.5+0.2459\times2.5+0.0243\times0 = 3.3697$$

Assume that in the closed mode the amount of flow should exceed $w_c = 2$, while in the open mode it should not exceed $w_o = 0.5$. Applying Equation (6.21) with acceptability functions (6.28) we obtain

$$E(U_c(z) \underset{F_c}{\otimes} 2) = 0.6641\times1(4\geq2)+0.0657\times1(1.5\geq2)+0.2459\times1(2.5\geq2)$$

$$+0.0243\times1(0\geq2) = 0.6641+0.2459 = 0.91$$

$$E(U_o(z) \underset{F_o}{\otimes} 0.5) = 0.7965\times1(0\leq0.5)+0.0234\times1(1.5\leq0.5)$$

$$+0.1749\times1(2.5\leq0.5)+0.0051\times1(4\leq0.5) = 0.7965$$

$$A(2, 0.5) = 0.5(E(U_c(z) \underset{F_c}{\otimes} 2) + E(U_o(z) \underset{F_o}{\otimes} 0.5))$$

$$= 0.5(0.91+0.7965) = 0.85325$$

Example 6.6

Consider a switching system with the configuration presented in Figure 6.13. Each one of the ten system elements is characterized by its availability and nominal performance rate in open and closed modes. In the case of a flow transmission system, the performance of an element is its transmitting capacity f. In the case of a system of electronic switches, the performance of an element is determined by its operation times in open mode t_o and in closed mode t_c. The parameters of the system elements are presented in Table 6.11.

In order to determine the system PD in the open and closed modes one has to obtain the u-function of the entire system using the composition operators over u-functions of the individual elements $u_{o1}(z)$-$u_{o10}(z)$ and $u_{c1}(z)$-$u_{c10}(z)$ respectively.

Table 6.11. Parameters of MSS elements

No. of element		Flow transmission model		
		f	p_c	p_o
		Task processing model		
	t_c	t_o	p_c	p_o
1	3.0	1.5	0.89	0.87
2	5.0	3.5	0.82	0.78
3	3.5	2.5	0.91	0.82
4	2.5	3.0	0.85	0.82
5	3.0	2.5	0.80	0.76
6	3.0	3.0	0.80	0.78
7	4.0	3.0	0.91	0.85
8	4.0	4.5	0.84	0.79
9	5.3	2.5	0.93	0.91
10	5.0	2.7	0.92	0.90

First, consider the system to be a combination of flow valves (flow transmission system with flow dispersion). The flow through the system can vary in the range of 0.0-7.0. In the closed mode the expected flow is $\varepsilon_c = 5.306$. The probability that the system provides the maximal flow in the closed mode is $\Pr\{G_c = 7\} = 0.30$. In the open mode the expected flow is $\varepsilon_o = 0.303$ and the probability that the system totally prevents the flow is $\Pr\{G_c = 0\} = 0.892$. The system failure is defined as its inability to provide at least the required constant level of flow w_c in its closed mode and to prevent the flow exceeding w_o in its open mode. The failure probabilities in both modes as functions of demand w are presented in Figure 6.14A. Note that $Q_c(w_c)$ is an increasing function (the greater the demand, the tougher the condition $G_c \geq w_c$), while $Q_o(w_o)$ is a decreasing function (the greater the demand, the easier the condition $G_o \leq w_o$). The entire system availability $A(w_o,w_c)$ as a function of maximal allowable flow in the open mode w_o and minimal required flow in the closed mode w_c is presented in Figure 6.15A.

Now consider the system to be a combination of electronic switches (task processing system). The probabilities that the system is able to operate in the open and closed modes (operation time is less than infinity) are $\Pr\{G_o<\infty\} = 0.892$ and $\Pr\{G_c<\infty\} = 0.990$. When $G_o<\infty$, the time needed by the system to disconnect its input from output in the open mode cannot be less than 3 and greater than 4.5. The conditional expected operation time is $\tilde{\varepsilon}_o = 3.02$. When $G_c<\infty$, the time needed by the system to connect its input with output in the closed mode cannot be less than 3 and greater than 5.3. The conditional expected operation time is $\tilde{\varepsilon}_c = 3.23$. When the system failure is defined as its inability to switch within the required time (w_o and w_c in open and closed modes respectively), the failure probabilities in both modes are functions of this time. The functions $Q_o(w_o)$ and $Q_c(w_c)$ are presented in Figure 6.14B. The entire system availability as a function of required switching times in the open and closed modes $A(w_o,w_c)$ is presented in Figure 6.15B.

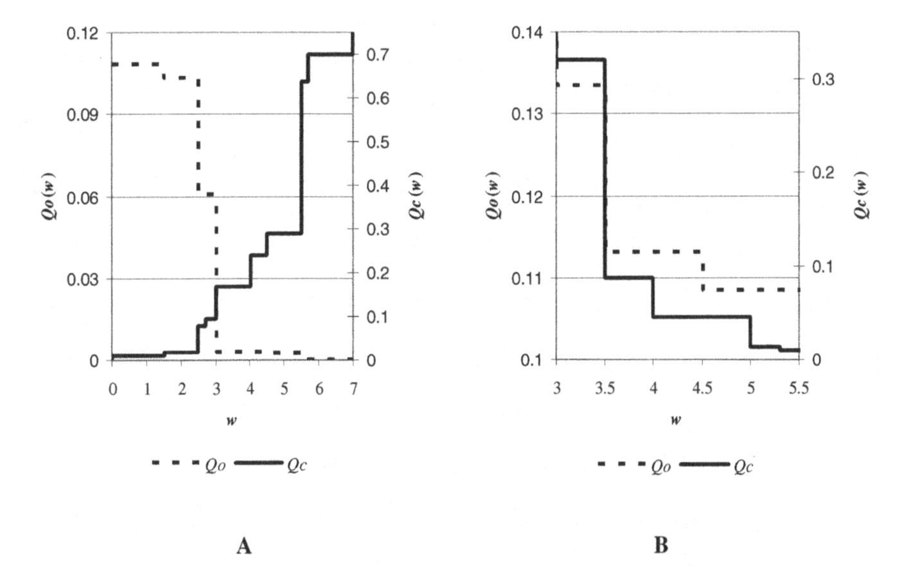

Figure 6.14. Failure probabilities as functions of demand.
A: flow transmission system; B: task processing system

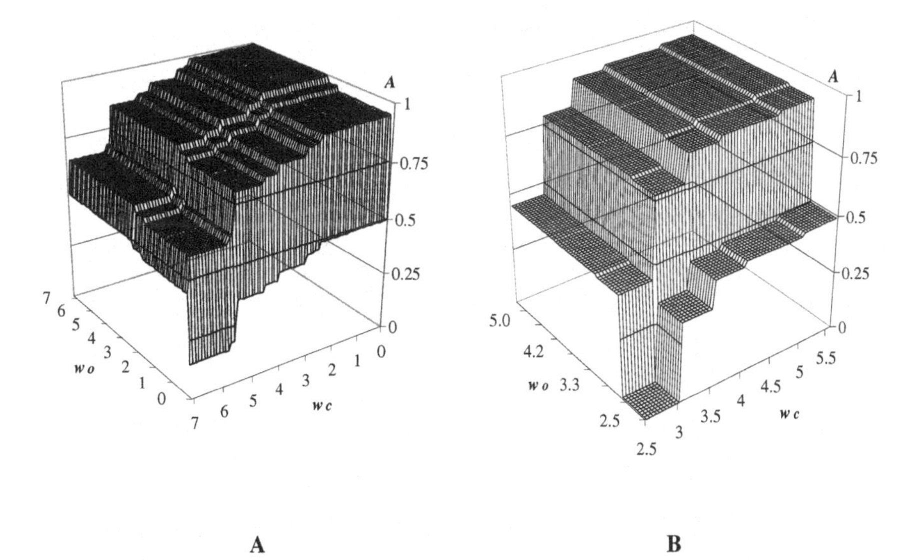

Figure 6.15. System availability as function of demands in open and closed modes.
A: flow transmission system; B: task processing system

The duality of roles of parallel and series connection of units in the two operation modes creates a situation in which any change in system configuration that increases system availability in an open mode can decrease it in a closed mode and *vice versa* [170, 171]. Therefore, the optimal system configuration should be found that provides the maximal overall system availability (6.21).

There exist two types of structure optimization problem when the systems with two failure modes are considered. The first one is an extension of the well-known redundancy optimization problem. In this problem, one has to determine the number of parallel elements (with identical functionality) for each system component when the system structure (topology of the reliability block diagram representing interaction of the components) is given. The algorithms for solving this problem were studied in [174, 175] for the binary-state systems without respect to the element performances. The second problem is to find the configuration (topology of the reliability block diagram) for a given set of elements that provides the greatest possible system availability. This problem was formulated and solved in [173] for the binary-state systems (also without respect to the element performances). In the following sections, algorithms for solving the two system optimization problems for MSSs with two failure nodes are presented.

6.2.3 Structure Optimization of Systems with Two Failure Modes

6.2.3.1 Problem Formulation

A system consists of N components connected according to a block diagram. Each component of type i contains a number of different switching elements connected in parallel. Different versions and numbers of elements may be chosen for any given system component. Element operation in open and closed modes is characterized by its availability and nominal performance rate.

For each component i there are B_i element versions available. A vector of parameters $g_{io}(b)$, $g_{ic}(b)$, $p_{io}(b)$ and $p_{ic}(b)$ can be specified for each version b of element of type i. The structure of system component i is defined by the numbers of parallel elements of each version n_{ib} for $1\le b\le B_i$. The vectors $\boldsymbol{n}_i=\{n(i,b)\}$ $(1\le i\le N,$ $1\le b\le B_i)$ define the entire system structure.

For a given set of vectors $\{\boldsymbol{n}_1, \ldots, \boldsymbol{n}_N\}$, the entire system fault probabilities $Q_o(w_o, \boldsymbol{n}_1, \ldots, \boldsymbol{n}_N)$ and $Q_c(w_c, \boldsymbol{n}_1, \ldots, \boldsymbol{n}_N)$ can be obtained for both modes. The requirement of providing the desired system availability in open and closed modes can be formulated as follows:

$$Q_o(w_o, \boldsymbol{n}_1, \ldots, \boldsymbol{n}_N)\le Q^*_o,\ \ Q_c(w_c, \boldsymbol{n}_1, \ldots, \boldsymbol{n}_N) \le Q^*_c, \tag{6.42}$$

where Q^*_o and Q^*_c are maximal allowable levels of system unavailability in open and closed modes respectively.

Having the given system structure, one can also determine the expected system performance in the both modes $\varepsilon_o(\boldsymbol{n}_1, \ldots, \boldsymbol{n}_N)$ and $\varepsilon_c(\boldsymbol{n}_1, \ldots, \boldsymbol{n}_N)$. While satisfying the availability requirements (6.42), one can desire to obtain expected system performance values as close to some specified values ε^*_o and ε^*_c as possible. The proximity between expected system performance and the desired level can be of different importance in open and closed modes.

Now consider two possible formulations of the problem of system structure optimization.

Formulation 1. Find system configuration $\{\boldsymbol{n}_1,\ldots, \boldsymbol{n}_N\}$ that provides maximal system availability:

$$A(w_o,w_c,\boldsymbol{n}_1,\ldots,\boldsymbol{n}_N)$$
$$=1-0.5(\mathrm{Q_o}(w_o,\boldsymbol{n}_1,\ldots,\boldsymbol{n}_N)+\mathrm{Q_c}(w_c,\boldsymbol{n}_1,\ldots,\boldsymbol{n}_N))\rightarrow \max \tag{6.43}$$

Formulation 2: find system configuration $\{\boldsymbol{n}_1, \ldots, \boldsymbol{n}_N\}$ that provides the maximal proximity of expected system performance to the desired levels for both modes, while satisfying the availability requirements:

$$\begin{gathered}\alpha\,|\,\varepsilon_o(\boldsymbol{n}_1,\ldots,\boldsymbol{n}_N)-\varepsilon_o^*\,|+(1-\alpha)\,|\,\varepsilon_c(\boldsymbol{n}_1,\ldots,\boldsymbol{n}_N)-\varepsilon_c^*\,|\rightarrow \min\\ \text{subject to }\mathrm{Q_o}(w_o,\boldsymbol{n}_1,\ldots,\boldsymbol{n}_N)\le Q_o^*,\ \mathrm{Q_c}(w_c,\boldsymbol{n}_1,\ldots,\boldsymbol{n}_N)\le Q_c^*\end{gathered} \tag{6.44}$$

where constant α reflects the relative importance of the open mode over the closed mode ($0\le\alpha\le1$). Note that for the task processing systems the measures ε_o and ε_c should be substituted by the corresponding conditional measures $\tilde{\varepsilon}_o$ and $\tilde{\varepsilon}_c$.

6.2.3.2 Implementing the Genetic Algorithm

The solution encoding is the same as described in Section 5.1.2.2, where the element a_j of the integer string $\boldsymbol{a}$ defines the number of parallel elements for each component i and version b (the relation between i, b and j is determined by Equation (5.10)).

In order to let the GA look for the solution meeting requirements (6.43) or (6.44), the following universal expression of solution quality (fitness) is used:

$$\begin{gathered}M-\alpha\,|\,\varepsilon_o(\boldsymbol{a})-\varepsilon^*{}_o\,|-(1-\alpha)\,|\,\varepsilon_c(\boldsymbol{a})-\varepsilon^*{}_c\,|\\ -\pi(\max\{0,\mathrm{Q_o}(\boldsymbol{a})-Q^*{}_o\}+\max\{0,\mathrm{Q_c}(\boldsymbol{a})-Q^*{}_c\})\end{gathered} \tag{6.45}$$

where π and M are constants much greater than the maximal possible value of system output performance.

The case when $Q^*{}_o=Q^*{}_c=0$ corresponds to formulation (6.43). Indeed, since π is sufficiently large, the value to be minimized in order to maximize the fitness is $\pi(Q_o+Q_c)$. On the other hand, when $Q^*{}_o=Q^*{}_c=1$ all availability limitations are removed and expected performance becomes the only factor in determining the system structure.

The solution decoding procedure determines n_{ib} for each system component i and each element version b from the string $\boldsymbol{a}$ and determines the performance measures of the system separately for open and closed modes according to the algorithm described in Section 5.1.2.2. Then it determines the solution fitness using expression (6.45).

Example 6.7

Consider a system of electronic switches consisting of four components connected in series [176]. Each component can contain a number of switches connected in parallel. The elements in each component should belong to a certain type. (For example, each component can operate in a different medium, which causes specific requirements on the switches.) Each element version is characterized by element parameters: availability and performance rates (operation times in open mode t_o and in closed mode t_c). The problem is to find the optimal system configuration by choosing elements for each component from the lists of versions presented in Table 6.12.

Table 6.12. Parameters of electronic switches

No. of component	Version of element	t_o	t_c	p_o	p_c
	1	5.80	3.00	0.81	0.76
	2	4.60	3.30	0.85	0.79
1	3	4.50	3.50	0.86	0.75
	4	4.00	3.10	0.84	0.76
	1	1.80	1.20	0.84	0.78
2	2	1.81	1.30	0.86	0.72
	3	1.85	1.10	0.89	0.70
	1	2.00	1.90	0.82	0.80
3	2	2.10	1.92	0.87	0.81
	3	2.10	1.89	0.89	0.73
	1	3.60	3.30	0.85	0.78
4	2	4.00	2.80	0.87	0.77

Table 6.13 contains the results obtained for a system that is considered to be in normal condition if the switching time in both modes is not greater than $w_o = w_c = 5$; the desired operation time is $\varepsilon^*_o = \varepsilon^*_c = 0$. The conditional expected value $\tilde{\varepsilon}$ is estimated for switching time distributed in the range of allowable values (0, 5). It is assumed that the operation speed is equally important in both modes: $\alpha = 0.5$. Three solutions were obtained for different levels of desired availability in both modes $Q^*_o = Q^*_c = 0$, $Q^*_o = Q^*_c = 0.035$ and $Q^*_o = Q^*_c = 0.05$.

Table 6.13. Solutions obtained for the system of electronic switches

Component	$Q^*_o=Q^*_c=0$	$Q^*_o=Q^*_c=0.035$	$Q^*_o=Q^*_c=0.05$
1	3*2,1*3	4*1,1*3,1*4	7*1,6*3,1*4
2	5*3	3*3	3*3
3	3*2	3*2	3*3
4	4*2	5*2	7*2
Q_o	0.030	0.035	0.050
Q_c	0.014	0.034	0.046
A	0.978	0.965	0.952
$\tilde{\varepsilon}_o$	2.213	2.076	1.997
$\tilde{\varepsilon}_c$	3.301	3.001	3.000

Observe that the solution maximizing the system availability (first formulation corresponding to $Q^*_o = Q^*_c = 0$) has a relatively small number of elements. Further growth of the number of elements decreases the system availability. One can see that, with the growth of availability requirements, the system availability increases by the price of the increase of expected switching time. The failure probability distributions in closed and open modes for the solutions obtained are presented in Figure 6.16. Note that the requirement to improve the conditional expected values of system performance contradicts the requirement to maximize the system availability defined as its ability to reach a threshold level of the performance.

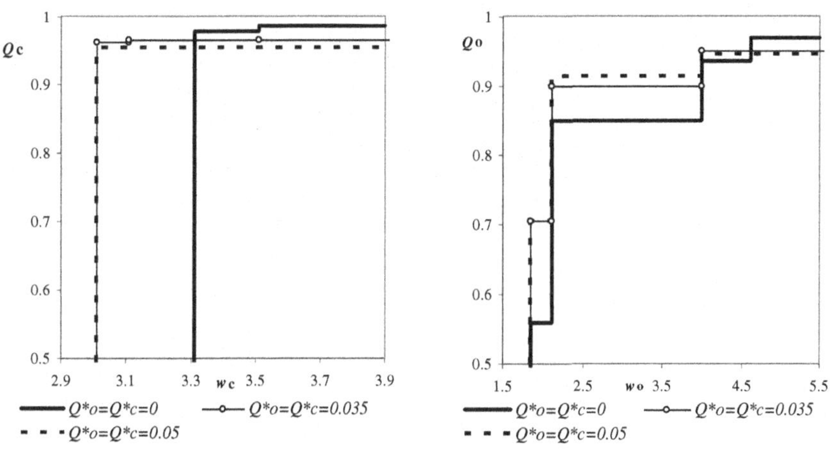

Figure 6.16. Failure probability distributions for the system of electronic switches

6.2.4 Optimal Topology of Systems with Two Failure Modes

6.2.4.1 Problem Formulation

The problem of optimizing a series-parallel MSS configuration is the following: find the series-parallel configuration of a given number of statistically independent units that provides maximal system availability when the units can experience two failure modes and are characterized by different performance rates and availability indices.

6.2.4.2 Implementing the Genetic Algorithm

According to its definition, any series-parallel system is either a single unit or it is two series-parallel subsystems connected in series or in parallel. Therefore, any such system can be represented by a binary tree. One of the possible representations was suggested in [173], where tree leaf nodes correspond to the primary units from which the configuration is built and the rest of the nodes are distinguished by the way the two children of the node are joined. As was shown in [173], such a binary tree can be easily represented by a symbolic string (post-order traversal) in which symbols from the set $\{1,\ldots,N\}$ correspond to unit numbers and symbols from the set {S, P} correspond to types of connection (S for series one and P for parallel one).

For example, the binary tree corresponding to the system presented in Figure6.17A can be represented by the following string: 12P3S45PP. The main disadvantage of this representation is that different strings can represent the same configuration, since the order of substrings representing the two child subtrees of any junction does not matter (observe that the system from Figure 6.17A can also be represented by the string 45P12P3SP). This causes situations in which the GA population is overwhelmed with different strings representing identical solutions. Such situations slow the algorithm convergence.

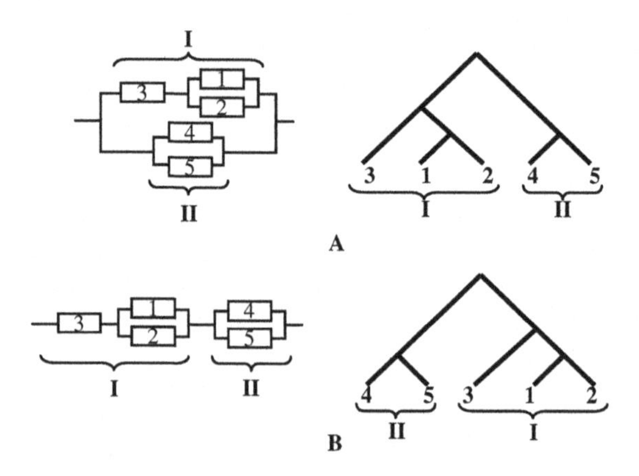

Figure 6.17. Binary tree representation of series-parallel system

In order to simplify the representation and to reduce the number of cases in which the same configuration is represented by different trees (strings), the following rule was introduced in [177] to determine the types of connection: for each node joining two child subtrees, determine the minimal number among the numbers of units belonging to the left child subtree x_L and right child subtree x_R. If $x_L<x_R$, then the subtrees are joined in parallel; if $x_L>x_R$, then they are joined in series.

Using this simplified representation one obtains a new configuration by swapping child subtrees of a given node (see Figure 6.17B) and does not need to distinguish nonleaf tree nodes (P or S), since they no longer determine the type of connection.

We use the following rule to represent the binary tree corresponding to a series-parallel configuration by a string: all numbers corresponding to units should appear in the string in the order that they appear in the tree from left to right. Each time that all of the numbers corresponding to subtrees connected by some node appear on the left-hand side of the given position in the string, the sign * representing the node should be inserted in this position.

Example 6.8

Consider the tree presented in Figure 6.17A. The corresponding string representation is 312**45**, where the underlined substring represents subsystem I and the double underlined substring represents subsystem II. Since for the root node $x_L = 1$ and $x_R = 4$, the subsystems I and II are connected in parallel. To make them connected in series, one just has to swap corresponding substrings and obtain string 45*312***, corresponding to the configuration presented in Figure 6.17B.

Note that, in the representation given, the last position of the string is always occupied by *, representing the root node of the tree. Therefore, this string element provides no information and can be removed. Since the total number of nonleaf nodes in the binary tree with N leafs is N–1, the string representing series-parallel configuration of N units should contain 2(N–1) elements.

Not every arbitrary string can represent a feasible solution. Indeed, consider string 3**1245*. Since each node sign * corresponds to two subtrees that the node connects, it should follow at least two unit numbers. In order to make an arbitrary solution feasible, in all cases where there are not enough numbers from the left-hand side of the node sign *, one has to find the closest number following the node sign * on the right-hand side and insert it immediately before the sign. (For the string given, such a procedure first produces string 31**245* and, when repeated, produces a feasible string 31*2*45*).

The simplest way to represent solution strings in the GA is by using permutations of integer numbers {1,…,2(N–1)} and by treating all the numbers greater than N as node signs * (for example, string 31827456 can be treated as 31*2*45*).

The following is a procedure for system availability evaluation based on decoding the system configuration from an arbitrary permutation $\boldsymbol{a}$ of integer numbers ranging from 1 to 2(N–1). The procedure enables the obtaining of u-functions $U_o(z)$ and $U_c(z)$ for the entire system, by applying composition operators over the corresponding u-functions of individual units in sequence determined by the system configuration encoded by a string $\boldsymbol{a} = (a_i,\ldots, a_{2N-2})$. To store the intermediate u-functions, a stack memory is used that allows binary subtrees (and corresponding u-functions) to be treated in order that is encoded by the string.

With each u-function $u(z)$ representing a subtree, we associate a number $x(u(z))$ equal to the smallest one from among the numbers of units belonging to the subtree.

The procedure performs the following steps:

1. Assigns number of string element $i = 1$.

2. If $a_i \le N$ (a_i corresponds to the number of the unit), assign $x(u_{a_i}(z)) = a_i$, place unit u-function $u_{a_i}(z)$ to the stack and go to step 5.

If $a_i > N$ (a_i corresponds to nonleaf node sign *) go to step 3.

3. If there are no fewer than two u-functions in the stack, go to step 4, else find string element a_j closest to a_i ($j>i$), which corresponds to the number of the unit ($a_j \le N$). Remove the element a_j from the string, shift all the elements a_i, …, a_{j-1} one position right and place element a_j into position i. Return to step 2.

4. Remove the upper u-function $u'(z)$ and second one from the top $u''(z)$ from the stack. Obtain the new u-function $u'''(z)$ either as $u'''(z) = u'(z) \otimes_{\varphi_{\text{ser}}} u''(z)$ (series connection) if $x(u'(z)) < x(u''(z))$ or as $u'''(z) = u'(z) \otimes_{\varphi_{\text{par}}} u''(z)$ (parallel connection) if $x(u'(z)) > x(u''(z))$.

Obtain index $x(u'''(z)) = \min\{x(u'(z)), x(u''(z))\}$. Place the new u-function $u'''(z)$ and the index $x(u'''(z))$ into the stack.

5. Increment i by one. If $i \le 2(N-1)$, return to step 2, else obtain the entire system u-function $U(z)$ as described in step 4.

Repeating steps 1-5 with the unit u-functions and composition operators corresponding to open and closed modes, one finally obtains the system u-functions $U_o(z)$ and $U_c(z)$ and determines the system's availability using Equation (6.21). The solution fitness is equal to the system's availability.

Since the solution of the optimization problem considered is represented by permutations of integer numbers (which corresponds to the sequencing problem), the corresponding fragment crossover operator and the mutation procedure that swaps two string elements (discussed in Section 1.3.2.6) are to be used. The following example illustrates the use of the fragment crossover operator, mutation procedure, and solution correction algorithm (used within the solution decoding procedure) for obtaining new feasible solutions from two parents.

Example 6.9

Consider two parent strings P1 and P2, representing configurations presented in Figure 6.18:

P1: $1\ 2\ \underline{3\ 4\ 5\ 6}\ 7\ 8$
P2: $4\ 2\ \underline{6\ 5\ 7\ 1}\ 8\ 3$

The fragment crossover operator is applied twice with the roles of the parents reversed. After applying the crossover with a randomly determined fragment, we obtain two offspring solutions O1 and O2 (the elements belonging to the fragment are underlined):

O1: $4\ 2\ \underline{1\ 5\ 6\ 7}\ 8\ 3$
O2: $1\ 2\ \underline{4\ 6\ 5\ 3}\ 7\ 8$

After applying the mutation procedure to O1 and O2, we obtain strings S1 and S2:

S1: $4\ 2\ \underline{3}\ 5\ 6\ 7\ 8\ \underline{1}$
S2: $1\ \underline{7}\ 4\ 6\ 5\ 3\ \underline{2}\ 8$

(the two randomly chosen positions are underlined).

Note that string S2 is infeasible (according to the feasibility rule presented in Section 3.2, the node sign 7 should follow at least two unit numbers representing two subtrees connected by the node). During the solution decoding procedure it is transformed into the string

S2: 1 4 7 5 6 3 2 8

The final solutions, as well as the intermediate ones, are also presented in Figure 6.18.

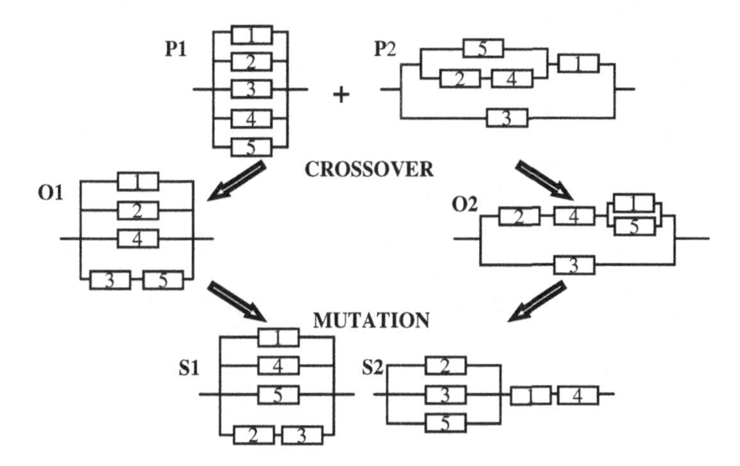

Figure 6.18. Examples of series-parallel configurations obtained by GA procedures

Example 6.10

Consider a set of 10 fluid flow valves. Each valve is characterized by its availability in open and closed modes (p_o, p_c), and by the nominal flow transmitting capacity f. These parameters are presented in Table 6.14. We want the flow to be not less than w_c in the closed mode and not greater than w_o in the open (disconnected) mode.

Three different system configurations were found by the GA for the different desired system transmitting capacities in open and closed modes: configuration A for $w_c = 5$, $w_o = 0.1$, configuration B for $w_c = 7$, $w_o = 0.1$, and configuration C for $w_c = 10$, $w_o = 3$. These configurations are presented in Figure 6.19. The probabilistic distributions of flows through the system in closed and open modes for the configurations obtained are presented in Figure 6.20 in the form of cumulative probabilities $\Pr\{G_c>w_c\}$ and $\Pr\{G_o<w_o\}$. Table 6.15 contains system fault probabilities Q_o and Q_c and availability index A obtained for each configuration for all the three demand combinations (w_c, w_o). Note that each configuration, though being the best for a certain combination (w_c, w_o), does not provide the greatest system availability for the two other combinations. Configuration A can not provide flow $f = w_c = 10$ in the closed mode even when all the units are available

Note that while the configurations obtained seem to be not realistic for pure switching systems they are relevant when one considers the configuration of different types of flow transmission equipment (pumps, filters, *etc.*) having alarm valves aimed at preventing the flow in the case of contingency.

Table 6.14. Parameters of fluid flow valves

No. of unit	f	p_c	p_o
1	2.0	0.86	0.82
2	2.0	0.92	0.88
3	2.5	0.95	0.89
4	2.5	0.95	0.89
5	3.0	0.90	0.86
6	3.0	0.90	0.86
7	4.0	0.87	0.83
8	4.0	0.84	0.80
9	5.0	0.87	0.81
10	5.0	0.82	0.80

Table 6.15. Reliability characteristics of the obtained solutions

Solution		w_c=5.0, w_o=0.1	w_c=7.0, w_o=0.1	w_c=10.0, w_o=3.0
	Q_c	0.059	0.489	1.000
A	Q_o	0.032	0.032	0.008
	A	**0.955**	0.739	0.000
	Q_c	0.017	0.045	0.208
B	Q_o	0.119	0.119	0.089
	A	0.932	**0.918**	0.852
	Q_c	0.002	0.004	0.042
C	Q_o	0.323	0.323	0.110
	A	0.837	0.836	**0.924**

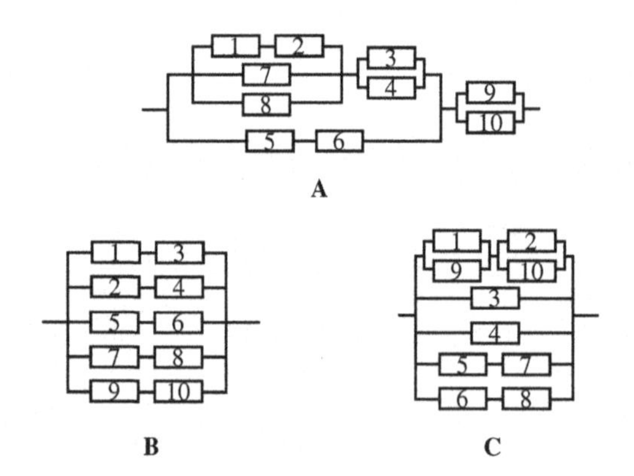

Figure 6.19. Optimal configurations obtained for the system of fluid flow valves

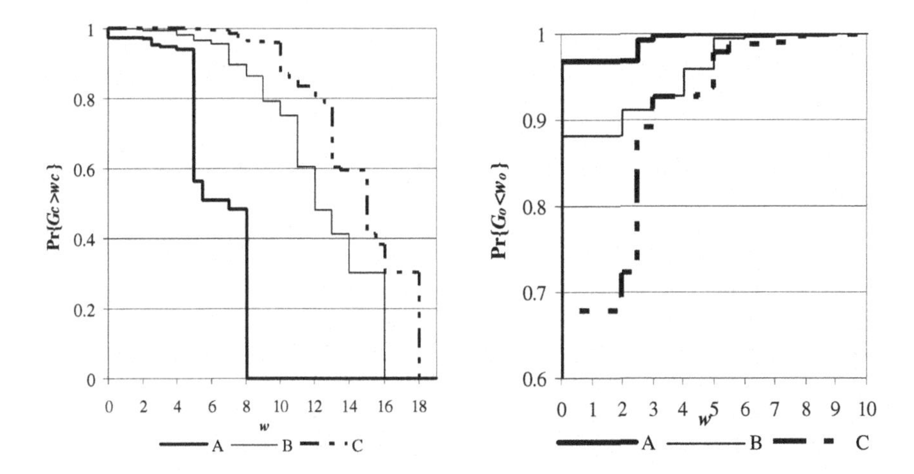

Figure 6.20. Cumulative probabilities $\Pr\{G_c>w_c\}$ and $\Pr\{G_o<w_o\}$for the fluid flow valves configurations obtained

6.3 Weighted Voting Systems

The weighted voting system (WVS) consists of n independent voting units that provide a binary decision or abstain from voting. Each unit has its own individual weight. The system accepts the proposition I if the cumulative weight of the units supporting this proposition is at least the prespecified fraction τ of the cumulative weight of all non-abstaining units. The system abstains if all n units abstain. In all other cases, the system rejects proposition I. The system fails if it does not accept the proposition that should be accepted, does not reject the proposition that should be rejected, or abstains from voting.

This can be modelled by considering the system input I being either 1 (proposition to be accepted) or 0 (proposition to be rejected) which is supplied to each unit. Each unit j produces its decision (unit output) $d_j(I)$ which can be 1, 0, or x (in the case of abstention). Inequality $d_j(I) \neq I$ means that the decision made by the unit is wrong. The above listed errors can be expressed as

1. $d_j(0) = 1$ (unit fails stuck-at-1)
2. $d_j(1) = 0$ (unit fails stuck-at-0)
3. $d_j(I) = x$ (unit fails stuck-at-x)

Accordingly, the reliability of each unit j can be characterized by the probabilities of these errors: $q_{01}^{(j)}$ for the first one, $q_{10}^{(j)}$ for the second one, $q_{1x}^{(j)}$ and $q_{0x}^{(j)}$ for the third one, where $q_{im}^{(j)}$ is $\Pr\{d_j(I) = m \mid I = i\}$ (note that stuck-at-x probabilities can be different for inputs $i = 0$ and $i = 1$).

To make a decision about the proposition acceptance, the system incorporates all of the unit decisions into a unanimous system output D in the following manner:

$$D(I)=\begin{cases}1, & \text{if} \sum\limits_{d_j(I)\neq x} \psi_j d_j(I) \geq \tau \sum\limits_{d_j(I)\neq x} \psi_j, \quad \sum\limits_{d_j(I)\neq x} \psi_j \neq 0 \\ 0, & \text{if} \sum\limits_{d_j(I)\neq x} \psi_j d_j(I) < \tau \sum\limits_{d_j(I)\neq x} \psi_j, \quad \sum\limits_{d_j(I)\neq x} \psi_j \neq 0 \\ x, & \text{if} \sum\limits_{d_j(I)\neq x} \psi_j = 0\end{cases} \tag{6.46}$$

where ψ_j is the nonnegative weight of an individual unit j which expresses its relative importance in the WVS and τ is a threshold factor which determines what fraction of the overall weight of voted units should correspond to those that approve the proposition to make it accepted by the entire system.

The entire system output distribution is characterized by WVS output probabilities $Q_{im} = \Pr\{D(I) = m \mid I = i\}$, where $m \in \{0, 1, x\}$. The system fails if $D(I) \neq I$. The entire WVS reliability can be defined as $R = \Pr\{D(I) = I\}$. One can see that the system reliability is a function of the reliabilities of its units. The reliability characteristics of the WVS units, as well as the probability distribution of the propositions, $P_0 = \Pr\{I = 0\}$ and $P_1 = \Pr\{I = 1\}$, can be elicited from historical statistics. In technical systems, probabilities of different kinds of error can be obtained for each unit with a high precision by intensive testing. The entire WVS reliability also depends on the unit weights and the threshold. The proper choice of these parameters can improve the WVS reliability without improving the reliability of the voting units.

6.3.1 Evaluating the Weighted Voting System Reliability

Let us define the total weight of WVS units supporting proposition I as Ψ_I^1:

$$\Psi_I^1 = \sum_{d_j(I)\neq x} \psi_j d_j(I) \tag{6.47}$$

and the total weight of units voting for the proposition rejection as Ψ_I^0:

$$\Psi_I^0 = \sum_{d_j(I)\neq x} \psi_j (1-d_j(I)) \tag{6.48}$$

The decision rule (6.46) can now be rewritten as follows:

$$D(I)=\begin{cases}1, & \text{if } \Psi_I^1 \geq \tau(\Psi_I^1+\Psi_I^0), \quad \Psi_I^1+\Psi_I^0 \neq 0 \\ 0, & \text{if } \Psi_I^1 < \tau(\Psi_I^1+\Psi_I^0), \quad \Psi_I^1+\Psi_I^0 \neq 0 \\ x, & \text{if } \Psi_I^1+\Psi_I^0 = 0\end{cases} \tag{6.49}$$

Following this expression, the condition $D(I) = 0$ can be rewritten as

$$\Psi_I^1 < \tau(\Psi_I^1 + \Psi_I^0) \tag{6.50}$$

or

$$(1-\tau)\Psi_I^1 - \tau\Psi_I^0 < 0 \tag{6.51}$$

This gives one a simple way of tallying the units' votes: each unit j adds a value of $(1-\tau)\psi_j$ to the total WVS score if it votes for the proposition's acceptance, a value of $-\tau\psi_j$ if it votes for proposition's rejection, and nothing if it abstains. The proposition is rejected if the total score is negative.

6.3.1.1 Universal Generating Function Technique for Weighted Voting System Reliability Evaluation

Using the UGF approach one can describe the distributions of the random output G_{ij} of an individual three-state voting unit j for input i as

$$u_{ij}(z) = \sum_{k=0}^{2} p_{jk} z^{g_{jk}} \tag{6.52}$$

where for $i = 1$

$$\begin{aligned} p_{j0} &= q_{10}^{(j)}, \quad g_{j0} = -\tau\psi_j \\ p_{j1} &= q_{11}^{(j)} = (1 - q_{10}^{(j)} - q_{1x}^{(j)}), \quad g_{j1} = (1-\tau)\psi_j \\ p_{j2} &= q_{1x}^{(j)}, \quad g_{j2} = 0 \end{aligned} \tag{6.53}$$

and for $i = 0$

$$\begin{aligned} p_{j0} &= q_{01}^{(j)}, \quad g_{j0} = (1-\tau)\psi_j \\ p_{j1} &= q_{00}^{(j)} = (1 - q_{01}^{(j)} - q_{0x}^{(j)}), \quad g_{j1} = -\tau\psi_j \\ p_{j2} &= q_{0x}^{(j)}, \quad g_{j2} = 0 \end{aligned} \tag{6.54}$$

In each voting unit, state 0 corresponds to an incorrect decision, state 1 corresponds to a correct decision, and state 2 corresponds to an abstention.

The total random WVS score G_i for input $I = i$ is equal to the sum of the random outputs of n individual voting units:

$$G_i = \sum_{j=1}^{n} G_{ij} \tag{6.55}$$

Therefore, the u-function of the system score $U_i(z)$ can be obtained using the following composition operator:

$$U_i(z) = \underset{+}{\otimes}(u_{i1}(z), \ldots, u_{in}(z)) \tag{6.56}$$

Since the function (6.55) possesses commutative and associative properties, the u-function of the entire WVS can be obtained recursively by the consecutive determination of u-functions of the arbitrary subsets of the elements. For example it can be obtained by the recursive procedure

$$\tilde{U}_{i1}(z) = u_{i1}(z),\ \ \tilde{U}_{im}(z) = \tilde{U}_{im\text{-}1}(z) \underset{+}{\otimes} u_{im}(z) \text{ for } 1<m\le n$$

$$U_i(z) = \tilde{U}_{in}(z) \tag{6.57}$$

In this procedure, $\tilde{U}_{im}(z)$ represents the score distribution of the WVS subsystem consisting of the first m voting units.

Note that, while the total number of different possible WVS states is 3^n, many of these states can result in the same values of score G_i. Therefore, the total number of terms $U_i(z)$ can be less than 3^n because of the like terms collection.

Using the criterion of proposition rejection as an acceptability function $F(G_i) = 1(G_i < 0)$ we obtain the proposition rejection probability Q_{i0}:

$$Q_{i0} = \Pr\{D(I) = 0 \mid I = i\} = E(F(G_i)) \tag{6.58}$$

which is equal to the sum of the coefficients of the terms with the negative exponents in $U_i(z)$.

Having Q_{10} one can easily obtain Q_{11} as

$$Q_{11}=1-Q_{10}-Q_{1x},\quad \text{where } Q_{1x} = \prod_{j=1}^{n} q_{1x}^{(j)} \tag{6.59}$$

Events $I = 0$ and $I = 1$ are mutually exclusive. Therefore, the entire WVS reliability $\Pr\{D(I) = I\}$ can be defined as

$$\Pr\{D(I) = 0 \mid I = 0\}\ \Pr\{I = 0\}+\Pr\{D(I) = 1 \mid I = 1\}\ \Pr\{I = 1\} \tag{6.60}$$

and calculated as follows:

$$R = P_0Q_{00}+P_1Q_{11} = P_0Q_{00}+P_1(1-Q_{10}-Q_{1x}) \tag{6.61}$$

Example 6.11

Consider a WVS with

$$n = 2,\ P_0 = P_1 = 0.5,\ \tau = 0.6$$

$$q_{01}^{(1)} = 0.02,\ q_{10}^{(1)} = 0.02,\ q_{0x}^{(1)} = q_{1x}^{(1)} = 0.01,\ \psi_1 = 5$$

$$q_{01}^{(2)} = 0.02,\ q_{10}^{(2)} = 0.05,\ q_{0x}^{(1)} = q_{1x}^{(2)} = 0.02,\ \psi_2 = 3$$

Then

$$(1-\tau)\psi_1 = 2,\ (1-\tau)\psi_2 = 1.2,\ -\tau\psi_1 = -3,\ -\tau\psi_2 = -1.8$$

$$u_{01}(z) = 10^{-2}(2z^2 + 97z^{-3} + z^0),\ u_{11}(z) = 10^{-2}(2z^{-3} + 97z^2 + z^0)$$

$$u_{02}(z) = 10^{-2}(2z^{1.2} + 96z^{-1.8} + 2z^0),\ u_{12}(z) = 10^{-2}(5z^{-1.8} + 93z^{1.2} + 2z^0)$$

u-functions for the entire WVS are

$$\tilde{U}_{01}(z) = u_{01}(z),\ \tilde{U}_{02}(z) = u_{01}(z) \underset{+}{\otimes} u_{02}(z)$$

$$= 10^{-4}(4z^{2.2} + 4z^2 + 192z^{0.2} + 2z^{1.2} + 2z^0 + \mathbf{96z^{-1.8}} + \mathbf{194z^{-0.2}} + \mathbf{194z^{-1.8}} + \mathbf{9312z^{-4.8}})$$

$$\tilde{U}_{11}(z) = u_{11}(z),\ \tilde{U}_{12}(z) = u_{11}(z) \underset{+}{\otimes} u_{12}(z)$$

$$= 10^{-4}(\mathbf{10z^{-4.8}} + \mathbf{4z^{-3}} + \mathbf{186z^{-1.8}} + \mathbf{5z^{-1.8}} + 2z^0 + 93z^{1.2} + 485z^{0.2} + 194z^2 + 9021z^{3.2})$$

The terms with negative exponents are marked in bold. To obtain Q_{10} and Q_{00} one should calculate the sums of the coefficients of the marked terms:

$$Q_{00} = 10^{-4}(96+194+194+9312) = 0.9796$$

$$Q_{10} = 10^{-4}(10+4+186+5) = 0.0205$$

In accordance with (6.59) $Q_{1x} = q_{1x}^{(1)} q_{1x}^{(2)} = 0.01\times0.02 = 0.0002$. In accordance with (6.61) the WVS reliability is

$$R = P_0Q_{00} + P_1(1-Q_{10}-Q_{1x}) = 0.5\times0.9796 + 0.5\times(1-0.0205-0.0002) = 0.97945$$

It should be noted that, owing to the additive property of the structure function, when adding a unit to an already evaluated system the new system does not need to be evaluated from scratch. Instead, the operator $\otimes_+$ should be applied to the u-functions of the evaluated system and to the new unit. Moreover, the associative property of the structure function allows the reliability of the WVS to be easily evaluated when it is combined from a number of subsystems for which corresponding u-functions are already obtained.

6.3.1.2 Simplification Technique

Consider the *u*-function $\tilde{U}_{im}(z)$ that represents the distribution of the score $\tilde{G}_{im}$ of the WVS subsystem λ_m consisting of first *m* voting units.

Let V_i be the sum of the weights of the units from *i* to *n*:

$$V_i = \sum_{j=i}^{n} \psi_j \tag{6.62}$$

One can see that V_{m+1} represents the sum of the weights of WVS units not belonging to λ_m.

The maximal possible value of the WVS score after the remainder of the units add their votes is $\tilde{G}_{im} + (1-\tau)V_{m+1}$ (if all of the units from $m+1$ to *n* vote for the proposition acceptance) and the minimal possible value of the WVS score is $\tilde{G}_{im} - \tau V_{m+1}$ (if all of the units from $m+1$ to *n* vote for the proposition rejection).

Therefore, if

$$\tilde{G}_{im} + (1-\tau)V_{m+1} < 0 \tag{6.63}$$

the proposition will be rejected independently of the states of the units $m+1, \ldots, n$. (We will refer to the *u*-function terms corresponding to the realizations of the score $\tilde{G}_{im}$ meeting condition (6.63) as 0-terms). Indeed, in each 0-term the realization of the score $\tilde{G}_{im}$ is low enough to prevent the total system score from being positive. Therefore, there is no need to continue the calculations by combining the states of the remainder of the units with the states corresponding to 0-terms. The sum of the probabilities of all of the possible combinations of the units $m+1, \ldots, n$ is equal to unity. Therefore, the total overall probability of the unit state combinations in which the score $\tilde{G}_{im}$ guarantees the proposition rejection is equal to the sum of the coefficients of the 0-terms in the *u*-function $\tilde{U}_{im}(z)$.

If

$$\tilde{G}_{im} - \tau V_{m+1} \geq 0 \tag{6.64}$$

then there is no chance that WVS will reject the proposition even if the units $m+1, \ldots, n$ vote for its rejection. (We will refer to the *u*-function terms corresponding to the realizations of the score $\tilde{G}_{im}$ meeting condition (6.64) as 1-terms.) Combining any 1-term of $\tilde{U}_{im}(z)$ with any terms corresponding to the not-yet-considered units cannot produce a term with a negative score. Therefore, this term cannot participate in determining the Q_{i0}. This means that all of the 1-terms can be removed from the *u*-function without affecting the resulting value of Q_{i0}.

The technique described allows one to evaluate the entire WVS reliability using the following algorithm.

1. For each voting element j, define the two u-functions $u_{0j}(z)$ and $u_{1j}(z)$ in the form (6.52) using Equations (6.53) and (6.54).
2. Assign $Q_{10} = Q_{00} = 0$, $\tilde{U}_{01}(z) = u_{01}(z)$, $\tilde{U}_{11}(z) = u_{11}(z)$.
3. For $i = 0$, 1 and $m = 2, \ldots, n$ (voting units can be ordered arbitrarily):
   - remove 1-terms and 0-terms from $\tilde{U}_{\text{im-1}}(z)$;
   - add the coefficients of the removed 0-terms to Q_{i0};
   - obtain $\tilde{U}_{\text{im}}(z) = \tilde{U}_{\text{im-1}}(z) \otimes_{+} u_{\text{im}}(z)$.
3. Add the coefficients of the negative terms in $U_i(z) = \tilde{U}_{in}(z)$ to Q_{i0}.
4. Calculate the fault probability of Q_{11} using Equation (6.59).
5. Calculate the WVS reliability R using Equation (6.61).

Example 6.12

Consider the WVS from Example 6.11 and apply to it the suggested simplification technique. For the given WVS:

$$V_2 = \psi_2 = 3, -\tau V_2 = -1.8, (1-\tau)V_2 = 1.2$$

First, assign

$$Q_{10} = Q_{00} = 0$$

$$\tilde{U}_{01}(z) = u_{01}(z) = 10^{-2}(\underline{2z^2} + \mathbf{97z^{-3}} + z^0)$$

$$\tilde{U}_{11}(z) = u_{11}(z) = 10^{-2}(\mathbf{2z^{-3}} + \underline{97z^2} + z^0)$$

The 1-terms in the u-functions are underlined; the 0-terms are marked in bold. The coefficients of the 0-terms are added to Q_{00} and Q_{10}:

$$Q_{00} = 0.97, Q_{10} = 0.02$$

After removal of the 0-terms and the 1-terms one obtains

$$\tilde{U}_{01}(z) = 0.01z^0$$

$$\tilde{U}_{11}(z) = 0.01z^0$$

The u-functions for two voting units are

$$\tilde{U}_{02}(z) = \tilde{U}_{01}(z) \underset{+}{\otimes} u_{02}(z) = 10^{-4}z^0(2z^{1.2} + 96z^{-1.8} + 2z^0)$$

$$= 10^{-4}(2z^{1.2} + \mathbf{96z^{-1.8}} + 2z^0)$$

$$\tilde{U}_{12}(z) = \tilde{U}_{11}(z) \underset{+}{\otimes} u_{12}(z) = 10^{-4}z^0(5z^{-1.8}+93z^{1.2}+2z^0)$$

$$= 10^{-4}(\mathbf{5z^{-1.8}}+93z^{1.2}+2z^0)$$

After adding the coefficients of the negative terms (marked in bold) from $\tilde{U}_{12}(z)$ to Q_{10} one obtains the same values of Q_{00} and Q_{10} as in Example 6.11:

$$Q_{00} = 0.97+0.0096 = 0.9796, \; Q_{10} = 0.02+0.0005 = 0.0205$$

6.3.2 Optimization of Weighted Voting System Reliability

While the reliabilities of the voting units usually cannot be changed when the WVS is built, the weights and the threshold can be chosen in such a way that maximizes the entire system reliability. The WVS optimization problem is, therefore, formulated as follows.

Find the units' weights and the threshold value that maximize the reliability of the WVS consisting of units with the given fault probabilities:

$$R(\psi_1,\ldots,\psi_n,\tau)\rightarrow\max \tag{6.65}$$

For an existing WVS with given weights the "tuning" problem can arise in which just the threshold value maximizing the system reliability should be found subject to changing conditions. For example, having information about the probability distribution between propositions that should be accepted or rejected (P_1 and P_0), one can modify the threshold value to achieve the greatest reliability.

Note that the reliability characteristics of WVS units, as well as the propositions' probability distributions, can be elicited from the historical statistics without respect to changes in WVS weights and threshold variation.

6.3.2.1 Implementing the Genetic Algorithm

The natural representation of a WVS weight distribution is by an n-length integer string in which the value in the jth position corresponds to the weight of the jth unit of the WVS. One can see that multiplying all the unit weights by the same value does not affect the WVS output defined by rule (6.46). Therefore, the unit weights can be normalized in such a way that the total weight V_1 is always equal to some constant c. The normalized weights from arbitrary integer string $\boldsymbol{a} = (a_1,\ldots,a_n)$ are obtained as follows:

$$\psi_j = a_j c / \sum_{m=1}^{n} a_m \tag{6.66}$$

The range in which the integer numbers are generated affects the precision of weights determination.

For each given weight distribution (determined by the string $\boldsymbol{a}$), the solution decoding procedure obtains the optimal value of the WVS threshold $\tau(\boldsymbol{a}, P_0, P_1)$ by solving the single-variable optimization problem

$$R(\boldsymbol{a}, P_0, P_1, \tau) \rightarrow \max \tag{6.67}$$

and uses the optimal value of the system reliability obtained as the solution fitness.

Example 6.13

Consider a target identification WVS consisting of five voting units making their decisions based on different properties of the target [75]. The voting unit weights for the system can be represented in the GA by an arbitrary integer vector of length 5. In this example we used c = 10 and generated the elements of vector $\boldsymbol{a}$ in the range (0, 100). For such parameters, the vector (25 10 55 62 38) produces, according to (6.66), weights $\psi_1 = 1.316$, $\psi_2 = 0.526$, $\psi_3 = 2.895$, $\psi_4 = 3.263$, and $\psi_5 = 2.0$.

The reliability indices of the voting units $q_{i0}^{(j)}$ $q_{i1}^{(j)}$ and $q_{ix}^{(j)}$ are presented in Table 6.16.

Table 6.16. Parameters of WVS units

No. of unit	1	2	3	4	5
$q_{01}^{(j)}$	0.02	0.06	0.07	0.08	0.18
$q_{0x}^{(j)}$	0.08	0.00	0.05	0.16	0.12
$q_{10}^{(j)}$	0.15	0.18	0.07	0.12	0.16
$q_{1x}^{(j)}$	0.20	0.00	0.05	0.16	0.12
ψ_i	2.4862	1.8784	2.4033	1.6851	1.5470

The optimal weights of units obtained by the GA-based optimization procedure for $P_0 = P_1 = 0.5$ are also presented in this table.

The optimal value of threshold is $\tau = 0.412$, for which the system reliability is $R = 0.982$.

To estimate the effect of the input probability distribution on WVS reliability, the optimal threshold values and corresponding system reliabilities were obtained for the WVS with the obtained weights, and for WVS with equal weights as a functions of P_1 ($P_0 = 1-P_1$). These functions are presented in Figure 6.21, where $\tau(P_1)$ and $R(P_1)$ correspond to a WVS with optimal weights obtained for $P_1 = 0.5$, and $\tau^*(P_1)$ and $R^*(P_1)$ correspond to a WVS with equal weights. Note that optimal weights obtained for $P_1 = 0.5$ are not optimal for $P_1 \neq 0.5$, but they provide greater WVS reliability than equal weights on the whole range $0 \le P_1 \le 1$.

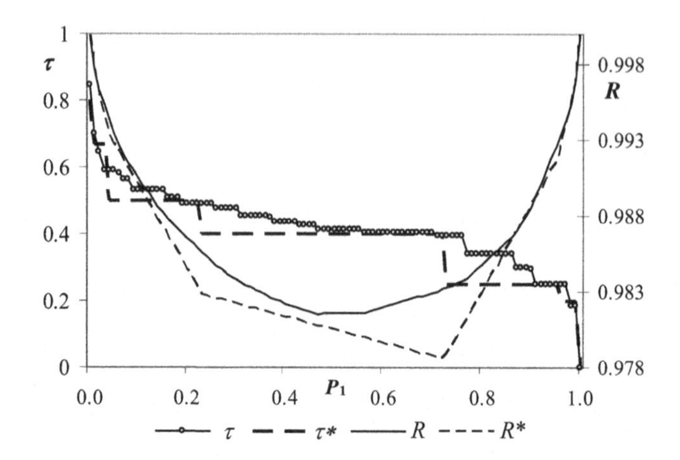

Figure 6.21. Optimal threshold value and WVS reliability as functions of P_1

6.3.3 Weighted Voting System Consisting of Voting Units with Limited Availability

In the WVS model considered in Section 6.3.1, all the voting units were assumed to be fully available (unit unavailability and stuck-at-x failure were not distinguished). In practice, one can deal with separate data concerning unit availability and probabilities of unit failure (wrong decision or abstention) when the unit is in its operating condition.

Two types of WVS can be defined with respect to their treatment of unavailable voting units. In the system of type 1, the unit stuck-at-x failure state and unit inoperable state cannot be distinguished by the system or the system cannot react to information about unit unavailability by changing its weights and threshold. The absence of a unit's output is interpreted by the WVS of this type as abstention from voting. In the system of type 2, the unavailable state of a unit and its abstention from voting can be distinguished and the WVS parameters can be adjusted to optimize its performance for each combination of available voting units.

In the system of type 1, the parameters (weights and threshold) can be chosen only once. The optimal WVS parameters obtained for the system with fully available units can be far away from optimality when the voting units have limited availability. In this section we demonstrate the incorporation of data about units availability into the procedure of parameters optimization for WVS of type 1. For the WVS of type 2 the parameter optimization procedure for fully available units presented in Section 6.3.2 should be applied each time the change of set of available units is detected.

Consider the voting unit j that can be in one of two states: $s_j = 1$ if the unit is available and $s_j = 0$ if it is unavailable. Let the operational availability of unit j be $\Pr\{s_j = 1\} = \alpha_j$. The unit can produce output $d_j(I) \neq x$ only if it is available

Therefore, for the given $I = i$ andfor each decision $m \in \{0, 1\}$of the individualunit j

$$\Pr\{d_j(I) = m \mid I = i\} = \Pr\{ d_j(I) = m \mid I = i, s_j = 1\} \Pr\{s_j = 1\}$$
$$= \alpha_j q_{im}^{(j)} \tag{6.68}$$

When the unit is not available ($s_j = 0$), its output is interpreted by the WVS of type 1 as $d_j(I) = x$. The same output can also be produced by the unit when it is available but indecisive. Therefore:

$$\Pr\{d_j(I) = x \mid I = i \} = \Pr\{ s_j = 0\}$$
$$+\Pr\{d_j(I) = x \mid I = i, s_j = 1\} \Pr\{s_j = 1\} = 1 - \alpha_j + \alpha_j q_{ix}^{(j)} \tag{6.69}$$

Since the output distribution of the available unit is represented by the u-function (6.52)-(6.54), and since each unit j adds the value of zero to the total WVS score when $d_j(I) = x$, one can obtain the u-function $\hat{u}_{ij}(z)$ representing the output distribution of the unit, which has the availability α_j, using the following operator κ:

$$\hat{u}_{ij}(z) = \kappa(u_{ij}(z)) = \alpha_j u_{ij}(z) + (1 - \alpha_j) z^0 \tag{6.70}$$

The u-function $\hat{u}_{ij}(z)$ has the same form as the u-function $u_{ij}(z)$ (6.52)-(6.54), except that its coefficients are

$$\hat{p}_{j0} = \alpha_j p_{j0}$$
$$\hat{p}_{j1} = \alpha_j p_{j1}$$
$$\hat{p}_{j2} = \alpha_j p_{j2} + 1 - \alpha_j \tag{6.71}$$

Using the algorithm presented in Section 6.3.1.2 over u-functions $\hat{u}_{ij}(z)$, one can obtain the reliability of WVS consisting of units with limited availability.

Example 6.14

Consider the WVS of type 1 consisting of four voting units with reliability indices presented in Table 6.17. The optimal weights of units ψ_j obtained for the WVS with fully available units for $P_0 = P_1 = 0.5$ are also presented in this table. The optimal value of the threshold is $\tau = 0.58$, for which the system reliability is $R = 0.891$. Taking into account the limited availability of voting units (availability indices α_j for the units are also presented in Table 6.17), one obtains much lower reliability $R_\alpha = 0.815$ for a WVS with the same weights and threshold. The reliability of a WVS consisting of units with limited availability can be improved if the unit availability values are included in the reliability estimation procedure while the optimization problem is solved. The optimal weights ψ^*_j obtained for the system are presented in Table 6.17. The optimal value of the threshold is $\tau^* = 0.4$, for which the system reliability is $R^*_\alpha = 0.846$. Including information about voting unit availability into the WVS parameters optimization problem enables the system reliability to be improved.

Table 6.17. Parameters of WVS units with limited availability

Unit no.	α_j	$q_{01}^{(j)}$	$q_{0x}^{(j)}$	$q_{10}^{(j)}$	$q_{1x}^{(j)}$	ψ_j	ψ^*_j	I_Rb_j
1	0.76	0.00	0.35	0.35	0.00	2.759	3.459	0.071
2	0.80	0.34	0.10	0.23	0.10	0.172	1.541	0.044
3	0.82	0.11	0.07	0.36	0.06	3.060	2.444	0.065
4	0.78	0.30	0.12	0.07	0.00	4.009	2.556	0.056

In order to find weaknesses in the WVS design and to suggest modifications for system upgrade or to determine the optimal voting unit maintenance policy one has to perform the unit availability importance analysis. According to the definition (4.71), the Birnbaum importance index for the WVS element j can be obtained as

$$I_Rb_j = R_{j1} - R_{j0} \tag{6.72}$$

where R_{j1} is theWVS reliability when the voting unit j is fully available and the remainder of the units have their availability α_j; R_{j0} is the WVS reliability when the voting unit j is unavailable.

An improvement in availability of the unit with the highest importance I_Rb_j causes the greatest increase in WVS reliability.

To determine the voting unit importance in the WVS of type 1, one has to apply the algorithm presented in Section 6.3.1.2 twice: the first time substituting $\alpha_j = 1$ in (6.71) to obtain R_{j1}, and the second time substituting $\alpha_j = 0$ in (6.71) to obtain R_{j0} and then to use Equation (6.72).

Example 6.15

The availability importance indices I_Rb_j of voting units of the WVS from Example 6.14 were obtained for optimal weights ψ^*_j and threshold τ^* [178]. These indices are presented in Table 6.17. The unit availability importance does not depend on the availability of this unit, but it depends strongly on the availability of the rest of units. This dependence is linear. Figure 6.22 presents dependencies of unit importance indices on the availability of voting unit 1. It should be noted that the relative importance of units can vary with variation of unit availability. For example, unit 4 is the most important one in the WVS for $\alpha_1<0.6$, but it becomes the least important one when $\alpha_1>0.92$.

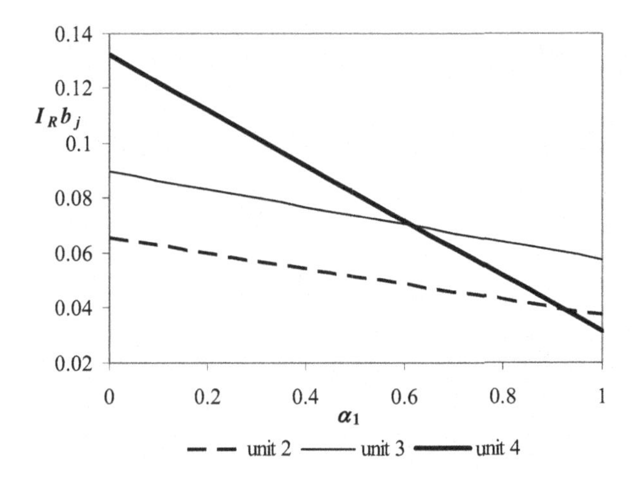

Figure 6.22. Voting unit availability importance as a function of availability α_1

In an adjustable WVS (WVS of type 2), the optimal weights and threshold are found for each combination of voting units available at the moment. Each hth combination is represented as the subset λ_h of the set Λ of all of the WVS units. The total number of possible combinations (subsets of Λ) in the WVS consisting of n units is 2^n. Let $\psi_j(\lambda_h)$ and $\tau(\lambda_h)$ be the optimal parameters of the WVS consisting of fully available voting units belonging to λ_h, and $R_\alpha(\lambda_h)$ is the reliability of this WVS with the optimal parameters. The entire reliability of the WVS of type 2 can be obtained as follows:

$$R = \sum_{h=1}^{2^n} [R_\alpha(\lambda_h) \prod_{e \in \lambda_h} \alpha_e \prod_{e \notin \lambda_h} (1-\alpha_e)] \tag{6.73}$$

In order to distinguish the availability of voting unit j, this expression can be rewritten as follows:

$$R = \alpha_j \sum_{h=1}^{2^{n-1}} [R_\alpha(\mu_h \cup \{j\}) \prod_{e \in \mu_h} \alpha_e \prod_{e \notin \mu_h} (1-\alpha_e)]$$

$$+(1-\alpha_j) \sum_{h=1}^{2^{n-1}} [R_\alpha(\mu_h) \prod_{e \in \mu_h} \alpha_e \prod_{i \notin \mu_h} (1-\alpha_e)] \tag{6.74}$$

where $\mu_h = \lambda_h \setminus \{j\}$. Using (6.74), one can determine the availability importance of the voting unit j as

$$I_R b_j = \partial R / \partial \alpha_j = \sum_{h=1}^{2^{n-1}} [R_\alpha(\mu_h \cup \{j\}) \prod_{e \in \mu_h} \alpha_e \prod_{e \notin \mu_h} (1-\alpha_e)]$$

$$- \sum_{h=1}^{2^{n-1}} [R_\alpha(\mu_h) \prod_{e \in \mu_h} \alpha_e \prod_{e \notin \mu_h} (1-\alpha_e)] \tag{6.75}$$

which is the same as

$$I_R b_j = \sum_{h=1}^{2^n} [\xi_h R_\alpha(\lambda_h) \prod_{e \in \lambda_h} \alpha_e \prod_{e \notin \lambda_h} (1-\alpha_e)] \tag{6.76}$$

where

$$\xi_h = \begin{cases} 1/(\alpha_j - 1), & j \notin \lambda_h \\ 1/\alpha_j, & j \in \lambda_h \end{cases} \tag{6.77}$$

For a WVS in which voting units have high availability, the terms multiplied by $(1-\alpha_e)$ can be neglected and, therefore, Equation (6.76) can be approximated as follows:

$$\widetilde{I}_R b_j = \frac{R_\alpha(\Lambda) - R_\alpha(\Lambda \setminus \{j\})}{\alpha_j} \prod_{e=1}^{n} \alpha_e \tag{6.78}$$

In some WVSs of type 2, only the threshold value can be adjusted according to different combinations of available units, whereas unit weights remain the same. Let ψ_1, ..., ψ_n be the constant unit weights and $\tau(\lambda_h)$ be the optimal threshold obtained for the given weights for the WVS consisting of fully available voting units belonging to λ_h. For the fixed weights and the optimal threshold $\tau(\lambda_h)$ one can obtain the reliability $R_\alpha^\tau(\lambda_h)$ of subsystem λ_h. Substituting in Equations (6.76) and (6.78) $R_\alpha(\lambda_h)$ with $R_\alpha^\tau(\lambda_h)$, one obtains the availability importance index $I_R B_j^\tau$ for unit j.

Example 6.16

Consider the WVS from Example 6.14. The maximal reliability values obtained by the optimization procedure for each possible combination of WVS units are presented in Table 6.18. Note that $R_\alpha^\tau(\lambda_h)$ (obtained for fixed weights ψ_j from Table 6.17) is always not greater than $R_\alpha(\lambda_h)$. Indeed, optimizing both weights and threshold results in better reliability than that obtained by optimizing just the threshold.

Table 6.18. WVS reliabilities for possible combinations of available voting units

λ_h	$R_\alpha(\lambda_h)$	$R_\alpha^\tau(\lambda_h)$	λ_h	$R_\alpha(\lambda_h)$	$R_\alpha^\tau(\lambda_h)$
∅	0	0	{2,3}	0.740	0.740
{1}	0.650	0.650	{2,4}	0.795	0.789
{2}	0.615	0.615	{3,4}	0.808	0.804
{3}	0.700	0.700	{1,2,3}	0.867	0.866
{4}	0.755	0.755	{1,2,4}	0.868	0.829
{1,2}	0.772	0.755	{1,3,4}	0.889	0.889
{1,3}	0.872	0.859	{2,3,4}	0.843	0.837
{1,4}	0.838	0.817	{1,2,3,4}	0.891	0.891

The voting unit availability importance indices I_Rb_j and $I_Rb_j^\tau$ are presented in Table 6.19. Observe that the relative importance of units differ for different types of WVS adjustment. For example, availability of unit 1 is most important for a WVS with adjustable weights and threshold, whereas in a WVS with adjustable threshold the most important is availability of unit 3.

Table 6.19. Unit availability importance indices for WVS with adjustable parameters

Unit no.	I_Rb_j	$I_Rb_j^\tau$
1	0.080	0.078
2	0.022	0.020
3	0.059	0.082
4	0.063	0.061

One can use Equation (6.78) to estimate the unit's availability importance only when the unit's availability is very high. For example, consider the availability importance indices obtained for the given WVS when the availability of all of its units is 0.99. Table 6.20 contains unit availability importance indices obtained using the exact expression (6.76) and the approximate expression (6.78). The indices take similar values. Table 6.21 contains the same indices obtained for a WVS with the availability of all of its units equal to 0.95. In this case, the difference between the values obtained by the exact and approximate expressions is much greater. Observe that, in both cases, substituting the exact availability importance values with their approximations does not violate the order of units when they are arranged according to their relative importance. Therefore, Equation (6.78) can be used to identify the most important element in the WVS.

Table 6.20. Exact and approximate values of unit availability importance indices ($\alpha_j = 0.99$)

Unit no.	I_Rb_j	$\tilde{I}_Rb_j$	$I_Rb_j^\tau$	$\tilde{I}_Rb_j^\tau$
1	0.0492	0.0464	0.0552	0.0527
2	0.0029	0.0023	0.0028	0.0023
3	0.0249	0.0230	0.0632	0.0609
4	0.0253	0.0232	0.0264	0.0244

Table 6.21. Exact and approximate values of unit availability importance indices ($\alpha_j = 0.95$)

Unit no.	$I_R b_j$	$\tilde{I}_R b_j$	$I_R b_j^\tau$	$\tilde{I}_R b_j^\tau$
1	0.0547	0.0410	0.0588	0.0466
2	0.0053	0.0021	0.0049	0.0021
3	0.0301	0.0203	0.0652	0.0538
4	0.0314	0.0205	0.0317	0.0215

6.3.4 Optimization of Weighted Voting Systems in the Presence of Common Cause Failures

When the voting units of a WVS are subject to CCFs caused by external impacts, the system's survivability can be enhanced by the proper separation of the units. In this section we consider the optimal unit separation problem that is analogous to the one considered in Section 5.2.1 for series-parallel systems. We assume that the units not separated from one another belong to the same CCG and can be destroyed by the same impact (total CCF).

Since the voting units have different decision probability distributions, the way in which they are partitioned into CCGs strongly affects the system's survivability (defined as the probability of making correct decisions). The way the units are separated and the values of the adjustable parameters of the WVS (weights and threshold) are interdependent factors affecting WVS survivability. Therefore, the WVS survivability maximization problem is to find the optimal separation of units, their weights, and the system threshold value.

6.3.4.1 Problem formulation

A WVS consists of n voting units with the given decision probability distributions $q_{i0}^{(j)}$, $q_{i1}^{(j)}$ $q_{ix}^{(j)}$. The units can be separated into B independent groups (see, for example, Figure 6.23), where B can vary from 1 (all of the units are gathered within a single group) to n (all of the units are separated from one another). It is assumed that all of the units belonging to the same group can be destroyed by the total CCF with probability v, which characterizes the WVS vulnerability. The destroyed units cannot produce positive or negative decisions and, therefore, are considered as abstaining ones.

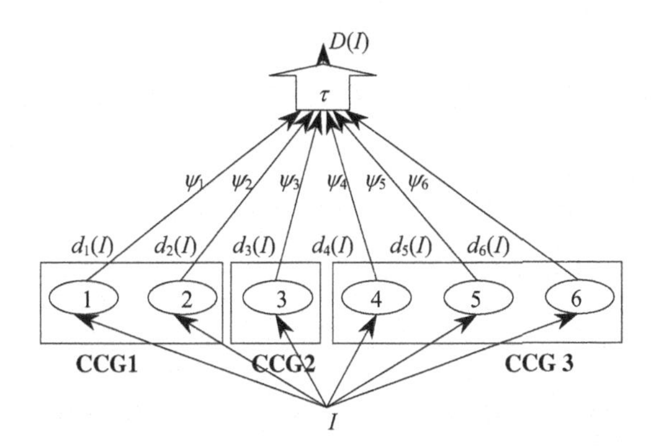

Figure 6.23. Example of WVS with separated voting units

The units' separation problem can be considered as a problem of partitioning a set Λ of n items into a collection of B mutually disjoint subsets λ_b ($1 \le b \le B$). Each set can contain from 0 to n elements. The partition of set Λ can be represented by the vector $\boldsymbol{x} = \{x_j: 1 \le j \le n\}$, where x_j is the number of the subset to which element j belongs. The weights of voting units in the WVS can be represented by the vector $\boldsymbol{\psi} = \{\psi_j: 1 \le j \le n\}$.

The WVS survivability optimization problem is formulated as follows. Find the vectors $\boldsymbol{x}$ (when no more than B different CCGs are allowed) and $\boldsymbol{\psi}$ and the threshold value τ that maximizes the system's survivability $S = \Pr\{D(I) = I\}$.

6.3.4.2 Evaluating Survivability of Weighted Voting Systems with Separated Common Cause Groups

Consider a separated group of voting units λ_b. Let the score distribution for this group be represented by the u-function $U_i^{\lambda_b}(z)$. Note that, since all of the units belonging to λ_b can be destroyed with the probability v, the probability of each state of the group (corresponding to a realization of its random score) should be multiplied by the probability of the group survival: $1-v$. If the group is destroyed, then the entire WVS considers all of the units belonging to λ_b as abstaining. This corresponds to the total score of group $\lambda_b = 0$.

The score of zero can be obtained when all of the units of group λ_b are indecisive or unavailable (because of internal causes) or when they are destroyed by the total CCF. Therefore, the overall probability that the score of separated group $\lambda_b = 0$ for input $I = i$ is

$$\Pr\{\bigcap_{j \in \lambda_b} d_j(I) = x \mid I = i\} = v + (1-v) \prod_{j \in \lambda_b} q_{ix}^{(j)} \tag{6.79}$$

To incorporate the group vulnerability into its score distribution one has to apply the operator ξ (4.58) over the u-function $U_i^{\lambda_b}(z)$

$$\xi(U_i^{\lambda_b}(z)) = (1-v)U_i^{\lambda_b}(z) + vz^0 \tag{6.80}$$

For the given distribution of voting units among CCGs one has to obtain the *u*-functions $U_i^{\lambda_b}(z)$ for each group λ_b:

$$U_i^{\lambda_b}(z) = \underset{+}{\otimes}(\breve{u}_{i1}(z),...,\breve{u}_{in}(z)) \tag{6.81}$$

where

$$\breve{u}_{ij}(z) = \begin{cases} u_{ij}(z), & \text{if } x_j = i \\ 1, & \text{if } x_j \neq i \end{cases} \tag{6.82}$$

The *u*-functions $U_i(z)$ for the entire WVS can then be obtained as

$$U_i(z) = \underset{+}{\otimes}(\xi(U_i^{\lambda_1}(z)), \xi(U_i^{\lambda_2}(z)),..., \xi(U_i^{\lambda_B}(z))) \tag{6.83}$$

After obtaining $U_0(z)$ and $U_1(z)$ one has to determine the system's survivability following steps 3-5 of the algorithm presented in Section 6.3.1.2.

6.3.4.3 Implementing the Genetic Algorithm

Let the WVS have n units that can be distributed among B groups. The system parameters are represented by the n-length integer string $\boldsymbol{a} = (a_1, ..., a_n)$ with values of elements ranging in the interval $(0,100B)$. In order to allow the value of a_j to represent both the weight of the jth unit and the number of the group to which it belongs, the following decoding procedure is used:

$$x_j = \lfloor a_j/100 \rfloor + 1, \ \psi_j' = \text{mod}_{100}(a_j) \tag{6.84}$$

The unit weights are further normalized in such a way that their total weight is always equal to some constant c:

$$\psi_j = \psi_j' c / \sum_{k=1}^{N} \psi_k' \tag{6.85}$$

In our GA we used $c = 10$.

Consider the example in which the parameters of a WVS consisting of $n = 6$ units and up to $B = 5$ groups are determined by the following string:

$\boldsymbol{a} = (264, 57, 74, 408, 221, 23)$

Using (6.84) we obtain: $x_1 = x_5 = 3$, $x_2 = x_3 = x_6 = 1$, $x_4 = 5$, vector of the unit weights before normalization $\psi' = (64, 57, 74, 8, 21, 23)$ and $\sum_{j=1}^{6}\psi'_j = 247$.

Now, using (6.85) with constant $c = 10$ we obtain the vector of the normalized unit weights:

$$\psi = (2.59, 2.31, 3.00, 0.32, 0.85, 0.93)$$

The WVS threshold τ is not determined by the string of system parameters. For each set of parameters determined by a solution string $\boldsymbol{a}$, WVS survivability S remains a function of the single argument τ. When WVS survivability is evaluated for a given set of parameters by the solution decoding procedure, this procedure determines the value of τ maximizing S. The maximal S obtained is considered to be a solution fitness, which is used to compare different solutions.

Example 6.17

Consider a WVS from [179] consisting of five voting units with the failure probabilities presented in Table 6.22. The solutions obtained for $P_0 = P_1 = 0.5$ and for vulnerability $v = 0.2$ are presented in Table 6.23. The solutions were obtained for each possible number of separated groups $1 \le B \le 5$. Table 6.23 contains voting unit weights and the WVS threshold for each solution obtained. It also contains for each unit the number of the group the unit belongs to. The values of the system's survivability are presented for each solution.

Table 6.22. Parameters of voting units

No. of unit	$q_{01}^{(j)}$	$q_{0x}^{(j)}$	$q_{10}^{(j)}$	$q_{1x}^{(j)}$
1	0.25	0.27	0.06	0.21
2	0.06	0.40	0.15	0.23
3	0.24	0.08	0.19	0.30
4	0.26	0.31	0.24	0.20
5	0.35	0.13	0.04	0.22

Table 6.23. Parameters of obtained solutions

No. of voting unit	$B = 1$		$B = 2$		$B = 3$		$B = 4$		$B = 5$	
	No. of group	Unit weight	No. of group	Unit weight	No. of group	Unit weight	No. of group	Unit weight	No. of group	Unit weight
1	1	2.381	1	1.969	1	2.263	1	2.413	1	2.302
2	1	2.275	2	2.563	2	2.514	2	2.297	2	2.474
3	1	1.905	2	1.875	2	1.844	3	1.919	3	1.856
4	1	1.085	2	1.531	1	1.034	3	1.047	4	0.997
5	1	2.354	1	2.063	3	2.346	4	2.326	5	2.371
τ	0.560		0.553		0.564		0.563		0.563	
S	0.710		0.824		0.843		0.850		0.852	

The WVS survivability as a function of group vulnerability v is presented in Figure 6.24 for each of the solutions obtained. It can be seen that, for $B = 1$, S is a linear function of v. For $B>1$ the dependencies are polynomial. It can also be seen from Figure 6.24 that the separation into two groups has the greatest effect on the system's survivability, whereas further separation leads to a smaller improvement of S. The growth of the group's vulnerability makes the separation more beneficial from the survivability improvement standpoint.

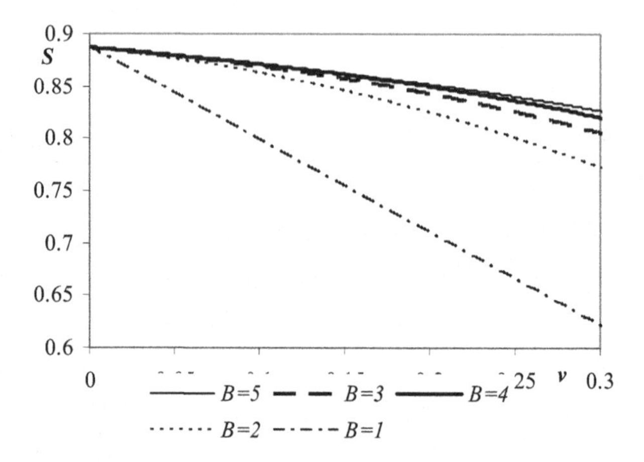

Figure 6.24. WVS survivability as a function of group vulnerability

6.3.5 Asymmetric Weighted Voting Systems

The reliability of a WVS consisting of a given set of voting units can be further improved by taking advantage of the knowledge about the statistical asymmetry of the units (asymmetric probabilities of making correct decisions with respect to the input I). In such a WVS, each voting unit j has two weights: ψ^0_j, which is assigned to the unit when it votes for the proposition rejection, and ψ^1_j, which is assigned to the unit when it votes for the proposition acceptance. As in the case of the regular (symmetric) voting systems, the proposition is rejected by the WVS if the total weight of the units voting for its acceptance is less than a prespecified fraction τ of the total weight of the non-abstaining units.

6.3.5.1 Evaluating the Reliability of Asymmetric Weighted Voting Systems

The decision rule in the asymmetric WVS takes the form

$$D(I)=x, \quad \text{if} \quad \sum_{d_j(I)\neq x} (\psi_j^0+\psi_j^1)=0 \qquad (6.86)$$

otherwise:

$$D(I)=\begin{cases}1, & \text{if} \sum_{d_j(I)\neq x}\psi_j^1 d_j(I)\geq\tau\sum_{d_j(I)\neq x}[\psi_j^1 d_j(I)+\psi_j^0(1-d_j(I))]\\ 0, & \text{if} \sum_{d_j(I)\neq x}\psi_j^1 d_j(I)<\tau\sum_{d_j(I)\neq x}[\psi_j^1 d_j(I)+\psi_j^0(1-d_j(I))]\end{cases} \tag{6.87}$$

From this rule we can obtain the condition that $D(I) = 0$:

$$(1-\tau)\sum_{d_j(I)\neq x}\psi_j^1 d_j(I)-\tau\sum_{d_j(I)\neq x}\psi_j^0(1-d_j(I))<0 \tag{6.88}$$

This provides a way for tallying the units' votes: each unit j adds the value of $(1-\tau)\psi^1_j$ to the total WVS score if it votes for proposition acceptance, a value of $-\tau\psi^0_j$ if it votes for proposition rejection, and nothing if it abstains. The proposition is rejected if the total score is negative.

One can define the terms of the u-functions (6.52) of the individual voting units as follows:

$$\begin{aligned}&p_{j0}=q_{10}^{(j)},\quad g_{j0}=-\tau\psi_j^0\\ &p_{j1}=q_{11}^{(j)}=(1-q_{10}^{(j)}-q_{1x}^{(j)}),\quad g_{j1}=(1-\tau)\psi_j^1\\ &p_{j2}=q_{1x}^{(j)},\quad g_{j2}=0\end{aligned} \tag{6.89}$$

for $i = 1$ and

$$\begin{aligned}&p_{j0}=q_{01}^{(j)},\quad g_{j0}=(1-\tau)\psi_j^1\\ &p_{j1}=q_{00}^{(j)}=(1-q_{01}^{(j)}-q_{0x}^{(j)}),\quad g_{j1}=-\tau\psi_j^0\\ &p_{j2}=q_{0x}^{(j)},\quad g_{j2}=0\end{aligned} \tag{6.90}$$

for $i = 0$ and obtain the system reliability applying steps 2-5 of the algorithm presented in Section 6.3.1.2.

It can be easily seen that the conditions (6.63) and (6.64) in the simplification technique should be replaced in the case of the asymmetric WVS by the conditions

$$\tilde{G}_{im}+(1-\tau)V^1{}_{m+1}<0 \tag{6.91}$$

and

$$\tilde{G}_{im}-\tau V^0{}_{m+1}\geq 0 \tag{6.92}$$

respectively, where

$$V^i{}_m = \sum_{j=m}^{n} \psi^i{}_m \text{, for } i = 0, 1 \tag{6.93}$$

Example 6.18

Given

$$n = 2, P_0 = P_1 = 0.5, \tau = 0.6$$

and that the parameters of the first voting unit are

$$q_{01}{}^{(1)} = 0.02, q_{10}{}^{(1)} = 0.02, q_{0x}{}^{(1)} = q_{1x}{}^{(1)} = 0.01, \psi^0{}_1 = 2, \psi^1{}_1 = 4$$

and the parameters of the second voting unit are

$$q_{01}{}^{(2)} = 0.02, q_{10}{}^{(2)} = 0.05, q_{0x}{}^{(2)} = q_{1x}{}^{(2)} = 0.02, \psi^0{}_2 = 3, \psi^1{}_2 = 1$$

then:

$$(1-\tau)\psi^1{}_1 = 1.6, -\tau\psi^0{}_1 = -1.2, (1-\tau)\psi^1{}_2 = 0.4, -\tau\psi^0{}_2 = -1.8$$

For the given weights of the second unit:

$$V^0{}_2 = \psi^0{}_2 = 3, V^1{}_2 = \psi^1{}_2 = 1$$

$$-\tau V^0{}_2 = -1.8, (1-\tau)V^1{}_2 = 0.4$$

The *u*-functions of the units are

$$u_{01}(z) = 10^{-2}(2z^{1.6} + \mathbf{97z^{-1.2}} + z^0), \quad u_{11}(z) = 10^{-2}(\mathbf{2z^{-1.2}} + 97z^{1.6} + z^0)$$

$$u_{02}(z) = 10^{-2}(2z^{0.4} + 96z^{-1.8} + 2z^0), \quad u_{12}(z) = 10^{-2}(5z^{-1.8} + 93z^{0.4} + 2z^0)$$

There are no 1-terms in $u_{01}(z)$ and $u_{11}(z)$. The 0-terms are marked in bold.

First, assign

$$Q_{00} = Q_{10} = 0$$

$$\tilde{U}_{01}(z) = u_{01}(z), \tilde{U}_{11}(z) = u_{11}(z)$$

After the 0-terms removal we obtain

$$Q_{00} = 0.97, Q_{10} = 0.02$$

$$\tilde{U}_{01}(z) = 0.02z^{1.6} + 0.01z^0, \tilde{U}_{11}(z) = 0.97z^{1.6} + 0.01z^0$$

The u-functions for two voting units are

$$\tilde{U}_{02}(z) = \tilde{U}_{01}(z) \underset{+}{\otimes} u_{02}(z) = 10^{-4}(2z^{1.6}+z^{0})(2z^{0.4}+96z^{-1.8}+2z^{0})$$

$$= 10^{-4}(4z^{2}+\mathbf{192z^{-0.2}}+4z^{1.6}+2z^{0.4}+\mathbf{96z^{-1.8}}+2z^{0}),$$

$$\tilde{U}_{12}(z) = \tilde{U}_{11}(z) \underset{+}{\otimes} u_{12}(z) = 10^{-4}(97z^{1.6}+1z^{0})(5z^{-1.8}+93z^{0.4}+2z^{0})$$

$$= 10^{-4}(\mathbf{485z^{-0.2}}+9021z^{2}+194z^{1.6}\mathbf{+5z^{-1.8}}+93z^{0.4}+2z^{0})$$

The terms with the negative exponents are marked in bold. Finally we obtain:

$$Q_{00} = 0.97+0.0192+0.0096 = 0.9988$$

$$Q_{10} = 0.02+0.0485+0.0005 = 0.0690$$

$$Q_{1x} = q_{1x}^{(1)} q_{1x}^{(2)} = 0.01\times 0.02 = 0.0002$$

$$R = P_0 Q_{00} + P_1(1-Q_{10}-Q_{1x})$$

$$= 0.5\times 0.9988+0.5\times(1-0.069-0.0002) = 0.9648$$

6.3.5.2 Optimization of Asymmetric Weighted Voting Systems

The parameter optimization problem for asymmetric WVSs can be formulated as follows:

$$R(\psi^{0}_{1}, \psi^{1}_{1}, \ldots, \psi^{0}_{n}, \psi^{1}_{n}, \tau) \rightarrow \max \tag{6.94}$$

The natural representation of a WVS weight distribution is by a $2n$-length integer string $\boldsymbol{a}$ in which the values in a_{2j-1} and a_{2j} correspond to the weights ψ^{0}_{j} and ψ^{1}_{j} respectively. The unit weights can be normalized in such a way that the total weight is always equal to some constant c. The normalized weights from arbitrary integer string $\boldsymbol{a} = (a_1, \ldots, a_{2n})$ are obtained as follows:

$$\psi_j^0 = a_{2j-1}c / \sum_{i=1}^{2n} a_i, \qquad \psi_j^1 = a_{2j}c / \sum_{i=1}^{2n} a_i \tag{6.95}$$

where c is a constant.

The solution decoding procedure determines the value of τ maximizing R for given unit weights. The obtained maximal $R(\boldsymbol{a})$ is considered as a solution fitness, which is used to compare different solutions.

Example 6.19

Consider a WVS consisting of five voting units with reliability indices presented in Table 6.24 [180]. The optimal weights of units obtained for a symmetric WVS ($\psi_j = \psi^0_j = \psi^1_j$) and for an asymmetric WVS when $P_0 = P_1 = 0.5$ are also presented in this table. The optimal values of the threshold, the decision probabilities, and the reliability of symmetric and asymmetric WVS obtained are presented in Table 6.25.

Table 6.24. Parameters of voting units

No. of unit	1	2	3	4	5
q_{01}	0.224	0.243	0.208	0.000	0.204
q_{0x}	0.209	0.077	0.073	0.249	0.168
q_{10}	0.287	0.219	0.103	0.197	0.133
q_{1x}	0.025	0.106	0.197	0.014	0.067
ψ (symmetric WVS)	1.155	1.763	1.915	3.040	2.128
ψ^1	0.402	1.796	2.848	0.526	0.557
ψ^0	0.372	0.124	0.031	2.693	0.650

Table 6.25. Parameters of optimal WVSs

	Symmetric WVS	Asymmetric WVS
τ	0.48	0.30
Q_{00}	0.927	0.958
Q_{01}	0.073	0.042
Q_{0x}	4.9E-05	4.9E-05
Q_{10}	0.054	0.051
Q_{11}	0.946	0.949
Q_{1x}	4.9E-07	4.9E-07
R	0.936	0.954

Observe that the asymmetric WVS is more reliable than the symmetric one. The system reliability as a function of the threshold value is presented in Figure 6.25A for the both WVSs with weights from Table 6.24 (note that for $\tau = 0$ $R = P_1(1-\prod_{j=1}^{n} q_{1x}^{(j)})$ and for $\tau = 1$, $R = P_0(1-\prod_{j=1}^{n} q_{01}^{(j)} - \prod_{j=1}^{n} q_{0x}^{(j)}) + P_1 \prod_{j=1}^{n} q_{11}^{(j)})$

For an existing WVS with given weights, the "turning" problem can arise in which just the threshold value maximizing the system reliability should be found subject to changing conditions. For example, based on information about the probability distribution P_i between propositions that should be accepted or rejected, one can modify the threshold value to achieve the greatest reliability. To estimate the effect of the input probability distribution on WVS reliability, the optimal threshold values, and the corresponding system reliabilities were obtained for the two WVSs as functions of P_1. These functions are presented in Figure 6.25B, where $\tau(P_1)$ and $R(P_1)$ correspond to the asymmetric WVS and $\tau^*(P_1)$ and $R^*(P_1)$ correspond to the symmetric WVS (weights of the both WVSs are optimal for $P_1 =$ 0.5).

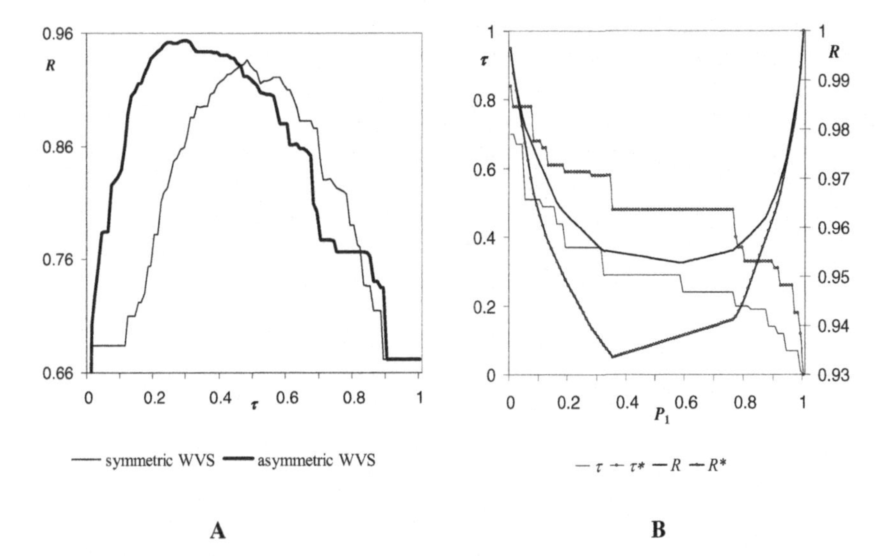

Figure 6.25. WVS reliability as a function of threshold and input reliability distribution

6.3.6 Weighted Voting System Decision-making Time

This section addresses the aspect of the WVS decision-making time. In many technical systems the time when the output (decision) of each voting unit is available is predetermined. For example, the decision time of a chemical analyzer is determined by the time of a chemical reaction. The decision time of a target detection radar system is determined by the time of the radio signal return and by the time of the signal processing by the electronic subsystem. In both these cases the variation of the decision times for a single voting unit is usually negligibly small.

On the contrary, the decision time of the entire WVS composed of voting units with different constant decision times can vary because in some cases the decisions of the slow voting units do not affect the decision of the entire system since this decision becomes evident after the faster units have voted. This happens when the total weight of the units voting for the proposition acceptance or rejection is enough to guarantee the system's decision independently of the decisions of the units that have not yet voted. In such situations, the voting process can be terminated without waiting for the slow units' decisions and the WVS decision can be made in a shorter time.

6.3.6.1 Determination of Weighted Voting System Decision Time Distribution

Assume that each voting unit j needs a fixed time t_j to produce its decision and all the WVS units are arranged in order of the decision time increase: $t_j<t_{j+1}$. In this case, u-functions $\widetilde{U}_{im}(z)$ represent the distribution of score $\widetilde{G}_{im}$ obtained by the voting of m fastest units. As was shown in Section 6.3.1.2, the 1-terms and 0-terms in the u-function $\widetilde{U}_{im}(z)$ correspond to combinations of decisions of the first m

units that guarantee the entire WVS decision (proposition acceptance and rejection respectively) independent of the decisions of the rest of the units. The sum of the coefficients of these terms in $\tilde{U}_{im}(z)$ is equal to π_{im}, the conditional probability that the WVS decision can be made at time t_m given the system input is i. By determining π_{im} as the sum of the coefficients of the removed 0-terms and 1-terms for each u-function $\tilde{U}_{im}(z)$ in step 3 of the algorithm presented in Section 6.3.1.2, we obtain the probabilities that the WVS decision time is equal to t_m.

Having the WVS decision time distribution represented by values of π_{im} and t_m for $m = 1, \ldots, n$ we obtain the expected WVS decision-making time as

$$\varepsilon = P_0 \sum_{m=1}^{n} \pi_{0m} t_m + P_1 \sum_{m=1}^{n} \pi_{1m} t_m \tag{6.96}$$

Example 6.20

Consider a WVS with parameters

$$n = 3,\ P_0 = P_1 = 0.5,\ \tau = 0.6$$

The parameters of the three voting units are:

$$q_{01}^{(1)} = 0.02,\ q_{10}^{(1)} = 0.02,\ q_{0x}^{(1)} = q_{1x}^{(1)} = 0.01,\ \psi^0{}_1 = 3,\ \psi^1{}_1 = 5,\ t_1=1$$

$$q_{01}^{(2)} = 0.02,\ q_{10}^{(2)} = 0.05,\ q_{0x}^{(2)} = q_{1x}^{(2)} = 0.02,\ \psi^0{}_2 = 4,\ \psi^1{}_2 = 3,\ t_2=2$$

$$q_{01}^{(3)} = 0.01,\ q_{10}^{(3)} = 0.03,\ q_{0x}^{(3)} = q_{1x}^{(3)} = 0.0,\ \psi^0{}_3 = 3,\ \psi^1{}_3 = 2,\ t_3=4$$

For the given parameters we have

$$(1-\tau)\psi^1{}_1 = 2,\ (1-\tau)\psi^1{}_2 = 1.2,\ (1-\tau)\psi^1{}_3 = 0.8$$

$$-\tau\psi^0{}_1 = -1.8,\ -\tau\psi^0{}_2 = -2.4,\ -\tau\psi^0{}_3 = -1.8$$

and

$$\tau V^0{}_2 = \tau(\psi^0{}_2 + \psi^0{}_3) = 0.6\times 7 = 4.2,\ \tau V^0{}_3 = \tau\psi^0{}_3 = 0.6\times 3 = 1.8,\ \tau V^0{}_4 = 0$$

$$(\tau-1)V^1{}_2 = (\tau-1)(\psi^1{}_2 + \psi^1{}_3) = -0.4\times 8 = -3.2$$

$$(\tau-1)V^1{}_3 = (\tau-1)\psi^1{}_3 = -0.4\times 2 = -0.8,\ (\tau-1)V^1{}_3 = 0$$

The u-functions for the individual voting units are

$$u_{01}(z) = 10^{-2}(2z^2 + z^0 + 97z^{-1.8}),\ u_{11}(z) = 10^{-2}(2z^{-1.8} + z^0 + 97z^2)$$

$$u_{02}(z) = 10^{-2}(2z^{1.2}+2z^0+96z^{-2.4}),\ u_{12}(z) = 10^{-2}(5z^{-2.4}+2z^0+93z^{1.2})$$

$$u_{03}(z) = 10^{-2}(1z^{0.8}+99z^{-1.8}),\ u_{13}(z) = 10^{-2}(3z^{-1.8}+97z^{0.8})$$

First, we assign

$$Q_{00} = Q_{11} = 0$$

$$\tilde{U}_{01}(z) = u_{01}(z),\ \tilde{U}_{11}(z) = u_{11}(z)$$

The u-functions $u_{01}(z)$ and $u_{11}(z)$ contain neither 1-terms nor 0-terms. This means that the WVS cannot make any decision based on voting of the first unit and

$$\pi_{01} = \pi_{11} = 0$$

We also can add nothing to Q_{00} and Q_{11}. The u-functions for the subsystem consisting of two units are

$$\tilde{U}_{02}(z) = \tilde{U}_{01}(z) \underset{+}{\otimes} u_{02}(z) = 10^{-4}(2z^2+z^0+97z^{-1.8})(2z^{1.2}+2z^0+96z^{-2.4})$$

$$= 10^{-4}(\underline{4z^{3.2}}+\underline{4z^2}+192z^{-0.4}+2z^{1.2}+2z^0+\mathbf{96z^{-2.4}}+194z^{-0.6}+\mathbf{194z^{-1.8}}+\mathbf{9312z^{-4.2}})$$

$$\tilde{U}_{12}(z) = \tilde{U}_{11}(z) \underset{+}{\otimes} u_{12}(z) = 10^{-4}(2z^{-1.8}+z^0+97z^2)(5z^{-2.4}+2z^0+93z^{1.2})$$

$$=10^{-4}(\mathbf{10z^{-4.2}}+\mathbf{4z^{-1.8}}+186z^{-0.6}+\mathbf{5z^{-2.4}}+2z^0+93z^{1.2}+485z^{-0.4}+\underline{194z^2}+\underline{9021z^{3.2}})$$

In these u-functions, the 1-terms are underlined and the 0-terms are marked in bold.

The sums of the coefficients of all of the marked terms are

$$\pi_{02} = 10^{-4}(4+4+96+194+9312) = 0.961$$

$$\pi_{12} = 10^{-4}(10+4+5+194+9021) = 0.9234$$

The sums of coefficients of 0-terms are

$$Q_{00} = 10^{-4}(96+194+9312) = 0.9602$$

$$Q_{10} = 10^{-4}(10+4+5) = 0.0019$$

After removing the marked terms, the u-functions take the form

$$\tilde{U}_{02}(z) = 10^{-4}(192z^{-0.4}+2z^{1.2}+2z^0+194z^{-0.6})$$

$$\tilde{U}_{12}(z) = 10^{-4}(186z^{-0.6}+2z^{0}+93z^{1.2}+485z^{-0.4})$$

The UGF for a subsystem consisting of three units is

$$\tilde{U}_{03}(z) = \tilde{U}_{02}(z) \underset{+}{\otimes} u_{03}(z)$$

$$= 10^{-6}(192z^{-0.4}+2z^{1.2}+2z^{0}+194z^{-0.6})(1z^{0.8}+99z^{-1.8}) = 10^{-6}(\underline{192z^{0.4}}$$

$$+\underline{2z^{2.0}}+\underline{2z^{0.8}}+\underline{194z^{0.2}}+\mathbf{19008z^{-2.2}}+\mathbf{198z^{-0.6}}+\mathbf{198z^{-1.8}}+\mathbf{19206z^{-2.4}})$$

$$\tilde{U}_{13}(z) = \tilde{U}_{12}(z) \underset{+}{\otimes} u_{13}(z)$$

$$= 10^{-6}(186z^{-0.6}+2z^{0}+93z^{1.2}+485z^{-0.4})(3z^{-1.8}+97z^{0.8}) = 10^{-6}(\mathbf{558z^{-2.4}}$$

$$\mathbf{+6z^{-1.8}}\mathbf{+279z^{-0.6}}\mathbf{+1455z^{-2.2}}+\underline{18042z^{0.2}}+\underline{194z^{0.8}}+\underline{9021z^{3.0}}+\underline{47045z^{0.4}})$$

In the final u-function, all of the terms are either 1-terms or 0-terms. Summing the coefficients of the terms we obtain

$$\pi_{03} = 10^{-6}(192+2+2+194+19008+198+198+19206) = 0.039$$

$$\pi_{13} = 10^{-6}(558+6+279+1455+18042+194+9021+47045) = 0.0766$$

and adding the coefficients of 0-terms to Q_{I0} we obtain

$$Q_{00} = 0.9602+10^{-6}(19008+198+198+19206) = 0.99881$$

$$Q_{10} = 0.0019+10^{-6}(558+6+279+1455) = 0.004198$$

Since $Q_{1x} = q_{1x}^{(1)}\, q_{1x}^{(2)}\, q_{1x}^{(3)} = 0$

$$Q_{11} = 1 - Q_{10} = 0.995802$$

The WVS reliability is

$$R = P_0 Q_{00} + P_1 Q_{11} = 0.5\times 0.99881+0.5\times 0.995802 = 0.997306$$

The expected decision time is

$$\varepsilon = 0.5(0.961\times 2 + 0.039\times 4) + 0.5(0.9234\times 2 + 0.0766\times 4) = 2.1156$$

6.3.6.2 Weighted Voting System Reliability Optimization Subject to Decision-time Constraint

The number of combinations of unit decisions that allow the entire system's decision to be obtained before the outputs of all of the units become available depends on the unit weights distribution and on the threshold value. By increasing the weights of the fastest units one makes the WVS more decisive in the initial stage of voting and, therefore, reduces the mean system decision time by the price of making it less reliable.

In applications where the WVS should make many decisions in a limited time, the expected system decision time is considered to be a measure of its performance. Since the units' weights and threshold affect both the WVS's reliability and its expected decision time, the problem of the optimal system turning can be formulated as follows: find the voting units' weights and the threshold that maximize the system reliability R while providing the expected decision time ε not greater than a prespecified value ε^*:

$$R(\psi^0{}_1, \psi^1{}_1, \ldots, \psi^0{}_n, \psi^1{}_n, \tau) \rightarrow \max$$

$$\text{subject to } \varepsilon(\psi^0{}_1, \psi^1{}_1, \ldots, \psi^0{}_n, \psi^1{}_n, \tau) \le \varepsilon^* \qquad (6.97)$$

The solution encoding for solving this problem by the GA is the same as in Section 6.3.5.2. The only difference is in the solution fitness formulation. In the constrained problem, the fitness of a solution defined by the integer string $\boldsymbol{a}$ is determined as $R(\boldsymbol{a})-\pi\max(\varepsilon-\varepsilon^*,0)$, where π is a penalty coefficient.

Example 6.21

A WVS consists of six voting units with the voting times and fault probabilities presented in Table 6.26. The optimal voting unit weights and thresholds and the parameters of the optimal WVS obtained for $\varepsilon^* = 35$ (when P_0=0.7, P_0=0.5, P_0=0.3) are presented in Tables 6.27 and 6.28.

Table 6.26. Parameters of voting units

No. of unit j	t_j	$q_{01}^{(j)}$	$q_{0x}^{(j)}$	$q_{10}^{(j)}$	$q_{1x}^{(j)}$
1	10	0.22	0.31	0.29	0.12
2	12	0.35	0.07	0.103	0.30
3	38	0.24	0.08	0.22	0.15
4	48	0.10	0.05	0.2	0.01
5	55	0.08	0.10	0.15	0.07
6	70	0.08	0.01	0.10	0.05

The system abstention probabilities do not depend on its weights and threshold. For any solution, $Q_{0x} = 0.868\times10^{-7}$ and $Q_{1x} = 1.89\times10^{-7}$.

It can be seen that for $P_0 \ne 0.5$ the WVS takes advantage of the knowledge about statistical asymmetry of the input and provides greater reliability than in the case where $P_0 = 0.5$. Observe that when $P_0>0.5$ the WVS provides Q_{00} greater than Q_{11}, and *vice versa* when $P_0<0.5$ $Q_{00}<Q_{11}$.

The R vs. ε trade-off curves for the WVS are presented in Figure 6.26. These curves are obtained by solving the optimization problem (6.97) for different values of time constraint ε^*.

Table 6.27. Optimal unit weights for $\varepsilon^* = 35$

	No. of unit j	1	2	3	4	5	6
p_0=0.7	ψ^1_j	0.018	0.240	0.564	0.300	0.388	0.476
	ψ^0_j	1.958	0.018	0.370	2.487	1.005	0.176
p_0=0.5	ψ^1_j	0.017	2.367	0.497	0.017	0.635	1.475
	ψ^0_j	2.281	0.360	0.189	1.561	0.566	0.034
p_0=0.3	ψ^1_j	0.019	2.597	1.243	0.019	0.019	0.742
	ψ^0_j	1.688	0.334	0.204	1.967	0.909	0.260

Table 6.28. Parameters of WVS optimal for $\varepsilon^* = 35$

	$P_0 = 0.7$	$P_0 = 0.5$	$P_0 = 0.3$
τ	0.76	0.50	0.45
Q_{00}	0.9798	0.9005	0.8477
Q_{01}	0.0202	0.0995	0.1523
Q_{10}	0.1611	0.0719	0.0283
Q_{11}	0.8389	0.9281	0.97166
R	0.9375	0.9143	0.9345
ε	34.994	34.987	34.994

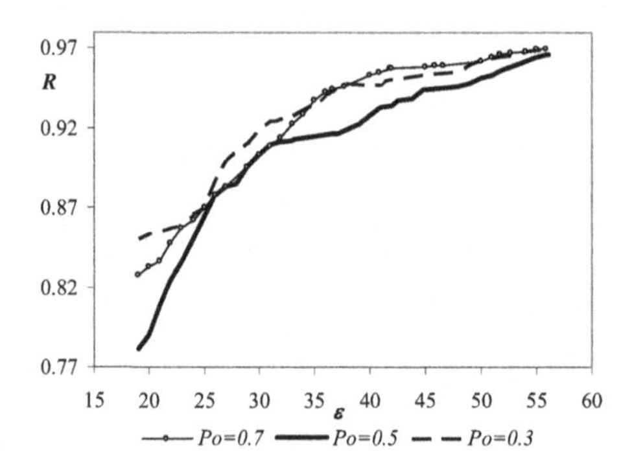

Figure 6.26. Reliability vs. expected decision time for $P_0 = 0.7$, $P_0 = 0.5$, $P_0 = 0.3$

6.3.7 Weighted Voting Classifiers

The weighted voting classifier (WVC) should classify objects belonging to a set of H classes. When an object belonging to some class I ($1 \le I \le H$) is presented to the system, its classification decision $D(I)$ is based on the classification decisions made by a set of n independent voting units. Each unit j while identifying objects from class I generates its individual classification decision $d_j(I) \in \{1, ..., H\}$. This decision can be correct $d_j(I) = I$ or incorrect $d_j(I) \neq I$. The unit can also abstain from voting $d_j(I) = 0$ (note that the abstention is always considered to be the wrong decision because $I \neq 0$). Each unit j has its individual weight ψ_j depending on the importance of its decision to the entire system.

Given the outputs of the individual units, the WVC can calculate for each classification decision h ($0 \le h \le H$) the sum of the weights of the units supporting this decision:

$$\Psi_I^h = \sum_{d_j(I)=h} \psi_j \tag{6.98}$$

The decision h' that obtained the greatest sum of the weights is determined as

$$\Psi_I^{h'} \ge \Psi_I^h \text{ for } 1 \le h \le H \tag{6.99}$$

(if there are several such decisions any of them can be chosen at random). The second best decision h'' can be determined as

$$\Psi_I^{h''} \ge \Psi_I^h \text{ for any } h \neq h' \tag{6.100}$$

There exist two different ways of making the entire WVC decision. The first one is based on a plurality voting rule. Using this rule, the entire WVC output is calculated as follows:

$$D(I) = \begin{cases} h', & \Psi_I^{h'} > \Psi_I^{h''} \\ 0, & \Psi_I^{h'} = \Psi_I^{h''} \end{cases} \tag{6.101}$$

which means that the WVC is able to classify the input if there exists an ultimate majority of weighted votes corresponding to some output h'. One can see that the system abstains from making a decision in two cases:

- all the units abstain from making a decision;
- more than one decision has the same support while the remaining decisions are supported less.

The second manner of decision making is based on a threshold voting rule. Using this rule, the WVC output is calculated as follows:

$$D(I) = \begin{cases} h', & \Psi_I^{h'} > \tau V_1 \\ 0, & \text{otherwise} \end{cases} \tag{6.102}$$

where, according to (6.62), V_1 is the sum of the weights of all of the voting units. If $\tau \geq 0.5$, then no more than one decision can satisfy the condition $\Psi_i^h > \tau V_1$. The greater τ, the less decisive the WVC. Indeed, the system is inclined to abstain when τ grows, since a lower number of combinations of voters outputs produces the winning decisions.

The entire WVC reliability can be defined as the probability that it makes the correct decisions: $R = \Pr\{D(I) = I\}$.

Each system unit has the probabilities of incorrect classification and abstention. It is natural that the probability of incorrect output depends on the class of the input object for each unit (for example, in target detecting systems some targets can be unrecognizable by speed detectors while highly recognizable by heat radiation detectors and *vice versa*). The same is true for the unit abstention probability $\Pr\{d_j(I) = 0\}$. Therefore, to define the probabilistic behavior of units one has to determine their fault probabilities $q_{ih}^{(j)}$ for $1 \leq i \leq H$, $0 \leq h \leq H$ $(h \neq i)$, $1 \leq j \leq n$, where $q_{ih}^{(j)} = \Pr\{d_j(I) = h \mid I = i\}$.

Conditional unit success (correct classification) probability, given the system input is $I = i$, can be determined, therefore, as

$$q_{ii}^{(j)} = \Pr\{d_j(I) = i \mid I = i\} = 1 - \sum_{0 \leq h \leq H, h \neq i} q_{ih}^{(j)} \tag{6.103}$$

The probability of correct classification of an object belonging to class i by the entire WVC $r_i = \Pr\{D(I) = I \mid I = i\}$ depends on the probabilities $q_{ih}^{(j)}$. Since the correct identifications of the objects belonging to different classes are mutually exclusive events, one can obtain the entire system reliability as

$$R = \sum_{i=1}^{H} P_i r_i \tag{6.104}$$

where the probabilities $P_i = \Pr\{I = i\}$ for $1 \leq i \leq H$ define the input probability distribution. (In the most common special case of evenly distributed input $R = \frac{1}{H} \sum_{i=1}^{H} r_i$).

The different states of WVC can be distinguished by the unit output distribution (UOD). WVC consisting of n voting units can have $(H+1)^n$ different states corresponding to different combinations of unit outputs (each unit can produce $H+1$ different outputs). Each WVC state can be characterized by a distribution of the weights of the units supporting different classification decisions named voting weight distribution (VWD).

Note that some different UODs can result in the same VWD. (For example, in a WVC with $n = 3$, $\psi_1 = \psi_2 = 1$ and $\psi_3 = 2$, UOD $d_1(I) = 1$, $d_2(I) = 1$ and $d_3(I) = 0$ results in the same VWD $\Psi_I^0 = \Psi_I^1 = 2$ as that of UOD $d_1(I) = 0$, $d_2(I) = 0$, $d_3(I) = 1$). From the entire WVC output point of view, these different UODs are indistinguishable and, therefore, can be treated as the same state.

To define the VWD of a classifier in state k one can use vector $g_k = \{g_k(h)\}$, $1 \le h \le H$, in which $g_k(h)$ is equal to Ψ_I^h in state k. Using the UGF approach one can describe distributions of random VWD $\boldsymbol{G}_j$ of an individual unit j as

$$u_{ij}(z) = \sum_{k=0}^{H} q_{ik}^{(j)} z^{\boldsymbol{g}_{jk}} \tag{6.105}$$

In this u-function each state k has the probability $q_{ik}^{(j)}$ and corresponds to the unit output $d_j(I) = k$ (when $I = i$) and, therefore, to VWD $\boldsymbol{g}_{jk}$ in which

$$g_{jk}(i) = \begin{cases} \psi_j, & i = k \\ 0, & i \ne k \end{cases} \tag{6.106}$$

(for $k = 0$, corresponding to element abstention, the vector contains only zeros).

Since the total weight of votes supporting any classification decision in the WVC is equal to the sum of weights of individual units supporting this decision, the resulting system VWD $\boldsymbol{G}$ can be obtained by summing the random VWDs $\boldsymbol{G}_j$ ($1 \le j \le n$) of individual voters. Therefore, the distribution $\boldsymbol{G}$ can be represented by the u-function

$$U_i(z) = \underset{+}{\otimes}(u_{i1}(z), \ldots, u_{in}(z)) = \sum_k p_{ik} z^{g_k} \tag{6.107}$$

(note that in this operator the exponents are obtained as sums of vectors, not scalar variables). Since the procedure of vector summation possesses commutative and associative properties, the u-function of the entire WVC can be obtained recursively by the consecutive determination of u-functions of arbitrary subsets of elements. For example, it can be obtained by the recursive procedure (6.57).

By following the decision rule (6.101) one can obtain the entire WVC output in each state k (for each term of $U_i(z)$) as

$$D_k(I) = \begin{cases} x, & g_k(x) > \max\limits_{1 \le h \le H, h \ne x} g_k(h) \\ 0, & \text{otherwise} \end{cases} \tag{6.108}$$

By following the decision rule (6.102) one can obtain the entire WVC output in each state k (for each term of $U_i(z)$) as

$$D_k(I)=\begin{cases}x, & g_k(x)>\tau V_1\\ 0, & \text{otherwise}\end{cases} \tag{6.109}$$

The correct classification corresponds to states in which $D_k(I)=i$ when $I = i$ and, therefore, to those u-function terms (further referred to as cc-terms) for which $g_k(i)>\max_{1\le h\le H,h\ne i} g_k(h)$ for plurality voting or $g_k(i)>\tau V_1$ for threshold voting.

Using the acceptability function

$$F_i(\mathbf{G})=\begin{cases}1, & D(I)=i\\ 0, & D(I)\ne i\end{cases} \tag{6.110}$$

over $U_i(z)$ representing all the possible WVC classification results, one can obtain the probability of successfully identifying the object of class i for plurality voting as follows:

$$r_i = E(F_i(\mathbf{G}))=\sum_k p_{ik} 1(g_k(i)>\max_{1\le h\le H,h\ne i} g_k(h)) \tag{6.111}$$

and for threshold voting as follows:

$$r_i = E(F_i(\mathbf{G}))=\sum_k p_{ik} 1(g_k(i)>\tau V_1) \tag{6.112}$$

Example 6.22

Consider a WVC consisting of two units ($n = 2$) that classifies objects belonging to three different classes ($H = 3$). The probabilities of wrong classification for each type of input object are presented in Table 2.29, as well as the probabilities of correct classification calculated in accordance with (6.103).

Table 6.29. Parameters of WVC units

		Unit 1			Unit 2		
		$i=1$	$i=2$	$i=3$	$i=1$	$i=2$	$i=3$
	h=1	0.94	0.01	0.06	0.68	0.3	0.01
q_{ih}	h=2	0.02	0.95	0.05	0.3	0.63	0.01
	h=3	0.04	0.02	0.85	0.01	0.05	0.97
	h=0	0.0	0.02	0.04	0.01	0.02	0.01

Note that unit 2 can scarcely distinguish objects of class 1 and 2 whereas specializing in the identification of objects of class 3. On the contrary, unit 1 specializes in recognizing objects of classes 1 and 2. Weights of units are $\psi_1 = 2$ and $\psi_2 = 1$. Input probability distribution is $P_1 = P_2 = P_3 = 1/3$. The threshold value is $\tau = 0.5$ ($\tau V_1 = 1.5$).

The u-functions of individual units are as follows:

$$u_{11}(z) = 10^{-2}(94z^{(200)}+2z^{(020)}+4z^{(002)})$$

$$u_{12}(\mathrm{z}) = 10^{-2}(68z^{(100)}+30z^{(010)}+z^{(001)}+z^{(000)})$$

for objects of class 1;

$$u_{21}(z) = 10^{-2}(z^{(200)}+95z^{(020)}+2z^{(002)}+2z^{(000)})$$

$$u_{22}(z) = 10^{-2}(30z^{(100)}+63z^{(010)}+5z^{(001)}+2z^{(000)})$$

for objects of class 2;

$$u_{31}(z) = 10^{-2}(6z^{(200)}+5z^{(020)}+85z^{(002)}+4z^{(000)})$$

$$u_{32}(z) = 10^{-2}(z^{(100)}+z^{(010)}+97z^{(001)}+z^{(000)})$$

for objects of class 3.

The u-functions for the entire WVC are as follows:

$$U_1(z)= u_{11}(z) \underset{+}{\otimes} u_{12}(z) = 10^{-4}(\underline{\mathbf{6392z^{(300)}}}+136z^{(120)}+272z^{(102)}+\underline{\mathbf{2820z^{(210)}}}$$

$$+60z^{(030)}+120z^{(012)}+\underline{\mathbf{94z^{(201)}}}+2z^{(021)}+4z^{(003)}+\underline{\mathbf{94z^{(200)}}}+2z^{(020)}+4z^{(002)})$$

for objects of class 1;

$$U_2(z) = u_{21}(z) \underset{+}{\otimes} u_{22}(z) = 10^{-4}(30z^{(300)}+\underline{\mathbf{2850z^{(120)}}}+60z^{(102)}+\underline{\underline{60z^{(100)}}}$$

$$+63z^{(210)}+\underline{\mathbf{5985z^{(030)}}}+126z^{(012)}+\underline{\mathbf{126z^{(010)}}}+5z^{(201)}+\underline{\mathbf{475z^{(021)}}}+10z^{(003)}$$

$$+\underline{\underline{10z^{(001)}}}+2z^{(200)}+\underline{\mathbf{190z^{(020)}}}+4z^{(002)}+\underline{\underline{4z^{(000)}}})$$

for objects of class 2;

$$U_3(z) = u_{31}(z) \underset{+}{\otimes} u_{32}(z) = 10^{-4}(6z^{(300)}+ 5z^{(120)}+\underline{\mathbf{85z^{(102)}}}+\underline{\underline{4z^{(100)}}} +6z^{(210)}$$

$$+5z^{(030)}+\underline{\mathbf{85z^{(012)}}}+\underline{\underline{4z^{(010)}}}+582z^{(201)}+485z^{(021)}+\underline{\mathbf{8245z^{(003)}}}+\underline{\underline{\mathbf{388z^{(001)}}}}$$

$$+6z^{(200)}+5z^{(020)}+\underline{\mathbf{85z^{(002)}}}+\underline{\underline{4z^{(000)}}})$$

for objects of class 3.

The cc-terms in these u-functions are marked in bold for the plurality voting rule and are underlined for the threshold voting rule. The terms corresponding to

WVC abstention are marked in italics for the plurality voting and are double underlined for the threshold voting. The probability of correct classification for each class can now be obtained as the sum of the coefficients of cc-terms for the corresponding u-function:

$$r_1 = 10^{-4}(6392+2820+94+94) = 0.94$$

$$r_2 = 10^{-4}(2850+5985+126+475+190) = 0.9626$$

$$r_3 = 10^{-4}(85+85+8245+388+85) = 0.8888$$

for plurality voting and

$$r_1 = 10^{-4}(6392+2820+94+94) = 0.94$$

$$r_2 = 10^{-4}(2850+5985+475+190) = 0.95$$

$$r_3 = 10^{-4}(85+85+8245+85) = 0.85$$

for threshold voting.

The entire WVC reliability is

$$R = (0.94+0.9626+0.8888)/3 = 0.9305$$

for plurality voting and

$$R = (0.94+0.95+0.85)/3 = 0.9133$$

for threshold voting. Note that the probability of recognizing objects of class 3 is much lower than the same for classes 1 and 2.

The probabilities of WVS abstaining are

$$\Pr\{D(I) = 0 \mid I = 1\} = 0.0$$

$$\Pr\{D(I) = 0 \mid I = 2\} = \Pr\{D(I) = 0 \mid I = 3\} = 0.0004$$

for plurality voting and

$$\Pr\{D(I) = 0 \mid I = 2\} = 10^{-4}(60+126+10+4) = 0.02$$

$$\Pr\{\mathrm{D}(I) = 0 \mid I = 3\} = 10^{-4}(4+4+388+4) = 0.04$$

for threshold voting. One can obtain the probability of wrong classifications for both types of system as

$$\Pr\{D(I) \neq I\} = \sum_{i=1}^{3} P_i(1 - r_i - \Pr\{D(I) = 0 \mid I = i\})$$

For the plurality voting WVC this index is equal to 0.069 and for the threshold voting it is equal to 0.067. One can see that the plurality voting WVC is more "decisive". It provides both correct and incorrect decisions with greater probability than the threshold voting classifier (even with the minimal possible threshold factor) and has a smaller abstention probability.

The description of the simplification technique used in the WVC reliability evaluation algorithm, as well as algorithms for solving the WVC optimization problem, can be found in [59, 181, 182].

6.4 Sliding Window Systems

The linear multi-state sliding window system (SWS) consists of n linearly ordered, statistically independent, multi-state elements (MEs). Each ME j has the random performance G_j and can be in one of k_j different states. Each state $i \in \{0, 1, \ldots, k_j-1\}$ of ME j is characterized by its fixed performance rate g_{ji} and probability $p_{jk} = \Pr\{G_j = g_{ji}\}$ (where $\sum_{i=0}^{k_j-1} p_{ji} = 1$). The SWS fails if the performance rates of any r consecutive MEs do not satisfy some condition. In terms of acceptability function, the failure criteria can be expressed as

$$F(G_1,\ldots,G_n) = \prod_{h=1}^{n-r+1} f(G_h,\ldots,G_{h+r-1}) = 0 \quad (6.113)$$

where F is the acceptability function for the entire SWS and f is the acceptability function for any group of r consecutive MEs. For example, if the sum of the performance rates of any r consecutive MEs should be not lower than the demand w, then Equation (6.113) takes the form

$$F(G_1,\ldots,G_n) = \prod_{h=1}^{n-r+1} 1(\sum_{m=h}^{h+r-1} G_m \geq w) = 0 \quad (6.114)$$

The special case of SWS where all of the n MEs are identical and have two states with performance rates of 0 and 1, $w = r-k+1$ and the acceptability function takes the form (6.114) is a k-out-of-r-from-n:F system.

6.4.1. Evaluating the Reliability of the Sliding Window Systems

The algorithm for evaluating the reliability of the sliding window system is very similar to that described in Section 2.4 for a k-out-of-r-from-n:F system.

6.4.1.1 Implementing the Universal Generating Function

The u-function representing p.m.f. of the random performance rate of ME j G_j takes the form:

$$u_j(z) = \sum_{i=0}^{k_j-1} p_{ji} z^{g_{j,i}} \tag{6.115}$$

The performance of a group consisting of r MEs numbered from h to $h+r-1$ is represented by the random vector $\boldsymbol{G}_h = (G_h, ..., G_{h+r-1})$ consisting of random performance values corresponding to all of the MEs belonging to the group.

Having the p.m.f. of independent random variables G_h, …, G_{h+r-1} one can obtain the p.m.f. of the random vector $\boldsymbol{G}_h$ by evaluating the probabilities of each combination of realizations of these values. Doing so by a recursive procedure, one can first obtain the p.m.f. of the r-length vector (0, …, 0, G_h) (corresponding to a single ME), then obtain the p.m.f. of r-length vector (0, …, 0, G_h, G_{h+1}) (corresponding to a pair of MEs), and so on until obtaining the p.m.f. of the vector (G_h, …, G_{h+r-1}).

Let the u-function $U_{-r+h}(z)$ represent the p.m.f. of a vector consisting of $r-h+1$ zeros and random values from G_1 to G_{h-1}. This u-function represents the PDs of MEs from 1 to $h-1$. In order to obtain the PD of a group of MEs from 1 to h, one has to evaluate all possible combinations of the realizations of a random vector (0,…,0, G_1, …, G_{h-1}) and a random variable G_h. Therefore, the u-function $U_1(z)$, representing the p.m.f. of the random vector $\boldsymbol{G}_1 = (G_1, ..., G_r)$, can be obtained by assigning

$$U_{1-r}(z) = z^{\boldsymbol{g}_0} \tag{6.116}$$

where the vector $\boldsymbol{g}_0$ consists of r zeros and the consecutive application of the shift operator $\underset{\leftarrow}{\otimes}$ (2.63):

$$U_{-r+h+1}(z) = U_{-r+h}(z) \underset{\leftarrow}{\otimes} u_h(z) \text{ for } h = 1, …, r \tag{6.117}$$

where the procedure $\boldsymbol{x} \leftarrow y$ over arbitrary r-length vector $\boldsymbol{x}$ and value y shifts all of the vector elements one position to the left, $x(s-1) = x(s)$ for $s = 2, …, r$ in sequence, and adds the value y to the right position, $x(r) = y$. (The first element of vector $\boldsymbol{x}$ disappears after applying the operator).

Having the PD of the first r MEs, one can obtain the PD of the next group of MEs (from 2 to r+1) by estimating all of the possible combinations of random vector values represented by $U_1(z)$, and random variable G_{r+1} represented by $u_{r+1}(z)$. Note that the performance G_1 does not influence the PD of this group. Therefore, in order to obtain the vector $\boldsymbol{G}_2$ one has to remove G_1 from vector $\boldsymbol{G}_1$ and replace it with G_{r+1}. By replacing the first element of the random vector with the new element corresponding to the following ME, one obtains vectors corresponding to the next groups of MEs.

By applying the shift operator $\underset{\leftarrow}{\otimes}$ further for $h = r+1, \ldots, n$ one obtains the u-functions for all of the possible groups of r consecutive MEs: $U_2(z), \ldots, U_{n-r+1}(z)$. The SWS contains exactly $n-r+1$ groups of r consecutive MEs, with each ME belonging to no more than r groups.

Let u-function $U_h(z) = \sum_{i=0}^{E_h-1} q_{hi} z^{\boldsymbol{g}_{hi}}$ for $1 \le h \le n-r+1$ represent the p.m.f. of vector $\boldsymbol{G}_h$. By summing the probabilities of all of the realizations $\boldsymbol{g}_{hi}$ of vector $\boldsymbol{G}_h$ producing zero values of the acceptability function $f(\boldsymbol{G}_h) = f(G_h, \ldots, G_{h+r-1})$, one can obtain the probability of failure Q_h of the hth group of r consecutive MEs:

$$Q_h = E(1 - f(\boldsymbol{G}_h)) = \theta_f(U_h(z)) = \sum_{i=0}^{E_{h-1}} q_{hi}(1 - f(\boldsymbol{g}_{hi})) \tag{6.118}$$

Consider the u-function $U_h(z)$. For each combination of values of $G_{h+1}, \ldots, G_{h+r-1}$, it contains exactly k_h different terms corresponding to different values of G_h, which takes all of the possible values of the performance rate of ME h. After applying the operator $\otimes_{\leftarrow}$, G_h disappears from the vector $\boldsymbol{G}_{h+1}$ and is replaced with G_{h+1}. This produces k_h terms in $U_{h+1}(z)$, corresponding to the same value of vector $\boldsymbol{G}_{h+1}$. Collecting these like terms, one obtains a single term for each vector $\boldsymbol{G}_{h+1}$. Therefore, the number of different terms in each u-function $U_h(z)$ is equal to $E_h = \prod_{i=h}^{h+r-1} k_i$.

By applying the operator θ_f (6.118) over $U_h(z)$ one can obtain the probability Q_h that the group consisting of MEs $h, \ldots, h+r-1$ fails. If for some combination of MEs' states the group fails, the entire SWS fails independently of the states of the MEs that do not belong to this group. Therefore, the terms corresponding to the group failure can be removed from $U_h(z)$, since they should not participate in determining further state combinations that cause system failures. This consideration lies at the base of the following algorithm for SWS availability evaluation:

1. Assign: $x = 0$; $U_{-r+1}(z) = z^{\boldsymbol{g}_0}$. Determine the u-functions of the individual MEs using (6.115).

2. Main loop. Repeat the following for $h = 1, \ldots, n$:

2.1. Obtain $U_{-r+h+1}(z) = U_{-r+h}(z) \underset{\leftarrow}{\otimes} u_h(z)$.

2.2. If $h \geq r$ add value $\theta_f(U_{h+1-r}(z))$ to x and remove all of the terms with the exponents producing the zero acceptability function from $U_{h+1-r}(z)$.

3. Obtain the SWS availability as $R = 1-x$.

Example 6.23

Consider an SWS with five MEs ($n = 5$) in which the sum of the performance rates of any three ($r = 3$) adjacent MEs should not be less than four. Each ME has two states: total failure (corresponding to a performance rate of zero) and functioning with a nominal performance rate. The nominal performance rates of the MEs from 1 to 5 are 1, 2, 3, 1 and 1 respectively.

The u-functions of the individual MEs are:

$$u_1(z) = p_{10}z^0 + p_{11}z^1,\ u_2(z) = p_{20}z^0 + p_{21}z^2,\ u_3(z) = p_{30}z^0 + p_{31}z^3$$

$$u_4(z) = p_{40}z^0 + p_{41}z^1,\ u_5(z) = p_{50}z^0 + p_{51}z^1$$

First, we assign

$$x = 0,\ U_{-2}(z) = z^{(0,0,0)}$$

Following step 2 of the algorithm, we obtain

$$U_{-1}(z) = U_{-2}(z) \underset{\leftarrow}{\otimes} u_1(z)) = z^{(0,0,0)} \underset{\leftarrow}{\otimes} (p_{10}z^0 + p_{11}z^1) = p_{10}z^{(0,0,0)} + p_{11}z^{(0,0,1)}$$

$$U_0(z) = U_{-1}(z) \underset{\leftarrow}{\otimes} u_2(z) = (p_{10}z^{(0,0,0)} + p_{11}z^{(0,0,1)}) \underset{\leftarrow}{\otimes} (p_{20}z^0 + p_{21}z^2)$$

$$= p_{10}p_{20}z^{(0,0,0)} + p_{11}p_{20}z^{(0,1,0)} + p_{10}p_{21}z^{(0,0,2)} + p_{11}p_{21}z^{(0,1,2)}$$

$$U_1(z) = U_0(z) \underset{\leftarrow}{\otimes} u_3(z)$$

$$= (p_{10}p_{20}z^{(0,0,0)} + p_{11}p_{20}z^{(0,1,0)} + p_{10}p_{21}z^{(0,0,2)} + p_{11}p_{21}z^{(0,1,2)}) \underset{\leftarrow}{\otimes} (p_{30}z^0 + p_{31}z^3)$$

$$= \mathbf{p_{10}p_{20}p_{30}z^{(0,0,0)}} + \mathbf{p_{11}p_{20}p_{30}z^{(1,0,0)}} + \mathbf{p_{10}p_{21}p_{30}z^{(0,2,0)}} + \mathbf{p_{11}p_{21}p_{30}z^{(1,2,0)}}$$

$$+ \mathbf{p_{10}p_{20}p_{31}z^{(0,0,3)}} + p_{11}p_{20}p_{31}z^{(1,0,3)} + p_{10}p_{21}p_{31}z^{(0,2,3)} + p_{11}p_{21}p_{31}z^{(1,2,3)}$$

The terms of $U_1(z)$ with exponents in which sums of elements are less than 4 are marked in bold. Following step 2.2 of the algorithm, we obtain

$$x = p_{10}p_{20}p_{30} + p_{11}p_{20}p_{30} + p_{10}p_{21}p_{30} + p_{11}p_{21}p_{30} + p_{10}p_{20}p_{31}$$

After removing the marked terms, $U_1(z)$ takes the form

$$U_1(z) = p_{11}p_{20}p_{31}z^{(1,0,3)}+p_{10}p_{21}p_{31}z^{(0,2,3)}+p_{11}p_{21}p_{31}z^{(1,2,3)}$$

Applying further the $\underset{\leftarrow}{\otimes}$ operator, we obtain

$$U_2(z)= U_1(z) \underset{\leftarrow}{\otimes} u_4(z)$$

$$= (p_{11}\,p_{20}\,p_{31}\,z^{(1,0,3)}+p_{10}\,p_{21}\,p_{31}\,z^{(0,2,3)}+p_{11}\,p_{21}\,p_{31}\,z^{(1,2,3)}) \underset{\leftarrow}{\otimes} (p_{40}z^{0}+p_{41}z^{1})$$

$$= \boldsymbol{p_{11}p_{20}p_{31}p_{40}z^{(0,3,0)}}+p_{10}p_{21}p_{31}p_{40}z^{(2,3,0)}+p_{11}p_{21}p_{31}p_{40}\,z^{(2,3,0)}$$

$$+p_{11}\,p_{20}\,p_{31}\,p_{41}\,z^{(0,3,1)}+p_{10}p_{21}\,p_{31}\,p_{41}\,z^{(2,3,1)}+p_{11}\,p_{21}\,p_{31}\,p_{41}z^{(2,3,1)}$$

Following step 2.2 of the algorithm, we modify x as follows:

$$x = p_{10}p_{20}p_{30}+p_{11}p_{20}p_{30}+p_{10}p_{21}p_{30}+p_{11}p_{21}p_{30}+p_{10}p_{20}p_{31}+p_{11}\,p_{20}p_{31}\,p_{40}$$

After removing the marked term and collecting like terms, $U_2(z)$ takes the form:

$$U_2(z) = p_{21}p_{31}p_{40}z^{(2,3,0)}+p_{11}p_{20}p_{31}p_{41}z^{(0,3,1)}+p_{21}p_{31}p_{41}z^{(2,3,1)}$$

Following steps 2.1 and 2.2 of the algorithm we obtain

$$U_3(z) = U_2(z) \underset{\leftarrow}{\otimes} u_5(z)$$

$$= (p_{21}p_{31}p_{40}z^{(2,3,0)}+p_{11}p_{20}p_{31}p_{41}z^{(0,3,1)}+p_{21}p_{31}p_{41}z^{(2,3,1)}) \underset{\leftarrow}{\otimes} (p_{50}z^{0}+p_{51}z^{1})$$

$$= \boldsymbol{p_{21}p_{31}p_{40}p_{50}z^{(3,0,0)}}+p_{11}p_{20}p_{31}p_{41}p_{50}z^{(3,1,0)}+p_{21}p_{31}p_{41}p_{50}z^{(3,1,0)}$$

$$+p_{21}p_{31}p_{40}p_{51}z^{(3,0,1)}+p_{11}p_{20}p_{31}p_{41}p_{51}z^{(3,1,1)}+p_{21}p_{31}p_{41}p_{51}z^{(3,1,1)}$$

After adding the coefficient of the marked term to x we have

$$x = p_{10}p_{20}p_{30}+p_{11}p_{20}p_{30}+p_{10}p_{21}p_{30}+p_{11}p_{21}p_{30}+p_{10}p_{20}p_{31}+p_{11}\,p_{20}p_{31}\,p_{40}$$

$$+p_{21}p_{31}\,p_{40}\,p_{50}$$

Finally:

$$R=1-x=1-p_{30}-p_{31}[p_{10}p_{20}+(p_{11}p_{20}+p_{21}p_{50})p_{40}]$$

6.4.1.2 Simplification Technique

Note that the first elements of vectors $\mathbf{g}_{hi}$ in the u-function $U_h(z)$ do not participate in determining $U_{h+1}(z)$ (according to the definition of the procedure $\mathbf{x}\leftarrow y$), which leads to producing k_h like terms in $U_{h+1}(z)$. In order to avoid excessive term multiplication procedures in operator $\otimes_{\leftarrow}$, one can perform a like term collection in $U_h(z)$. To do this, one can, after step 2.2 of the algorithm, replace the first elements in all vectors of $U_h(z)$ with zeros and collect like terms.

The algorithm can be further simplified if the SWS acceptability function takes the form (6.114). Consider the sth term $q_{hs}z^{\mathbf{g}_{hs}}$ of a u-function $U_h(z)$ after replacing the first element $g_{hs}(1)$ of vector $\mathbf{g}_{hs}$ with zero. If $\tilde{g}_{h+r}$ is the greatest possible value of the performance rate of the $(h+r)$th ME and

$$\sum_{i=2}^{r} g_{hs}(i) < w - \tilde{g}_{h+r} \tag{6.119}$$

any combination of the term $q_{hs}z^{\mathbf{g}_{hs}}$ with terms of $u_{h+r}(z)$ produces terms corresponding to SWS failure. This means that, in the u-function $U_{h+1}(z)$, all of the terms with coefficients $q_{hs}p_{h+r,i}$ should be removed and the sum of the corresponding coefficients should be added to x. Since $\sum_{i=0}^{k_{h+r}-1} p_{h+r,i} = 1$, the sum of these coefficients is equal to q_{hs}. In order to avoid k_{h+r} redundant term multiplication procedures, one can remove the term $q_{hs}z^{\mathbf{g}_{hs}}$ meeting condition (6.119) from $U_h(z)$ and add its coefficient to x.

In order to reduce the algorithm computation complexity considerably using the considerations described above, one has to apply to any newly obtained u-function $U_m(z)$ (in step 2 of the algorithm) for $m = 0, \ldots, n-r$ the following operator φ, which:

- replaces all of the first elements of vectors $\mathbf{g}_{hs}$ with zeros;
- collects like terms in the u-function;
- removes the terms meeting (6.119) and adds the coefficients of the replaced terms to x.

Example 6.24

Consider the Example 6.23 and apply to it the simplification technique. First, we obtain

$$w-\tilde{g}_3 = 4-3 = 1,\ w-\tilde{g}_4 = w-\tilde{g}_5 = 4-1 = 3$$

The operator φ applied to

$$U_0(z) = \mathbf{p_{10}p_{20}z^{(0,0,0)}} + p_{11}p_{20}z^{(0,1,0)} + p_{10}p_{21}z^{(0,0,2)} + p_{11}p_{21}z^{(0,1,2)}$$

removes the term $p_{10}p_{20}z^{(0,0,0)}$, since $0+0+0<w-\tilde{g}_3 = 1$, and adds $p_{10}p_{20}$ to x. After applying the operator, $U_0(z)$ takes the form

$$\varphi(U_0(z)) = p_{11}p_{20}z^{(0,1,0)}+p_{10}p_{21}z^{(0,0,2)}+p_{11}p_{21}z^{(0,1,2)}$$

Following steps 2.1 and 2.2 of the algorithm, we obtain

$$U_1(z) = (\varphi(U_0(z)) \underset{\leftarrow}{\otimes} u_3(z))$$

$$= (p_{11}p_{20}z^{(0,1,0)}+p_{10}p_{21}z^{(0,0,2)}+p_{11}p_{21}z^{(0,1,2)} \underset{\leftarrow}{\otimes} (p_{30}z^{0}+p_{31}z^{3})$$

$$= \mathbf{p_{11}p_{20}p_{30}z^{(1,0,0)}+p_{10}p_{21}p_{30}z^{(0,2,0)}+p_{11}p_{21}p_{30}z^{(1,2,0)}}+p_{11}p_{20}p_{31}z^{(1,0,3)}$$

$$+p_{10}p_{21}p_{31}z^{(0,2,3)}+p_{11}p_{21}p_{31}z^{(1,2,3)}$$

Applying the operator φ that removes the terms meeting condition (6.119) (marked in bold) one obtains

$$x = p_{10}p_{20}+p_{11}p_{20}p_{30}+p_{10}p_{21}p_{30}+p_{11}p_{21}p_{30},$$

$$\varphi(U_1(z)) = p_{11}p_{20}p_{31}z^{(0,0,3)}+p_{21}p_{31}z^{(0,2,3)}$$

Further:

$$U_2(z) = (\varphi(U_1(z)) \underset{\leftarrow}{\otimes} u_4(z)) = p_{11}p_{20}p_{31}z^{(0,0,3)}+p_{21}p_{31}z^{(0,2,3)}) \underset{\leftarrow}{\otimes} (p_{40}z^{0}+p_{41}z^{1})$$

$$= \mathbf{p_{11}p_{20}p_{31}p_{40}z^{(0,3,0)}}+p_{21}p_{31}p_{40}z^{(2,3,0)}+p_{11}p_{20}p_{31}p_{41}z^{(0,3,1)}+p_{21}p_{31}p_{41}z^{(2,3,1)}$$

Applying the operator φ over $U_2(z)$ one obtains

$$x = p_{10}p_{20}+p_{11}p_{20}p_{30}+p_{10}p_{21}p_{30}+p_{11}p_{21}p_{30}+p_{11}p_{20}p_{31}p_{40}$$

$$\varphi(U_2(z)) = p_{21}p_{31}p_{40}z^{(0,3,0)}+(p_{11}p_{20}+p_{21})p_{31}p_{41}z^{(0,3,1)}$$

For the last group of MEs:

$$U_3(z) = (\varphi(U_2(z)) \underset{\leftarrow}{\otimes} u_5(z))$$

$$= (p_{21}p_{31}p_{40}z^{(0,3,0)}+(p_{11}p_{20}+p_{21})p_{31}p_{41}z^{(0,3,1)}) \underset{\leftarrow}{\otimes} (p_{50}z^{0}+p_{51}z^{1})$$

$$= \mathbf{p_{21}p_{31}p_{40}p_{50}z^{(3,0,0)}}+(p_{11}p_{20}+p_{21})p_{31}p_{41}p_{50}z^{(3,1,0)}$$

$$+p_{21}p_{31}p_{40}p_{50}z^{(3,0,1)}+(p_{11}p_{20}+p_{21})p_{31}p_{41}p_{50}z^{(3,1,1)}$$

$$x = p_{10}p_{20}+p_{11}p_{20}p_{30}+p_{10}p_{21}p_{30}+p_{11}p_{21}p_{30}+p_{11}p_{20}p_{31}p_{40}+p_{21}p_{31}p_{40}p_{50}$$

Finally:

$$R = 1-x$$

$$= 1-p_{10}p_{20}-p_{11}p_{20}p_{30}-p_{10}p_{21}p_{30}-p_{11}p_{21}p_{30}-p_{11}p_{20}p_{31}p_{40}-p_{21}p_{31}p_{40}p_{50}$$

Taking into account that $p_{10}p_{20} = p_{10}p_{20}p_{30}+p_{10}p_{20}p_{31}$, one obtains the same result:

$$R = 1-p_{30}-p_{31}[p_{10}p_{20}+(p_{11}\,p_{20}+p_{21}\,p_{50})p_{40}]$$

Using the SWS reliability evaluation procedure described, one can analyze the effect of demand variation on the overall system reliability.

Example 6.25

Consider the two following SWSs [183]. The first one consists of 10 identical three-state MEs. The probabilities of the MEs' states are $p_{j0} = 0.1$, $p_{j1} = 0.3$, $p_{j2} = 0.6$. The corresponding performance rates are $g_{j0} = 0$, $g_{j1} = 1$, $g_{j2} = 3$.

The SWS reliability, as a function of constant demand w, is presented in Figure 6.27A for different r ($2 \le r \le 10$). Note that, because the cumulative performance of groups of MEs takes a finite number of discrete values, the $R(w)$ is a step function. One can see that the greater the r, the greater the SWS reliability for the same w. This is natural, because the growth of r provides growing redundancy in each group.

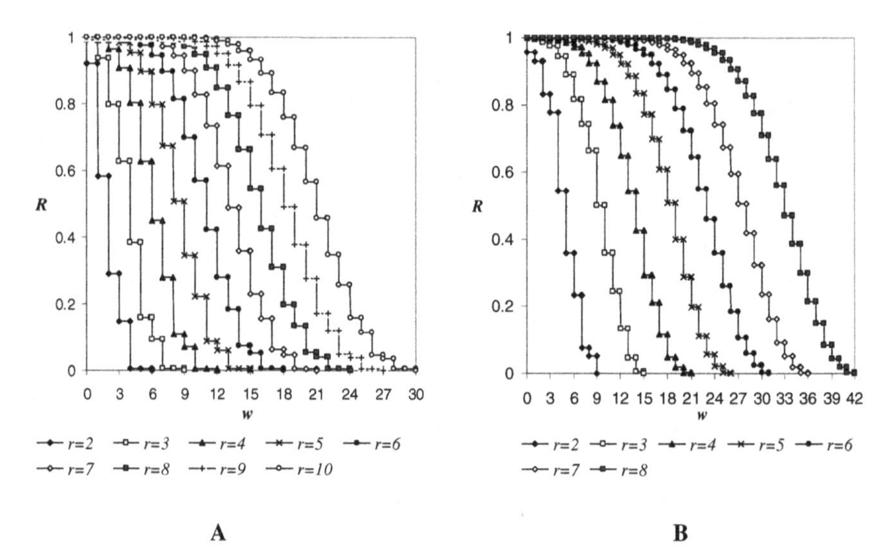

Figure 6.27. Reliability of SWS as a function of w and r
(A: SWS with 10 identical MEs. B: SWS with eight different MEs)

The second SWS consists of eight different MEs. The number of states of these MEs varies from two to five. The performance distributions of the MEs are presented in Table 6.30. The SWS reliability as a function of constant demand w is presented in Figure 6.27B for different r ($2\leq r\leq 8$). Observe that the functions $R(w)$ for the second SWS have more steps than the functions $R(w)$ for the first one. Indeed, different MEs produce a greater variety of levels of cumulative group performance rates than the identical ones.

Table 6.30. Parameters of SWS elements

No. of ME	1		2		3		4		5		6		7		8	
No. of state	p	G	p	g	p	g	p	g	p	g	p	g	p	g	p	g
0	0.03	0	0.10	0	0.17	0	0.05	0	0.08	0	0.01	0	0.20	0	0.05	0
1	0.22	2	0.10	1	0.83	6	0.25	3	0.20	1	0.22	4	0.10	3	0.25	4
2	0.75	5	0.40	2	-	-	0.40	5	0.15	2	0.77	5	0.10	4	0.70	6
3	-	-	0.40	4	-	-	0.30	6	0.45	4	-	-	0.60	5	-	-
4	-	-	-	-	-	-	-	-	0.12	5	-	-	-	-	-	-

6.4.2 Multiple Sliding Window Systems

The existence of multiple failure criteria is a common situation for complex systems, especially for consecutive-type systems. In this section we consider an extension of the linear SWS model to a multi-criteria case. In this multiple sliding window system (MSWS) a vector $\boldsymbol{r} = (r_i\colon 1\leq i\leq Y)$ is defined such that $r_i<r_{i+1}$, and $1\leq r_i\leq n$ for any i. The system fails if for any i ($1\leq i\leq Y$) at least one of the functions f_i over the performance rates of any r_i consecutive MEs is equal to zero. The entire MSWS acceptability function takes the form

$$F(G_1,...,G_n)=\prod_{i=1}^{Y}\prod_{h=1}^{n-r_i+1} f_i(G_h,...,G_{h+r_i-1})=0 \tag{6.120}$$

The introduction of the linear MSWS model is motivated by the following examples.

Example 6.26

Consider a sequence of service stations in which each station should process the same sequence of n different tasks. Each station i can process r_i incoming tasks simultaneously according to the first-in-first-out rule using a limited resource w_i. Each incoming task can have different states and the amount of the resource needed to process the task is different for each state of each task. The total resource needed to process r_i consecutive tasks should not exceed the available amount of the resource w_i. The system fails if in at least one of the stations there is no available resource to process r_i tasks simultaneously.

The simplest example of such a model is a transportation system in which n randomly ordered containers are carried by consecutive conveyors characterized by a different length and allowable load. The number of containers r_i that are loaded onto each conveyor i is defined by its length. The transportation system fails if the total load of any one of the conveyors is greater than its maximal allowed load w_i.

An example of the transportation system is presented in Figure 6.28. The system consists of Y = 3 conveyors and transports n = 14 randomly ordered containers of four types (each type m is characterized by its weight g_m). The first conveyor can simultaneously carry r_1 = 2 containers, the second and third conveyors can carry r_2 = 6 and r_3 = 3 containers respectively. The maximal allowable loads of conveyors 1, 2 and 3 are w_1, w_2 and w_3 respectively. The system fails if the total weight of any two adjacent containers is greater than w_1, or if the total weight of any six adjacent containers is greater than w_2, or if the total weight of any three adjacent containers is greater than w_3. The weight of the jth container in the line can be represented by a random value G_j: $G_j \in \{g_1, g_2, g_3, g_4\}$. The acceptability function for each conveyor i can be determined as

$$f_i(G_h,...,G_{h+r_i-1}) = 1(\sum_{j=h}^{h+r_i-1} G_j \le w_i),\ 1 \le i \le 3$$

for any group of r_i adjacent containers starting with hth one ($r_1 = 2$, $r_2 = 6$, $r_3 = 3$).

The system reliability (defined as its expected acceptability) takes the form

$$R = E(\prod_{i=1}^{3} \prod_{h=1}^{15-r_i} 1(\sum_{j=h}^{h+r_i-1} G_j \le w_i)\ ),\ \text{where}\ r_1{=}2,\ r_2{=}6,\ r_3{=}3$$

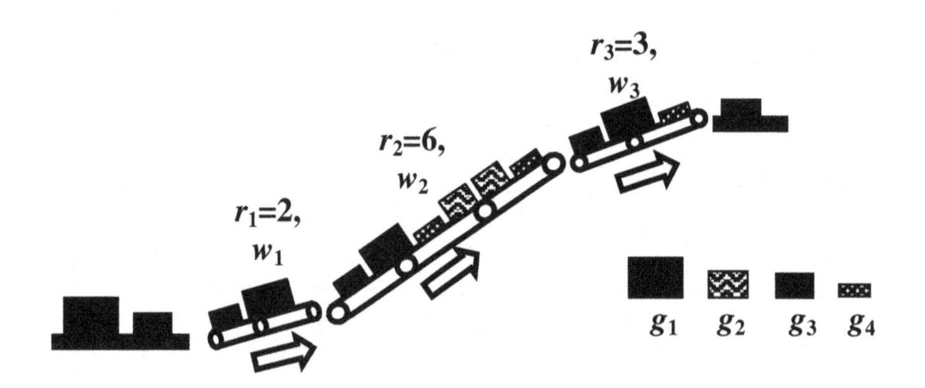

Figure 6.28. Example of transportation MSWS

Example 6.27

Consider a heating system that should provide a certain temperature along several lines with moving parts placed at different distances from the heaters. The temperature at each point of the line i is determined by a cumulative effect of r_i closest heaters. Each heater consists of several electrical heating elements. The heating effect of each heater depends on the availability of its heating elements and, therefore, can vary discretely (if the heaters are different, then the number of different levels of heat radiation and the intensity of the radiation at each level are specific to each heater). In order to provide the temperature, which is not less than some specified value at each point of line i, any r_i adjacent heaters should be in states where the sum of their radiation intensity is greater than the minimum allowed level w_i. The system fails if any group of r_i adjacent heaters provides the cumulative radiation intensity lower than w_i.

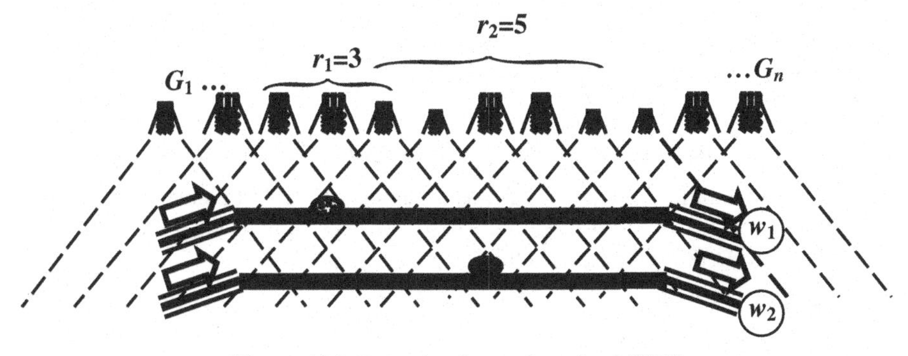

Figure 6.29. Example of manufacturing MSWS

In the example presented in Figure 6.29 there are 12 heaters providing random radiation intensity G_j ($1 \le j \le 12$). The parts located at any point of the close conveyor are heated by three adjacent heaters. The cumulative heating intensity along this conveyor should not be lower than w_1. The parts located at any point of the remote conveyor are heated by five adjacent heaters. The cumulative heating intensity along this conveyor should not be lower than w_2. The system fails if any three adjacent heaters fail to provide the desired heating intensity w_1 or if any five adjacent heaters fail to provide the desired heating intensity w_2. The acceptability function for each conveyor i can be defined as

$$f_i(G_h,...,G_{h+r_i-1}) = 1(\sum_{j=h}^{h+r_i-1} G_j \ge w_i),\ 1 \le i \le 2$$

which corresponds to any group of r_i adjacent heaters starting with the hth one ($r_1 = 3$, $r_2 = 5$).

The system reliability takes the form

$$R = E(\prod_{i=1}^{2} \prod_{h=1}^{13-r_i} 1(\sum_{j=h}^{h+r_i-1} G_j \ge w_i)),\ \text{where } r_1 = 3,\ r_2 = 5$$

A variety of other systems also fit the model: quality control systems that detect deviations from given values of parameters in product samples, combat systems that should provide certain fire density along a defence line, *etc.*

6.4.2.1 Evaluating the Multiple Sliding Window System Reliability

Let the u-function $U_{-r_Y+1}(z)$ take the form (6.116) where the vector $\mathbf{g}_0$ consists of r_Y zeros. According to the algorithm presented in Section 6.4.1.1, by applying the operator (6.117) for $h = 1, \ldots, n$ one obtains distributions for all of the possible random vectors of performance rates of r_Y consecutive MEs.

Note that the vectors of length r_Y considered also contain all of the possible vectors of the smaller length. For any $r_i<r_Y$ the last r_i elements of vectors in the exponents of u-functions $U_{-r_Y+1+r_i}(z)$, $U_{-r_Y+2+r_i}(z)\ldots$, $U_{-r_Y+1+n}(z)$ represent all of the possible vectors of performance rates of r_i consecutive MEs. Therefore, in each u-function $U_{-r_Y+1+h}(z)$ obtained by the recursive operator (6.117) for $r_1\leq h\leq n$, one can obtain the failure probability of groups of r_i consecutive MEs for any $r_1\leq h\leq r_Y$ satisfying the condition $r_i\leq h$ by applying operators $\theta_{f_i}(U_{-r_Y+1+h}(z))$ in which acceptability functions f_i take as arguments r_i the last elements of vectors from the exponents of the u-function. These considerations lead to the following algorithm for MSWS reliability evaluation:

1. Assign: $x = 0$; $U_{-r_Y+1}(z) = z^{\mathbf{g}_0}$. Determine the u-functions of the individual MEs using (6.115).
2. Main loop. Repeat the following for $h = 1, \ldots, n$:
   2.1. Obtain $U_{-r_Y+h+1}(z) = U_{-r_Y+h}(z) \underset{\leftarrow}{\otimes} u_h(z)$.
   2.2. For $i = 1, \ldots, Y$: if $h\geq r_i$ add value $\theta_{f_i}(U_{-r_Y+h+1}(z))$ to x and remove from $U_{-r_Y+h+1}(z)$ terms with the exponents in which the last r_i elements produce zero acceptability function f_i.
3. Obtain the SWS availability as $R = 1-x$.

Alternatively, the system reliability can be obtained as the sum of the coefficients of the last u-function $U_{-r_Y+n+1}(z)$.

Example 6.28

Consider an MSWS with $n = 5$, $Y = 2$, $r_1 = 3$, $r_2 = 4$, $f_1(x_1, x_2, x_3) = 1(\sum_{j=1}^{3} x_j \geq 5)$, $f_2(x_1, x_2, x_3, x_4) = 1(\sum_{j=1}^{4} x_j \geq 6)$. Each ME has two states: total failure (corresponding to a performance rate of zero) and functioning with a nominal performance rate. The nominal performance rates of the MEs are 2, 2, 3, 1, and 2.

In the initial step of the algorithm, a value of zero is assigned to x. The u-functions of the individual MEs are

$$u_1(z) = p_{10}z^0+p_{11}z^2,\ u_2(z) = p_{20}z^0+p_{21}z^2,\ u_3(z) = p_{30}z^0+p_{31}z^3$$

$$u_4(z) = p_{40}z^0+p_{41}z^1,\ u_5(z) = p_{50}z^0+p_{51}z^2$$

Since, in the MSWS considered, $r_Y = r_2 = 4$, the initial u-function takes the form

$$U_{-3}(z) = z^{(0,0,0,0)}$$

Following step 2 of the algorithm we obtain:
for $h = 1$

$$U_{-2}(z) = U_{-3}(z) \underset{\leftarrow}{\otimes} u_1(z) = z^{(0,0,0,0)} \underset{\leftarrow}{\otimes} (p_{10}z^0+p_{11}z^2) = p_{10}z^{(0,0,0,0)}+p_{11}z^{(0,0,0,2)}$$

for $h = 2$

$$U_{-1}(z) = U_{-2}(z) \underset{\leftarrow}{\otimes} u_2(z)) = (p_{10}z^{(0,0,0,0)}+p_{11}z^{(0,0,0,2)}) \underset{\leftarrow}{\otimes} (p_{20}z^0+p_{21}z^2)$$

$$= p_{10}p_{20}z^{(0,0,0,0)}+p_{11}p_{20}z^{(0,0,2,0)}+p_{10}p_{21}z^{(0,0,0,2)}+p_{11}p_{21}z^{(0,0,2,2)}$$

for $h = 3$

$$U_0(z) = U_{-1}(z) \underset{\leftarrow}{\otimes} u_3(z)$$

$$= (p_{10}p_{20}z^{(0,0,0,0)}+p_{11}p_{20}z^{(0,0,2,0)}+p_{10}p_{21}z^{(0,0,0,2)}+p_{11}p_{21}z^{(0,0,2,2)}) \underset{\leftarrow}{\otimes} (p_{30}z^0+p_{31}z^3)$$

$$= \mathbf{p_{10}p_{20}p_{30}z^{(0,0,0,0)}}+\mathbf{p_{11}p_{20}p_{30}z^{(0,2,0,0)}}+\mathbf{p_{10}p_{21}p_{30}z^{(0,0,2,0)}}+\mathbf{p_{11}p_{21}p_{30}z^{(0,2,2,0)}}$$

$$+\mathbf{p_{10}p_{20}p_{31}z^{(0,0,0,3)}}+p_{11}p_{20}p_{31}z^{(0,2,0,3)}+p_{10}p_{21}p_{31}z^{0,0,2,3)}+p_{11}p_{21}p_{31}z^{(0,2,2,3)}$$

In this step, operator θ_{f_1} should be applied to $U_0(z)$. The terms of $U_0(z)$ with $f_1 = 0$ are marked in bold. The value of $\theta_{f_1}(U_0(z))$ is added to x:

$$x = p_{10}p_{20}p_{30}+p_{11}p_{20}p_{30}+p_{10}p_{21}p_{30}+p_{11}p_{21}p_{30}+p_{10}p_{20}p_{31}$$

After removing the marked terms, $U_0(z)$ takes the form

$$U_0(z) = p_{11}p_{20}p_{31}z^{(0,2,0,3)}+p_{10}p_{21}p_{31}z^{0,0,2,3)}+p_{11}p_{21}p_{31}z^{(0,2,2,3)}$$

Proceeding for $h = 4$ we obtain

$$U_1(z) = U_1(z) \underset{\leftarrow}{\otimes} u_4(z) = (p_{11}p_{20}p_{31}z^{(0,2,0,3)}+p_{10}p_{21}p_{31}z^{(0,0,2,3)}$$

$$+p_{11}p_{21}p_{31}z^{(0,2,2,3)}) \underset{\leftarrow}{\otimes} (p_{40}z^0+p_{41}z^1) = \underline{\mathbf{p_{11}p_{20}p_{31}p_{40}z^{(2,0,3,0)}}}$$

$$+\underline{p_{10}p_{21}p_{31}p_{40}z^{(0,2,3,0)}}+p_{11}p_{21}p_{31}p_{40}z^{(2,2,3,0)}+\mathbf{p_{11}p_{20}p_{31}p_{41}z^{(2,0,3,1)}}$$

$$+p_{10}p_{21}p_{31}p_{41}z^{(0,2,3,1)}+p_{11}p_{21}p_{31}p_{41}z^{(2,2,3,1)}$$

Both operators θ_{f_1} and θ_{f_2} should be applied to $U_1(z)$. The terms of $U_1(z)$ with $f_1 = 0$ are marked in bold; the terms with $f_2 = 0$ are underlined. One can see that in the first term both $f_1 = 0$ and $f_2 = 0$, in the second term only $f_2 = 0$, and in the fourth term only $f_1 = 0$. First, the value of

$$\theta_{f_1}(U_1(z)) = p_{11}p_{20}p_{31}p_{40}+p_{11}p_{20}p_{31}p_{41}$$

is added to x and the terms with $f_1 = 0$ are removed. Then, in the remaining u-function $U_1(z)$, the value of

$$\theta_{f_2}(U_1(z)) = p_{10}p_{21}p_{31}p_{40}$$

is added to x and the terms with $f_2 = 0$ are removed.

After removing all of the marked terms, $U_1(z)$ takes the form

$$U_1(z) = p_{11}p_{21}p_{31}p_{40}z^{(2,2,3,0)}+p_{10}p_{21}p_{31}p_{41}z^{(0,2,3,1)}+p_{11}p_{21}p_{31}p_{41}z^{(2,2,3,1)}$$

Finally, for $h = 5$

$$U_2(z) = U_1(z) \underset{\leftarrow}{\otimes} u_5(z)) = (p_{11}p_{21}p_{31}p_{40}z^{(2,2,3,0)}+p_{10}p_{21}p_{31}p_{41}z^{(0,2,3,1)}$$

$$+p_{11}p_{21}p_{31}p_{41}z^{(2,2,3,1)}) \underset{\leftarrow}{\otimes} (p_{50}z^{0}+p_{51}z^{2}) = \underline{\mathbf{p_{11}p_{21}p_{31}p_{40}p_{50}z^{(2,3,0,0)}}}$$

$$+\mathbf{(p_{10}+p_{11})p_{21}p_{31}p_{41}p_{50}z^{(2,3,1,0)}}+p_{11}p_{21}p_{31}p_{40}p_{51}z^{(2,3,0,2)}$$

$$+(p_{10}+p_{11})p_{21}p_{31}p_{41}p_{51}z^{(2,3,1,2)}$$

The terms of $U_2(z)$ with $f_1 = 0$ are marked in bold and the terms with $f_2 = 0$ are underlined. After adding the value of

$$\theta_{f_1}(U_2(z)) = p_{11}p_{21}p_{31}p_{40}p_{50}+(p_{10}+p_{11})p_{21}p_{31}p_{41}p_{50}$$

to x and removing the corresponding terms from $U_2(z)$, this u-function does not contain terms with $f_2 = 0$. Now x is equal to the system unreliability and $R = 1-x$.

The final u-function $U_2(z)$ takes the form

$$U_2(z) = p_{11}p_{21}p_{31}p_{40}p_{51}z^{(2,3,0,2)}+(p_{10}+p_{11})p_{21}p_{31}p_{41}p_{51}z^{(2,3,1,2)}$$

$$= p_{11}p_{21}p_{31}p_{40}p_{51}z^{(2,3,0,2)}+p_{21}p_{31}p_{41}p_{51}z^{(2,3,1,2)}$$

The system reliability can also be obtained as the sum of the coefficients of the resulting u-function:

$$R = p_{11}p_{21}p_{31}p_{40}p_{51} + p_{21}p_{31}p_{41}p_{51} = p_{21}p_{31}p_{51}(1-p_{10}p_{40})$$

Example 6.29

An MSWS with $n = 10$, $Y = 3$, $r_1 = 3$, $r_2 = 5$, $r_3 = 7$ consists of identical two-state elements. Total failure of the elements corresponds to a performance rate of 0, and a normal state corresponds to performance rate of 1. The reliability of each element j is $p_{j1} = 0.8$. The system fails if the total performance of any r_i adjacent elements is less then w_i. The graphs of the MSWS reliability as a function of the demands w_1, w_2 and w_3 are presented in Figure 6.30.

When $w_i = r_i$ the system becomes a series one and its reliability is equal to $0.8^{10} = 0.1074$. Observe that the variation of demands w_i do not necessarily influence the system's reliability because of failure criteria superposition. For example, satisfying one of the system success conditions

$$\prod_{h=1}^{11-r} 1(\sum_{m=h}^{h+r_i-1} G_m \geq w_i) = 1$$

for $r_1 = 3$, $w_1 = 2$ guarantees satisfying this condition for $r_2 = 5$, $w_2 = 3$. Therefore, the reliability of the MSWS with $w_1 = 2$ does not depend on w_2 if $w_2 \leq 3$.

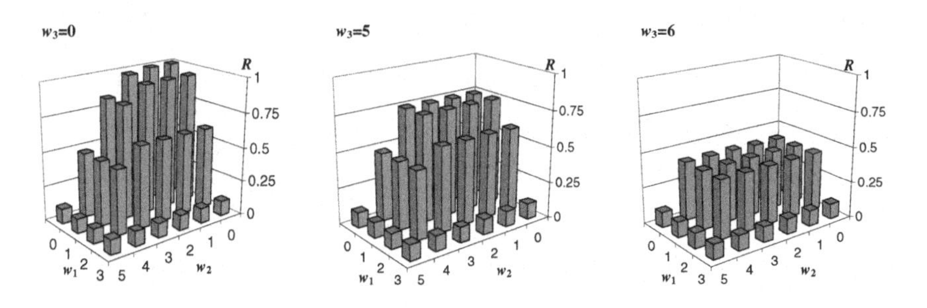

Figure 6.30. Reliability of MSWS as a function of demands

Satisfying the system success condition for $r_2 = 5$, $w_2 = 4$ guarantees satisfying this condition for $r_1 = 3$, $w_1 = 2$. Therefore, the reliability of the MSWS with $w_2 = 4$ does not depend on w_1 if $w_1 \leq 2$.

Satisfying the system success condition for $r_3 = 7$, $w_3 = 6$ guarantees satisfying this condition for both $r_1 = 3$, $w_1 = 2$ and $r_2 = 5$, $w_2 = 4$. Therefore, the reliability of the MSWS with $w_3 = 6$ does not depend on w_1 if $w_1 \leq 2$ and does not depend on w_2 if $w_2 \leq 4$.

6.4.3 Element Reliability Importance in Sliding Window System

The elements' importance measures and the methods of their evaluation for SWSs are the same as for the series-parallel systems. In order to evaluate the importance measures one has to apply the technique described Section 4.5 using the algorithm for SWS reliability evaluation instead of the series-parallel block diagram method.

For SWSs consisting of identical elements it may be important to know how the improvement of all of the elements' reliability influences the entire system's reliability. In order to obtain this importance measure one has to calculate the values of the system reliability for the different values of element reliability, simultaneously changing parameters of the u-functions of all of the elements.

Example 6.30

Consider an MSWS with $n = 10$, $Y = 2$, $r_1 = 3$, $r_2 = 5$. The MSWS consists of identical two-state elements. Total failure of the elements corresponds to a performance rate of 0, and anormal state corresponds to a performance rate of 1. The system fails if the total performance of any r_i adjacent elements is less than w_i.

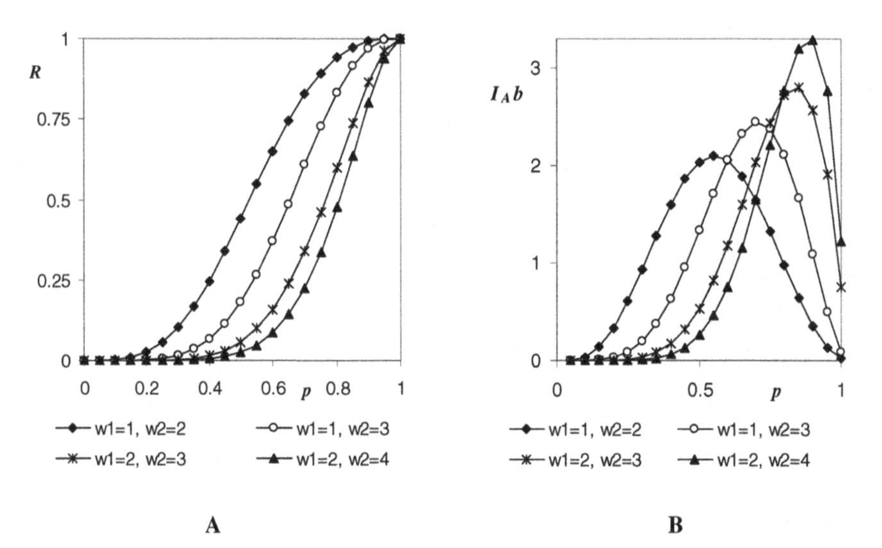

Figure 6.31. SWS reliability (A) and elements' reliability importance (B) as functions of the elements' reliability

In Figure 6.31A one can see the reliability of the MSWS considered as a function of p for different combinations of w_1 and w_2. For the same p, the system reliability decreases with the growth of w_1 and w_2. The elements' Birnbaum reliability importance indices $I_Ab = dR/dp$ as functions of p are presented in Figure 6.31B. One can see that, until a certain level of p corresponding to a maximal I_Ab, the more reliable the elements the greater the entire system benefits from further improvement of the elements' reliability. After achieving the maximal value of I_Ab, the influence of the element's reliability improvement on the system's reliability is drastically reduced (this means that further improvement of the elements' reliability is less justified).

With the growth of w_1 and w_2, the element's reliability corresponding to the maximal reliability importance moves toward the greater values.

In SWSs consisting of nonidentical elements, different elements play different roles in providing for the system's reliability. Evaluating the relative influence of the element's reliability on the reliability of the entire system provides useful information for tracing system bottlenecks.

Example 6.31

Consider an SWS with n = 10 and r = 3 [184]. The parameters of the two-state system elements are presented in Table 6.31. Total failure of any element j corresponds to a performance rate of 0, and a normal state corresponds to a performance rate of g_{j1}. The element's reliability is p_{j1}. The system fails if the cumulative performance of any three adjacent elements is less than the demand w. The system's reliability as a function of demand w is presented in Figure 6.32A.

Table 6.31. Parameters of SWS elements

No. of element j	1	2	3	4	5	6	7	8	9	10
p_{j1}	0.87	0.90	0.83	0.95	0.92	0.89	0.80	0.85	0.82	0.95
g_{j1}	200	200	400	300	100	400	100	200	300	200

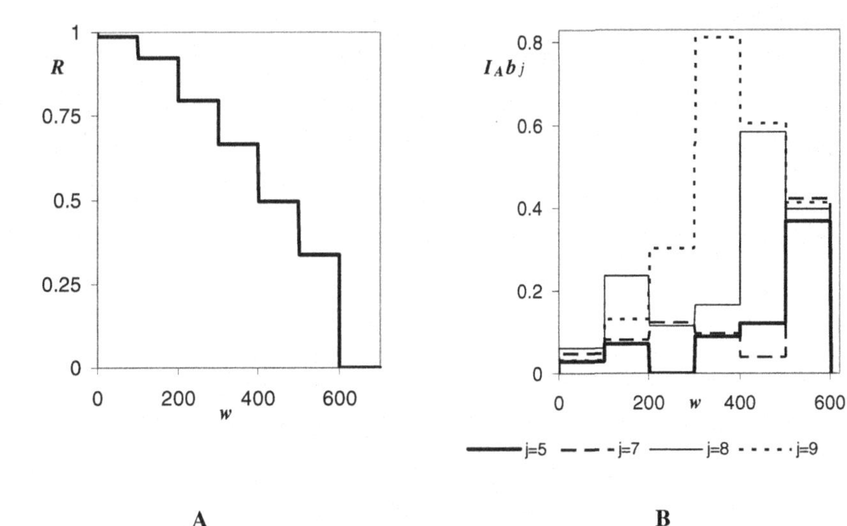

Figure 6.32. SWS reliability (A) and the elements' reliability importance (B) as functions of demand

The reliability importance indices for several elements as functions of the system's demand are presented in Figure 6.32B. Observe that the relative importance of the elements changes with the demand variation. For example, when

$100<w<200$, element 8 is the most important one, whereas when $200<w<300$ this element becomes less important than element 9. This means that in making a decision about the system's reliability enhancement one has to take into account the range of the possible demand levels.

One can see that for some w the importance of the elements can be equal to zero. This means that these elements have no influence on the entire SWS availability and can be removed. Indeed, consider element 5 when $200<w<300$. This element belongs to three triplets with the following nominal performance rates $\{g_{31} = 400, g_{41} = 300, g_{51} = 100\}$, $\{g_{41} = 300, g_{51} = 100, g_{61} = 400\}$, $\{g_{51} = 100, g_{61} = 400, g_{71} = 100\}$. The cumulative performance rate of the first triplet is greater than w if at least one of elements 3 and 4 works and is less than w if both of these elements fail. The cumulative performance rate of the second triplet is greater than w if at least one of elements 4 and 6 works and is less than w if both of these elements fail. The cumulative performance rate of the third triplet is greater than w if element 6 works and is less than w if this element fails. The state of element 5 does not affect the value of the acceptability function for any one of the three triplets.

6.4.4 Optimal Element Sequencing in Sliding Window Systems

Having a given set of MEs, one can achieve considerable reliability improvement of the linear SWS by choosing the elements' proper arrangement along a line. Indeed, it can be easily seen that the order of the tasks' arrivals to the service system (Example 6.26) or allocation of heaters along a line (Example 6.27) can strongly affect the system's entire reliability. For the set of MEs with a given performance rate distribution, the only factor affecting the entire SWS reliability (for fixed r and w) is the sequence of MEs. Papastavridis and Sfakianakis [185] first considered the optimal element arrangement problem for SWSs with binary elements having a different reliability. In this section, the optimal element arrangement problem is considered for the general SWS model. This problem is formulated as follows: find the sequence of MEs in the SWS that maximizes the system reliability.

6.4.4.1 Implementing the Genetic Algorithm

In order to represent the sequence of n MEs in the SWS in the GA one can consider a line with n consequent positions and use a string $\boldsymbol{a} = (a_1, \ldots, a_n)$ in which a_j is equal to the number of the position occupied by ME j (see Section 1.3.2.4). One can see that the total number of different arrangement solutions (number of different possible vectors $\boldsymbol{a}$) is equal to $n!$ (number of possible permutations in a string of n different numbers).

The solution decoding procedure should apply the algorithm for SWS reliability determination for the given sequence of MEs represented by string $\boldsymbol{a}$. The solution's fitness is equal to the value of system reliability $R(\boldsymbol{a}, r, w)$ obtained.

Example 6.32

Consider an SWS with n = 10 [186]. The parameters of the system MEs are presented in Table 6.32. Three element sequencing solutions were obtained by the GA for the SWS with $r = 3$ (for $w = 6$, $w = 8$ and $w = 10$) and three solutions were obtained for the same SWS with $r = 5$ (for $w = 10$, $w = 15$, $w = 20$). These solutions are presented in Table 6.33. The system's reliability as a function of demand w is presented in Figure 6.33 for the ME sequences obtained. One can see that the greater r, the greater the SWS reliability for the same w. This is natural, because the growth of r provides a growing redundancy in each group.

Table 6.32. Performance distributions of SWS elements

No. of ME	1		2		3		4		5		6		7		8		9		10	
State	p	g	p	g	p	g	p	g	p	g	p	g	p	g	p	g	p	g	p	g
0	0.03	0	0.10	0	0.17	0	0.05	0	0.08	0	0.01	0	0.20	0	0.05	0	0.20	0	0.05	0
1	0.22	2	0.10	1	0.83	6	0.25	3	0.20	1	0.22	4	0.10	3	0.25	4	0.10	3	0.25	2
2	0.75	5	0.40	2	-	-	0.40	5	0.15	2	0.77	5	0.10	4	0.70	6	0.15	4	0.70	6
3	-	-	0.40	4	-	-	0.30	6	0.45	4	-	-	0.60	5	-	-	0.55	5	-	-
4	-	-	-	-	-	-	-	-	0.12	5	-	-	-	-	-	-	-	-	-	-

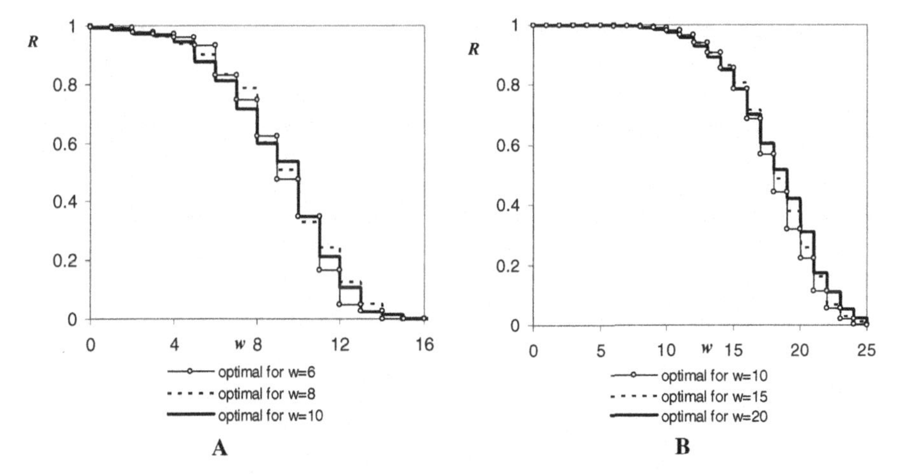

Figure 6.33. Reliability of SWS with the optimal element arrangements as function of demand. A: for $r = 3$; B: for $r = 5$

Table 6.33. Parameters of the solutions obtained

r	w	R	Sequence of SWS elements									
3	6	0.931	2	1	6	5	4	8	7	10	3	9
	8	0.788	5	1	8	9	6	4	7	3	10	2
	10	0.536	5	9	3	1	4	7	10	8	6	2
5	10	0.990	2	5	1	4	6	8	10	3	7	9
	15	0.866	9	7	3	10	1	6	8	4	5	2
	20	0.420	2	5	4	8	3	6	10	7	1	9

Note that, for solutions which provide the greatest SWS reliability for a certain w, the reliability for the rest of the values of w is less than for the solutions optimal for those values of w. Indeed, the optimal allocation provides the greatest system probability of meeting just the specified demand by the price of reducing the probability of meeting greater demands.

6.4.5 Optimal Uneven Element Allocation in Sliding Window Systems

While the problem of the optimal ordering of tasks' arrivals to the service system (Example 6.26) presumes the arrival of one task at a time (only one task can be in each position in the service line), in the problem of the optimal arrangement of heaters (Example 6.27) we can assume that n positions are distributed along a line and the heaters may be allocated unevenly at these positions (several heaters can be gathered at the same position while some positions remain empty).

In many cases such uneven allocation of the MEs in an SWS results in greater system reliability than the even allocation.

Example 6.33

Consider a simple case in which four MEs should be allocated within an SWS with four positions. Each ME j has two states: a failure state with a performance of 0 and a normal state with a performance of 1. The probability of a normal state is p_j, the probability of a failure is $q_j = 1-p_j$. For $r = 3$ and $w = 2$, the system succeeds if each three consecutive positions contain at least two elements in a normal state. Consider two possible allocations of the MEs within the SWS (Figure 6.34):

A. MEs are evenly distributed among the positions.

B. Two MEs are allocated at second position and two MEs are allocated at third position.

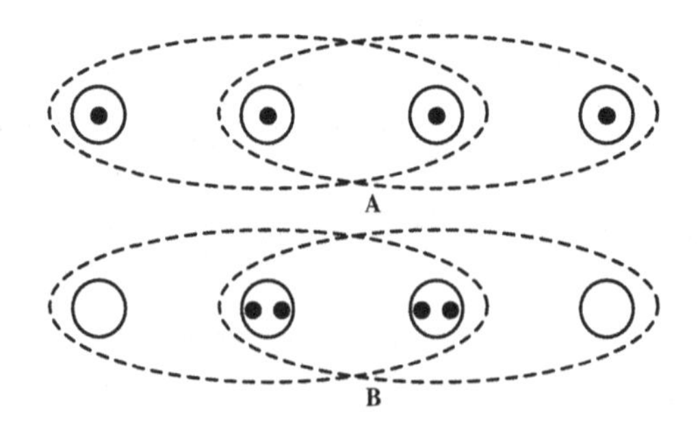

Figure 6.34. Two possible allocations of MEs in an SWS

In case A, the SWS succeeds either if no more than one ME fails or if MEs in the first and fourth positions fail and MEs in the second and third positions are in a normal state. Therefore, the system reliability is

$$R_A = p_1p_2p_3p_4+q_1p_2p_3p_4+p_1q_2p_3p_4+p_1p_2q_3p_4+p_1p_2p_3q_4+q_1p_2p_3q_4$$

For identical MEs with $p_j = p$

$$R_A = p^4+4qp^3+q^2p^2$$

In case B, the SWS succeeds if at least two MEs are in a normal state. The system reliability in this case is

$$R_B = p_1p_2p_3p_4+q_1p_2p_3p_4+p_1q_2p_3p_4+p_1p_2q_3p_4+p_1p_2p_3q_4$$
$$+q_1q_2p_3p_4+q_1p_2q_3p_4+q_1p_2p_3q_4+p_1q_2q_3p_4+p_1q_2p_3q_4+p_1p_2q_3q_4$$

For identical MEs

$$R_B = p^4+4qp^3+6q^2p^2$$

One can see that the uneven allocation B is more reliable:

$$R_B - R_A=5q^2p^2=5(1-p)^2p^2$$

Consider now the same system when w=3. In case A the system succeeds only if it does not contain any failed ME:

$$R_A = p_1p_2p_3p_4$$

In case B it succeeds if it contains no more than one failed element:

$$R_B = p_1p_2p_3p_4+q_1p_2p_3p_4+p_1q_2p_3p_4+p_1p_2q_3p_4+p_1p_2p_3q_4$$

For identical MEs:

$$R_A = p^4,\ R_B = p^4+4qp^3 \text{ and } R_B-R_A = p^4+4qp^3-p^4 = 4(1-p)p^3$$

Observe that, even for w = 4, when in case A the system is unable to meet the demand (R_A = 0) because $w>r$, in case B it still succeeds with probability $R_B = p_1p_2p_3p_4$.

In this section we consider a general optimal allocation problem in which the number of MEs m is not necessarily equal to the number of positions n ($m \neq n$) and an arbitrary number of elements can be allocated at each position (some positions may be empty):

The SWS consists of n consecutively ordered positions. At each position any group of MEs can be allocated. The allocation problem can be considered as a problem of partitioning a set of m items into a collection of n mutually disjoint subsets. This partition can be represented by the integer string $\boldsymbol{a}$ = (a_j: 1≤j≤m),

$1 \le a_j \le n$, where a_j is the number of the position at which ME j locates. It is assumed that the SWS has the acceptability function (6.114). The total performance of the group of the MEs located at the same position is equal to the sum of the performances of these MEs. The empty position can be represented by an element with the constant performance of zero.

For any given integer string $\boldsymbol{a}$, the GA determines the solution fitness (equal to the SWS reliability) using the following procedure:

1. Assign $\tilde{u}_i(z) = z^0$ for each $i = 1, \dots, n$, corresponding to SWS positions.

Determine u-functions $u_j(z)$ for each individual ME j ($1 \le j \le m$) in the form (6.115) in accordance with their performance distributions.

2. According to the given string $\boldsymbol{a}$ for each $j = 1, \dots, m$ modify $\tilde{u}_{a_j}(z)$ as follows:

$$\tilde{u}_{a_j}(z) = \tilde{u}_{a_j}(z) \underset{+}{\otimes} u_j(z) \tag{6.121}$$

3. Apply the algorithm for SWS reliability evaluation described in Section 6.4.1 over n u-functions $\tilde{u}_i(z)$.

Example 6.34

Consider an SWS with $n = 10$ positions in which $m = 10$ identical binary MEs are to be allocated [187]. The performance distribution of each ME j is $p_{j1} = \Pr\{G_j = 1\} = 0.9$, $p_{j0} = \Pr\{G_j = 0\} = 0.1$.

Table 6.34 presents allocation solutions obtained for different r and w (number of identical elements in each position). The reliability of the SWS corresponding to the allocations obtained is compared with its reliability corresponding to the case when the MEs are evenly distributed among the positions. One can see that the reliability improvement achieved by the free allocation increases with the increase of r and w. On the contrary, the number of occupied positions in the best solutions obtained decreases when r and w grow. Figure 6.35 presents the SWS reliability as a function of demand w for $r = 2$, $r = 3$ and $r = 4$ for even ME allocation and for unconstrained allocation obtained by the GA.

Table 6.34. Solutions of ME allocation problem (SWS with identical MEs)

Position	$r=2$, $w=1$	$r=3$, $w=2$	$r=4$, $w=3$
1			
2	2	1	
3		3	
4	2		5
5		3	
6	2		
7			5
8	2	3	
9			
10	2		
	Reliability		
Free allocation	0.951	0.941	0.983
Even allocation	0.920	0.866	0.828
Improvement	3.4%	8.7%	18.7%

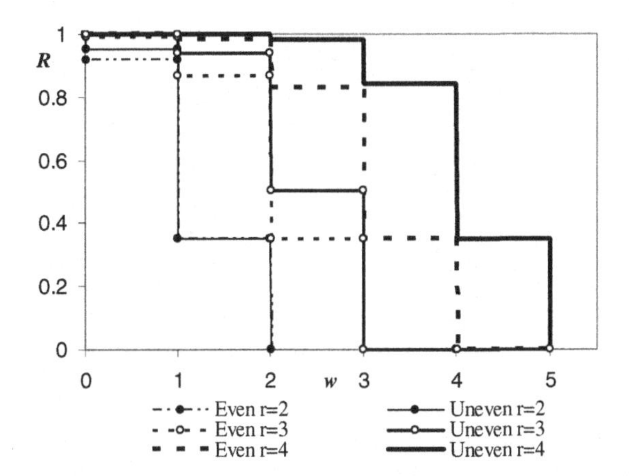

Figure 6.35. Reliability of SWS with identical MEs for different r and ME allocations

Example 6.35

Consider the SWS allocation problem from Example 6.32, in which $n = m = 10$, $r = 3$ and $w = 10$. The best ME allocation solutions obtained by the GA are presented in Table 6.35 (list of elements located at each position).

Table 6.35. Solutions of ME allocation problem (SWS with different MEs)

Position	$m = 10$ Even allocation	$m = 10$ Uneven allocation	$m = 9$	$m = 8$	$m = 7$
1	5				
2	9				1
3	3	6, 7, 10	2, 5, 8, 9	3, 6	7
4	1				
5	4	2, 5	7		3
6	7	1, 4	1, 4	5, 7, 8	6
7	10				
8	8	3, 8, 9	3	4	4, 5
9	6		6	1, 2	2
10	2				
Reliability	0.536	0.765	0.653	0.509	0.213

The best even allocation solution obtained in Example 6.32 improves considerably when the even allocation constraint is removed. One can see that the best unconstrained allocation solution obtained by the GA in which only 4 out of 10 positions are occupied by the MEs provides a 42% reliability increase over even allocation. The system reliability as a function of demand for the even and unconstrained allocations obtained is presented in Figure 6.36.

Table 6.35 also presents the best allocations of the first m MEs from Table 6.32 (for $m = 9$, $m = 8$ and $m = 7$). Observe that uneven allocation of nine MEs in the SWS still provides greater reliability than does even allocation of 10 MEs.

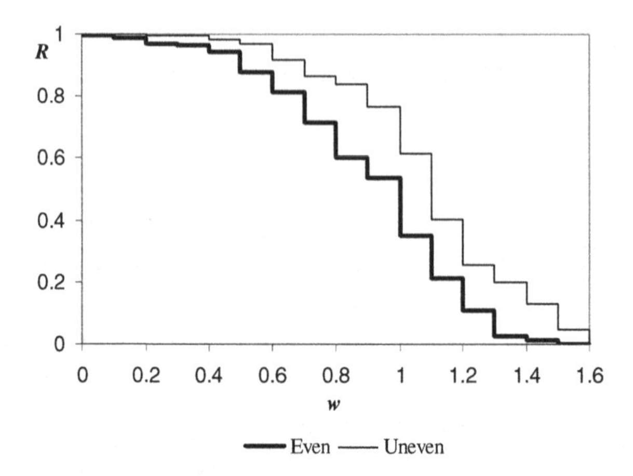

Figure 6.36. Reliability of SWS with different MEs

6.4.6 Sliding Window System Reliability Optimization in the Presence of Common Cause Failures

In many cases, when SWS elements are subject to CCFs, the system can be considered as consisting of mutually disjoint CCGs with total CCFs. The origin of CCFs can be outside the system's elements they affect (external impact), or they can originate from the elements themselves, causing other elements to fail. Usually, the CCFs occur when a group of elements share the same resource (energy source, space, *etc.*)

Example 6.36

Consider the manufacturing heating system from Example 6.27 and assume that the power to the heaters is supplied by B independent power sources (Figure 6.37). Each heater is connected to one of these sources. The heaters supplied from the same source compose the CCG. Each source has a certain failure probability. When the source fails, all of the heaters connected to this source (belonging to the corresponding CCG) are in a state of total failure. Therefore, the failure of any power source causes the CCF in the heating system.

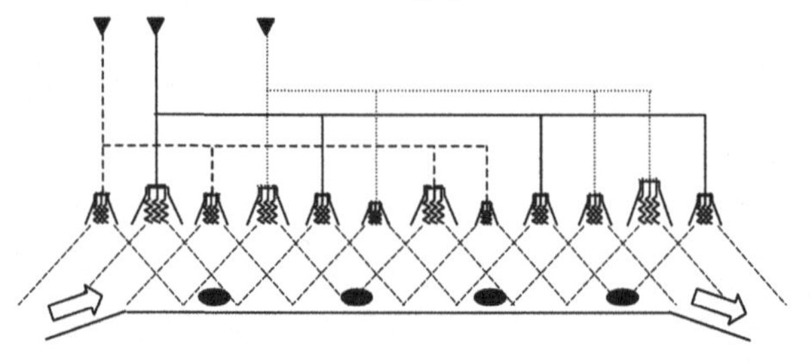

Figure 6.37. Example of SWS with several CCGs

6.4.6.1 Evaluating Sliding Window System Reliability in the Presence of Common Cause Failures

The u-function $u_j(z)$ representing the p.m.f. of the random performance rate of ME j G_j takes the form (6.115) only when the element is not the subject of CCF.

We assume that the performance rate of element j when it is a subject of CCF is equal to zero. The u-function corresponding to this case takes the form $u_j^0(z) = z^0$. The u-functions $u_j(z)$ and $u^0_j(z)$ represent, therefore, conditional performance distributions of the performance rate of element j.

Let the entire SWS consists of a set Λ of n ordered MEs and has B independent CCGs. Each CCG can be in two states (normal state and failure). The failure probability of CCG i is f_i. It can be seen that the total number of the combinations of CCG states is 2^B. Each CCG i can be defined by a subset $\lambda_i \subseteq \Lambda$ such that

$$\bigcup_{i=1}^{B} \lambda_i = \Lambda,\ \lambda_i \cap \lambda_e = \varnothing \text{ for } i \neq e \tag{6.122}$$

Let binary variable s_i define the state of CCG i such that $s_i = 1$ corresponds to the normal state of the group and $s_i = 0$ corresponds to failure of the group. When $s_i = 1$ the performance of each ME j belonging to λ_i is a random value having the distribution determined by its u-function $u_j(z)$. When $s_i = 0$ the performance of each ME belonging to λ_i is equal to zero, which corresponds to the u-function $u^0_j(z)$. One can connect the state s_i of each individual CCG with a number of state combination h in the following way:

$$s_i(h)=\text{mod}_2\lfloor h/2^{i-1} \rfloor \tag{6.123}$$

When h varies from 0 to 2^B-1 one obtains all the possible combinations of states of CCGs using Equation (6.123) for $1 \le i \le B$.

The probability of each CCG state combination h is

$$q_h = \prod_{i=1}^{B} (f_i)^{1-s_i(h)} (1-f_i)^{s_i(h)} \tag{6.124}$$

If one defines the u-function of each ME $1 \le j \le n$ as

$$\tilde{u}_j(z) = \begin{cases} u_j^0(z), & \text{if} \quad j \in \lambda_i, \quad s_i = 0 \\ u_j(z), & \text{otherwise} \end{cases} \tag{6.125}$$

and applies the algorithm for SWS reliability evaluation (described in Section 6.4.1) over u-functions $\tilde{u}_j(z)$, then one obtains the conditional probability of the SWS success r_h when the CCG state combination is h. Since all of the 2^B

combinations of CCG states are mutually exclusive, in order to calculate the unconditional probability of the SWS success (SWS reliability) one can apply the following equation:

$$R = \sum_{h=0}^{2^B-1} q_h r_h \tag{6.126}$$

Now we can evaluate the SWS reliability using the following algorithm:

1. Assign $R = 0$. For each j ($1 \le j \le n$) determine two u-functions $u_j(z)$ (in accordance with Equation (6.115) and $u^0_j(z) = z^0$.

2. Repeat the following for $h = 0, \ldots, 2^B-1$:

2.1. For $i = 1, \ldots, B$ determine $s_i(h)$ using Equation (6.123).

2.2. Determine q_h (the probability of CCG state combination h) using Equation (6.124).

2.3. For each $i = 1, \ldots, B$ determine the numbers of elements belonging to the CCG i and define the u-functions of these elements $\tilde{u}_j(z)$ in accordance with Equation (6.125).

2.4. Determine r_k (the conditional SWS reliability for CCG state combination h) applying the procedure described in Section 6.4.1 over u-functions $\tilde{u}_j(z)$ ($1 \le j \le n$) for a given demand w.

2.5. Add the product $q_h r_h$ to R.

6.4.6.2 Optimal Distribution of Multi-state Elements among Common Cause Groups

The way the MEs are distributed among CCGs strongly affects the SWS reliability. Consider a simple example in which SWS with $r = 2$ consists of four MEs composing two CCGs. Each ME has two states with performance rates of 0 and 1. The system demand is 1. When $\lambda_1 = \{1, 2\}$ and $\lambda_2 = \{3, 4\}$ each CCF causes the system's failure. When $\lambda_1 = \{1, 3\}$ and $\lambda_2 = \{2, 4\}$ the SWS can succeed in the case of a single CCF if both MEs not belonging to the failed CCG are in the operational state.

The elements' distribution problem can be considered as a problem of partitioning a set Λ of n MEs into a collection of B mutually disjoint subsets λ_i ($1 \le i \le B$). Each set can contain from 0 to n elements. The partition of set Λ can be represented in the GA by the integer string $\boldsymbol{a} = \{a_j: 1 \le j \le n\}$, $0 \le a_j \le B$, where a_j is the number of the subset to which ME j belongs: $j \in \lambda_{a_j}$.

Example 6.36

Consider the SWS with $n = 10$ from Example 6.32 [188]. The parameters of the system's MEs are presented in Table 6.32. It is assumed that the failure probability f_i of each CCG i is equal to 0.2.

First, distribution solutions were obtained by the GA for a SWS with fixed B. The solutions that provide the greatest SWS reliability for a certain demand w

provide a reliability for the remainder of the values of w that is less than the reliability of the optimal solutions for those values of w. Indeed, the optimal allocation provides the greatest system probability of meeting just the specified demand by the price of reducing the probability of meeting other demands. Therefore, the optimal distribution solution depends on system demand. The solutions obtained for different demands for SWS with $r = 3$ and $r = 5$ when $B = 2$ are presented in Table 6.36.

When $r = 3$, solution A is optimal for demand $0<w<3$ and solution B is optimal for demand $3\le w\le 5$. When $r = 5$, solution D is optimal for demand $7\le w\le 11$ and solution E is optimal for demand $12\le w\le 16$. The system reliabilities, as functions of demand for the solutions obtained, are presented in Figure 6.38. The solutions C (for $r = 3$) and F (for $r = 5$), in which adjacent elements belong to different CCGs, are presented for comparison. These solutions provide lower SWS reliability than the optimal ones.

Table 6.36. Solutions of ME grouping problem obtained for $B = 2$

r		Distribution of MEs
	A	{1, 4, 6, 7, 9, 10} {2, 3, 5, 8}
3	B	{1, 2, 4, 5, 7, 8, 10} {3, 6, 9}
	C	{1, 3, 5, 7, 9} {2, 4, 6, 8, 10}
	D	{1, 2, 5, 6, 7, 10} {3, 4, 8, 9}
5	E	{1, 2, 3, 6, 7, 8} {4, 5, 9, 10}
	F	{1, 3, 5, 7, 9} {2, 4, 6, 8, 10}

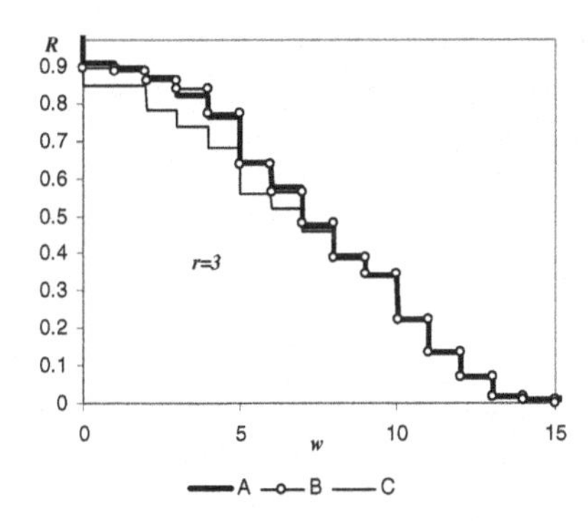

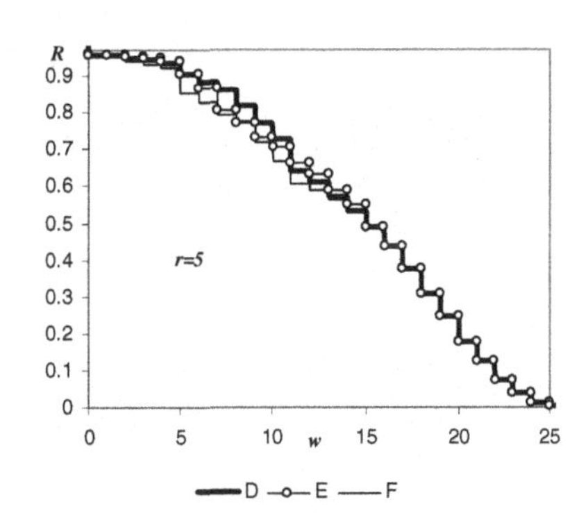

Figure 6.38. SWS reliability for the best ME distribution solutions obtained for $B = 2$

Observe that with the growth of w the difference of system reliability provided by the different distribution solutions becomes negligible. This is because, when w is great, the system becomes intolerant to any common supply failure without regard to the structure of the failed CCG.

The problem of choosing the optimal number of CCGs is also of interest. The increase in the number of CCGs reduces the damage caused by a single CCG to the system, but, on the other hand, it increases the probability that at least one CCG has failed. Therefore, when the system does not tolerate the loss of even a small portion of its elements (which happens when w is great), the increase of B decreases the system's overall reliability. The smaller the system demand w, the greater the benefit from the elements' distribution among different CCGs. Table 6.37 presents the solutions obtained for the given SWS for different B. For each $B>1$ the solution that provides the system's reliability greater than the reliability of a system with a single CCG for the greatest w is chosen (this means that for the given B, the solution presented is better than the solution with $B = 1$ for a given demand w^*, but, when the demand is greater than w^*, no distribution solution with B CCGs exists that outperforms the single CCG solution).

Table 6.37. Solutions of ME grouping problem obtained for different B

r	B	Distribution of MEs
	1	{1, 2, 3, 4, 5, 6, 7, 8, 9, 10}
3	2	{1, 2, 3, 6, 7, 8} {4, 5, 9, 10}
	3	{1, 4, 7, 10} {2, 5, 8} {3, 6, 9}
	1	{1, 4, 6, 7, 9, 10} {2, 3, 5, 8}
	2	{1, 4, 5, 6, 9, 10} {3, 4, 7, 8}
5	3	{1, 5, 6, 10} {2, 4, 7, 9} {3, 8}
	4	{1, 5, 6, 10} {2, 7} {3, 8} {4, 9}
	5	{1, 6} {2, 7} {3, 8} {4, 9} {5, 10}

The SWS reliabilities as functions of demand for the solutions obtained are presented in Figure 6.39. The single CCG solution is the worst for the small demands and the best for the great demands. On the contrary, the solutions with $B = r$, in which any r adjacent elements belong to different CCGs, provide the greatest system reliability for small demands and provide the lowest system reliability for great demands. The solutions with $1<B<r$ provide intermediate values of system reliability. Therefore, when the number of CCGs is not fixed, the greatest reliability solution is either with $B = r$ for low demands or with $B = 1$ for great demands.

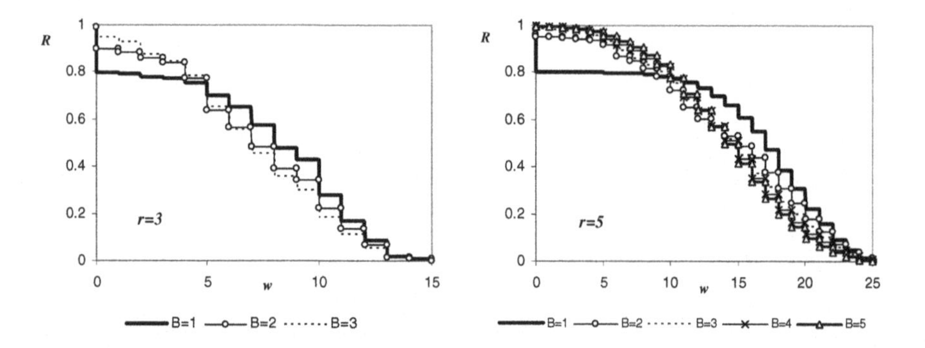

Figure 6.39. SWS reliability for the best ME distribution solutions obtained for different B

7. Universal Generating Function in Analysis and Optimization of Consecutively Connected Systems and Networks

7.1 Multi-state Consecutively Connected Systems

7.1.1 Connectivity Model

The linear multi-state consecutively connected system (LMCCS) consists of $n+1$ consecutively ordered positions (nodes) C_j, $j\in[1, n+1]$. The first node C_1 is the source and the last one C_{n+1} is the sink. At some nodes C_j ($1\le j\le n$) MEs are allocated (some nodes remain empty). These elements provide connections (arcs) between the node in which they are allocated and further nodes.

Each ME e_i has k_j states, where state S_i of e_i is a discrete random variable with p.m.f.

$$\Pr\{S_i = h\} = p_{ih}, \quad \sum_{h=0}^{k_j-1} p_{ih} = 1 \tag{7.1}$$

$S_i = h$ ($1\le h\le k_j-1$) for element e_i allocated at node C_j implies that arcs exist from C_j to each of $C_{j+1}, C_{j+2}, \ldots, C_{\alpha(j+h)}$, where $\alpha(x)=\min\{x, n+1\}$. $S_i = 0$ implies the total failure state of e_i (no arcs created by e_i exist from C_j). Note that, although different MEs can have a different number of states, one can define the same number of states for all of the MEs without a loss of generality. Indeed, if ME e_i has k_i states and ME e_m has k_m states ($k_i\le k_m$), one can consider both MEs as having $k = \max\{k_i, k_m\}$ states while assigning $p_{ih} = 0$ for $k_i\le h<k$.

The system fails if and only if there is no path from C_1 to C_{n+1}. Therefore, the system reliability can be defined as the probability that it can provide the connection between C_1 to C_{n+1}.

Let the random value $G_i^{(j)}$ be the number of the most remote node to which exists an arc that ME e_i located at C_j provides. The u-function $u_{ij}(z)$ representing the p.m.f. of $G_i^{(j)}$ takes the form

$$u_{ij}(z) = \sum_{h=0}^{k-1} p_{ih} z^{\alpha(j+h)} \tag{7.2}$$

The absence of any ME at node C_j implies that no paths exist from C_j to any further node. This means that any arc reaching C_j has no continuation with probability 1. In this case, the corresponding u-function takes the form

$$u_{0j}(z) = z^j \tag{7.3}$$

If node C_j contains several MEs, then the number of the most remote node to which an arc from C_j exists is determined by the ME, which is in the maximal state. Therefore, the random value $G^{(j)}$ (the number of the most remote node connected to C_j) can be obtained using the structure function

$$G^{(j)} = \max\{G_{i_1}^{(j)},...,G_{i_m}^{(j)}\} \tag{7.4}$$

where $i_1,\ldots,i_m$ are numbers of MEs located at C_j. The u-function $U_j(z)$ that represents the p.m.f. of $G^{(j)}$ can, therefore, be obtained as

$$U_j(z) = \underset{\max}{\otimes}(u_{i_1 j}(z),...,u_{i_m j}(z)) = \sum_{h=j}^{\alpha(j+k-1)} q_{jh} z^h \tag{7.5}$$

For any $u_{ij}(z)$

$$u_{0j}(z) \underset{\max}{\otimes} u_{ij}(z) = u_{ij}(z) \tag{7.6}$$

After collecting like terms, the resulting u-function $U_j(z)$ (as well as u-functions $u_{ij}(z)$ for individual MEs) can have no more than $\min\{k, n-j+2\}$ terms (corresponding to possible values of $G^{(j)} \in \{j, j+1, \ldots, \min\{j+k-1, n+1\}\}$).

Consider now the paths starting from C_1 that are provided by elements allocated in subsequent nodes. Let the random value $\tilde{G}^{(j)}$ be the number of the farthest node of a path from C_1 provided by the MEs located at nodes from C_1 to C_j. Let $\tilde{U}_j(z)$ represents the p.m.f. of $\tilde{G}^{(j)}$. All the paths provided by the MEs located at C_1 are single arc paths and, therefore, $\tilde{G}^{(1)} = G^{(1)}$. The p.m.f. of $\tilde{G}^{(1)}$ can be represented by the u-function $\tilde{U}_1(z)$, which is equal to $U_1(z)$.

The paths provided by the MEs located at C_1, …, C_j can be continued by arcs provided by MEs located at C_{j+1} only if $\tilde{G}^{(j)} > j$ (the path reaches C_{j+1}). If this condition is satisfied, then the most remote node of a path from C_1 provided by the MEs located at $C_1,\ldots,C_{j+1}$ can be determined as $\tilde{G}^{(j+1)} = \max\{\tilde{G}^{(j)}, G^{(j+1)}\}$. This can be expressed by the following function:

$$\tilde{G}^{(j+1)} = \begin{cases} \max\{\tilde{G}^{(j)}, G^{(j+1)}\}, \tilde{G}^{(j)} > j \\ j, \qquad\qquad \tilde{G}^{(j)} \le j \end{cases} \tag{7.7}$$

When $\tilde{G}^{(j)} \le j$, the path from C_1 to nodes with numbers greater than j does not exist. Therefore, the corresponding state does not participate in the ME state combinations providing the system's success. In order to consider only the combinations of states of elements located at C_1, …, C_j corresponding to cases in which paths from C_1 to C_{j+1} exist ($\tilde{G}^{(j)} > j$), we introduce the following φ operator which eliminates the term with $\tilde{G}^{(j)} = j$ from the u-function $\tilde{U}_j(z)$:

$$\varphi(\tilde{U}_j(z)) = \varphi(\sum_{h=j}^{\alpha(j+k-1)} q_{jh} z^h) = \sum_{h=j+1}^{\alpha(j+k-1)} q_{jh} z^h \tag{7.8}$$

Note that $\varphi(\tilde{U}_j(z)) = 0$ if $\tilde{U}_j(z)$ does not contain terms with $\tilde{G}^{(j)} > j$.

Now, having the p.m.f. of random variables $\tilde{G}^{(j)}$ and $G^{(j+1)}$, represented by $\tilde{U}_j(z)$ and $U_{j+1}(z)$ respectively, one can determine the u-function $\tilde{U}_{j+1}(z)$ representing the p.m.f. of $\tilde{G}^{(j+1)}$ using the operator

$$\tilde{U}_{j+1}(z) = \begin{cases} \varphi(\tilde{U}_j(z)) \underset{\max}{\otimes} U_{j+1}(z) \;\text{ if } \tilde{U}_j(z) \neq 0 \\ z^j \quad \text{if } \tilde{U}_j(z) = 0 \end{cases} \tag{7.9}$$

Applying Equation (7.9) recursively, one obtains $\tilde{U}_n(z)$ that contains two terms corresponding to $\tilde{G}^{(n)} = n$ and $\tilde{G}^{(n)} = n+1$. $\varphi(\tilde{U}_n(z))$ has only one term corresponding to the probability that the path from C_1 to C_{n+1} exists. The coefficient of this term is equal to the reliability of the LMCCS.

The following procedure determines the system reliability:

1. Assign $U_j(z) = u_{0j}(z) = z^j$ for each $j = 1, …, n$.
2. For each ME e_i located at C_j determine $u_{ij}(z)$ using Equation (7.2) and modify $U_j(z)$: $U_j(z) = U_j(z) \underset{\max}{\otimes} u_{ij}(z)$.
3. Assign $\tilde{U}_1(z) = U_1(z)$ and apply in sequence Equation (7.9) for $j = 1, …, n-1$.
4. Obtain the coefficient of the resulting single term u-function $\varphi(\tilde{U}_n(z))$ as the system reliability.

Example 7.1

Consider a system with four nodes and three MEs. Each ME has three states: state 0 (total failure), state 1 in which the ME is able to connect the node at which it is located with the next node, and state 2 in which the ME is able to connect the node it is located at with the next two nodes. The first and second MEs are located at C_1 and the third one is located at C_3. No MEs are located at C_2.

The u-functions of the individual MEs located at C_1 are

$$u_{11}(z) = p_{10}z^1+p_{11}z^2+p_{12}z^3,\ u_{21}(z) = p_{20}z^1+p_{21}z^2+p_{22}z^3$$

The second node is empty, which corresponds to the u-function $u_{02}(z) = z^2$. The u-function of the ME located at C_3 is

$$u_{33}(z) = p_{30}z^3+(p_{31}+p_{32})z^4$$

The u-functions representing the p.m.f. of random values $G^{(1)}$, $G^{(2)}$ and $G^{(3)}$ for the groups of MEs allocated at the same nodes are, after simplification:

$$U_1(z) = u_{11}(z) \underset{\max}{\otimes} u_{21}(z) = p_{10}p_{20}z^1+[p_{10}p_{21}+p_{11}(1-p_{22})]z^2$$

$$+(p_{12}+p_{22}-p_{12}p_{22})z^3,\ \ U_2(z) = u_{02}(z),\ U_3(z) = u_{33}(z)$$

The u-function representing the p.m.f. of $\tilde{G}^{(1)}$ is

$$\tilde{U}_1(z) = U_1(z),\ \ \varphi(\tilde{U}_1(z)) = [p_{10}p_{21}+p_{11}(1-p_{22})]z^2+(p_{12}+p_{22}-p_{12}p_{22})z^3$$

The u-function representing the p.m.f. of $\tilde{G}^{(2)}$ is

$$\tilde{U}_2(z) = \varphi(\tilde{U}_1(z)) \underset{\max}{\otimes} U_2(z) = [p_{10}p_{21}+p_{11}(1-p_{22})]z^2+(p_{12}+p_{22}-p_{12}p_{22})z^3$$

$$\varphi(\tilde{U}_2(z)) = (p_{12}+p_{22}-p_{12}p_{22})z^3$$

The u-function representing the p.m.f. of $\tilde{G}^{(3)}$ is

$$\tilde{U}_3(z) = \varphi(\tilde{U}_2(z)) \underset{\max}{\otimes} U_3(z) = (p_{12}+p_{22}-p_{12}p_{22})p_{30}z^3$$

$$+(p_{12}+p_{22}-p_{12}p_{22})(p_{31}+p_{32})z^4$$

$$\varphi(\tilde{U}_3(z)) = (p_{12}+p_{22}-p_{12}p_{22})(p_{31}+p_{32})z^4$$

Finally, the system availability obtained from $\varphi(\tilde{U}_3(z))$ is

$$R_L = (p_{12}+p_{22}-p_{12}p_{22})(p_{31}+p_{32})$$

In order to reduce the problem of evaluating the reliability of the circular multi-state consecutively connected system (CMCCS) into the problem of evaluating the reliability of the LMCCS, we use the following definition of a system's failure (Malinowski and Preuss [83]): a CMCCS fails if there exists at least one node C_j, $1 \le j \le n$, such that no arcs (signals) originated from any other node can reach C_j directly.

Assume that each node C_j contains a single ME e_j that has k_j states (any set of MEs located at the same node can be replaced by an equivalent ME with the u-function obtained using Equation (7.5)). Consider a set σ_i of consecutive MEs e_i, e_{i+1}, …, e_n (Figure 7.1). Let the random variable Y_i represent the number of MEs located beyond σ_i and reached directly by any arc originated from σ_i. It can be seen that $0 \le Y_i \le h_{i,\max}$, where

$$\begin{aligned} h_{i,\max} &= \max\{k_n - 1, k_{n-1} - 2, \ldots, k_i - (n-i) - 1\} \\ &= \max_{i \le j \le n} \{k_j - (n-j) - 1\} \end{aligned} \tag{7.10}$$

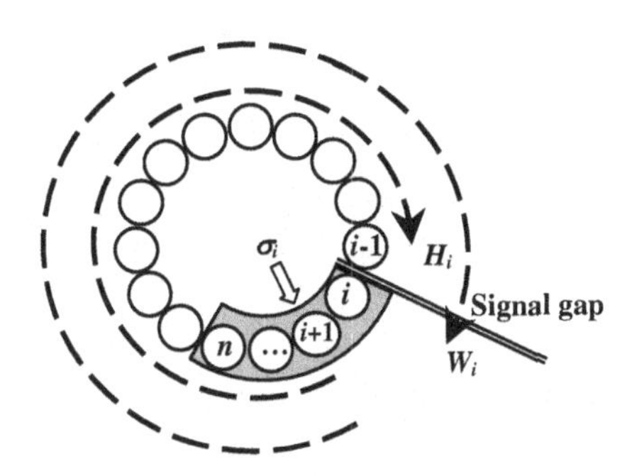

Figure 7.1. Circular multi-state consecutively connected system

In order to obtain the p.m.f. of the variable Y_i one has to determine the probabilities $p_h^{(i)} = \Pr\{Y_i = h\}$ for $0 \le h \le h_{i,\max}$. Let $q_{i,j}$ be the probability that the arcs originated from e_i can reach no more than j next MEs:

$$q_{i,j} = p_{i,0} + p_{i,1} + \ldots + p_{i,j} = \sum_{m=0}^{j} p_{i,m} \tag{7.11}$$

The probability $Q_j^{(i)} = \Pr\{Y_i \le h\}$ that any arc originated from the set σ_i can reach no more than h MEs beyond this set takes the form

$$Q_h^{(i)} = q_{n,h} q_{n-1,h+1} \dots q_{i,n-i+h} = \prod_{j=i}^{n} q_{j,n-j+h} \tag{7.12}$$

The probabilities $p_h^{(i)}$ can be obtained using the following recursive procedure:

$$p_0^{(i)} = Q_0^{(i)} = \prod_{j=i}^{n} q_{j,n-j} = \prod_{j=i}^{n} \sum_{m=0}^{n-j} p_{j,m}$$

$$p_h^{(i)} = \Pr\{Y_i = h\} = \Pr\{Y_i \le h\} - \sum_{j=0}^{h-1} \Pr\{Y_i = j\}$$

$$= Q_h^{(i)} - \sum_{j=0}^{h-1} p_j^{(i)} = \prod_{j=i}^{n} q_{j,n-j+h} - \sum_{j=0}^{h-1} p_j^{(i)}$$

$$= \prod_{j=i}^{n} \sum_{m=0}^{n-j+k} p_{j,m} - \sum_{j=0}^{h-1} p_j^{(i)} \quad \text{for } 1 \le h \le h_{i,\max} \tag{7.13}$$

Consider, now, an equivalent CMCCS in which the set of MEs σ_i is replaced by a single equivalent element with the random state represented by the variable Y_i. According to the definition of system failure, it can be represented as a sum of n disjoint events E_i corresponding to the signal gap between MEs e_{i-1} and e_i (Figure 7.1). Let W_i for $1 \le i \le n$ be an event when the signal from σ_i reaches (directly or indirectly) MEs $e_1,\dots, e_i$ and H_i $2 \le i \le n$ be an event when the signal from σ_i reaches (directly or indirectly) MEs $e_1,\dots, e_{i-1}$. It can be seen that

$$\begin{aligned} &E_1 = \overline{W_1} \\ &H_i = W_i \cup E_i, \ \ W_i \cap E_i = \varnothing \ \text{ for } 2 \le i \le n \end{aligned} \tag{7.14}$$

Therefore:

$$1 - R_C = \sum_{i=1}^{n} \Pr\{E_i\} = 1 - \Pr\{W_1\} + \sum_{i=2}^{n} \Pr\{E_i\}$$

$$= 1 - \Pr\{W_1\} + \sum_{i=2}^{n} (\Pr\{H_i\} - \Pr\{W_i\}) \tag{7.15}$$

Note that, according to the definitions of the events W_i and H_i, $\Pr\{W_i\}$ is equal to the reliability $R_L(L_i^W)$ of LMCCS L_i^W with source ME σ_i, MEs $e_1,\dots, e_{i-1}$ as intermediate elements, and sink right after element i–1, $\Pr\{H_i\}$ is equal to the reliability $R_L(L_i^H)$ of LMCCS L_i^H with source ME σ_i, MEs $e_1,\dots, e_{i-2}$ as intermediate elements, and sink right after element i–2. Therefore:

$$R_C = R_L(L_1^W) - \sum_{i=2}^{n}[R_L(L_i^H) - R_L(L_i^W)] \quad (7.16)$$

7.1.2 Retransmission Delay Model

In the previous section, the only condition for the system's success was its ability to keep its source and sink connected. For communication systems, this means that any signal leaving the source can reach the receiver without respect to the signal transmission time. However, in digital telecommunication systems the retransmission process is usually associated with a certain delay. When this is so, the total time T of the signal propagation from the source to the receiver can vary depending on the combination of states of MEs (retransmitters). Even when the retransmission delays of the individual MEs are constant values, the retransmission delay of the entire system is a random variable because of the random connectivity provided by the MEs. The entire system is considered to be in working condition if T is not greater than a certain specified level w. Otherwise, the network fails. In this section we present an algorithm for the evaluation of the reliability of LMCCSs consisting of MEs with fixed retransmission delays.

7.1.2.1 Evaluating System Reliability and Conditional Expected Delay

Consider a system with a given allocation of MEs. Each ME e_i receiving the signal can retransmit it to further nodes after a fixed delay τ_i. The signal generated by e_i located at some node reaches all of the S_i next nodes immediately, where S_i is a state of e_i. Since the states of each ME are random values, the total time T of the signal propagation from C_1 to C_{n+1} is also a random value with the distribution

$$\Pr\{T = t_k\} = Q_k, \quad \sum_{k=1}^{K} Q_k = 1 \quad (7.17)$$

where K is the total number of system states characterized by different values of T (different combinations of the individual MEs' states can result in the same T). Some combinations of MEs' states lead to the absence of the connection between C_1 and C_{n+1} ($t_k = \infty$ relates to this specific state).

The system reliability $R(w)$ is defined as a probability that the signal generated at C_1 can be delivered to C_{n+1} at a time not greater than w. Having the distribution (7.17) one can obtain this reliability as

$$R(w) = E(1(T \le w)) = \sum_{t_k \le w} Q_k \quad (7.18)$$

The conditional expected time of signal delivery $\tilde{\varepsilon}$, can be obtained as

$$\tilde{\varepsilon} = \sum_{t_k < \infty} t_k Q_k \Big/ \sum_{t_k < \infty} Q_k \quad (7.19)$$

Consider an ME e_i allocated at node C_j. In order to represent the delay distribution of the signal generated by this ME we use the random vector $\boldsymbol{G}_i^{(j)} = \{G_i^{(j)}(1),...,G_i^{(j)}(n+1)\}$ so that $G_i^{(j)}(h)$ represents the random time of the signal arrival to node C_h since it has arrived at C_j. When the ME e_i is in state S_i ($0 \le S_i < k_i$), it provides a signal retransmission from C_j to a set of nodes $\{C_{j+1},...,C_{\alpha(j+Si)}\}$ with delay τ_i. In order to represent the times taken to retransmit the signal entering the node C_j to the rest of nodes when ME e_i is state S_i, we determine vector $\boldsymbol{g}_{iS_i}^{(j)}$ (realization of $\boldsymbol{G}_i^{(j)}$ corresponding to state S_i of e_i) as follows:

$$\boldsymbol{g}_{iS_i}^{(j)}(h) = \begin{cases} \infty, & 1 \le h \le j \\ \tau_i, & j < h \le \alpha(j+S_i) \\ \infty, & h > \alpha(j+S_i) \end{cases} \tag{7.20}$$

(∞ can be represented by any number greater than w).

The u-function

$$u_{ij}(z) = \sum_{k=0}^{k_j-1} p_{ik} z^{\boldsymbol{g}_{ik}^{(j)}} \tag{7.21}$$

represents all of the possible states of the ME e_i located at C_j by relating the probability of each state S_i to the value of a random vector $\boldsymbol{G}_i^{(j)}$ in this state.

The absence of any ME at node C_j implies that no connections exist between C_j and any other node. This means that any signal reaching C_j cannot be retransmitted in this node. In this case, the corresponding u-function takes the form

$$u_{\infty j}(z) = z^{\boldsymbol{G}_\infty}, \quad \text{where } G_\infty(h) = \infty \text{ for } 1 \le h \le n+1. \tag{7.22}$$

Consider two MEs e_i and e_f allocated at the same node C_j. The signal reaching C_j can be retransmitted by e_i to each node h by the time $G_i^{(j)}(h)$. The same signal can be retransmitted by e_f to each node h by the time $G_f^{(j)}(h)$. The time of arrival of the signal retransmitted at node j to any node h can be obtained as $\min\{G_i^{(j)}(h), G_f^{(j)}(h)\}$. Therefore, the delay distribution for the signal generated by the MEs e_i and e_f can be represented by the random vector $\min\{\boldsymbol{G}_i^{(j)}, \boldsymbol{G}_f^{(j)}\}$.

The structure function for a set of m MEs $\{e_{i_1}, ..., e_{i_m}\}$ located at the same node C_j takes the form

$$\boldsymbol{G}^{(j)} = \min\{\boldsymbol{G}_{i_1}^{(j)},...,\boldsymbol{G}_{i_m}^{(j)}) \tag{7.23}$$

The u-function corresponding to the p.m.f. of a random vector $\boldsymbol{G}^{(j)}$ that represents the delays of the signal generated by a subsystem of MEs located at C_j

can be obtained by the composition operator $\underset{\min}{\otimes}$ over the u-functions $u_{i_l j}(z)$ ($1 \le l \le m$).

For any $u_{ij}(z)$:

$$u_{\infty j}(z) \underset{\min}{\otimes} u_{ij}(z) = u_{ij}(z) \tag{7.24}$$

The operator $\underset{\min}{\otimes}$ can be applied recursively to obtain the u-function for an arbitrary group of MEs allocated at C_j.

Assume that a signal generated at C_m reaches C_j with delay $G^{(m)}(j)$. The signal generated (retransmitted) at C_j reaches any node h when the time $G^{(m)}(j)+G^{(j)}(h)$ has passed since the signal has arrived at C_m.

Consider two adjacent nodes C_m and C_{m+1}. After the signal arrives at C_m it can be retransmitted to the rest of the nodes directly with delays determined by the vector $\boldsymbol{G}^{(m)}$ or through C_{m+1} with delays determined by the vector $\boldsymbol{G}^{(m)}+\boldsymbol{G}^{(m+1)}$. If there are different ways of signal propagation to a certain node, which are characterized by different times, then the minimal time determines the signal delay. Therefore, in order to determine the random delay for a signal retransmitted by the two groups of MEs allocated at two adjacent nodes C_m and C_{m+1}, one can use the function $f(\boldsymbol{G}^{(m)},\boldsymbol{G}^{(m+1)})$ over vectors $\boldsymbol{G}^{(m)}$ and $\boldsymbol{G}^{(m+1)}$, where for each individual term $1 \le h \le n$

$$f(G^{(m)},G^{(m+1)})(h) = \min\{G^{(m)}(h), G^{(m)}(m+1) + G^{(m+1)}(h)\} \tag{7.25}$$

Let $\tilde{U}_m(z)$ be the u-function that determines the p.m.f. of a random vector $\tilde{\boldsymbol{G}}^{(m)}$ representing delays of a signal retransmitted by all of the MEs located at C_1, …, C_m (since the signal has arrived at C_1) and $U_{m+1}(z)$ be the u-function representing the p.m.f. of delays for signal retransmitted by all the MEs located at C_{m+1} (since the signal has arrived at C_{m+1}). Using the composition operator $\otimes_f$ with function (7.25) one can obtain $\tilde{U}_{m+1}(z)$ as

$$\tilde{U}_{m+1}(z) = \tilde{U}_m(z) \underset{f}{\otimes} U_{m+1}(z) \tag{7.26}$$

Recursively applying this expression for $m = 1, \ldots, m = n-1$ one obtains the resulting u-function taking the form

$$\tilde{U}_n(z) = \sum_{k=1}^{K} Q_k z^{\tilde{g}_k^{(n)}} \tag{7.27}$$

This u-function relates the probabilities of all of the possible system states Q_k with the times of signal propagation from source to receiver $t_k=\tilde{g}_k^{(n)}(n+1)$ (the last elements of the vectors $\tilde{\boldsymbol{g}}_k^{(n)}$) corresponding to these states.

Defining the acceptability function as $F(\tilde{G}^{(n)}(n+1), w)=1(\tilde{G}^{(n)}(n+1)\leq w)$ one can determine the system reliability as

$$R(w)=\sum_{\tilde{g}_k^{(n)}(n+1)\leq w} Q_k \tag{7.28}$$

and the conditional expected delay as

$$\tilde{\varepsilon}=\sum_{\tilde{g}_k^{(n)}(n+1)<\infty} \tilde{g}_k^{(n)}(n+1)Q_k \Big/ \sum_{\tilde{g}_k^{(n)}(n+1)<\infty} Q_k \tag{7.29}$$

7.1.2.2 Simplification Technique

When the u-function $\tilde{U}_m(z)$ is obtained, the values $\tilde{G}^{(m)}(1),\ldots,\tilde{G}^{(m)}(m)$ of the random vector $\tilde{\boldsymbol{G}}^{(m)}$ are no longer used for determining $\tilde{U}_{m+1}(z)$ (and all of the $\tilde{U}_s(z)$ for $s>m$) according to (7.25). Indeed, when determining $\tilde{U}_{m+1}(z)$ one needs to know only the probabilities that the signal reaches nodes $C_{m+1},\ldots,C_n$ and the corresponding delays. It does not matter through which paths the signal reaches these nodes. For example, if in different system states the signal reaches C_{m+1} through a number of different combinations of paths (represented by different terms in $\tilde{U}_m(z)$) resulting in the same delay, then one does not have to distinguish among these combinations. The only thing one has to know is the sum of the probabilities of the states in which combinations of paths with the given minimal delay exist. This means that one can reduce the number of terms in $\tilde{U}_m(z)$ by replacing all of the values $\tilde{g}_k^{(m)}(1),\ldots,\tilde{g}_k^{(m)}(m)$ in vectors $\tilde{\boldsymbol{g}}_k^{(m)}$ of the u-function with ∞ symbols and collecting the like terms.

If, in some term of the u-function $\tilde{U}_m(z)$ $\tilde{G}^{(m)}(m+1)=\ldots=\tilde{G}^{(m)}(n+1)=\infty$, the signal cannot reach any node from C_{m+1} to C_{n+1} independently of the states of MEs located in these nodes, then this state does not contribute to the signal propagation to C_{n+1} and the corresponding term can be removed from the u-function $\tilde{U}_m(z)$.

Taking into account the above-mentioned considerations, one can drastically simplify the u-functions $\tilde{U}_m(z)$ for $1\leq m\leq n$ using the following operator $\varphi(\tilde{U}_m(z))$ which

- assigns ∞ to $\tilde{g}_k{}^{(m)}(1),\ldots,\tilde{g}_k{}^{(m)}(m)$ in each term of $\tilde{U}_m(z)$;

- removes all of the terms in which vectors $\tilde{\boldsymbol{g}}_k^{(m)}$ contain only ∞ symbols;
- collects the like terms in the resulting u-function.

In order to determine the system reliability and the conditional expected delay, one can apply the same four-step procedure described in Section 7.1.1 by assigning in the first step $U_j(z) = u_{\infty j}(z)$ and substituting the composition and simplification operators with those defined for systems with delays.

Example 7.2

Consider an LMCCS with four nodes and three MEs. The delays of the MEs are τ_1, τ_2 and τ_3 respectively. The maximal allowable system delay is w. The first ME is allocated at C_1 (C_1 receives a signal from a fully reliable source), the second and third MEs are allocated at C_3. Each ME can provide a connection with the further allocated nodes with the following given probabilities: $\Pr\{S_1 = k\} = p_{1k}$ for $0 \le k \le 4$, $\Pr\{S_2 = k\} = p_{2k}$ and $\Pr\{S_3 = k\} = p_{3k}$ for $0 \le k \le 1$, $\Pr\{S_2 > 1\} = \Pr\{S_3 > 1\} = 0$.

According to (7.20) and (7.21), the u-functions of individual MEs are

$$u_{11}(z) = p_{10}z^{(*,*,*,*)} + p_{11}z^{(*,\tau_1,*,*)} + p_{12}z^{(*,\tau_1,\tau_1,*)} + p_{13}z^{(*,\tau_1,\tau_1,\tau_1)},$$

$$u_{23}(z) = p_{20}z^{(*,*,*,*)} + p_{21}z^{(*,*,*,\tau_2)},\ u_{33}(z) = p_{30}z^{(*,*,*,*)} + p_{31}z^{(*,*,*,\tau_3)}.$$

For $1 \le j \le 3$, $u_{\infty j}(z) = z^{(*,*,*,*)}$. (In this example, symbol * stands for ∞.)

Using the operator $\otimes_{\min}$ we obtain u-functions for groups of MEs located in each node:

$$U_1(z) = u_{\infty 1}(z) \underset{\min}{\otimes} u_{11}(z) = u_{11}(z);\ U_2(z) = u_{\infty 2}(z) = z^{(*,*,*,*)}$$

$$U_3(z) = u_{\infty 3}(z) \underset{\min}{\otimes} u_{23}(z) \underset{\min}{\otimes} u_{33}(z) = p_{20}p_{30}z^{(*,*,*,*)} + p_{21}p_{30}z^{(*,*,*,\tau_2)}$$

$$+ p_{20}p_{31}z^{(*,*,*,\tau_3)} + p_{21}p_{31}z^{(*,*,*,\min\{\tau_2,\tau_3\})}$$

Following the consecutive procedure (7.26), we obtain

$$\tilde{U}_1(z) = U_1(z),\ \varphi(\tilde{U}_1(z)) = p_{11}z^{(*,\tau_1,*,*)} + p_{12}z^{(*,\tau_1,\tau_1,*)} + p_{13}z^{(*,\tau_1,\tau_1,\tau_1)}$$

$$\tilde{U}_2(z) = \varphi(\tilde{U}_1(z)) \underset{f}{\otimes} U_2(z) = p_{11}z^{(*,\tau_1,*,*)} + p_{12}z^{(*,\tau_1,\tau_1,*)} + p_{13}z^{(*,\tau_1,\tau_1,\tau_1)}$$

$$\varphi(\tilde{U}_2(z)) = p_{12}z^{(*,*,\tau_1,*)} + p_{13}z^{(*,*,\tau_1,\tau_1)}$$

$$\tilde{U}_3(z) = \varphi(\tilde{U}_2(z)) \underset{f}{\otimes} U_3(z) = p_{12}(p_{20}p_{30}z^{(*,*,\tau_1,*)} + p_{21}p_{30}z^{(*,*,\tau_1,\tau_1+\tau_2)}$$

$$+ p_{20}p_{31}z^{(*,*,\tau_1,\tau_1+\tau_3)} + p_{21}p_{31}z^{(*,*,\tau_1,\tau_1+\min\{\tau_2,\tau_3\})}) + p_{13}(p_{20}p_{30}z^{(*,*,\tau_1,\tau_1)}$$

$$+p_{21}p_{30}z^{(*,*,\tau_1,\tau_1)}+p_{20}p_{31}z^{(*,*,\tau_1,\tau_1)}+p_{21}p_{31}z^{(*,*,\tau_1,\tau_1)})$$

The operator φ applied to $\tilde{U}_3(z)$ first replaces the third element of each vector with *, then removes the term $p_{12}p_{20}p_{30}z^{(*,*,*,*)}$ and collects the like terms as follows:

$$\varphi(\tilde{U}_3(z)) = p_{13}(p_{20}p_{30}+p_{21}p_{30}+p_{20}p_{31}+p_{21}p_{31})z^{(*,*,*,\tau_1)}$$

$$+ p_{12}p_{21}p_{30}z^{(*,*,*,\tau_1+\tau_2)}+p_{12}p_{20}p_{31}z^{(*,*,*,\tau_1+\tau_3)}$$

$$+ p_{12}p_{21}p_{31}z^{(*,*,*,\tau_1+\min\{\tau_2,\tau_3\})} = p_{13}z^{(*,*,*,\tau_1)} + xz^{(*,*,*,\tau_1+\tau_2)} + yz^{(*,*,*,\tau_1+\tau_3)}$$

where

$$x = p_{12}p_{21},\ y = p_{12}p_{20}p_{31} \text{ if } \tau_3>\tau_2$$

$$x = p_{12}p_{21}p_{30},\ y = p_{12}p_{31} \text{ if } \tau_2>\tau_3$$

Now using (7.28) we obtain

$$R(w) = 0 \text{ if } w<\tau_1$$

$$R(w) = p_{13} \text{ if } \tau_1\le w<\tau_1+\min\{\tau_2, \tau_3\}$$

$$R(w) = p_{13}+p_{12}p_{31} \text{ if } \tau_1+\tau_3\le w<\tau_1+\tau_2,\ (\tau_2>\tau_3)$$

$$R(w) = p_{13}+p_{12}p_{21} \text{ if } \tau_1+\tau_2\le w<\tau_1+\tau_3,\ (\tau_3>\tau_2)$$

$$R(w) = p_{13}+p_{12}(p_{21}+p_{20}p_{31}) \text{ if } \tau_1+\max\{\tau_2, \tau_3\}\le w$$

and using (7.29) we obtain

$$\tilde{\varepsilon} = \tau_1+(p_{12}p_{21}\tau_2+p_{12}p_{20}p_{31}\tau_3)/[p_{13}+p_{12}(p_{21}+p_{20}p_{31})] \text{ if } \tau_3>\tau_2$$

$$\tilde{\varepsilon} = \tau_1+(p_{12}p_{31}\tau_3+p_{12}p_{21}p_{30}\tau_2)/[p_{13}+p_{12}(p_{21}+p_{20}p_{31})] \text{ if } \tau_2>\tau_3$$

Note that the operator φ reduces the number of terms in $\tilde{U}_1(z)$ from four to three, in $\tilde{U}_2(z)$ from three to two, and in $\tilde{U}_3(z)$ from eight to three.

7.1.3 Optimal Element Allocation in a Linear Multi-state Consecutively Connected System

The problem of optimal ME allocation in an LMCCS was first formulated by Malinowski and Preuss in [189]. In this problem, m MEs should be allocated in nodes C_1, ..., C_n in such a way that maximizes the LMCCS reliability. In [189] only the systems with $m = n$ were considered in which only one ME is located in each node.

In many cases, even for $m = n$, greater reliability can be achieved if some of the MEs are gathered in the same node, thus providing redundancy, rather than if all the MEs are evenly distributed between all the nodes.

Example 7.3

Consider the simplest case in which two identical MEs should be allocated within an LMCCS with $n = m = 2$. Each ME has three states: state 0 (total failure), state 1 in which the ME is able to connect the node in which it is located with the next node, and state 2 in which the ME is able to connect node it is located in with the next two nodes. The probabilities of being in each state do not depend on the ME's allocation and are p_0, p_1 and p_2 respectively. The LMCCS succeeds if C_1 is connected with C_3 (connectivity model). There are two possible allocations of the MEs within the LMCCS (Figure 7.2A):

I. Both MEs are located in the first node.

II. The MEs are located in the first and second nodes.

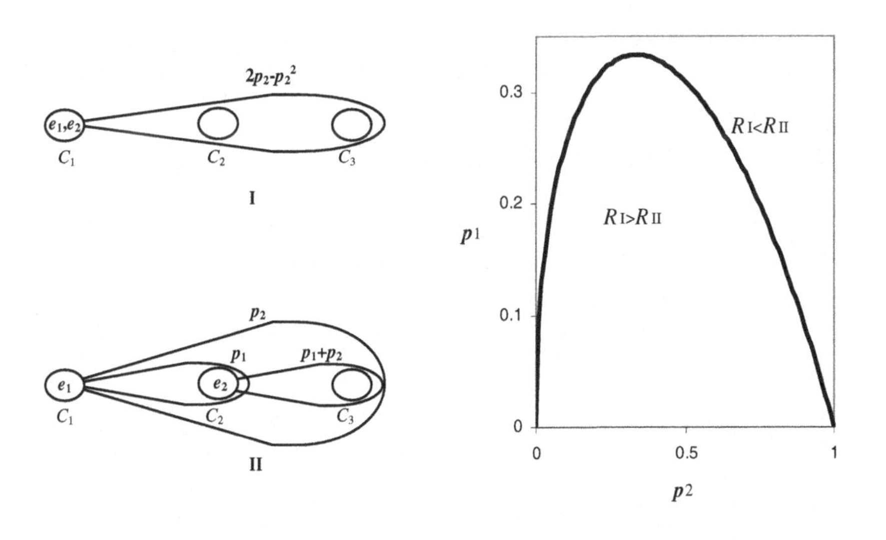

Figure 7.2. Two possible allocations of MEs in LMCCS with $n = m = 2$

In case I, the LMCCS succeeds if at least one of the MEs is in state 2 and the system reliability is

$$R_{\mathrm{I}} = 2p_2 - p_2^2 \tag{7.30}$$

In case II, the LMCCS succeeds either when the ME located in the first node is in state 2 or if it is in state 1 and the second element is not in state 0. The system reliability in this case is

$$R_{\mathrm{II}} = p_2 + p_1(p_1 + p_2) \tag{7.31}$$

By comparing (7.30) and (7.31), one can decide which allocation of the elements is preferable for any given p_1 and p_2. Figure 7.2B presents the decision curve $R_{\mathrm{I}} = R_{\mathrm{II}}$ on the plane (p_1, p_2). Solution I is preferable for combinations of p_1 and p_2 located below the curve, whereas solution I provides lower system reliability than solution II for combinations of p_1 and p_2 located above the curve.

In a general optimal allocation problem the LMCCS consists of n+1 consecutively ordered nodes. At each one of the first n nodes any group of MEs from a given set of m different MEs can be allocated. The allocation problem can be considered as a problem of partitioning a set of m items into a collection of n mutually disjoint subsets. This partition can be represented by the integer string $\boldsymbol{a}$ = (a_j: $1 \le j \le m$), $1 \le a_j \le n$, where a_j is the number of the node at which ME j is located.

7.1.3.1 Connectivity Model

The LMCCS connectivity model defines the system's reliability as the probability that the path from C_1 to C_{n+1} exists.This probability should be maximized. For any given integer string $\boldsymbol{a}$, the GA determines the solution fitness (equal to the LMCCS's reliability) using the following procedure:

1. Define the u-functions $u_j(z)$ of individual MEs e_j ($1 \le j \le m$) in accordance with their state distribution.
2. Assign $U_i(z) = u_{0i}(z) = z^i$ for each $1 \le i \le n$ corresponding to LMCCS nodes.
3. According to the given string $\boldsymbol{a}$, for each j = 1, …, m modify $U_{a_j}(z)$ as follows: $U_{a_j}(z) = U_{a_j}(z) \underset{\max}{\otimes} u_j(z)$.
4. Assign $\tilde{U}_1(z) = U_1(z)$ and apply in sequence Equation (7.9) for j = 1, …, n−1.
5. Obtain the coefficient of the resulting single term u-function $\varphi(\tilde{U}_n(z))$ as the system reliability.

Example 7.4

Consider the ME allocation problem presented in [189], in which $n = m = 13$ and reliability characteristics of MEs do not depend on their allocation. All the MEs have two states with nonzero probabilities. The state probability distributions of the MEs are presented in Table 7.1.

First, we solve the allocation problem when allocation of no more than one ME is allowed at each node. This is done by imposing a penalty on the solutions in which more than one ME is allocated in the same node. The best solution that provides the system reliability $R = 0.592$ is presented in Table 7.2. The LMCCS reliability considerably improves when all of the limitations on the ME allocation are removed. The best solution obtained by the GA in which only 5 out of 13 nodes are occupied by the MEs provides the system reliability $R = 0.723$ [190] (see Table 7.2).

Table 7.1. State distributions of MEs

ME	No. of state				
	0	1	2	3	4
e_1	0.70	0.00	0.00	0.30	0.00
e_2	0.65	0.00	0.00	0.00	0.35
e_3	0.60	0.00	0.00	0.40	0.00
e_4	0.55	0.00	0.45	0.00	0.00
e_5	0.50	0.50	0.00	0.00	0.00
e_6	0.45	0.55	0.00	0.00	0.00
e_7	0.40	0.00	0.00	0.00	0.60
e_8	0.35	0.00	0.00	0.65	0.00
e_9	0.30	0.00	0.70	0.00	0.00
e_{10}	0.25	0.75	0.00	0.00	0.00
e_{11}	0.20	0.00	0.00	0.80	0.00
e_{12}	0.15	0.00	0.85	0.00	0.00
e_{13}	0.10	0.00	0.00	0.00	0.90

Table 7.2. Solutions of the ME allocation problem

Node	MEs	
	Even allocation	Uneven allocation
C_1	13	4,9,12
C_2	6	-
C_3	5	13
C_4	1	-
C_5	12	-
C_6	3	-
C_7	11	1,3,11
C_8	2	-
C_9	8	-
C_{10}	7	2,7,8
C_{11}	4	-
C_{12}	9	-
C_{13}	10	5,6,10
R	0.59201	0.72319

7.1.3.2 Retransmission Delay Model

For the LMCCS retransmission delay model one can consider two possible formulations of the optimal ME allocation problem.

1. Find allocation $\boldsymbol{a}$ maximizing the LMCCS reliability for the given w:

$$R(w, \boldsymbol{a}) \rightarrow \max \quad (7.32)$$

2. Find allocation $\boldsymbol{a}$ minimizing the conditional expected delay while providing desired LMCCS reliability R^* for the given w:

$$\tilde{\varepsilon}\,(\mathbf{a}) \rightarrow \min \text{ subject to } R(w, \boldsymbol{a}) \geq R^* \quad (7.33)$$

The same solution representation that was used for the connectivity model can be used for the retransmission delay model in the GA. The reliability and conditional expected delay can be evaluated for any ME allocation solution using the procedure described in Section 7.1.2. The solution fitness can be defined as

$$M - \pi_1 \tilde{\varepsilon}(\boldsymbol{a}) - \pi_2 (1 + R^* - R(w, \boldsymbol{a})) \times 1(R(w, \boldsymbol{a}) < R^*) \quad (7.34)$$

where π_1 and π_2 are the penalty coefficients and M is a constant value. Note that $\pi_1 = 0$ and $R^* = 1$ correspond to formulation (7.32), while $\pi_1 = 1$ and $0<R^*<1$ correspond to formulation (7.33).

Example 7.5

Consider an LMCCS in which $n = m = 8$ and reliability characteristics of MEs do not depend on their allocation [191]. The state probability distributions of the MEs are presented in Table 7.3, as well as delay times for each ME. This example corresponds to the problem of optimal allocation of eight retransmitters with different technical characteristics among eight potential nodes with identical signal propagation conditions.

The best ME allocation obtained for the first formulation of the problem for $w = 3$ (solution A) is presented in Figure 7.3A (in each figure, MEs are presented in their maximal possible state corresponding to the maximal signal span). The probability that the system delay is not greater than 3 for solution A is $R(3) = 0.808$ and conditional expected delay is $\tilde{\varepsilon} = 2.755$. Note that in this solution some nodes remain empty while others contain more then one ME. To compare this solution with the best possible even ME allocation, the solution was obtained for the constrained allocation problem in which allocation of no more than one ME in each node is allowed (solution B). This solution is presented in Figure 7.3B. One can see that the reliability of the even allocation solution $R(3) = 0.753$ is smaller than the free allocation solution. Observe that while solution A has greater reliability than solution B, the latter has a lower conditional expected delay of $\tilde{\varepsilon} =$ 2.657.

The solution of formulation 2 of the allocation problem for $R^* = 0.75$ and $w = 3$ is presented in Figure 7.3C (solution C). In this solution, $\tilde{\varepsilon} = 2.368$ and $R(3) = 0.751>R^*$. The even ME allocation solution obtained for formulation 2 coincides with solution B.

The LMCCS reliability as a function of the maximal allowable delay is presented in Figure 7.4 for all of the solutions obtained. One can see that not only the reliability of LMCCS for a given w depends on ME allocation, but also the minimal possible delay (minimal w for which $R(w)>0$) and the maximal finite delay (maximal w for which $R(w)<R(\infty)$).

The example presented shows that the proper allocation of retransmitters can improve the reliability of the radio relay system without any additional investment in retransmitter reliability improvement.

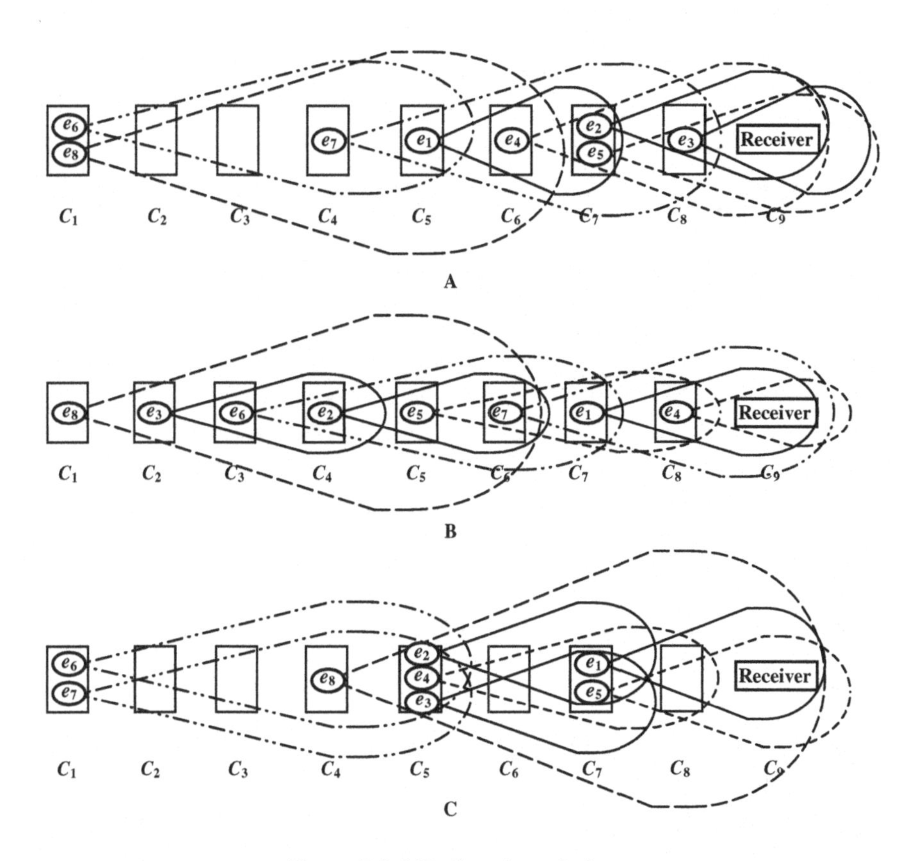

Figure 7.3. ME allocation solutions

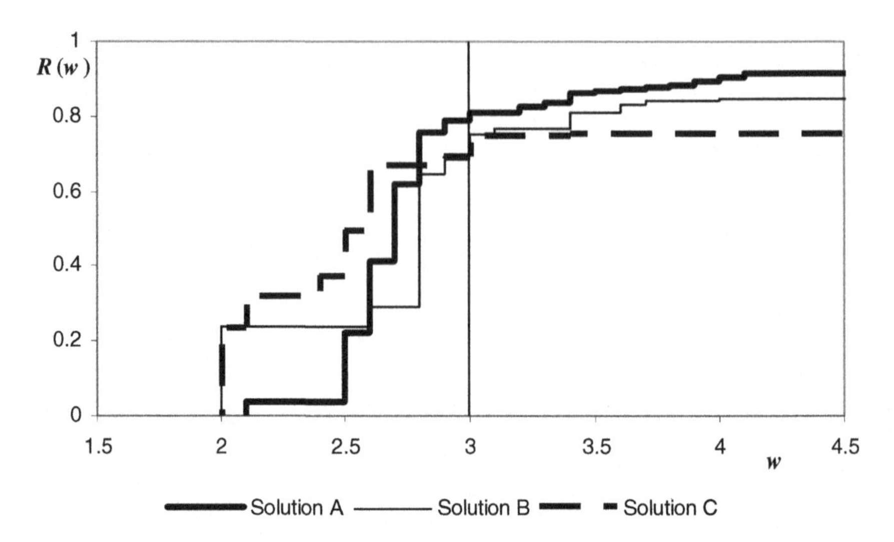

Figure 7.4. LMCCS reliability as function of maximal allowable delay

7.1.3.3 Optimal Element Allocation in the Presence of Common Cause Failures

In many cases, when LMCCS elements are subject to CCFs, the system can be considered as consisting of mutually disjoint common cause groups with total CCFs. Usually, the CCFs occur when a group of elements are allocated at the same node. The vulnerability of each node strongly affects the optimal allocation of MEs.

Example 7.6

Consider the simplest LMCCS model from Example 7.3 and assume that each node of this system (with all the MEs it contains) can be destroyed with probability v.

In case I, the LMCCS survives if node C_1 is not destroyed and at least one of the MEs is in state 2. The system survivability is

$$S_{\mathrm{I}} = s(2p_2 - p_2^2) \tag{7.35}$$

where $s = 1-v$ is the node survivability.

In case II, the LMCCS survives either when the node C_1 is not destroyed and the ME located in C_1 is in state 2, or, if both nodes C_1 and C_2 are not destroyed, the ME located in C_1 is in state 1 and the ME located in C_2 is not in state 0. The system survivability in this case is

$$S_{\mathrm{II}} = sp_2 + s^2 p_1(p_1 + p_2) \tag{7.36}$$

By comparing (7.35) and (7.36), one can decide which allocation of the elements is preferable for any given v, p_1 and p_2Condition $S_{\mathrm{I}} \geq S_{\mathrm{II}}$ can be rewritten as

$$s(2p_2-p_2^2)\geq sp_2+s^2p_1(p_1+p_2) \tag{7.37}$$

and finally as

$$s\leq p_2(1-p_2)/(p_1(p_1+p_2)) \tag{7.38}$$

Figure 7.5 presents the maximal values of s for which $S_I \geq S_{II}$ as a function of variables p_1 and p_2. For given combinations of p_1 and p_2, the values of s located below the curve correspond to cases when the solution I is preferable; the values of s located above the curve correspond to cases when solution I provides lower system survivability than solution II.

The end point of each curve $s(p_2)$ belongs to line $s = p_2$. Indeed, since $p_0+p_1+p_2 = 1$, the maximal possible value of p_1 for each given p_2 is $1-p_2$. Substituting p_1 with $1-p_2$ in Equation (7.38) one obtains $s\leq p_2$.

When several MEs are gathered within a single node C_j that has vulnerability v, this means that all of these elements can be destroyed with probability v. Let the p.m.f. of the number of the most remote node to which an arc from C_j exists be represented by the u-function $U_j(z)$ obtained by Equation (7.5). In order to incorporate the node vulnerability into the model, the probabilities q_{jh} in Equation (7.5) should be considered as conditional probabilities that the node C_j is connected with the set of nodes $C_{j+1},\dots, C_{j+h}$ under the assumption that the node C_j is not destroyed. The unconditional probability that node C_j is connected with nodes $C_{j+1},\dots, C_{j+h}$ is, therefore, equal to $(1-v)q_{jh} = sq_{jh}$. If the node C_j is destroyed, then its MEs cannot provide connection with any other node (this state corresponds to term z^j in Equation (7.5)).

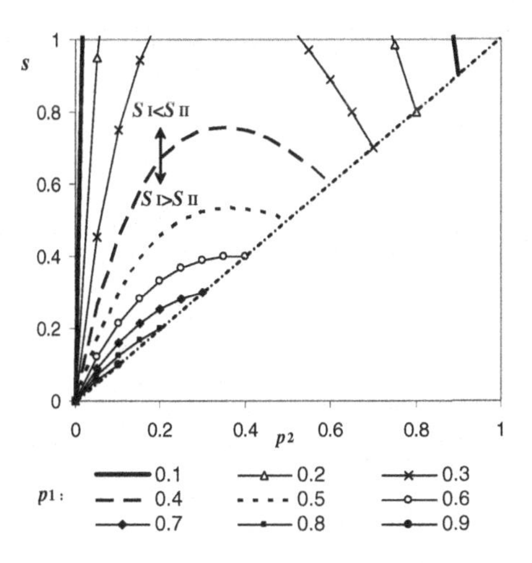

Figure 7.5. Comparison of possible ME allocations in an LMCCS with vulnerable nodes

Therefore, the u-function of MEs located at vulnerable node C_j can be obtained using the following operator over the u-function $U_j(z)$:

$$\xi(U_j(z)) = sU_j(z) + vz^j = s\sum_{h=j+1}^{\alpha(j+k-1)} q_{jh}z^h + (v + sq_{jj})z^j \qquad (7.39)$$

The survivability of an LMCCS with vulnerable nodes can be obtained using the algorithm presented in Section 7.1.1, in which after step 2 the operator (7.39) is applied over u-functions $U_1(z), \ldots, U_n(z)$.

Example 7.7

In order to illustrate the procedure, consider the LMCCS from Example 7.3 (case I). The u-functions of the individual MEs e_1 and e_2 located at C_1 are respectively

$$u_{11}(z) = p_{10}z^1 + p_{11}z^2 + p_{12}z^3 \text{ and } u_{21}(z) = p_{20}z^1 + p_{21}z^2 + p_{22}z^3$$

The second node is empty, which corresponds to the u-function $u_{02}(z) = z^2$.

The u-functions for groups of MEs located at nodes 1 and 2 are

$$U_1(z) = u_{11}(z) \underset{\max}{\otimes} u_{21}(z) = p_{10}p_{20}z^1 + [p_{10}p_{21} + p_{11}(1-p_{22})]z^2$$

$$+(p_{12}+p_{22}-p_{12}p_{22})z^3$$

$$\xi(U_1(z)) = (v+sp_{10}p_{20})z^1 + s[p_{10}p_{21}+p_{11}(1-p_{22})]z^2 + s(p_{12}+p_{22}-p_{12}p_{22})z^3$$

$$U_2(z) = u_{02}(z),\ \xi(U_2(z)) = \xi(z^2) = (v+s)z^2 = z^2$$

The u-function representing the p.m.f. of $\tilde{G}^{(1)}$ is

$$\tilde{U}_1(z) = \xi(U_1(z))$$

$$\varphi(\tilde{U}_1(z)) = s[p_{10}p_{21}+p_{11}(1-p_{22})]z^2 + s(p_{12}+p_{22}-p_{12}p_{22})z^3$$

The u-function representing the p.m.f. of $\tilde{G}^{(2)}$ is:

$$\tilde{U}_2(z) = \varphi(\tilde{U}_1(z)) \underset{\max}{\otimes} \xi(U_2(z)) = s[p_{10}p_{21}+p_{11}(1-p_{22})]z^2$$

$$+s[p_{12}+p_{22}-p_{12}p_{22}]z^3,\ \varphi(\tilde{U}_2(z)) = s[p_{12}+p_{22}-p_{12}p_{22}]z^3$$

Finally, the LMCCS survivability obtained from $\varphi(\tilde{U}_2(z))$ is

$$S_{\mathrm{I}} = s[p_{12}+p_{22}-p_{12}p_{22}]$$

If $p_{12} = p_{22} = p_2$, then $S_{\mathrm{I}} = s(2p_2 - p_2^2)$.

The definition of the optimal ME allocation problem for an LMCCS with vulnerable nodes and the optimization technique do not differ from those suggested for an LMCCS without CCF (see Section 7.1.3.1).

Example 7.8

Consider the LMCCS with $n = m = 13$ from Example 7.4 and assume that each node has the same vulnerability v. The ME allocation solutions obtained in [192] for $v = 0$, $v = 0.06$, and $v = 0.3$ are presented in Table 7.3.

Each solution provides the greatest possible system survivability only for the given value of v. The optimal ME allocation changes when v varies. One can see the survivability of the system with ME allocations obtained for $v = 0$, $v = 0.06$, and $v = 0.3$ as a function of the variable node vulnerability in Figure 7.6.

The greater the node vulnerability, the greater the number of occupied nodes in the optimal solution. Indeed, by increasing the ME separation, the system tries to compensate its increasing vulnerability.

Table 7.3. ME allocation solutions for an LMCCS with vulnerable nodes

	MEs		
Node	$v = 0$	$v = 0.06$	$v = 0.3$
C_1	4,9,12	13	13
C_2	-	-	-
C_3	13	-	-
C_4	-	-	2
C_5	-	1,2,11	11
C_6	-	-	-
C_7	1,3,11	-	3,4
C_8	-	5,6,12	1,6,12
C_9	-	4,9	5,9
C_{10}	2,7,8	7,10	7,10
C_{11}	-	3,8	8
C_{12}	-	-	-
C_{13}	5,6,10	-	-
S when $v = 0$	0.723	0.708	0.648
S when $v = 0.06$	0.559	0.572	0.541
S when $v = 0.3$	0.164	0.193	0.207

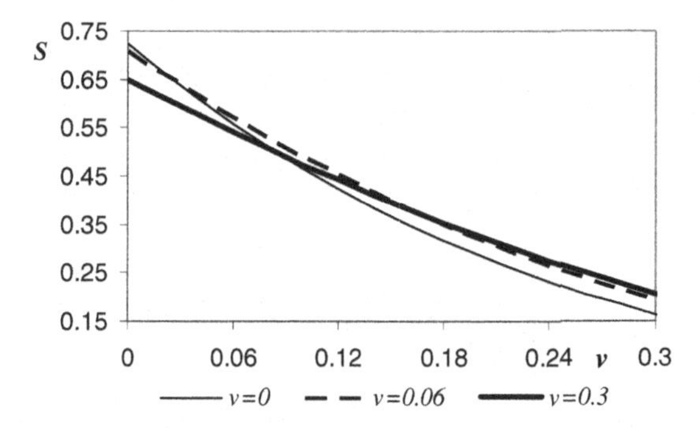

Figure 7.6. LMCCS survivability as a function of node vulnerability v

7.2 Multi-state Networks

In LMCCSs, when a connection between C_i and C_j is available, all connections between C_i and C_m for $i<m<j$ are also available. By removing this constraint, one obtains a single-source-single-sink multi-state acyclic network (MAN), which is a generalization of an LMCCS. In the MAN, all the nodes are ordered and each node C_i (if it is not a sink) is connected by links with a set of nodes Θ_i, which have numbers greater than i. It can be seen that the MANs have no cycles. By allowing the network to have several sinks (terminal nodes), one obtains the most general definition of the MAN.

7.2.1 Connectivity Model

A MAN can be represented by an acyclic directed graph with n nodes C_i ($1\le i\le n$), $n*$ of which compose a set Π of sinks (leaf nodes). The nodes are numbered in such a way that for any arc (C_j, C_e) $e>j$ and last $n*$ numbers are assigned to the leaf nodes: $\Pi = \{C_{n-n*+1},\dots, C_n\}$ (such numbering is always possible in anacyclic directed graph). The existence of arc (C_j, C_e) means that a signal or flow can pass directly from node C_j to node C_e. One can define for each nonleaf node C_j a set of nodes Θ_j directly following C_j: $C_h\in\Theta_j$ if (C_j, C_h) exists (see Figure 7.7).

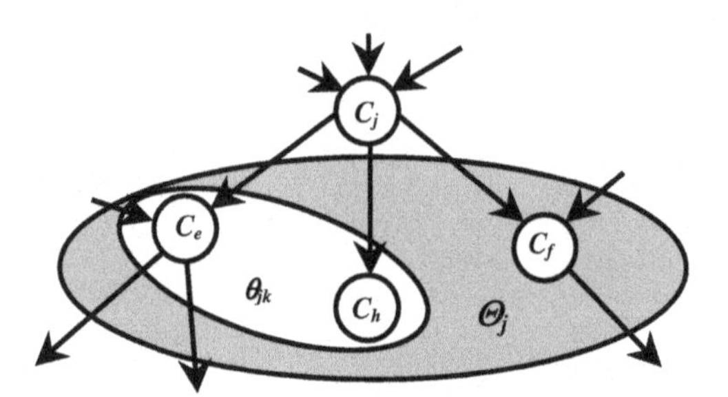

Figure 7.7. Fragment of a MAN

The ME located in each nonleaf node C_j ($1\le j\le n-n*$) provides a connection between C_j and the nodes belonging to the set Θ_j. In each state k, this ME provides a connection to some subset θ_{jk} of Θ_j. (In the case of total failure, the MEs cannot provide connection with any node: $\theta_{jk} = \varnothing$; in the case of a fully operational state: $\theta_{jk} = \Theta_j$). Each ME e_i located at C_j can have k_i different states and each state k has probability $p_{i\theta_{jk}}$, such that $\sum_{k=1}^{k_i} p_{i\theta_{jk}} = 1$. The states of all MEs are independent.

A signal or flow can be retransmitted by the ME located at C_i only if it reaches this node. The MAN reliability R_Π is defined as a probability that a signal generated at the root node C_1 reaches all the $n*$ leaf nodes $C_{n-n*+1},\dots,C_n$. (In some cases the MAN reliability R_π is defined as the probability that a signal reaches a subset π of the set of leaf nodes Π).

Consider an ME e_i located at node C_j. In each state k ($0 \le k < k_i$), the ME provides a signal transmission from C_j to a set of nodes θ_{jk}. In order to represent the random set of MAN nodes that receive a signal from e_i, we determine random vector $\boldsymbol{G}_i^{(j)} = \{G_i^{(j)}(1), \ldots, G_i^{(j)}(n)\}$, so that $G_i^{(j)}(h) = 1$ if the signal from e_i can reach C_h and $G_i^{(j)}(h) = 0$ otherwise. The value of the vector $\boldsymbol{G}_i^{(j)}$ in each state k of e_i takes the form [96]:

$$g_{ik}^{(j)}(h) = \begin{cases} 1, & C_h \in \theta_{jk} \\ 0, & C_h \notin \theta_{jk} \end{cases} \tag{7.40}$$

The u-function

$$u_{ij}(z) = \sum_{k=0}^{k_i-1} p_{i\theta_{jk}} z^{\boldsymbol{g}_{ik}^{(j)}} \tag{7.41}$$

represents the p.m.f. of random vector $\boldsymbol{G}_i^{(j)}$. When C_j is empty, the signal cannot be retransmitted from C_j to any other node. This situation is represented by the u-function $u_{0j}(z) = z^{\boldsymbol{G}_0}$, where $\boldsymbol{G}_0$ is the zero vector.

Consider a set of m MEs $\{e_{i_1}, \ldots, e_{i_m}\}$ located at the same node C_j. The signal that reached C_j can be retransmitted to node C_h if at least one of the MEs provides a connection with C_h. This corresponds to the existence of $G_{i_h}^{(j)}(h) = 1$ for at least one ME e_{i_b} ($1 \le b \le m$). Therefore, the structure function for a set of m MEs $\{e_{i_1}, \ldots, e_{i_m}\}$ located at the same node C_j takes the form

$$\boldsymbol{G}^{(j)} = \max(\boldsymbol{G}_{i_1}^{(j)}, \ldots, \boldsymbol{G}_{i_m}^{(j)}) \tag{7.42}$$

The u-function, representing the p.m.f. of random vector $\boldsymbol{G}^{(j)}$ (corresponding to nodes connected to C_j by all the MEs located at C_j), can be obtained by the composition operator $\underset{\max}{\otimes}$ over u-functions $u_{i_b j}(z)$ ($1 \le b \le m$). For any $u_{ij}(z)$:

$$u_{0j}(z) \underset{\max}{\otimes} u_{ij}(z) = u_{ij}(z) \tag{7.43}$$

Let $\tilde{\boldsymbol{G}}^{(h)}$ represent the set of nodes connected to C_1 by MEs located at C_1, C_2, …, C_h. Assume that MEs located at the first h nodes provide connection with C_{h+1} (which corresponds to $\tilde{G}^{(h)}(h+1) = 1$). If MEs located at C_{h+1} provide connection between C_{h+1} and arbitrary node C_s (which corresponds to $\tilde{G}^{(h+1)}(s) = 1$), the signal generated at C_1 reaches C_s. Therefore, in this case, the set of nodes connected to C_1 by MEs located at the first h+1 nodes is the union of sets represented by $\tilde{\boldsymbol{G}}^{(h)}$ and $\boldsymbol{G}^{(h+1)}$. This can be expressed as follows:

$$\tilde{\boldsymbol{G}}^{(h+1)} = \max\{\tilde{\boldsymbol{G}}^{(h)},\ \boldsymbol{G}^{(h+1)}\} \tag{7.44}$$

If a signal generated at C_1 does not reach C_{h+1} through the first h nodes (which corresponds to $\tilde{G}^{(h)}(h+1) = 0$), the MEs located at C_{h+1} cannot transmit the signal in any of its states. Therefore, these MEs do not affect the state of the MAN. The set of nodes receiving the signal remains one that is represented by the vector $\tilde{\boldsymbol{G}}^{(h)}$. In the general case, the following function ω can be used in order to determine the random vector $\tilde{\boldsymbol{G}}^{(h+1)}$:

$$\tilde{\boldsymbol{G}}^{(h+1)} = \omega(\tilde{\boldsymbol{G}}^{(h)}, \boldsymbol{G}^{(h+1)}) = \begin{cases} \tilde{\boldsymbol{G}}^{(h)}, & \tilde{G}^{(h)}(h+1) = 0 \\ \max\{\tilde{\boldsymbol{G}}^{(h)}, \boldsymbol{G}^{(h+1)}\}, & \tilde{G}^{(h)}(h+1) = 1 \end{cases} \tag{7.45}$$

Recursively applying the operator $\underset{\omega}{\otimes}$ over u-functions $\tilde{U}_h(z)$ representing the p.m.f. of random vectors $\tilde{\boldsymbol{G}}^{(h)}$:

$$\tilde{U}_{h+1}(z) = \tilde{U}_h(z) \underset{\omega}{\otimes} U_{h+1}(z) \text{ for } h = 1,\dots,n-n^*-1 \tag{7.46}$$

where $\tilde{U}_1(z) = U_1(z)$ one finally obtains the u-function representing the distribution of the MAN states when all the MEs are considered (as the p.m.f. of random vector $\tilde{\boldsymbol{G}}^{(n-n^*)}$).

One can obtain the MAN reliability (the probability that the signal reaches any subset $\pi \subset \Pi$ of leaf nodes) by defining the acceptability function

$$F_\pi(\tilde{\boldsymbol{G}}^{(n-n^*)}) = \begin{cases} 1, & \text{if } \tilde{G}^{(n-n^*)}(j) = 1 \text{ for any } j: C_j \in \pi \\ 0, & \text{otherwise} \end{cases} \tag{7.47}$$

and evaluating

$$R_\pi = E(F_\pi(\tilde{\boldsymbol{G}}^{(n-n^*)})) = \delta(\tilde{U}_{n-n^*}(z)) \tag{7.48}$$

where operator δ produces the sum of the coefficients of those terms in $\tilde{U}_{n-n^*}(z)$ that have exponents satisfying the condition $F(\tilde{\boldsymbol{G}}^{(n-n^*)}) = 1$.

Using the same considerations as in Section 7.1.2.2, one can define a u-function simplification operator φ that

- zeros $\tilde{g}_k^{(h)}(1), \ldots, \tilde{g}_k^{(h)}(h)$ in each term k of $\tilde{U}_h(z)$;
- removes all the terms of $\tilde{U}_h(z)$ in which $\tilde{\boldsymbol{g}}_k^{(h)}$ contain only zeros;
- collects like terms in the resulting u-function.

Example 7.9

Consider the MAN with $n = 5$ and $n* = 2$ presented in Figure 7.8A. Assume that ME e_1 is located at C_1 and provides a connection between C_1 and any subset of $\{C_2, C_3\}$. ME e_2 is located at C_2 and provides a connection between C_2 and any subset of $\{C_3, C_5\}$. ME e_3 is also located at C_2 and provides a connection between C_2 and C_3. ME e_4 is located at C_3 and provides a connection between C_3 and C_4.

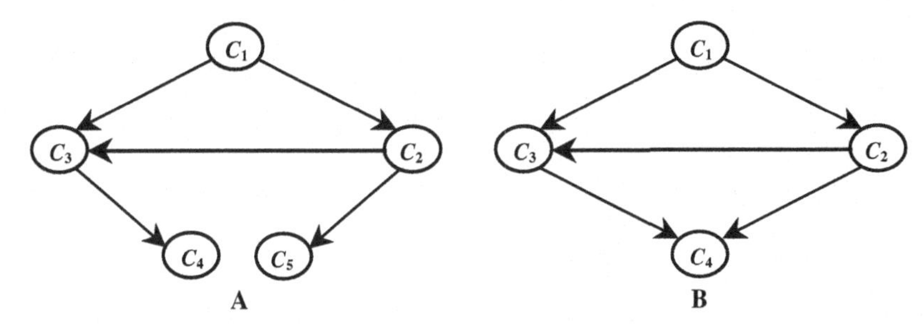

Figure 7.8. Two simple MANs

According to (7.40) and (7.41), the u-functions of individual MEs located at nodes C_1, C_2 and C_3 are

$$u_{11}(z) = p_{1\varnothing}z^{(0,0,0,0,0)}+p_{1\{2\}}z^{(0,1,0,0,0)}+p_{1\{3\}}z^{(0,0,1,0,0)}+p_{1\{2,3\}}z^{(0,1,1,0,0)}$$

$$u_{22}(z) = p_{2\varnothing}z^{(0,0,0,0,0)}+p_{2\{3\}}z^{(0,0,1,0,0)}+p_{2\{5\}}z^{(0,0,0,0,1)}+p_{2\{3,5\}}z^{(0,0,1,0,1)}$$

$$u_{32}(z) = p_{3\varnothing}z^{(0,0,0,0,0)}+p_{3\{3\}}z^{(0,0,1,0,0)},\ u_{43}(z) = p_{4\varnothing}z^{(0,0,0,0,0)}+p_{4\{4\}}z^{(0,0,0,1,0)}$$

The u-functions of groups of MEs located at the same nodes are

$$U_1(z) = u_{11}(z),\ U_3(z) = u_{43}(z)$$

$$U_2(z)=u_{22}(z) \underset{\max}{\otimes} u_{32}(z)=p_{2\varnothing}p_{3\varnothing}z^{(0,0,0,0,0)}+p_{2\{3\}}p_{3\varnothing}z^{(0,0,1,0,0)}+p_{2\{5\}}p_{3\varnothing}z^{(0,0,0,0,1)}$$

$$+p_{2\{3,5\}}p_{3\varnothing}z^{(0,0,1,0,1)}+p_{2\varnothing}p_{3\{3\}}z^{(0,0,1,0,0)}+p_{2\{3\}}p_{3\{3\}}z^{(0,0,1,0,0)}+p_{2\{5\}}p_{3\{3\}}z^{(0,0,1,0,1)}$$

$$+p_{2\{3,5\}}p_{3\{3\}}z^{(0,0,1,0,1)}=q_{2\varnothing}z^{(0,0,0,0,0)}+q_{2\{3\}}z^{(0,0,1,0,0)}+q_{2\{5\}}z^{(0,0,0,0,1)}+q_{2\{3,5\}}z^{(0,0,1,0,1)}$$

where

$$q_{2\varnothing} = p_{2\varnothing}p_{3\varnothing},\ q_{2\{3\}} = p_{2\{3\}}p_{3\varnothing}+p_{2\varnothing}p_{3\{3\}}+p_{2\{3\}}p_{3\{3\}},\ q_{2\{5\}} = p_{2\{5\}}p_{3\varnothing}$$

$$q_{2\{3,5\}} = p_{2\{3,5\}}p_{3\varnothing}+p_{2\{5\}}p_{3\{3\}}+p_{2\{3,5\}}p_{3\{3\}}$$

Since $p_{3\varnothing}+p_{3\{3\}} = 1$, the equations for $q_{2\{3\}}$ and $q_{2\{3,5\}}$ can be simplified:

$$q_{2\{3\}} = p_{2\{3\}}+p_{2\varnothing}p_{3\{3\}},\quad q_{2\{3,5\}} = p_{2\{3,5\}}+p_{2\{5\}}p_{3\{3\}}.$$

Following the consecutive procedure (7.46), we obtain

$$\tilde{U}_1(z) = U_1(z),\ \varphi(\tilde{U}_1(z)) = p_{1\{2\}}z^{(0,1,0,0,0)}+p_{1\{3\}}z^{(0,0,1,0,0)}+p_{1\{2,3\}}z^{(0,1,1,0,0)}$$

$$\tilde{U}_2(z) = \varphi(\tilde{U}_1(z)) \underset{\omega}{\otimes} U_2(z) = p_{1\{2\}}q_{2\varnothing}z^{(0,1,0,0,0)}+p_{1\{2\}}q_{2\{3\}}z^{(0,1,1,0,0)}$$

$$+p_{1\{2\}}q_{2\{5\}}z^{(0,1,0,0,1)}+p_{1\{2\}}q_{2\{3,5\}}z^{(0,1,1,0,1)}+p_{1\{3\}}z^{(0,0,1,0,0)}+p_{1\{2,3\}}q_{2\varnothing}z^{(0,1,1,0,0)}$$

$$+p_{1\{2,3\}}q_{2\{3\}}z^{(0,1,1,0,0)}+p_{1\{2,3\}}q_{2\{5\}}z^{(0,1,1,0,1)}+p_{1\{2,3\}}q_{2\{3,5\}}z^{(0,1,1,0,1)}$$

and, after simplification:

$$\varphi(\tilde{U}_2(z)) = (p_{1\{2\}}q_{2\{3\}}+p_{1\{3\}}+p_{1\{2,3\}}q_{2\varnothing}+p_{1\{2,3\}}q_{2\{3\}})z^{(0,0,1,0,0)}$$

$$+p_{1\{2\}}q_{2\{5\}}z^{(0,0,0,0,1)}+(p_{1\{2\}}q_{2\{3,5\}}+p_{1\{2,3\}}q_{2\{5\}}+p_{1\{2,3\}}q_{2\{3,5\}})z^{(0,0,1,0,1)}$$

(Operator φ reduces the number of different terms in this u-function from nine to three.).

$$\tilde{U}_3(z) = \varphi(\tilde{U}_2(z)) \underset{\omega}{\otimes} U_3(z) = (p_{1\{2\}}q_{2\{3\}}+p_{1\{3\}}+p_{1\{2,3\}}q_{2\varnothing}$$

$$+p_{1\{2,3\}}q_{2\{3\}})p_{4\varnothing}z^{(0,0,1,0,0)}+p_{1\{2\}}q_{2\{5\}}p_{4\varnothing}z^{(0,0,0,0,1)}+(p_{1\{2\}}q_{2\{3,5\}}+p_{1\{2,3\}}q_{2\{5\}}$$

$$+p_{1\{2,3\}}q_{2\{3,5\}})p_{4\varnothing}z^{(0,0,1,0,1)}+(p_{1\{2\}}q_{2\{3\}}+p_{1\{3\}}+p_{1\{2,3\}}q_{2\varnothing}$$

$$+p_{1\{2,3\}}q_{2\{3\}})p_{4\{4\}}z^{(0,0,1,1,0)}+p_{1\{2\}}q_{2\{5\}}p_{4\{4\}}z^{(0,0,0,0,1)}+(p_{1\{2\}}q_{2\{3,5\}}+p_{1\{2,3\}}q_{2\{5\}}$$

$$+p_{1\{2,3\}}q_{2\{3,5\}})p_{4\{4\}}\, z^{(0,0,1,1,1)}$$

and, after simplification:

$$\varphi(\tilde{U}_3(z)) = [p_{1\{2\}}q_{2\{5\}}+(p_{1\{2\}}q_{2\{3,5\}}+p_{1\{2,3\}}q_{2\{5\}}+p_{1\{2,3\}}q_{2\{3,5\}})p_{4\varnothing}]z^{(0,0,0,0,1)}$$

$$+(p_{1\{2\}}q_{2\{3\}}+p_{1\{3\}}+p_{1\{2,3\}}q_{2\varnothing}+p_{1\{2,3\}}q_{2\{3\}})p_{4\{4\}}z^{(0,0,0,1,0)}$$

$$+(p_{1\{2\}}q_{2\{3,5\}}+p_{1\{2,3\}}q_{2\{5\}}+p_{1\{2,3\}}q_{2\{3,5\}})p_{4\{4\}}z^{(0,0,0,1,1)}$$

The coefficient in the term with the vector $\tilde{G}^{(3)} = (0, 0, 0, 1, 1)$ in $\varphi(\tilde{U}_3(z))$ is the probability that the signal reaches both C_4 and C_5, which is equal to the MAN availability

$$R_{\{4,5\}} = (p_{1\{2\}}q_{2\{3,5\}}+p_{1\{2,3\}}q_{2\{5\}}+p_{1\{2,3\}}q_{2\{3,5\}})p_{4\{4\}}$$

One can also obtain the probabilities that the signal reaches nodes C_4 and C_5 by summing the coefficients of the terms with $\tilde{G}^{(3)}(4) = 1$ and $\tilde{G}^{(3)}(5) = 1$ respectively:

$$R_{\{4\}} = (p_{1\{2\}}q_{2\{3\}}+p_{1\{3\}}+p_{1\{2,3\}}q_{2\varnothing}+p_{1\{2,3\}}q_{2\{3\}})p_{4\{4\}}+(p_{1\{2\}}q_{2\{3,5\}}$$

$$+p_{1\{2,3\}}q_{2\{5\}}+p_{1\{2,3\}}q_{2\{3,5\}})p_{4\{4\}}$$

$$R_{\{5\}} = p_{1\{2\}}q_{2\{5\}}+(p_{1\{2\}}q_{2\{3,5\}}+p_{1\{2,3\}}q_{2\{5\}}+p_{1\{2,3\}}q_{2\{3,5\}})p_{4\varnothing}+$$

$$+(p_{1\{2\}}q_{2\{3,5\}}+p_{1\{2,3\}}q_{2\{5\}}+p_{1\{2,3\}}q_{2\{3,5\}})p_{4\{4\}}$$

7.2.2 Model with Constant Transmission Characteristics of Arcs

The aim of the networks considered is a transmission of information or material flow from the source to the sinks. The internode links (arcs) are associated with flow transmission media (lines, pipes, channels, *etc.*) and the nodes are associated with communication centres. The transmission process efficiency depends on the performance of the network elements. The most common performance characteristics are transmission speed and transmission capacity of the network arcs.

In this section, we consider two different types of MAN: minimal transmission time networks (task processing networks without work sharing) and maximal flow path networks (flow transmission networks without flow dispersion).

Examples of the minimal transmission time MAN are computer networks or networks of cellular telephones. The time of the signal transmission between each pair of nodes in a MAN depends on the type of equipment and exchange protocols. However, for each given retransmitter and line, it can be exactly estimated and considered as a constant. The total time of the signal propagation from transmitter to receiver can vary depending on line availability. The whole network is considered to be in working condition if the signal propagation time is not greater than a certain specified level w. Otherwise, the network fails.

Examples of the maximal flow path MAN are transportation networks or continuous production networks dealing with indivisible flows. The aim of the network is to provide the transmission of the maximal amount of an indivisible product from source to sink through a single path chosen from available network links. Each link has limited transmission capacity. The transmission capacity of a path is determined by the capacity of its bottleneck (the minimal transmission capacity among links belonging to the path). The transmission capacity and availability of each link can be exactly estimated. The capacity of the maximal flow path from source to sink can vary, depending on link availability. The whole network is considered to be in working condition if it contains an available path through which it can transmit the flow that is not less than the certain specified level w. Otherwise, the network fails.

In order to represent the multi-state nature of a MAN with fully reliable nodes and unreliable capacitated links, we assume that each node is an ME that provides connections to other nodes depending on network link availability. The model of the MAN with constant transmission characteristics of arcs differs from the MAN connectivity model by defining for each arc (C_i, C_j) its nominal performance v_{ij} and availability a_{ij}.

Consider an ME located at C_i. In each state k $(0 \le k < k_i)$, C_i is connected with a set of nodes θ_{ik}. In order to represent the random performance of connections between the node C_i and the rest of MAN nodes, we determine vector $\boldsymbol{G}^{(i)}$, which in each state k of the ME takes the value $\boldsymbol{g}_{ik}$ determined as follows [193]:

$$g_{ik}(j) = \begin{cases} v_{ij}, & if \;\; C_j \in \theta_{ik} \\ *, & if \;\; C_j \notin \theta_{ik} \end{cases} \tag{7.49}$$

where * stands for absence of connection. The u-function

$$u_i(z) = \sum_{k=0}^{k_i-1} p_{i\theta_{ik}} z^{g_{ik}} \tag{7.50}$$

represents the p.m.f. of random vector $\boldsymbol{G}^{(i)}$.

Consider the most complex (from the combinatorial standpoint) case of a multi-state node in which there exist M_i statistically independent arcs originated from C_i. Each arc (C_i, C_j) has availability a_{ij}. In this case, the total number of different ME states is $k_i = 2^{M_i}$. The probability of a state k in which only the arcs between C_i and nodes belonging to set θ_{ik} are available can be obtained as

$$p_{i\theta_{ik}} = \prod_{m \in \theta_{ik}} a_{im} \prod_{h \in \Theta_i \backslash \theta_{ik}} (1 - a_{ih}) \tag{7.51}$$

Therefore, the u-function for the ME C_i takes the form

$$u_i(z) = \sum_{k=0}^{2^{M_i}-1} p_{i\theta_{ik}} z^{g_{ik}} \tag{7.52}$$

where

$$p_{i\theta_{ik}} = \prod_{j=1}^{M_i} a_{ij}^{\beta_{jk}} (1-a_{ij})^{1-\beta_{jk}}, \quad g_{ik}(j) = \begin{cases} v_{ij}, & \text{if } \beta_{jk}=1, \ C_j \in \Theta_i \\ *, & \text{otherwise} \end{cases} \tag{7.53}$$

and $\beta_{jk} = \text{mod}_2 \lfloor k/2^{j-1} \rfloor$.

In many cases, the total number of different ME states is less than 2^{M_i} because of the presence of mutual dependence among arcs' states. (For example, if signal propagation conditions allow a signal generated at C_i to reach C_j, they always allow the signal to reach a closer node C_m.)

Consider an ME located at C_i. The random performance of connection between C_i and C_{i+1} is equal to $G^{(i)}(i+1)$ and the random performance of connection between C_i and arbitrary C_e ($e>i+1$) is equal to $G^{(i)}(e)$. The random performance of connection between C_{i+1} and C_e is $G^{(i+1)}(e)$. Therefore, two paths from C_i to C_e can exist: (C_i, C_e) and (C_i, C_{i+1}), (C_{i+1}, C_e). In order to replace the two MEs located at C_i and C_{i+1} with a single equivalent ME, one has to replace all of the connections among the two MEs and the rest of the network nodes with new connections having the same transmission performances (see Figure 7.9).

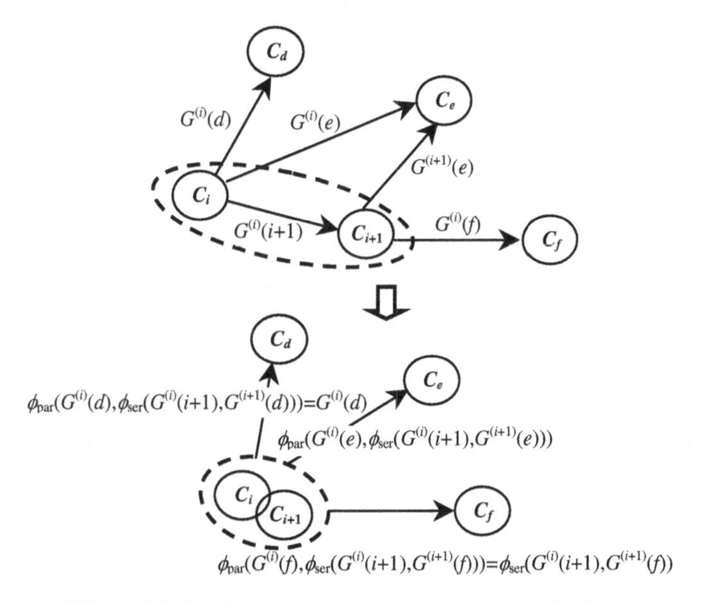

Figure 7.9. Transformation of two MEs into an equivalent one

The performance of the path (C_i, C_{i+1}), (C_{i+1}, C_e) is determined by the performances of consecutively connected arcs (C_i, C_{i+1}) and (C_{i+1}, C_e) (*i.e.* $G^{(i)}(i+1)$ and $G^{(i+1)}(e)$ respectively) and can be determined by a function

$$\phi_{ser}(G^{(i)}(i+1), G^{(i+1)}(e)) \tag{7.54}$$

corresponding to series connection of arcs. Note that, if at least one of the arcs is not available, the entire path does not exist. This should be expressed by the following property of the function:

$$\phi_{ser}(X, *) = \phi_{ser}(*, X) = * \text{ for any } X \tag{7.55}$$

If there are two parallel (alternative) paths from C_i to C_e (path (C_i, C_e) with performance $G^{(i)}(e)$ and path (C_i, C_{i+1}), (C_{i+1}, C_e) with performance $\phi_{ser}(G^{(i)}(i+1), G^{(i+1)}(e))$, then the performance of connection between C_i and C_e can be determined by the function

$$\phi(G^{(i)}, G^{(i+1)}) = \phi_{par}(G^{(i)}(e), \phi_{ser}(G^{(i)}(i+1), G^{(i+1)}(e)) \tag{7.56}$$

corresponding to parallel connection of paths. If one of the paths is not available, then the performance of the entire two-path connection is equal to the performance of the second path:

$$\phi_{par}(X, *) = X, \ \phi_{par}(*, X) = X \ \text{ for any } X \tag{7.57}$$

If arc (C_i, C_{i+1}) is not available (which corresponds to $G^{(i)}(i+1) = *$), then, according to (7.55) and (7.57), for any e

$$\phi_{par}(G^{(i)}(e), \phi_{ser}(G^{(i)}(i+1), G^{(i+1)}(e)) = G^{(i)}(e) \tag{7.58}$$

Different functions ϕ_{ser} and ϕ_{par} which meet conditions (7.55) and (7.57) can be defined according to the physical nature of the network. To obtain the performance of all of the connections, one has to apply the function (7.56) to the entire vectors $\boldsymbol{G}^{(i)}$ and $\boldsymbol{G}^{(i+1)}$. The resulting vector $\tilde{\boldsymbol{G}}^{(i+1)} = \phi(\boldsymbol{G}^{(i)}, \boldsymbol{G}^{(i+1)})$, in which each element $\tilde{G}^{(i+1)}(e)$ is determined using (7.56), represents the performance of arcs between the equivalent ME (two-ME subsystem) and any other ME in the MAN.

By applying a composition operator $\underset{\phi}{\otimes}$ with the function (7.56) over u-functions of individual MEs $u_i(z)$ and $u_{i+1}(z)$, one obtains the u-function $\tilde{U}_{i+1}(z)$ representing the p.m.f. of random vector $\tilde{\boldsymbol{G}}^{(i+1)}$. The two MEs with u-functions $u_i(z)$ and $u_{i+1}(z)$ can now be replaced in the MAN by the equivalent ME with u-function $\tilde{U}_{i+1}(z)$.

One can obtain the u-function for the entire MAN containing all of the MEs by defining $\tilde{U}_1(z) = u_1(z)$ and consecutively applying the equation

$$\tilde{U}_{i+1}(z) = \tilde{U}_i(z) \underset{\phi}{\otimes} u_{i+1}(z) \tag{7.59}$$

for $i = 1, \ldots, n-n^*-1$. Each u-function $\tilde{U}_i(z)$ represents the distribution of performance of connections between C_1 (direct or through $C_2, \ldots, C_i$) and the rest of the nodes.

Using the same considerations as in Section 7.1.2.2, one can define the u-function simplification operator φ that:

- assigns * symbols to $\tilde{g}_k^{(i)}(1),\ldots, \tilde{g}_k^{(i)}(i)$ in each term k of $\tilde{U}_i(z)$;
- removes all the terms of $\tilde{U}_i(z)$ in which $\tilde{\mathbf{g}}_k^{(i)}$ contain only * symbols;
- collects like terms in the resulting u-function.

This operator should be applied to each u-function obtained such that

$$\tilde{U}_{i+1}(z) = \varphi(\tilde{U}_i(z)) \underset{\phi}{\otimes} u_{i+1}(z) \tag{7.60}$$

Finally, we obtain the u-function

$$\tilde{U}_{n-n^*}(z) = \sum_{k=1}^{K} Q_k z^{\tilde{g}_k^{(n-n^*)}} \tag{7.61}$$

that represents the performance distribution of connections between an equivalent node (that replaced nodes $C_1, \ldots, C_{n-n^*}$) and the sink nodes $C_{n-n^*+1},\ldots, C_n$ of the MAN.

7.2.2.1 Minimal Transmission Time Multi-state Acyclic Network

When the arc performance is associated with the transmission time ($v_{ij} = t_{ij}$), the absence of an arc or its unavailability corresponds to infinite transmission time. Therefore, the sign * should be replaced with ∞ (* can also be represented by any number greater than the allowable network delay w).

The time of a signal transmission from C_i to C_e through any C_j ($i<j<e$) is equal to the sum of times of signal transmission between C_i and C_j and between C_j and C_e. Therefore, the ϕ_{ser} function takes the form

$$\phi_{\text{ser}}(G^{(i)}(i+1), G^{(i+1)}(e)) = G^{(i)}(i+1)+G^{(i+1)}(e) \tag{7.62}$$

When alternative ways of signal propagation from node C_i to node C_e exist, the least one of the transmission times determines the moment when the signal reaches node C_e. Therefore, the ϕ_{par} function takes the form

$$\phi_{\text{par}}(G^{(i)}(e),\phi_{\text{ser}}(G^{(i)}(i+1),G^{(i+1)}(e))) = \min\{G^{(i)}(e),G^{(i)}(i+1)+G^{(i+1)}(e)\} \tag{7.63}$$

Since the condition of normal MAN functioning is that the signal generated at C_1 reaches all the nodes from $\pi \subset \Pi$ with the delay not exceeding w, one can define the acceptability function as

$$F_\pi(\tilde{\boldsymbol{G}}^{(n-n^*)},w)=\begin{cases}1, & \text{if } \tilde{G}^{(n-n^*)}(j)\le w \text{ for any } j:C_j\in\pi\\ 0, & \text{otherwise}\end{cases} \quad (7.64)$$

and apply the operator δ_w (3.15) over the u-function $\tilde{U}_{n-n^*}(z)$ in order to obtain the network reliability:

$$R_\pi(w)=E(F_\pi(\tilde{\boldsymbol{G}}^{(n-n^*)},w))=\delta_w(\tilde{U}_{n-n^*}(z)) \quad (7.65)$$

The conditional expected transmission time for any node $C_j\in\Pi$ is determined from $\tilde{U}_{n-n^*}(z)$ as

$$\tilde{\varepsilon}_j=\sum_{\tilde{g}_k^{(n-n^*)}(j)<\infty}\tilde{g}_k^{(n-n^*)}(j)Q_k\Big/\sum_{\tilde{g}_k^{(n-n^*)}(j)<\infty}Q_k \quad (7.66)$$

7.2.2.2 Maximal Flow Path Multi-state Acyclic Network

When the arc performance is associated with transmission capacity, which determines the maximal flow that the arc can transmit ($v_{ij}=f_{ij}$), the absence of an arc or its unavailability corresponds to zero transmission capacity. Therefore, the sign * should be replaced with zero.

The amount of flow that can be transmitted through a path consisting of two consecutive links is determined by the link having the minimal capacity, which becomes the bottleneck of the path. Therefore, the ϕ_{ser} function takes the form

$$\phi_{ser}(G^{(i)}(i+1),G^{(i+1)}(e))=\min\{G^{(i)}(i+1),G^{(i+1)}(e)\} \quad (7.67)$$

When alternative ways of the flow transmission from the node C_i to the node C_e exist, the path with the greatest capacity should be chosen to maximize the flow. Therefore, the ϕ_{par} function takes the form

$$\phi_{par}(G^{(i)}(e),\phi_{ser}(G^{(i)}(i+1),G^{(i+1)}(e))$$

$$=\max\{G^{(i)}(e),\min\{G^{(i)}(i+1),G^{(i+1)}(e)\}\} \quad (7.68)$$

Since the condition of normal MAN functioning is that the capacities of flow paths from C_1 to nodes from $\pi\subset\Pi$ are not less then w, one can define the acceptability function as

$$F_\pi(\tilde{\boldsymbol{G}}^{(n-n^*)},w)=\begin{cases}1, & \text{if } \tilde{G}^{(n-n^*)}(j)\ge w \text{ for any } j:C_j\in\pi\\ 0, & \text{otherwise}\end{cases} \quad (7.69)$$

and obtain the MAN availability using Equation (7.65).

The expected capacity of the flow path from C_1 to $C_j \in \Pi$ is determined from $\tilde{U}_{n-n^*}(z)$ as

$$\varepsilon_j = \sum_{k=1}^{K} \tilde{g}_k^{(n-n^*)}(j) Q_k \tag{7.70}$$

Example 7.10

Consider the simple four-node MAN presented in Figure 7.8B. The ME located at C_1 has four states: total failure (no arcs available), connection with C_2, connection with C_3, and connection with C_2 and C_3. The probabilities of the states are $p_{1\varnothing}$, $p_{1\{2\}}$, $p_{1\{3\}}$ and $p_{1\{2,3\}}$ respectively. The ME located at C_2 also has four states: total failure, connection with C_3, connection with C_4, and connection with C_3 and C_4. The probabilities of the states are $p_{2\varnothing}$, $p_{2\{3\}}$, $p_{2\{4\}}$ and $p_{2\{3,4\}}$ respectively. The ME located at C_3 has only two states: total failure and connection with C_4. The probabilities of the states are $p_{3\varnothing}$ and $p_{3\{4\}}$ respectively. Assume that the MAN is of minimal transmission time type. In this case, the performance of arc (C_i, C_j) is transmission time t_{ij}. The u-functions corresponding to the network nodes are

$$u_1(z) = p_{1\varnothing} z^{(*,*,*,*)} + p_{1\{2\}} z^{(*,\, t_{12},*,*)} + p_{1\{3\}} z^{(*,*,\, t_{13},*)} + p_{1\{2,3\}} z^{(*,\, t_{12},\, t_{13},*)}$$

$$u_2(z) = p_{2\varnothing} z^{(*,*,*,*)} + p_{2\{3\}} z^{(*,*,\, t_{23},*)} + p_{2\{4\}} z^{(*,*,*,\, t_{24})} + p_{2\{3,4\}} z^{(*,*,\, t_{23},\, t_{24})}$$

$$u_3(z) = p_{3\varnothing} z^{(*,*,*,*)} + p_{3\{4\}} z^{(*,*,*,\, t_{34})}$$

where * stands for ∞.

Using (7.60) recursively we obtain

$$\tilde{U}_1(z) = u_1(z), \quad \varphi(\tilde{U}_1(z)) = p_{1\{2\}} z^{(*,\, t_{12},*,*)} + p_{1\{3\}} z^{(*,*,\, t_{13},*)} + p_{1\{2,3\}} z^{(*,\, t_{12},\, t_{13},*)}$$

$$\tilde{U}_2(z) = \varphi(\tilde{U}_1(z)) \underset{\phi}{\otimes} u_2(z) = p_{1\{2\}} p_{2\varnothing} z^{(*,\, t_{12},*,*)} + p_{1\{2\}} p_{2\{3\}} z^{(*,\, t_{12},\, t_{12}+t_{23},*)}$$

$$+ p_{1\{2\}} p_{2\{4\}} z^{(*,\, t_{12},*,\, t_{12}+t_{24})} + p_{1\{2\}} p_{2\{3,4\}} z^{(*,\, t_{12}, t_{12}+t_{23},\, t_{12}+t_{24})} + p_{1\{3\}} z^{(*,*,\, t_{13},*)}$$

$$+ p_{1\{2,3\}} p_{2\varnothing} z^{(*,\, t_{12}, t_{13},*)} + p_{1\{2,3\}} p_{2\{3\}} z^{(*,\, t_{12},\, \min\{t_{13}, t_{12}+t_{23}\},*)}$$

$$+ p_{1\{2,3\}} p_{2\{4\}} z^{(*,\, t_{12}, t_{13},\, t_{12}+t_{24})} + p_{1\{2,3\}} p_{2\{3,4\}} z^{(*,\, t_{12},\, \min\{t_{13}, t_{12}+t_{23}\}, t_{12}+t_{24})}$$

$$\varphi(\tilde{U}_2(z)) = p_{1\{2\}} p_{2\{3\}} z^{(*,*,\, t_{12}+t_{23},*)} + p_{1\{2\}} p_{2\{4\}} z^{(*,*,*,\, t_{12}+t_{24})}$$

$$+ p_{1\{2\}} p_{2\{3,4\}} z^{(*,*,\, t_{12}+t_{23},\, t_{12}+t_{24})} + (p_{1\{3\}} + p_{1\{2,3\}} p_{2\varnothing}) z^{(*,*,\, t_{13},*)}$$

$$+p_{1\{2,3\}}p_{2\{3\}}\, z^{(*,*,\, \min\{t_{13},\, t_{12}+t_{23}\},*)}+p_{1\{2,3\}}p_{2\{4\}}z^{(*,*,\, t_{13},\, t_{12}+t_{24})}$$

$$+p_{1\{2,3\}}p_{2\{3,4\}}z^{(*,*,\, \min\{t_{13},t_{12}+t_{23}\},\, t_{12}+t_{24})}$$

$$\tilde{U}_3(z)= \varphi(\tilde{U}_2(z)) \underset{\phi}{\otimes} u_3(z) = p_{1\{2\}}p_{2\{3\}}p_{3\varnothing}z^{(*,*,t_{12}+t_{23},*)}$$

$$+p_{1\{2\}}p_{2\{4\}}p_{3\varnothing}z^{(*,*,*,\, t_{12}+t_{24})}+p_{1\{2\}}p_{2\{3,4\}}p_{3\varnothing}z^{(*,*,\, t_{12}+t_{23},\, t_{12}+t_{24})}$$

$$+(p_{1\{3\}}+p_{1\{2,3\}}p_{2\varnothing})p_{3\varnothing}z^{(*,*,\, t_{13},*)}+p_{1\{2,3\}}p_{2\{3\}}p_{3\varnothing}z^{(*,*,\, \min\{t_{13},t_{12}+t_{23}\},*)}$$

$$+p_{1\{2,3\}}p_{2\{4\}}p_{3\varnothing}z^{(*,*,\, t_{13},\, t_{12}+t_{24})}+p_{1\{2,3\}}p_{2\{3,4\}}p_{3\varnothing}z^{(*,*,\, \min\{t_{13},t_{12}+t_{23}\},\, t_{12}+t_{24})}$$

$$+p_{1\{2\}}p_{2\{3\}}p_{3\{4\}}z^{(*,*,\, t_{12}+t_{23},\, t_{12}+t_{23}+t_{34})}+p_{1\{2\}}p_{2\{4\}}p_{3\{4\}}z^{(*,*,*,\, t_{12}+t_{24})}$$

$$+p_{1\{2\}}p_{2\{3,4\}}p_{3\{4\}}z^{(*,*,\, t_{12}+t_{23},\, \min\{t_{12}+t_{24},t_{12}+t_{23}+t_{34}\})}$$

$$+(p_{1\{3\}}+p_{1\{2,3\}}p_{2\varnothing})p_{3\{4\}}z^{(*,*,t_{13},t_{13}+t_{34})}+p_{1\{2,3\}}p_{2\{4\}}p_{3\{4\}}z^{(*,*,t_{13},\min\{t_{12}+t_{24},t_{13}+t_{34}\})}$$

$$+p_{1\{2,3\}}p_{2\{3\}}p_{3\{4\}}z^{(*,*,\, \min\{t_{13},t_{12}+t_{23}\},\, \min\{t_{13},t_{12}+t_{23}\}+t_{34})}$$

$$+p_{1\{2,3\}}p_{2\{3,4\}}p_{3\{4\}}z^{(*,*,\, \min\{t_{13},t_{12}+t_{23}\},\, \min\{t_{12}+t_{24},\, \min\{t_{13},t_{12}+t_{23}\}+t_{34}\})}$$

$$\varphi(\tilde{U}_3(z)) =[p_{1\{2\}}(p_{2\{4\}}+p_{2\{3,4\}}p_{3\varnothing})+p_{1\{2,3\}}(p_{2\{4\}}+p_{2\{3,4\}})p_{3\varnothing}]z^{(*,*,*,t_{12}+t_{24})}$$

$$+p_{1\{2\}}p_{2\{3\}}p_{3\{4\}}z^{(*,*,*,\, t_{12}+t_{23}+t_{34})}+p_{1\{2\}}p_{2\{3,4\}}p_{3\{4\}}z^{(*,*,*,\, \min\{t_{12}+t_{24},t_{12}+t_{23}+t_{34}\})}$$

$$+(p_{1\{3\}}+p_{1\{2,3\}}p_{2\varnothing})p_{3\{4\}}z^{(*,*,*,t_{13}+t_{34})}+p_{1\{2,3\}}p_{2\{3\}}p_{3\{4\}}z^{(*,*,*,\, \min\{t_{13},t_{12}+t_{23}\}+t_{34})}$$

$$+p_{1\{2,3\}}p_{2\{4\}}p_{3\{4\}}z^{(*,*,*,\, \min\{t_{12}+t_{24},t_{13}+t_{34}\})}$$

$$+p_{1\{2,3\}}p_{2\{3,4\}}p_{3\{4\}}z^{(*,*,*,\, \min\{t_{12}+t_{24},\, \min\{t_{13},t_{12}+t_{23}\}+t_{34}\})}$$

Note that operator φ reduces the number of different terms in the u-function $\tilde{U}_3(z)$ from 14 to 7. Being used with numerical data, the operator reduces this number to three by resolving the min function. Indeed, three possible finite network transmission times exist: $t_{12}+t_{24}$, $t_{13}+t_{34}$ and $t_{12}+t_{23}+t_{34}$.

Assume that t_{12} = 1, t_{13} = 4, t_{23} = t_{24} = t_{34} = 2. The final u-function takes the form

$$\varphi(\tilde{U}_3(z)) = [p_{1\{2\}}(p_{2\{4\}}+p_{2\{3,4\}}p_{3\varnothing})+p_{1\{2,3\}}(p_{2\{4\}}+p_{2\{3,4\}})p_{3\varnothing}]z^{(*,*,*,3)}$$

$$+p_{1\{2\}}p_{2\{3\}}p_{3\{4\}}z^{(*,*,*,5)}+p_{1\{2\}}p_{2\{3,4\}}p_{3\{4\}}z^{(*,*,*,3)}$$

$$+(p_{1\{3\}}+p_{1\{2,3\}}p_{2\varnothing})p_{3\{4\}}z^{(*,*,*,6)}+p_{1\{2,3\}}p_{2\{3\}}p_{3\{4\}}z^{(*,*,*,5)}$$

$$+p_{1\{2,3\}}p_{2\{4\}}p_{3\{4\}}z^{(*,*,*,3)}+p_{1\{2,3\}}p_{2\{3,4\}}p_{3\{4\}}z^{(*,*,*,3)}$$

$$=(p_{1\{2\}}+p_{1\{2,3\}})(p_{2\{4\}}+p_{2\{3,4\}})z^{(*,*,*,3)}+(p_{1\{2\}}+p_{1\{2,3\}})p_{2\{3\}}p_{3\{4\}}z^{(*,*,*,5)}$$

$$+(p_{1\{3\}}+p_{1\{2,3\}}p_{2\varnothing})p_{3\{4\}}z^{(*,*,*,6)}$$

The network availability for different values of desired delay w can be obtained from this u-function using the operator δ_w:

$$A(w)=0 \text{ for } w<3$$

$$A(3)=(p_{1\{2\}}+p_{1\{2,3\}})(p_{2\{4\}}+p_{2\{3,4\}})$$

$$A(5)=(p_{1\{2\}}+p_{1\{2,3\}})(p_{2\{4\}}+p_{2\{3,4\}}+p_{2\{3\}}p_{3\{4\}})$$

$$A(6)=(p_{1\{2\}}+p_{1\{2,3\}})(p_{2\{4\}}+p_{2\{3,4\}}+p_{2\{3\}}p_{3\{4\}})+(p_{1\{3\}}+p_{1\{2,3\}}p_{2\varnothing})p_{3\{4\}}$$

The conditional expected delay can be obtained using Equation (7.66):

$$\tilde{\varepsilon}_j=[3(p_{1\{2\}}+p_{1\{2,3\}})(p_{2\{4\}}+p_{2\{3,4\}})+5(p_{1\{2\}}+p_{1\{2,3\}})p_{2\{3\}}p_{3\{4\}}$$

$$+6(p_{1\{3\}}+p_{1\{2,3\}}p_{2\varnothing})p_{3\{4\}}]/A(6)$$

Now, assume that the MAN is of maximal flow path type. In this case, the performance of arc (C_i, C_j) is transmission capacity f_{ij}. The u-functions corresponding to the network nodes are

$$u_1(z)=p_{1\varnothing}z^{(0,0,0,0)}+p_{1\{2\}}z^{(0,f_{12},0,0)}+p_{1\{3\}}z^{(0,0,f_{13},0)}+p_{1\{2,3\}}z^{(0,f_{12},f_{13},0)}$$

$$u_2(z)=p_{2\varnothing}z^{(0,0,0,0)}+p_{2\{3\}}z^{(0,0,f_{23},0)}+p_{2\{4\}}z^{(0,0,0,f_{24})}+p_{2\{3,4\}}z^{(0,0,f_{23},f_{24})}$$

$$u_3(z)=p_{3\varnothing}z^{(0,0,0,0)}+p_{3\{4\}}z^{(0,0,0,f_{34})}$$

The same operators as in the previous example should be applied to the u-functions. The only difference is that the summation function and min function should be replaced with a min function and max function respectively. Finally, one obtains

$$\varphi(\tilde{U}_3(z))=[p_{1\{2\}}(p_{2\{4\}}+p_{2\{3,4\}}p_{3\varnothing})+p_{1\{2,3\}}(p_{2\{4\}}+p_{2\{3,4\}})p_{3\varnothing}]z^{(0,0,0,\min\{f_{12},f_{24}\})}$$

$$+p_{1\{2\}}p_{2\{3\}}p_{3\{4\}}z^{(0,0,0,\min\{f_{12},f_{23},f_{34}\})}+(p_{1\{3\}}+p_{1\{2,3\}}p_{2\varnothing})p_{3\{4\}}z^{(0,0,0,\min\{f_{13},f_{34}\})}$$

$$+p_{1\{2\}}p_{2\{3,4\}}p_{3\{4\}}z^{(0,0,0,\max\{\min\{f_{12},f_{24}\},\min\{f_{12},f_{23},f_{34}\}\})}$$

$$+p_{1\{2,3\}}p_{2\{3\}}p_{3\{4\}}z^{(0,0,0,\min\{\max\{f_{13},\min\{f_{12},f_{23}\}\},f_{34}\})}$$

$$+p_{1\{2,3\}}p_{2\{4\}}p_{3\{4\}}z^{(0,0,0,\ \max\{\min\{f_{12},f_{24}\},\ \min\{f_{13},f_{34}\}\})}$$

$$+p_{1\{2,3\}}p_{2\{3,4\}}p_{3\{4\}}z^{(0,0,0,\ \max\{\min\{f_{12},f_{24}\},\ \min\{\max\{f_{13},\ \min\{f_{12},f_{23}\}\},f_{34}\}\})}$$

For $f_{12} = 1,\ f_{13} = 4,\ f_{23} = f_{24} = f_{34} = 2$:

$$\varphi(\tilde{U}_3(z)) = [p_{1\{2\}}(p_{2\{4\}}+p_{2\{3,4\}}p_{3\varnothing})+p_{1\{2,3\}}(p_{2\{4\}}+p_{2\{3,4\}})p_{3\varnothing}]z^{(0,0,0,1)}$$

$$+p_{1\{2\}}p_{2\{3\}}p_{3\{4\}}z^{(0,0,0,1)}+p_{1\{2\}}p_{2\{3,4\}}p_{3\{4\}}z^{(0,0,0,1)}+(p_{1\{3\}}+p_{1\{2,3\}}p_{2\varnothing})p_{3\{4\}}z^{(0,0,0,2)}$$

$$+p_{1\{2,3\}}p_{2\{3\}}p_{3\{4\}}z^{(0,0,0,2)}+p_{1\{2,3\}}p_{2\{4\}}p_{3\{4\}}z^{(0,0,0,2)}+p_{1\{2,3\}}p_{2\{3,4\}}p_{3\{4\}}z^{(0,0,0,2)}$$

$$= [p_{1\{2\}}(p_{2\{4\}}+p_{2\{3,4\}}+p_{2\{3\}}p_{3\{4\}})+p_{1\{2,3\}}p_{3\varnothing}(p_{2\{4\}}+p_{2\{3,4\}})]z^{(0,0,0,1)}$$

$$+(p_{1\{3\}}+p_{1\{2,3\}})p_{3\{4\}}z^{(0,0,0,2)}$$

The availability of the network for different values of desired flow capacity w can be obtained from this u-function using the operator δ_w:

$$A(1) = p_{1\{2\}}(p_{2\{4\}}+p_{2\{3,4\}}+p_{2\{3\}}p_{3\{4\}})+p_{1\{2,3\}}p_{3\varnothing}(p_{2\{4\}}+p_{2\{3,4\}})$$
$$+(p_{1\{3\}}+p_{1\{2,3\}})p_{3\{4\}}$$

$$A(2) = (p_{1\{3\}}+p_{1\{2,3\}})p_{3\{4\}}$$

$$A(w) = 0 \text{ for } w>2$$

Finally, the expected network transmission capacity is obtained using Equation (7.70):

$$\varepsilon_4 = p_{1\{2\}}(p_{2\{4\}}+p_{2\{3,4\}}+p_{2\{3\}}p_{3\{4\}})$$
$$+p_{1\{2,3\}}p_{3\varnothing}(p_{2\{4\}}+p_{2\{3,4\}})+2(p_{1\{3\}}+p_{1\{2,3\}})p_{3\{4\}}$$

7.2.3 Optimal Element Allocation in a Multi-state Acyclic Network

The formulation of the ME allocation problem for a MAN is similar to the problem formulation for an LMCCS. An example of this problem is the allocation of a set of radio relay stations with different characteristics among different positions when the positions are not allocated along a line, but form a network. The optimal allocation should provide the greatest possible probability of the successful signal propagation from the root node (the position where a transmitter is located) to all of the terminal nodes (positions where the receivers are located). The connectivity model of a MAN (Section 7.2.1) is suited to this optimization problem.

7.2.3.1 Allocation Problem for a Multi-state Acyclic Network without Common Cause Failures

The MAN consists of n nodes from which n^* are terminal nodes. At each one of the first $n-n^*$ nodes, any group of MEs from a given set of m different elements should be allocated in a way that provides the greatest possible MAN reliability R_Π. The state distribution of each ME can depend on the node where it is allocated. For any ME allocated at any node this state distribution is given.

Like in the ME allocation problem for an LMCCS, any allocation of m MEs in $n-n^*$ nodes of a MAN can be represented by the integer string $\boldsymbol{a} = (a_j\text{: } 1\le j\le m)$, $1\le a_j\le n-n^*$, where a_j is the number of the node at which ME j locates. The fitness of any solution $\boldsymbol{a}$ generated by the GA is equal to the MAN reliability $R_\Pi(\boldsymbol{a})$ obtained using the algorithm described in Section 7.2.1.

Example 7.11

Consider the MAN with $n = 10$ and $n^* = 3$ presented in Figure 7.10. The list of possible states of identical MEs (represented by sets θ_{ik}) and corresponding probabilities $p_{i\theta_{ik}}$ is presented in Table 7.4, where θ_{ik} and $p_{i\theta_{ik}}$ correspond to the ME located at node C_i. The system fails if a signal generated at position C_1 does not reach at least one of the positions from the set $\{C_8, C_9, C_{10}\}$.

Table 7.4. State distribution of the network multi-state nodes

θ_{1k}	$P_{1\theta_{1k}}$	θ_{2k}	$p_{2\theta_{2k}}$	θ_{3k}	$p_{3\theta_{3k}}$	θ_{4k}	$p_{4\theta_{4k}}$	θ_{5k}	$p_{5\theta_{5k}}$	θ_{6k}	$p_{6\theta_{6k}}$	θ_{7k}	$p_{7\theta_{7k}}$
{2,3,4}	0.75	{4,6,8}	0.65	{4,5}	0.85	{6,7,10}	0.62	{6,7}	0.83	{8,9}	0.8	{9,10}	0.60
{2,3}	0.10	{4,6}	0.08	{4}	0.06	{6,7}	0.08	{6}	0.04	{8}	0.06	{9}	0.35
{3,4}	0.08	{4,8}	0.05	{5}	0.04	{6,10}	0.06	{7}	0.07	{9}	0.10	{10}	0.02
{2}	0.02	{6,8}	0.08	∅	0.05	{7,10}	0.02	∅	0.06	∅	0.04	∅	0.03
{3}	0.01	{4}	0.05			{6}	0.05						
∅	0.04	{6}	0.02			{7}	0.05						
		{8}	0.05			{10}	0.07						
		∅	0.02			∅	0.05						

Table 7.5 presents ME allocation solutions obtained by the GA for different m. This table contains numbers of identical MEs located at each node and the resulting MAN reliability.

In order to solve the allocation problem for nonidentical MEs, we modify the ME state probabilities in the following way:

$$p^j{}_{i\theta_{ik}} = \rho(j)\, p_{i\theta_{ik}} \quad \text{for } \theta_{ik}\neq\varnothing \text{ and } p^j{}_{i\varnothing} = 1-\rho(j) + \rho(j)\, p_{i\varnothing}$$

where $p_{i\theta_{ik}}$ are presented in Table 7.4, $p^j{}_{i\theta_{ik}}$ are state probabilities of ME e_j located at node C_i and $\rho(j) = 1.02-0.02j$ for $1\le j\le m=7$. Such a modification provides a unique state distribution for each element being allocated in each node.

The ME allocation solution for this problem (solution A) is presented in Table 7.6, which contains lists of MEs located in each node. The best solution obtained is compared with the solution of the constrained allocation problem in which allocation of no more than one ME in each node is allowed (solution B) and with the solution in which each ME j is located at C_j (solution C). The free ME allocation, when MEs occupy just three nodes out of seven, provides greater reliability than does the allocation in which the number of occupied nodes is equal to the number of MEs.

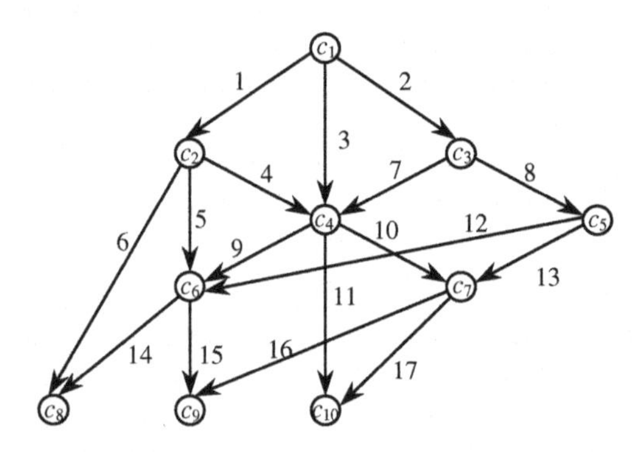

Figure 7.10. Example of a MAN

Table 7.5. Allocation solutions obtained for a MAN with identical MEs

m	Nodes							$R_{\{8,9,10\}}$
	1	2	3	4	5	6	7	
3	1	-	-	1	-	1	-	0.4515
4	1	-	-	2	-	1	-	0.6115
5	1	1	-	2	-	1	-	0.7665
6	2	-	-	2	-	2	-	0.8693
7	2	1	-	2	-	1	1	0.9309
8	2	1	-	3	-	2	-	0.9669
9	2	1	-	3	-	2	1	0.9830
10	2	2	-	3	-	2	1	0.9899
11	3	1	-	4	-	3	-	0.9946
12	3	2	-	4	-	2	1	0.9975
13	3	2	-	4	-	2	2	0.9986
14	3	2	-	5	-	3	1	0.9993
15	4	2	-	5	-	3	1	0.9996

Table 7.6. Allocation solutions obtained for MAN with different MEs

Solution	Nodes							$R_{\{8,9,10\}}$
	1	2	3	4	5	6	7	
A	2,4	-	-	5,6,7	-	1,3	-	0.8868
B	1	5	6	2	7	3	4	0.7611
C	1	2	3	4	5	6	7	0.7328

7.2.3.2 Allocation Problem for a Multi-state Acyclic Network in the Presence of Common Cause Failures

As in an LMCCS, the CCFs in a MAN usually occur when a group of elements are allocated at the same node. Each group of MEs can be considered to be a group with total CCF, since the destruction of any node by an external impact causes destruction of all of the MEs located at this node. The vulnerability of each node (CCF probability) v strongly affects the optimal allocation of MEs.

Example 7.12

Consider the simplest case in which two identical MEs should be allocated within a MAN with $n = 3$, $n^* = 1$. When allocated at node C_1, each ME can have four states:

- total failure, *i.e.* ME does not connect node C_1 with any other node (probability of this state is $p_{1\varnothing}$);
- ME connects C_1 with C_2 (probability of this state is $p_{1\{2\}}$);
- ME connects C_1 with C_3 (probability of this state is $p_{1\{3\}}$);
- ME connects C_1 with both C_2 and C_3 (probability of this state is $p_{1\{2,3\}}$).

When allocated at node C_2 each ME can have two states:

- total failure, *i.e.* ME does not connect node C_2 with any other node (probability of this state is $p_{2\varnothing}$;
- ME connects C_2 with C_3 (probability of this state is $p_{2\{3\}}$).

The probability that each node survives during the system operation time is $s = 1-v$.

Let $p_{1\varnothing} = p_{2\varnothing}$. There are two possible allocations of the MEs within the MAN (Figure 7.11):

I. Both MEs are located in the first position.

II. The MEs are located in the first and second positions.

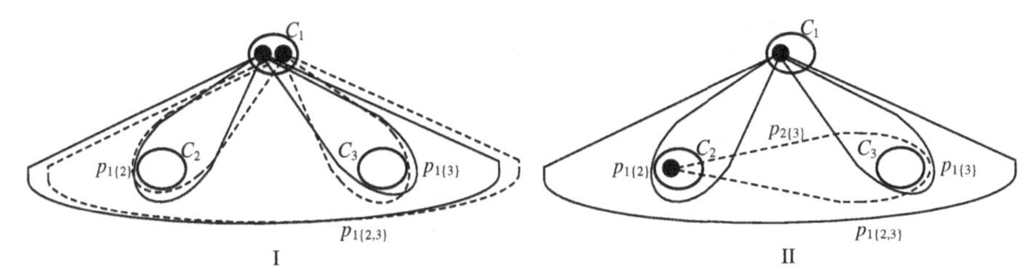

Figure 7.11. Two possible allocations of MEs in the simplest MAN

In case I, the MAN survives if node C_1 is not destroyed and at least one of the MEs is in state {3} or {2,3} and the system survivability is

$$S_{\mathrm{I}} = s\{2(p_{1\{3\}}+p_{1\{2,3\}})-(p_{1\{3\}}+p_{1\{2,3\}})^2\}$$

In case II, the MAN survives either when the node C_1 is not destroyed and the ME located in C_1 is in state {3} or {2,3}, or if both nodes C_1 and C_2 are not destroyed, the ME located in C_1 is in state {2} and the ME located in C_2 is in state {3}. The system survivability in this case is

$$S_{\mathrm{II}} = s\{p_{1\{3\}}+p_{1\{2,3\}}\}+s^2 p_{1\{2\}}p_{2\{3\}}$$

Since $p_{1\varnothing} = p_{2\varnothing}$, one can rewrite the last expression as

$$S_{\mathrm{II}} = s\{p_{1\{3\}}+p_{1\{2,3\}}\}+s^2(1-p_{1\{3\}}-p_{1\{2,3\}}-p_{1\varnothing})(1-p_{1\varnothing})$$

By comparing the expressions for S_{I} and S_{II}, one can decide which allocation of the elements is preferable for any given s, $p_{1\varnothing}$ and $p_{1\{3\}}+p_{1\{2,3\}}$. Condition $S_{\mathrm{I}}\geq S_{\mathrm{II}}$ can be rewritten as

$$(p_{1\{3\}}+p_{1\{2,3\}})(1-p_{1\{3\}}-p_{1\{2,3\}})\geq s(1-p_{1\{3\}}-p_{1\{2,3\}}-p_{1\varnothing})(1-p_{1\varnothing})$$

and finally as

$$s \leq (p_{1\{3\}}+p_{1\{2,3\}})(1-p_{1\{3\}}-p_{1\{2,3\}})/(1-p_{1\{3\}}-p_{1\{2,3\}}-p_{1\varnothing})(1-p_{1\varnothing})$$

Figure 7.12 presents the maximal values of s for which $S_{\mathrm{I}}\geq S_{\mathrm{II}}$ as a function of variables $p_{1\varnothing}$ and $p_{1\{3\}}+p_{1\{2,3\}}$. For given combinations of $p_{1\varnothing}$ and $p_{1\{3\}}+p_{1\{2,3\}}$ the values of s located below the curve correspond to cases when the solution I is preferable, whereas the values of s located above the curve correspond to cases when solution I provides lower system survivability than solution II.

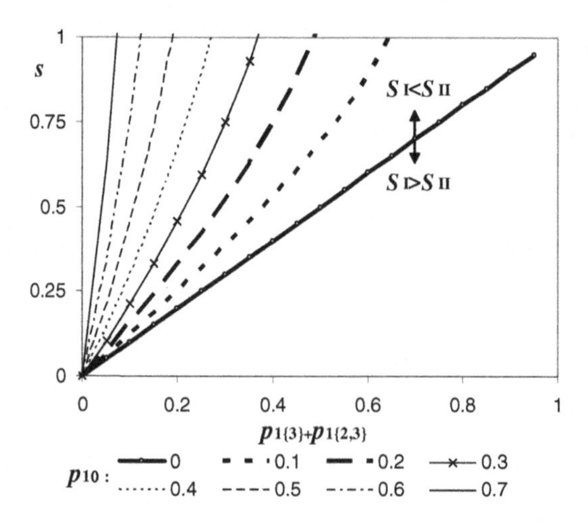

Figure 7.12. Comparison of two possible allocations of MEs in the simplest MAN

The same solution encoding scheme that was used in the previous section can be applied when implementing the GA for solving the ME's allocation problem in the presence of CCFs. For any ME allocation, the same algorithm for MAN reliability evaluation should be applied. The only difference lies in implementing the operator

$$\xi(U_j(z)) = (1-v)U_j(z) + vz^{G_0} \tag{7.71}$$

over each u-function $U_j(z)$ corresponding to the group of MEs located at node C_j (in analogy with Equation (7.39)). Here, $\boldsymbol{G}_0$ is the zero vector.

Example 7.13

Consider the MAN with identical MEs from Example 7.11 and assume that the vulnerability of each node is $v = 0.1$.

Table 7.7 presents the ME allocation solutions obtained by the GA in [194] for $m = 6$ and $m = 14$. This table contains the numbers of the identical MEs located at each position and the resulting MAN survivability. When $v = 0$ (no CCF), the solutions which are optimal for $v = 0.1$ provide network survivability very close to the survivability of a MAN with ME allocation optimal for $v = 0$. When $v = 0.1$, the solutions which are optimal for $v = 0.1$ provide greater MAN survivability than solutions which are optimal for $v = 0$ (survivability increase is 2.4% for $m = 6$ and 5.7% for $m = 14$). Figure 7.13 presents the survivability of MANs with the ME allocations obtained as a function of single node vulnerability v. One can see that the greater the node vulnerability, the greater the difference between MAN survivability provided by ME allocations optimal for $v = 0$ and $v = 0.1$.

Table 7.7. Allocation solutions obtained for a MAN with identical MEs

m	Solution for	Nodes 1	2	3	4	5	6	7	$R_{\{8,9,10\}}$ when $v = 0$	$R_{\{8,9,10\}}$ when $v = 0.1$
6	$v = 0$	2	-	-	2	-	2	-	0.8693	0.6337
	$v = 0.1$	1	1	-	2	-	1	1	0.8628	0.6491
14	$v = 0$	3	2	-	5	-	3	1	0.9993	0.7882
	$v = 0.1$	2	2	1	3	1	2	3	0.9965	0.8333

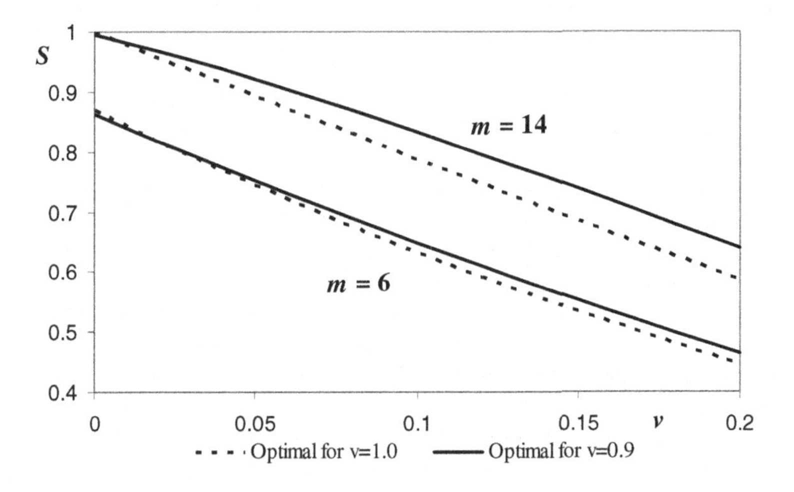

Figure 7.13. MAN survivability as a function of node vulnerability

7.2.4 Optimal Reliability Enhancement of a Multi-state Acyclic Network

Rapid changes in digital network technology provide increases of transmission capacity and transmission speed of communication lines. Replacement of old lines with new, high-performance ones can considerably reduce the signal delivery time in digital communication networks. Subject to budget limitations, a question arises regarding which lines should be replaced to obtain the desired reliability improvement effect.

Usually, line modernization does not affect its availability but improves its performance. The performance (transmission time) of an available line is a constant value that depends on data exchange protocols and properties of the communication media. Such networks can be represented by the model of a minimal transmission time network with constant transmission characteristics of arcs, as described in Section 7.2.2.1. This model uses multi-state nodes in order to represent the multi-state nature of the network with unreliable lines.

Reduction of the signal transmission time t_{ij} for a line connecting nodes C_i and C_j can affect the overall network availability $A(w)$, defined as a probability that the signal delivery delays in the network are not greater than w. If for each line connecting nodes C_i and C_j the values of transmission times before and after the line replacement are given (t_{ij} and t'_{ij} respectively), as well as the replacement cost c_{ij}, one can find a subset of lines for which replacement of times t_{ij} with times t'_{ij} produces the value of index $A(w)$ not less than the specified value A^*, while the total cost associated with this replacement is minimal.

In order to formulate the optimization problem, let us introduce an integer function $b(i, j)$ that produces a unique integer number $1 \le b(i, j) \le B$ for each pair of nodes connected by line (C_i, C_j) (where B is the total number of lines in the MAN).

Let us also define a binary vector $\boldsymbol{a}$ in which $a(b(i, j)) = 1$ means that line (C_i, C_j) is replaced and $a(b(i, j)) = 0$ means that the line is not replaced.

For any given line replacement solution $\boldsymbol{a}$, the total replacement cost is

$$C_{\text{tot}}(\boldsymbol{a}) = \sum_{j=1}^{B} c_{ij} a(b(i, j)) \tag{7.72}$$

and the transmission time of each line (C_i, C_j) is

$$\tau_{ij} = a(b(i, j)) t'_{ij} + [1 - a(b(i, j))] t'_{ij}. \tag{7.73}$$

Having the availabilities and the transmission times of each line, one can determine the entire MAN availability $A(\boldsymbol{a}, w)$ for any given allowable delay w. The optimization problem takes the form

$$C_{\text{tot}}(\boldsymbol{a}) \to \min \text{ subject to } A(\boldsymbol{a}, w) \ge A^* \tag{7.74}$$

Any binary vector $\boldsymbol{a}$ can represent a MAN modernization solution in the GA. For each given vector $\boldsymbol{a}$, the GA solution decoding procedure determines the system protection cost using Equation (7.72) and evaluates the system survivability $A(\boldsymbol{a}, w)$ using the algorithm described in Section 7.2.2.1. In order to let the GA look for the solution with minimal total cost and with $A(\boldsymbol{a}, w)$ not less than the required value A^*, the solution fitness is evaluated as follows:

$$M - C(\boldsymbol{a}) - \pi(1 + A^* - A(\boldsymbol{a}, w)) \times 1(A(\boldsymbol{a}, w) < A^*) \tag{7.75}$$

Example 7.14

Consider the data transmission network presented in Figure 7.10. The list of possible states of MEs located at nodes C_1, …, C_7 represented by sets of nodes θ_{ik} and corresponding state probabilities $p_{i\theta_{ik}}$ is presented in Table 7.8.

The signal transmission times for each line before and after the line replacement are presented in Table 7.9, as well as the replacement costs.

Table 7.8. State distribution of the network multi-state nodes

θ_{1k}	$p_{1\theta_{1k}}$	θ_{2k}	$p_{2\theta_{2k}}$	θ_{3k}	$p_{3\theta_{3k}}$	θ_{4k}	$p_{4\theta_{4k}}$	θ_{5k}	$p_{5\theta_{5k}}$	θ_{6k}	$p_{6\theta_{6k}}$	θ_{7k}	$p_{7\theta_{7k}}$
{2,3,4}	0.85	{4,6,8}	0.65	{4,5}	0.85	{6,7,10}	0.75	{6,7}	0.85	{8,9}	0.84	{9,10}	0.70
{2,3}	0.05	{4,6}	0.16	{4}	0.06	{6,7}	0.03	{6}	0.04	{8}	0.06	{9}	0.25
{3,4}	0.03	{4,8}	0.05	{5}	0.04	{6,10}	0.06	{7}	0.07	{9}	0.06	{10}	0.02
{2,4}	0.00	{6,8}	0.01	∅	0.05	{7,10}	0.02	∅	0.04	∅	0.04	∅	0.03
{2}	0.02	{4}	0.05			{6}	0.01						
{3}	0.01	{6}	0.02			{7}	0.01						
{4}	0.00	{8}	0.04			{10}	0.07						
∅	0.04	∅	0.02			∅	0.05						

Table 7.9. Line transmission times and enhancement costs

No. of line $b(i, j)$	Nodes connected i, j	t_{ij} (s)	t'_{ij} (s)	c_{ij} (10^6 \$)	No. of line $b(i, j)$	Nodes connected i, j	t_{ij} (s)	t'_{ij} (s)	c_{ij} (10^6 \$)
1	1, 2	0.20	0.10	1.2	10	4, 7	0.12	0.06	0.6
2	1, 3	0.34	0.15	1.4	11	4, 10	0.70	0.45	2.0
3	1, 4	0.52	0.30	1.0	12	5, 6	0.37	0.18	1.1
4	2, 4	0.45	0.18	1.8	13	5, 7	0.18	0.10	0.8
5	2, 6	0.55	0.25	1.9	14	6, 8	0.39	0.12	2.2
6	2, 8	0.84	0.40	0.5	15	6, 9	0.34	0.13	1.8
7	3, 4	0.29	0.20	0.7	16	7, 9	0.73	0.53	1.3
8	3, 5	0.12	0.06	0.8	17	7, 10	0.60	0.23	2.3
9	4, 6	0.33	0.13	2.1					

Before line replacement, the network cannot provide the signal delivery to all its terminal nodes in a time less than 1.22 ($A(1.22) = 0.684$, $A(w) = 0$ for $w<1.22$). The conditional expected signal delivery times (given the signal arrives at finite time) for the individual terminal nodes are $\tilde{\varepsilon}_8 = 1.074$, $\tilde{\varepsilon}_9 = 1.132$ and $\tilde{\varepsilon}_{10} = 1.227$.

The solutions obtained in [195] by the GA for $w \in \{0.8, 1.0, 1.2\}$ and $A^* \in \{0.7, 0.75, 0.8, 0.85\}$ are presented in Table 7.10. For each solution, the list of lines to be replaced, the total replacement cost, and the resulting system availability are presented.

One can see that the solution obtained for w = 1.0 and A^* = 0.75 coincides with one for w = 1.0 and A^* = 0.80, and the solution obtained for w = 1.2 and A^* = 0.70 coincides with one for w = 1.2 and A^* = 0.75.

Table 7.10. Parameters of the solutions obtained

w	A^* = 0.7	A^* = 0.75	A^* = 0.8	A^* = 0.85	$A_{max}(w)$
0.8	$A(w)$=0.721 C_{tot} = 9.6 {2,3,6,7,11,14,15}	$A(w)$=0.762 C_{tot} = 11.2 {2,3,6,11,14,15,17}	$A(w)$=0.807 C_{tot} = 14.2 {1,2,3,4,6,11,14,15,17}	-	0.811
1.0	$A(w)$=0.723 C_{tot} = 3.7 {2,3,6,8}	$A(w)$=0.801 C_{tot} = 5.0 {2,3,6,8,16}	$A(w)$=0.801 C_{tot} = 5.0 {2,3,6,8,16}	$A(w)$=0.869 C_{tot} = 7.5 {1,2,3,4,8,16}	0.888
1.2	$A(w)$=0.773 C_{tot} = 1.0 {3}	$A(w)$=0.773 C_{tot} = 1.0 {3}	$A(w)$=0.812 C_{tot} = 1.4 {2}	$A(w)$=0.868 C_{tot} = 2.4 {2,3}	0.888

The greatest possible network availability $A_{max}(w)$ obtained by replacing all the lines (the total cost of such replacement is 23.5) is presented in Table 7.10 for each value of w. Since $A_{max}(0.8)$ = 0.811, the algorithm cannot find a solution for w = 0.8 and A^* = 0.85.

The $A_{\{j\}}$ and $\tilde{\varepsilon}_j$ indices for individual terminal nodes C_8, C_9, and C_{10}, corresponding to the solutions obtained, are presented in Tables 7.11 and 7.12 respectively. Solutions providing greater system availability do not necessarily provide lower conditional expected signal delivery times. For example, for w = 1.2, A^* = 0.7 and A^* = 0.8 the conditional expected signal delivery times $\tilde{\varepsilon}_j$ are greater for more reliable solutions ($\tilde{\varepsilon}_j$ for A^* = 0.8 are greater than $\tilde{\varepsilon}_j$ for A^* = 0.7 for j = 8, 9, 10). This illustrates the fact that the availability and the conditional expected performance indices have different natures and cannot be used interchangeably.

Table 7.11. Availability of signal delivery to individual terminal nodes

	A^* = 0.7			A^* = 0.75			A^* = 0.8			A^* = 0.85			$A_{max}(w)$		
w	$A_{\{8\}}$	$A_{\{9\}}$	$A_{\{10\}}$	$A_{\{8\}}$	$A_{\{9\}}$	$A_{\{10\}}$	$A_{\{8\}}$	$A_{\{9\}}$	$A_{\{10\}}$	$A_{\{8\}}$	$A_{\{9\}}$	$A_{\{10\}}$	$A_{\{8\}}$	$A_{\{9\}}$	$A_{\{10\}}$
0.8	0.917	0.810	0.841	0.917	0.810	0.888	0.928	0.855	0.919	-	-	-	0.929	0.857	0.922
1.0	0.879	0.810	0.879	0.879	0.909	0.879	0.879	0.909	0.879	0.912	0.946	0.917	0.929	0.947	0.922
1.2	0.908	0.922	0.817	0.908	0.922	0.817	0.925	0.930	0.851	0.928	0.943	0.903	0.929	0.947	0.922

Table 7.12. Expected signal delivery time for individual terminal nodes

	A^* = 0.7			A^* = 0.75			A^* = 0.8			A^* = 0.85			$A_{max}(w)$		
w	$\tilde{\varepsilon}_8$	$\tilde{\varepsilon}_9$	$\tilde{\varepsilon}_{10}$	$\tilde{\varepsilon}_8$	$\tilde{\varepsilon}_9$	$\tilde{\varepsilon}_{10}$	$\tilde{\varepsilon}_8$	$\tilde{\varepsilon}_9$	$\tilde{\varepsilon}_{10}$	$\tilde{\varepsilon}_8$	$\tilde{\varepsilon}_9$	$\tilde{\varepsilon}_{10}$	$\tilde{\varepsilon}_8$	$\tilde{\varepsilon}_9$	$\tilde{\varepsilon}_{10}$
0.8	0.641	0.806	0.780	0.641	0.806	0.693	0.561	0.785	0.667	-	-	-			
1.0	0.700	0.956	1.004	0.700	0.932	1.004	0.700	0.932	1.004	0.950	0.923	0.983	0.479	0.523	0.598
1.2	1.034	1.019	1.032	1.034	1.019	1.032	1.040	1.025	1.095	1.026	0.996	1.014			

8. Universal Generating Function in Analysis and Optimization of Fault-tolerant Software

8.1 Reliability and Performance of Fault-tolerant Software

The NVP approach presumes the execution of n functionally equivalent software versions that receive the same input and send their outputs to a voter, which is aimed at determining the system's output. The voter produces an output if at least k out of n outputs agree. Otherwise, the system fails. As shown in Section 3.2.10, the RBS approach can also be considered as NVP with $k = 1$ when the system performance (task execution time) is considered.

In many cases, the information about the version's reliability and the execution time are available from separate testing and/or reliability prediction models [196]. This information can be incorporated into a fault-tolerant program model in order to obtain an evaluation of its reliability and performance.

8.1.1 Fault-tolerant Software Performance Model

According to the generally accepted model [197], the software system consists of C components. Each component performs a subtask and the sequential execution of the components performs a major task.

It is assumed that n_c functionally equivalent versions are available for each component c. Each version i has an estimated reliability r_{ci} and constant execution time τ_{ci}. Failures of versions for each component are statistically independent, as well as the total failures of the different components.

The software versions in each component c run on parallel hardware units. The total number of units is h_c. The units are independent and identical. The availability of each unit is a_c. The number H_c of units available at the moment determines the amount of available computational resources and, therefore, the number of versions that can be executed simultaneously $L_c(H_c)$. No hardware unit can change its state during the software execution.

The versions of each component c start their execution in accordance with a predetermined ordered list. L_c first versions from the list start their execution simultaneously (at time zero). If the number of terminated versions is less than k_c, after termination of each version a new version from the list starts its execution immediately. If the number of terminated versions is not less than k_c, after termination of each version the voter compares the outputs. If k_c outputs are identical, the component terminates its execution (terminating all the versions that are still executed), otherwise a new version from the list is executed immediately.

If after termination of n_c versions the number of identical outputs is less than k_c the component and the entire system fail.

In the case of component success, the time of the entire component execution T_c is equal to the termination time of the version that has produced the k_cth correct output (in most cases, the time needed by the voter to make the decision can be neglected). It can be seen that the component execution time is a random variable depending on the reliability and the execution time of the component versions and on the availability of the hardware units. We assume that if the component fails, then its execution time is equal to infinity.

The examples of time diagrams (corresponding to component 1 with $n_1 = 5$, $k_1 = 3$ and component 2 with $n_2 = 3$, $k_2 = 2$) for a given sequence of versions execution (the versions are numbered according to this sequence) and different values of L_c are presented in Figure 8.1.

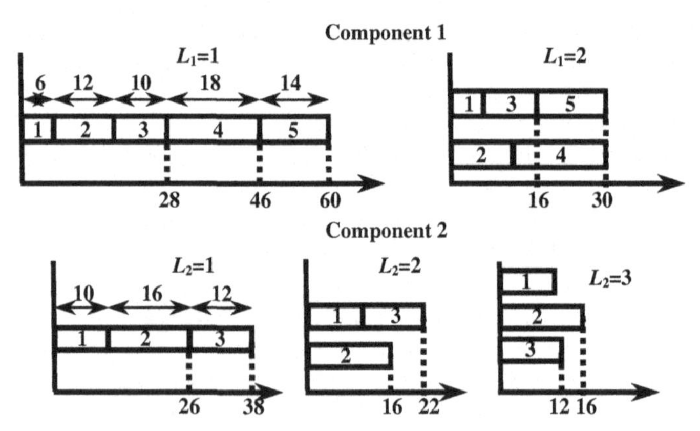

Figure 8.1. Time diagrams for software components with different numbers of versions executed simultaneously

The sum of the random execution times of each component gives the random task execution time for the entire system T. In order to estimate both the system's reliability and its performance, different measures can be used, depending on the application.

In applications where the execution time of each task is of critical importance, the system's acceptability function is defined (according to the performability concept [198, 199]) as $F(T, w) = 1(T<w)$, where w is a maximal allowed system execution time. The system's reliability $R(w) = E(F(T, w))$ in this case is the probability that the correct output is produced in time less than w. The conditional expected system execution time $\tilde{\varepsilon}(w) = E(T \times 1(T < w)) / R(w)$ is considered to be a measure of the system's performance. This index, defined according to Equation

(3.7), determines the system's expected execution time given that the system does not fail.

In applications where the system's average productivity (the number of executed tasks) over a fixed mission time is of interest [200], the system's acceptability function is defined as $F(T) = 1(T<\infty)$, the system's reliability is defined as the probability that it produces correct outputs regardless of the total execution time (this index can be referred to as $R(\infty)$), and the conditional expected system execution time $\tilde{\varepsilon}(\infty)$ is considered to be a measure of the system's performance.

8.1.1.1 Number of Versions that Can Be Simultaneously Executed

The number of available hardware units in component c can vary from 0 to h_c. Given that all of the units are identical and have availability a_c, one can easily obtain probabilities $Q_c(x) = \Pr\{H_c = x\}$ for $0 \le x \le h_c$:

$$Q_c(x) = \Pr\{H_c = x\} = \binom{h_c}{x} a_c{}^x (1-a_c)^{h_c - x} \tag{8.1}$$

The number of available hardware units x determines the number of versions that can be executed simultaneously: $l_c(x)$. Therefore:

$$\Pr\{L_c = l_c(x)\} = Q_c(x) \tag{8.2}$$

The pairs $Q_c(x)$, $l_c(x)$ for $1 \le x \le h_c$ determine the p.m.f. of the discrete random value L_c.

8.1.1.2 Version Termination Times

In each component c, a sequence where each version starts its execution is defined by the numbers of versions. This means that each version i starts its execution not earlier than versions 1, …, i–1 and not later than versions i+1, …, n_c. If the number of versions that can run simultaneously is l_c, then we can assume that the software versions run on l_c independent processors. Let α_m be the time when processor m terminates the execution of a version and is ready to run the next version from the list of not executed versions. Having the execution time of each version τ_{ci} ($1 \le i \le n_c$), one can obtain the termination time $t_{ci}(l_c)$ for each version i using the following simple algorithm:

1. Assign $\alpha_1 = \ldots = \alpha_{l_c} = 0$ (all of the processors are ready to run the software versions at time 0).
2. For $i = 1, \ldots, n_c$ repeat:
   2.1. Find any m ($1 \le m \le l_c$): $\alpha_m = \min\{\alpha_1, \ldots, \alpha_{l_c}\}$ (m is the number of the earliest processor that is ready to run a new version from the list).
   2.2. Obtain $t_{ci}(l_c) = \alpha_m + \tau_{ci}$ and assign $\alpha_m = t_{ci}(l_c)$.

Times $t_{ci}(l_c)$, $1 \le i \le n_c$, correspond to intervals between the beginning of component execution and the moment when the versions produce their outputs. Observe that the versions that start execution earlier can terminate later: $j<y$ does not guarantee that $t_{cj}(l_c) \le t_{cy}(l_c)$. In order to obtain the sequence, in which the versions produce their outputs, the termination times should be sorted in increasing order $t_{cm_1}(l_c) \le t_{cm_2}(l_c) \le \ldots \le t_{cm_{n_c}}(l_c)$, which gives the order of versions $m_1, m_2, \ldots, m_{n_c}$, corresponding to times of their termination.

The ordered list $m_1, m_2, \ldots, m_{n_c}$ determines the sequence of version outputs in which they arrive at the voter. Now one can consider the component c as a system in which the n_c versions are executed consecutively according to the order $m_1, m_2, \ldots, m_{n_c}$ and produce their outputs at times $t_{cm_1}(l_c)$, $t_{cm_2}(l_c)$, $\ldots$, $t_{cm_{n_c}}(l_c)$.

8.1.1.3 The Reliability and Performance of Components and the Entire System

Let r_{cm_i} be the reliability of the version that produces ith output in component c (r_{cm_i} is equal to the probability that this output is correct). Consider the probability that k out of n first versions of component c succeed. This probability can be obtained as

$$[\prod_{i=1}^{n}(1-r_{cm_i})][\sum_{i_1=1}^{n-k+1}\frac{r_{cm_{i_1}}}{(1-r_{cm_{i_1}})}\sum_{i_2=i_1+1}^{n-k+2}\frac{r_{cm_{i_2}}}{(1-r_{cm_{i_2}})}\ldots\sum_{i_k=i_{k-1}+1}^{n}\frac{r_{cm_{i_k}}}{(1-r_{cm_{i_k}})}] \quad (8.3)$$

(according to Equation (2.16) for k-out-of-n systems). The component c produces the correct output directly after the end of the execution of j versions ($j \ge k_c$) if the m_jth version succeeds and exactly k_c-1 out of the first executed $j-1$ versions succeed.

The probability of such event $p_{cj}(l_c)$ is

$$p_{cj}(l_c) = r_{cm_j}[\prod_{i=1}^{j-1}(1-r_{cm_i})][\sum_{i_1=1}^{j-k_c+1}\frac{r_{cm_{i_1}}}{(1-r_{cm_{i_1}})}\sum_{i_2=i_1+1}^{j-k_c+2}\frac{r_{cm_{i_2}}}{(1-r_{cm_{i_2}})} \ldots \sum_{i_{k_c-1}=i_{k_c-2}+1}^{j-1}\frac{r_{cm_{i_{k_c-1}}}}{(1-r_{cm_{i_{k_c-1}}})}] \quad (8.4)$$

Observe that $p_{cj}(l_c)$ is the conditional probability that the component execution time is $t_{cm_j}(l_c)$ given l_c versions can be executed simultaneously:

$$p_{cj}(l_c) = \Pr\{T_c = t_{cm_j}(l_c) \mid L_c = l_c\} \quad (8.5)$$

Having the p.m.f. of L_c we can now obtain for $1 \le x \le h_c$

$$\Pr\{T_c = t_{cm_j}(l_c(x))\} = \Pr\{T_c = t_{ck_j}(l_c(x)) \mid L_c = l_c(x)\}\Pr\{L_c = l_c(x)\}$$
$$= p_{cj}(l_c(x))Q_c(x) \quad (8.6)$$

The pairs $t_{cm_j}(l_c(x))$, $p_{cj}(l_c(x))Q_c(x)$, obtained for $1 \le x \le h_c$ and $k_c \le j \le n_c$, determine the p.m.f. of version execution time T_c.

Since the events of successful component execution termination for different j and x are mutually exclusive, we can express the probability of component c success as

$$R_c(\infty) = \Pr\{T_c < \infty\} = \sum_{x=1}^{h_c}[Q_c(x) \sum_{j=k_c}^{n_c} p_{cj}(l_c(x))] \quad (8.7)$$

Since failure of any component constitutes the failure of the entire system, the system's reliability can be expressed as

$$R(\infty) = \prod_{c=1}^{C} R_c(\infty) \quad (8.8)$$

From the p.m.f. of execution times T_c for each component c one can obtain the p.m.f. of the execution time of the entire system, which is equal to the sum of the execution times of components:

$$T = \sum_{c=1}^{C} T_c \quad (8.9)$$

8.1.1.4 Using Universal Generating Function for Evaluating the Execution Time Distribution of Components

In order to obtain the execution time distribution for a component c for a given l_c in the form $p_{cj}(l_c)$, $t_{cm_j}(l_c)$ $(k_c \le j \le n_c)$ one can determine the realizations $t_{cm_j}(l_c)$ of the execution time $T_c(l_c)$ using the algorithm presented in Section 8.1.1.2 and the corresponding probabilities $p_{cj}(l_c)$ using Equation (8.4). However, the probabilities $p_{cj}(l_c)$ can be obtained in a much simpler way using a procedure based on the UGF technique [201].

Let the random binary variable s_{cm_i} be an indicator of the success of version m_i in component c such that $s_{cm_i} = 1$ if the version produces the correct output and $s_{cm_i} = 0$ if it produces the wrong output. The p.m.f. of s_{cm_i} can be represented by the u-function

$$u_{cm_i}(z) = r_{cm_i} z^1 + (1 - r_{cm_i}) z^0 \quad (8.10)$$

It can be easily seen that using the operator $\underset{+}{\otimes}$ we can obtain the u-function

$$U_{cj}(z,l_c)=\underset{+}{\otimes}(u_{cm_1}(z),...,u_{cm_j}(z)) \tag{8.11}$$

that represents the p.m.f. of the number of correct outputs in component c after the execution of a group of first j versions (the order of elements m_1, m_2, ..., m_{n_c} and, therefore, $U_{cj}(z, l_c)$ depend on l_c). Indeed, the resulting polynomial relates the probabilities of combinations of correct and wrong outputs (the product of corresponding probabilities) with the number of correct outputs in these combinations (the sum of success indicators). Observe that after collecting the like terms (corresponding to obtaining the overall probability of a different combination with the same number of correct outputs) $U_{cj}(z, l_c)$ takes the form

$$U_{cj}(z,l_c)=\sum_{k=0}^{j}\pi_{jk}z^k \tag{8.12}$$

where π_{jk} is the probability that the group of first j versions produces k correct outputs.

Note that $U_{cj}(z, l_c)$ can be obtained by using the recurrent expression

$$U_{cj}(z,l_c)=U_{cj-1}(z,l_c)\underset{+}{\otimes}\ [r_{cm_j}z^1+(1-r_{cm_j})z^0] \tag{8.13}$$

According to its definition, $p_{cj}(l_c)$ is the probability that the group of first j versions produces k_c correct outputs and the group of first j–1 versions produces k_c – 1 correct outputs given that l_c versions can be executed simultaneously. The coefficient π_{jk_c} in polynomial $U_{cj}(z, l_c)$ is equal to the conditional probability that the group of first j versions produces k_c correct outputs given that l_c versions can be executed simultaneously.

In order to let the coefficient π_{jk_c} in polynomial $U_{cj}(z, l_c)$ be equal to $p_{cj}(l_c)$, the term with the exponent equal to k_c should be removed from $U_{c\,j\text{-}1}(z, l_c)$ before applying Equation (8.13) (excluding the combination in which j–1 first versions produce k_c correct outputs while the m_jth version fails).

If after the execution of j first versions the number of correct outputs produced is k and $k+n_c-j<k_c$, then the required number of correct outputs k_c cannot be obtained even if all the n_c-j subsequent versions produce correct outputs. Therefore, the terms $\pi_{jk}z^k$ with $k<k_c-n_c+j$ can be removed from $U_{cj}(z, l_c)$.

The above considerations lie at the base of the following algorithm for determining all of the probabilities $p_{cj}(l_c)$ ($k_c\le j\le n_c$):

1. For the given l_c, determine the order of version termination m_1, m_2, ..., m_{n_c} using the algorithm from Section 8.1.1.2.

2. Determine the u-function of each version of component c according to Equation (8.10).

3. Define $U_{c0}(z, l_c) = 1$. For $j = 1, 2, \ldots, n_c$:

3.1 Obtain $U_{cj}(z, l_c)$ using Equation (8.13) and, after collecting like terms, represent it in the form (8.12).

3.2. Remove from $U_{cj}(z, l_c)$ all the terms $\pi_{jk} z^k$ for which $k < k_c - n_c + j$.

3.3. If $j \geq k_c$, assign: $p_{cj}(l_c) = \pi_{jk_c}$ and remove term $\pi_{jk_c} z^{k_c}$ from $U_{cj}(z, l_c)$.

8.1.1.5 Execution Time Distribution for the Entire System

Having the pairs $p_{cj}(l_c(x))$, $t_{cm_j}(l_c(x))$ for each possible realization $l_c(x)$ of L_c ($1 \leq x \leq h_c$) and probabilities $\Pr\{L_c = l_c(x)\} = Q_c(x)$, one can obtain the p.m.f. of random execution times T_c for each component by applying Equation (8.6). If the conditional p.m.f. $p_{cj}(l_c(x))$, $t_{cm_j}(l_c(x))$ are represented by the u-function

$$\tilde{u}_c(z, l_c(x)) = \sum_{j=k_c}^{n_c} p_{cj}(l_c(x)) z^{t_{cm_j}(l_c(x))} \tag{8.14}$$

then the u-function representing the p.m.f. of the random value T_c takes the form:

$$\tilde{U}_c(z) = \sum_{x=1}^{h_c} Q_c(x) \tilde{u}_c(z, l_c(x)) \tag{8.15}$$

Since the random system execution time T is equal to the sum of the execution times of all of the C components, one can obtain the u-function $\tilde{U}(z)$ representing the p.m.f. of T as

$$\tilde{U}(z) = \underset{+}{\otimes}(\tilde{U}_1(z), \ldots, \tilde{U}_C(z)) = \prod_{c=1}^{C} \left(\sum_{x=1}^{h_c} Q_c(x) \tilde{u}_c(z, l_c(x)) \right) \tag{8.16}$$

8.1.1.6 Different Components Executed on the Same Hardware

Now consider the case where all of the software components are consecutively executed on the same hardware consisting of h parallel identical modules with the availability a. The number of available parallel hardware modules H is random with p.m.f. $Q(x) = \Pr\{H = x\}$, $1 \leq x \leq h$, defined in the same way as in Equation (8.1).

When $H = x$, the number of versions that can be executed simultaneously in each component c is $l_c(x)$. The u-functions representing the p.m.f. of the corresponding component execution times T_c are $\tilde{u}_c(z, l_c(x))$ defined by Equation (8.14). The u-function $\hat{U}(z, x)$ representing the conditional p.m.f. of the system execution time T (given the number of available hardware modules is x) can be obtained for any x ($1 \leq x \leq h$) as

$$\hat{U}(z,x) = \underset{+}{\otimes}(\tilde{u}_1(z,l_1(x)),...,\tilde{u}_C(z,l_C(x))) = \prod_{c=1}^{C} \tilde{u}_c(z,l_c(x)) \tag{8.17}$$

Having the p.m.f. of the random value H we obtain the u-function $\tilde{U}(z)$ representing the p.m.f. of T as:

$$\tilde{U}(z) = \sum_{x=1}^{H} Q(x)\hat{U}(z,x) \tag{8.18}$$

Example 8.1

Consider a system consisting of two components. The first component consists of $h_1 = 2$ hardware units with availability $a_1 = 0.9$ on which $n_1 = 5$ software versions with $k_1 = 3$ are executed. The second component consists of $h_2 = 3$ hardware units with availability $a_2 = 0.8$ on which and $n_2 = 3$ software versions with $k_2 = 2$ are executed. The parameters of versions r_{ci} and τ_{ci} are presented in Table 8.1.

Table 8.1. Parameters of software versions

Componen			$c = 1$				$c = 2$	
Version	1	2	3	4	5	1	2	3
r_{ci}	0.7	0.6	0.8	0.6	0.9	0.8	0.8	0.7
τ_{ci}	6	12	10	18	14	10	16	12
$t_{ci}(1)$	6	18	28	46	60	10	26	38
$t_{ci}(2)$	6	12	16	30	30	10	16	22
$t_{ci}(3)$	-	-	-	-	-	10	16	12

One software version can be executed on each hardware unit: $l_c(h_c) = h_c$.

The terminations times $t_{ci}(l_c)$ obtained for the different possible values of L_1 and L_2 using the algorithm described in Section 8.1.1.2 are also presented in this table. The version execution diagrams for different values of L_1 and L_2 are presented in Figure 8.1.

For the given parameters, the u-functions of the software versions are

$$u_{11}(z) = 0.3z^0+0.7z^1;\ u_{12}(z) = 0.4z^0+0.6z^1;\ u_{13}(z) = 0.2z^0+0.8z^1$$

$$u_{14}(z) = 0.4z^0+0.6z^1;\ u_{15}(z) = 0.1z^0+0.9z^1$$

for component 1 and

$$u_{21}(z) = 0.2z^0+0.8z^1;\ u_{22}(z) = 0.2z^0+0.8z^1;\ u_{23}(z) = 0.3z^0+0.7z^1$$

for component 2.

The order of version termination in the first component is 1, 2, 3, 4, 5 for both $L_1 = 1$ and $L_1 = 2$ (see Table 8.1).

According to the algorithm presented in Section 8.1.1.4, determine the u-functions for the groups of versions and corresponding probabilities $p_{1j}(1)$ and $p_{1j}(2)$.

$$U_{10}(z, 1) = U_{10}(z, 2) = 1; \; U_{11}(z, 1) = U_{11}(z, 2) = u_{11}(z) = 0.3z^0+0.7z^1$$

$$U_{12}(z, 1) = U_{12}(z, 2) = u_{11}(z) \underset{+}{\otimes} u_{12}(z) = (0.3z^0+0.7z^1)(0.4z^0+0.6z^1)$$

$$= 0.12z^0+0.46z^1+0.42z^2$$

$$U_{13}(z, 1) = U_{13}(z, 2) = U_{12}(z, 1) \underset{+}{\otimes} u_{13}(z) = (0.12z^0+0.46z^1+0.42z^2)$$

$$\times(0.2z^0+0.8z^1) = 0.024z^0+0.188z^1+0.452z^2+0.336z^3$$

Remove the term $0.024z^0$ from $U_{13}(z, 1)$ according to step 3.2 of the algorithm. Remove the term $0.336z^3$ from $U_{13}(z, 1)$ and $U_{13}(z, 2)$ and obtain $p_{13}(1) = p_{13}(2) = 0.336$ according to step 3.3 of the algorithm:

$$U_{14}(z, 1) = U_{14}(z, 2) = U_{13}(z, 1) \underset{+}{\otimes} u_{14}(z) = (0.188z^1+0.452z^2)$$

$$\times(0.4z^0+0.6z^1) = 0.0752z^1+0.2936\, z^2+0.2712z^3$$

Remove the term $0.0752z^1$ from $U_{14}(z, 1)$ according to step 3.2 of the algorithm. Remove the term $0.2712z^3$ from $U_{14}(z, 1)$ and $U_{14}(z, 2)$ and obtain $p_{14}(1) = p_{14}(2) = 0.2712$ according to step 3.3 of the algorithm:

$$U_{15}(z, 1) = U_{15}(z, 2) = U_{14}(z, 1) \underset{+}{\otimes} u_{15}(z)$$

$$= 0.2936z^2\,(0.1z^0+0.9z^1) = 0.02936z^2+0.11z^2+0.26424z^3$$

Finally, obtain $p_{15}(1) = p_{15}(2) = 0.26424$.

Having the probabilities $p_{1j}(1)$, $p_{1j}(2)$ and the corresponding termination times $t_{1j}(1)$, $t_{1j}(2)$, define the u-functions representing the component execution time distributions

$$\tilde{u}_1(z, 1) = 0.336z^{28}+0.271z^{46}+0.264z^{60}$$

$$\tilde{u}_1(z, 2) = 0.336z^{16}+0.271z^{30}+0.264z^{30} = 0.336z^{16}+0.535z^{30}$$

The p.m.f. of L_1 is

$$Q_1(1) = \Pr\{L_1 = 1\} = \Pr\{h_1 = 1\} = 2a_1(1-a_1) = 0.18$$

$$Q_1(2) = \Pr\{L_1 = 2\} = \Pr\{h_1 = 2\} = a_1^{\,2} = 0.81$$

According to Equation (8.15), obtain

$$\tilde{U}_1(z) = 0.18\,\tilde{u}_1(z, 1) + 0.81\,\tilde{u}_1(z, 2)$$

$$= 0.06z^{28} + 0.049z^{46} + 0.048z^{60} + 0.272z^{16} + 0.434z^{30}$$

The order of version termination in the second component is 1, 2, 3 for both $L_2 = 1$ and $L_2 = 2$ and 1, 3, 2 for $L_2 = 3$ (see Table 8.1).

According to the algorithm presented in Section 8.1.1.4, determine the u-functions for the groups of versions and corresponding probabilities $p_{2j}(1)$, $p_{2j}(2)$ and $p_{2j}(3)$. For $L_2 = 1$ and $L_2 = 2$:

$$U_{20}(z, 1) = U_{20}(z, 2) = 1;\ U_{21}(z, 1) = U_{21}(z, 2) = u_{21}(z) = 0.2z^0 + 0.8z^1$$

$$U_{22}(z, 1) = U_{22}(z, 2) = u_{21}(z) \underset{+}{\otimes} u_{22}(z)$$

$$= (0.2z^0 + 0.8z^1)^2 = 0.04z^0 + 0.32z^1 + 0.64z^2$$

Remove the term $0.04z^0$ from $U_{22}(z, 1)$ according to step 3.2 of the algorithm. Remove the term $0.64z^2$ from $U_{22}(z, 1)$ and $U_{22}(z, 2)$ and obtain $p_{22}(1) = p_{22}(2) = 0.64$ according to step 3.3 of the algorithm:

$$U_{23}(z, 1) = U_{23}(z, 2) = U_{22}(z, 1) \underset{+}{\otimes} u_{23}(z) = 0.32z^1\,(0.3z^0 + 0.7z^1)$$

$$= 0.096z^1 + 0.224z^2$$

Obtain $p_{23}(1) = p_{23}(2) = 0.224$.

Having the probabilities $p_{2j}(1)$, $p_{2j}(2)$ and the corresponding termination times $t_{2j}(1)$, $t_{2j}(2)$, define the u-functions representing the component execution time distributions

$$\tilde{u}_2(z, 1) = 0.64z^{26} + 0.224z^{38},\ \tilde{u}_2(z, 2) = 0.64z^{16} + 0.224z^{22}$$

For $L_2 = 3$:

$$U_{20}(z, 3) = 1;\ U_{21}(z, 3) = u_{21}(z) = 0.2z^0 + 0.8z^1;\ U_{22}(z, 3) = u_{21}(z) \underset{+}{\otimes} u_{23}(z)$$

$$= (0.2z^0 + 0.8z^1)(0.3z^0 + 0.7z^1) = 0.06z^0 + 0.38z^1 + 0.56z^2$$

Remove the term $0.06z^0$ from $U_{22}(z, 1)$ according to step 3.2 of the algorithm. Remove the term $0.56z^2$ from $U_{22}(z)$ and obtain $p_{22}(3) = 0.56$ according to step 3.3 of the algorithm:

$$U_{23}(z, 3) = U_{22}(z, 3) \underset{+}{\otimes} u_{22}(z) = 0.38z^1(0.2z^0 + 0.8z^1)$$

$$= 0.076z^1 + 0.304z^2$$

From $U_{23}(z, 3)$ obtain $p_{23}(3) = 0.304$.

The u-function representing the corresponding component execution time distribution takes the form

$$\tilde{u}_2(z, 3) = 0.56z^{12} + 0.304z^{16}$$

The p.m.f. of L_2 is

$$Q_2(1) = \Pr\{L_2 = 1\} = \Pr\{h_2 = 1\} = 3a_2(1-a_2)^2 = 0.096$$

$$Q_2(2) = \Pr\{L_2 = 2\} = \Pr\{h_2 = 2\} = 3a_2^2(1-a_2) = 0.384$$

$$Q_2(3) = \Pr\{L_2 = 3\} = \Pr\{h_2 = 3\} = a_2^3 = 0.512$$

According to Equation (8.15), obtain

$$\tilde{U}_2(z) = 0.096\tilde{u}_2(z, 1) + 0.384\tilde{u}_2(z, 2) + 0.512\tilde{u}_2(z, 3)$$

$$= 0.287z^{12} + 0.401z^{16} + 0.086z^{22} + 0.0614z^{26} + 0.0215z^{38}$$

The u-function representing the execution time distribution for the entire system takes the form

$$\tilde{U}(z) = \tilde{U}_1(z) \underset{+}{\otimes} \tilde{U}_2(z) = (0.272z^{16} + 0.434z^{30} + 0.06z^{28} + 0.049z^{46}$$

$$+ 0.048z^{60})(0.287z^{12} + 0.401z^{16} + 0.086z^{22} + 0.0614z^{26} + 0.0215z^{38})$$

$$= 0.078z^{28} + 0.109z^{32} + 0.023z^{38} + 0.017z^{40} + 0.141z^{42} + 0.024z^{44} + 0.174z^{46}$$

$$+ 0.005z^{50} + 0.037z^{52} + 0.007z^{54} + 0.027z^{56} + 0.014z^{58} + 0.020z^{62} + 0.001z^{66}$$

$$+ 0.012z^{68} + 0.020z^{72} + 0.019z^{76} + 0.004z^{82} + 0.001z^{84} + 0.003z^{86} + 0.001z^{98}$$

Having this u-function, one can obtain the system reliability and conditional expected execution time for different time constraints w:

$$R(\infty) = 0.739;\quad \tilde{\varepsilon}(\infty) = \frac{1}{0.739}(32.94) = 44.559$$

For $w = 50$:

$$R(50) = 0.078 + 0.109 + 0.023 + 0.017 + 0.141 + 0.024 + 0.174 = 0.567$$

$$\tilde{\varepsilon}(50) = \frac{1}{0.567}(0.078\times28+0.109\times32+0.023\times38+0.017\times40$$
$$+\,0.141\times42+0.024\times44+0.174\times46) = 22.24/0.567 = 39.24$$

Example 8.2

Consider a system consisting of five components. The number of parallel hardware units, the availability of these units, and the parameters n_c and k_c of the fault-tolerant programs for each component are presented in Table 8.2. Components 1, 3 and 5 are of the NVP type and components 2 and 4 are of the RBS type ($k_2 = k_4 = 1$). The reliability and execution times of the software versions are presented in Table 8.3.

Table 8.2. Parameters of system components

Component	1	2	3	4	5
h_c	3	4	2	3	2
a_c	0.80	0.95	0.90	0.85	0.95
n_c	5	2	3	3	5
k_c	3	1	2	1	3
$l_c(1)$	2	0	1	1	3
$l_c(2)$	4	1	3	2	5
$l_c(3)$	5	1	-	3	-
$l_c(4)$	-	2	-	-	-

Table 8.3. Parameters of software versions

	Version	1	2	3	4	5
$c = 1$	r_{1i}	0.86	0.77	0.98	0.93	0.91
	τ_{1i}	10	12	25	22	15
$c = 2$	r_{2i}	0.85	0.92	-	-	-
	τ_{2i}	30	45	-	-	-
$c = 3$	r_{3i}	0.87	0.94	0.98	-	-
	τ_{3i}	6	6	10	-	-
$c = 4$	r_{4i}	0.95	0.85	0.85	-	-
	τ_{4i}	25	15	20	-	-
$c = 5$	r_{5i}	0.68	0.79	0.90	0.90	0.94
	τ_{5i}	10	10	15	20	25

The entire system's execution time varies in the range of $81\le T\le 241$. The system's reliability and conditional expected execution time obtained by the algorithm described in Sections 8.1.1.1-8.1.1.5 are $R(\infty) = 0.77$ and $\tilde{\varepsilon}(\infty) = 96.64$ respectively. The functions $R(w)$ and $\tilde{\varepsilon}(w)$ are presented in Figure 8.2A.

Now consider the same fault-tolerant software system running on a single hardware block consisting of three parallel units with availability $a = 0.9$. The functions $l_c(x)$ for each software component are presented in Table 8.4. Note that the system can operate only when $H\ge 2$, since a single hardware unit has not enough resources for execution of the second software component.

Table 8.4. Number of versions executed simultaneously

Component	1	2	3	4	5
$l_c(1)$	2	0	3	1	1
$l_c(2)$	4	1	3	2	2
$l_c(3)$	5	1	3	3	3

The entire system execution time varies in the range $81 \le T \le 195$. The system reliability and expected execution time (without respect to execution time constraints) are respectively $R(\infty) = 0.917$ and $\tilde{\varepsilon}(\infty) = 103.18$. The functions $R(w)$ and $\tilde{\varepsilon}(w)$ are presented in Figure 8.2B.

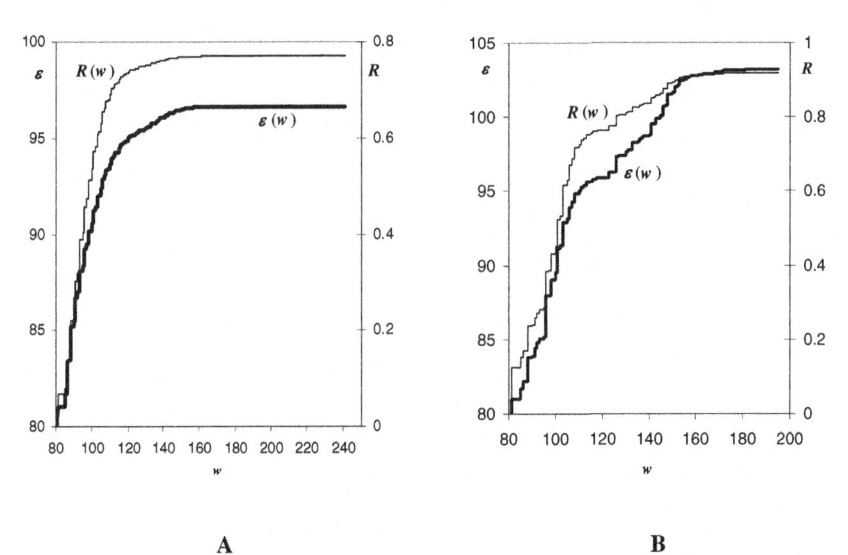

Figure 8.2. $R(w)$ and $\tilde{\varepsilon}(w)$ functions for a fault-tolerant system running on different hardware components (A) and on a single hardware component (B)

8.2 Optimal Version Sequencing in Fault-tolerant Programs

In programs consisting of versions with different parameters, the sequence of the version execution affects the distribution of the system's task execution time. The influence of the sequence of the version's execution on this distribution is demonstrated in Figure 8.3.

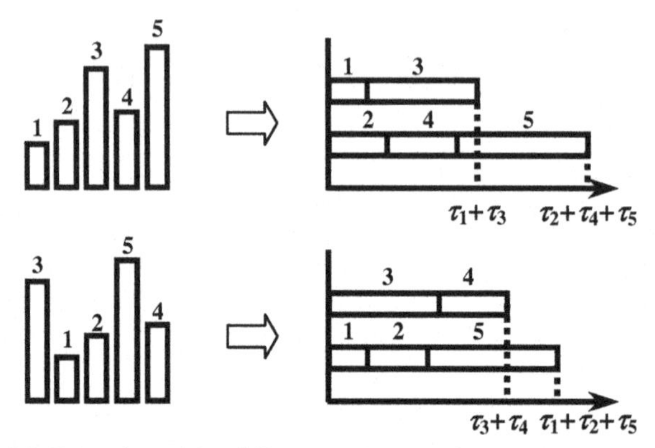

Figure 8.3. Execution of the different sequences of versions in a component with $n_c = 5$, $k_c = 4$, $L_c = 2$

While the sequence in which the versions in each component start their execution does not affect the $R(\infty)$ index, it can have a considerable influence on both $R(w)$ and $\tilde{\varepsilon}(w)$ when the task execution time is constrained. Therefore, two different optimization problems can be formulated in which the sequences maximizing the system's reliability $R(w)$ or minimizing its conditional expected execution time $\tilde{\varepsilon}(w)$ are to be determined.

To apply the GA, one has to represent the sequences of versions in each component in the form of strings. These sequences can be represented by C substrings corresponding to different components. Each substring c should be of length n_c and contain a permutation of integer numbers ranging from 1 to n_c.

The solution encoding scheme with different substrings is complicated and requires the development of sophisticated procedures for string generation, crossover, and mutation that preserve the feasibility of solutions. In order to simplify the GA procedures, an encoding method was developed in which the single permutation defines the sequences of the versions in each one of the C components.

The solution encoding string consists of $n = \sum_{c=1}^{C} n_c$ different integer numbers ranging from 1 to n. Each number j belonging to the interval

$$\sum_{c=1}^{m-1} n_c + 1 \le j \le \sum_{c=1}^{m} n_c \tag{8.19}$$

corresponds to version $j - \sum_{c=1}^{m-1} n_c$ of component m. The relative order in which the numbers corresponding to the versions of the same component appear in the string determines the sequence of their execution.

Example 8.3

Consider a system consisting of three components with $n_1 = 3$, $n_2 = 5$ and $n_3 = 4$. The solution encoding strings should consist of 3+5+4 = 12 integers. Numbers 1, 2, 3 correspond to component 1, numbers 4, 5, 6, 7, 8 correspond to component 2 (they are marked in bold), numbers 9, 10, 11, 12 correspond to component 3 (they are marked in italic). In the solution string $\boldsymbol{a}$ = (**4**, *9*, **7**, 1, *12*, 2, 3, *10*, **6**, **8**, **5**, *11*) the numbers corresponding to different components appear in the following relative order:

for component 1: 1, 2, 3
for component 2: **4, 7, 6, 8, 5**
for component 3: *9, 12, 10, 11*

This corresponds to the following sequences of versions execution:

in component 1: 1, 2, 3
in component 2: 1, 4, 3, 5, 2
in component 3: 1, 4, 2, 3

The solution decoding procedure determines the sequence of versions in each component according to string $\boldsymbol{a}$. Then it calculates the version termination times using the procedure presented in Section 8.1.1.2 and the probabilities p_{cj} using the algorithm presented in Section 8.1.1.3. After obtaining the components' execution time distribution, the time distribution of the entire system is calculated as described in Sections 8.1.1.5 and 8.1.1.6. Finally, the indices $R(w)$ and $\tilde{\varepsilon}(w)$ are obtained from the system's execution time distribution. The solution fitness can be defined as $R(\boldsymbol{a},w)$ or $M-\tilde{\varepsilon}(\boldsymbol{a},w)$, depending on the optimization problem formulation.

Example 8.4

Consider a fault-tolerant software system consisting of five components and running on a single reliable unit [202]. The parameters of the components are presented in Table 8.5. This table contains the values of n_c and k_c for each component c and the reliability and execution time for each version.

First consider the system in which $L_1 = 1$, $L_2 = 2$, $L_3 = 4$, $L_4 = 1$, $L_5 = 3$. The overall system reliability that does not depend on version sequencing is $R(\infty) = 0.92$. The solutions with minimal conditional expected execution time $\tilde{\varepsilon}(\infty)$ and with maximal system reliability $R(w)$ for $w = 300$ are presented in Table 8.6. The table contains minimal and maximal possible system execution times for each solution, values of indices $\tilde{\varepsilon}(\infty)$ and $R(w)$, and the corresponding version sequences.

It can be seen that the minimal possible system execution time $T_{\min}$ (achieved when in each component the first k_c versions succeed) can be obtained when the versions in each component are ordered according to the increased execution time. Such a solution is also presented in Table 8.6. Observe that the solution that provides the smallest minimal execution time is not optimal, neither in terms of $\tilde{\varepsilon}(\infty)$ nor in terms of $R(w)$.

Table 8.5. Parameters of system components for the numerical example

No. of component	n_c	k_c		Versions				
				1	2	3	4	5
1	4	1	τ	17	20	32	75	
			r	0.71	0.85	0.89	0.98	-
2	3	2	τ	28	55	58	-	-
			r	0.85	0.85	0.93	-	-
3	5	3	τ	17	20	38	41	63
			r	0.80	0.80	0.86	0.98	0.98
4	3	2	τ	17	20	32	-	-
			r	0.75	0.93	0.97	-	-
5	3	1	τ	30	54	70	-	-
			r	0.70	0.80	0.89	-	-

The poor solution corresponding to the greatest possible $\tilde{\varepsilon}(\infty)$ is presented in Table 8.6 for comparison. The system's reliabilities $R(w)$ as functions of the maximal allowable execution time w are presented in Figure 8.4A for all of the solutions obtained (numbered according to Table 8.6).

Table 8.6. Parameters of solutions obtained for the fault-tolerant system with $L_1 = 1, L_2 = 2, L_3 = 4, L_4 = 1, L_5 = 3$

No.	Problem formulation	Sequence of versions	T_{min}	T_{max}	$\tilde{\varepsilon}(\infty)$	$R(300)$
1	$\tilde{\varepsilon}(\infty)\rightarrow$min	2134|132|54321|213|132	183	429	211.91	0.914
2	$R(300)\rightarrow$max	2314|312|43521|321|123	198	429	220.22	0.915
3	Increasing τ	1234|123|12345|123|123	177	449	213.84	0.909
4	$\tilde{\varepsilon}(\infty)\rightarrow$max	4312|213|52134|132|231	247	432	277.67	0.776

Now consider the same system with $L_1 = L_2 = L_3 = L_4 = L_5 = 1$, which corresponds to consecutive execution of versions one at a time. The maximal possible time of system execution in this case does not depend on the versions sequence and is equal to the sum of execution times of all of the versions. The overall system reliability that does not depend on either version sequencing or on L_c is still $R(\infty) = 0.92$. The same four types of solution that were obtained in the previous example are presented in Table 8.7 (the maximal allowable execution time in this case is $w = 430$).

Table 8.7. Parameters of solutions obtained for the fault-tolerant system with $L_1 = L_2 = L_3 = L_4 = L_5 = 1$

No.	Problem formulation	Sequence of versions	T_{min}	T_{max}	$\tilde{\varepsilon}(\infty)$	$R(430)$
1	$\tilde{\varepsilon}(\infty)\rightarrow$min	2134|3122|12435|213|123	251	687	313.02	0.886
2	$R(430)\rightarrow$max	2314|132|41235|231|123	266	687	321.33	0.892
3	increasing τ	1234|123|12345|123|123	242	687	316.19	0.886
4	$\tilde{\varepsilon}(\infty)\rightarrow$max	4312|321|53421|312|321	449	687	469.62	0

Observe that the minimal possible system execution time for the solution with the greatest possible $\tilde{\varepsilon}(\infty)$ is greater than w. Therefore, the corresponding $R(w) = 0$. The system reliabilities $R(w)$ as functions of maximal allowable execution time w are presented in Figure 8.4B for all of the solutions obtained.

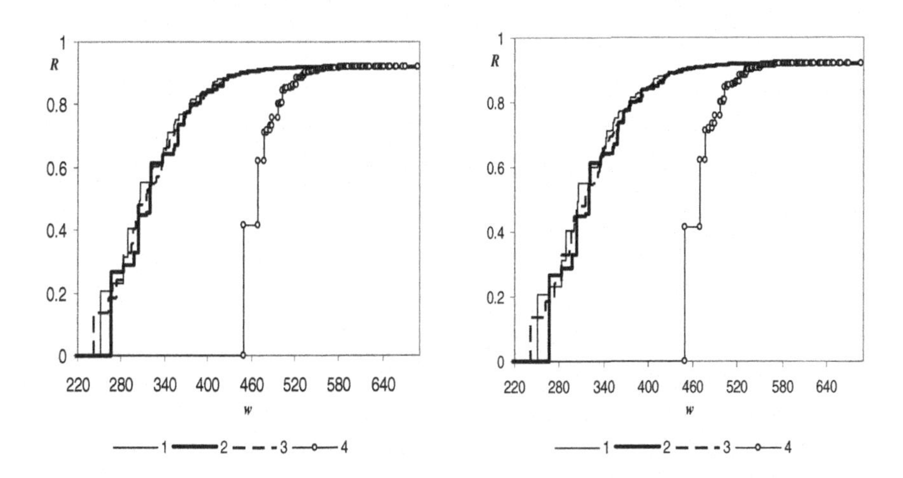

Figure 8.4. $R(w)$ functions for solutions obtained for the system with $L_1 = 1$, $L_2 = 2$, $L_3 = 4$, $L_4 = 1$, $L_5 = 3$ (A) and for the system with $L_1 = L_2 = L_3 = L_4 = L_5 = 1$ (B)

8.3 Optimal Structure of Fault-tolerant Software Systems

When a fault-tolerant software system is designed, one has to choose software versions for each component and find the sequence of their execution in order to achieve the system's greatest reliability subject to cost constraints. The versions are chosen from a list of the available products. Each version is characterized by its reliability, execution time, and cost. The total cost of the system's software is defined according to the cost of its versions. The cost for each version can be the purchase cost if the versions are commercial and the off-the-shelf cost, or it can be an estimate based upon the version's size, complexity, and performance.

Assume that B_c functionally equivalent versions are available for each component c and that the number k_c of the versions that should agree in each component is predetermined. The choice of the versions and the sequence of their execution in each component determine the system's entire reliability and performance.

The permutation $\boldsymbol{x}^*_c$ of B_c different integer numbers ranging from 1 to B_c determines the order of the version that can be used in component c. Let $y_{cb} = 1$ if the version b is chosen to be included in component c and $y_{cb} = 0$ otherwise. The binary vector $\boldsymbol{y}_c = \{y_{c1},\dots,y_{cB_c}\}$ determines the subset of versions chosen for component c. Having the vectors $\boldsymbol{x}^*_c$ and $\boldsymbol{y}_c$ one can determine the execution order $\boldsymbol{x}_c$ of the versions chosen by removing from $\boldsymbol{x}^*_c$ any number b for which $y_{cb} = 0$. The total number of versions in component c (equal to the length of vector $\boldsymbol{y}_c$ after removing the unchosen versions) is determined as

$$n_c = \sum_{b=1}^{B_c} y_{cb} \tag{8.20}$$

The system structure optimization problem can now be formulated as find vectors $\boldsymbol{x}_c$ for $1 \le c \le C$ that maximize $R(w)$ subject to cost constraint

$$\Omega = \sum_{c=1}^{C} \sum_{b \in \boldsymbol{x}_c} \omega_{cb} \le \Omega^* \tag{8.21}$$

where ω_{cb} is the cost of version b used in component c, Ω is the entire system cost and Ω^* is the maximal allowable system cost. Note that the length of vectors $\boldsymbol{x}_c$ can vary depending on the number of versions chosen.

In order to encode the variable-length vectors $\boldsymbol{x}_c$ in the GA using the constant-length integer strings one can use (B_c+1)-length strings containing permutations of numbers $1,\ldots, B_c, B_c+1$. The numbers that appear before B_c+1 determine the vector $\boldsymbol{x}_c$. For example, for $B_c = 5$ the permutations (2, 3, 6, 5, 1, 4) and (3, 1, 5, 4, 2, 6) correspond to $\boldsymbol{x}_c = (2, 3)$ and $\boldsymbol{x}_c = (3, 1, 5, 4, 2)$ respectively. Any possible vector $\boldsymbol{x}_c$ can be represented by the corresponding integer substring containing the permutation of B_c+1 numbers. By combining C substrings corresponding to different components one obtains the integer string $\boldsymbol{a}$, that encodes the entire system structure.

As in the version sequencing problem (Section 8.2), the encoding method is used in which the single permutation defines the sequences of the versions chosen in each of the C components. The solution encoding string is a permutation of $n = \sum_{c=1}^{C}(B_c+1)$ integer numbers ranging from 1 to n. Each number j belonging to the interval $\sum_{c=1}^{m-1}(B_c+1)+1 \le j \le \sum_{c=1}^{m}(B_c+1)$ corresponds to version $j - \sum_{c=1}^{m-1}(B_c+1)$ of component m. The relative order in which the numbers corresponding to the versions of the same component appear in the string determines the structure of this component.

Example 8.5

Consider, for example, a system consisting of three components with $B_1 = 3$, $B_2 = 5$ and $B_3 = 4$. The solution encoding strings consist of (3+1)+(5+1)+(4+1) = 15 integers. Numbers from 1 to 4 correspond to component 1, numbers from 5 to 10 correspond to component 2, and numbers from 11 to 15 correspond to component

3. In the solution string $\boldsymbol{a}$ = (5, 9, 7, 1, 12, 2, 15, 3, 10, 6, 14, 8, 4, 13, 5, 11), the numbers corresponding to different components appear in the following relative order:

for component 1: 1, 2, 3, **4**
for component 2: 5, 9, 7, **10**, 8
for component 3: 12, **15**, 14, 13, 11

Only the numbers located before the largest number in each substring (marked in bold) represent the component's structure. This gives the following substrings:

for component 1: 1, 2, 3; for component 2: 5, 9, 7; for component 3: 12

These substrings correspond to the following sequences of versions execution:

in component 1: 1, 2, 3; in component 2: 1, 5, 3; in component 3: 2

The solution decoding procedure determines the sequence of versions in each component. Then it determines the system reliability $R(w)$ as described in Section 8.1 and calculates the system cost using Equation (8.21). The solution fitness is evaluated as $R(w)-\pi\max(\Omega-\Omega^*, 0)$, where π is a penalty coefficient.

Example 8.6

Consider a fault-tolerant software system consisting of four components running on fully available hardware [203]. The parameters of the versions that can be used in these components are presented in Table 8.8. This table contains the values of k_c and L_c for each component and the cost, reliability, and execution time for each version.

Table 8.8. Parameters of fault-tolerant system components and versions

No. of component	k_c	L_c		Versions							
				1	2	3	4	5	6	7	8
			τ	17	10	20	32	30	75	-	-
1	1	1	r	0.71	0.85	0.85	0.89	0.95	0.98	-	-
			c	5	15	7	8	12	6	-	-
			τ	28	55	35	55	58	-	-	-
2	2	2	r	0.82	0.82	0.88	0.90	0.93	-	-	-
			c	11	8	18	10	16	-	-	-
			τ	17	20	38	38	48	50	41	63
3	3	4	r	0.80	0.80	0.86	0.90	0.90	0.94	0.98	0.98
			c	4	3	4	6	5	4	9	6
			τ	17	10	20	32	-	-	-	-
4	2	1	r	0.75	0.85	0.93	0.97	-	-	-	-
			c	12	16	17	17	-	-	-	-
			τ	30	54	40	65	70	-	-	-
5	1	3	r	0.70	0.80	0.80	0.80	0.89	-	-	-
			c	5	9	11	7	12	-	-	-

Two sets of solutions were obtained for the maximal allowable system operation times w = 250 and w = 300. For each value of w, four different solutions were obtained for different cost constraints. These solutions are presented in Tables 8.9 and 8.10. The tables contain the system cost and reliability for each solution, the expected conditional execution time, minimal and maximal possible system execution times, and the corresponding sequences of the versions chosen.

The functions $R(w)$ for the solutions obtained are presented in Figure 8.5.

Table 8.9. Parameters of solutions obtained for $w = 250$

Ω*	Sequence of versions	Ω	T_{min}	T_{max}	$\tilde{\varepsilon}(\infty)$	$R(250)$
160	231|541|37162|324|214	159	166	307	188.34	0.913
140	34|241|64231|234|123	140	173	301	194.43	0.868
120	5|431|31562|43|21	119	205	249	217.07	0.752
100	3|241|4562|43|41	100	205	270	220.52	0.598

Table 8.10. Parameters of solutions obtained for $w = 300$

Ω*	Sequence of versions	Ω	T_{min}	T_{max}	$\tilde{\varepsilon}(\infty)$	$R(300)$
160	341|4521|85632|324|41	160	188	369	210.82	0.951
140	53|541|28361|431|51	140	208	335	231.02	0.889
120	6|241|61372|241|31	120	240	307	252.87	0.813
100	4|142|2386|43|41	100	219	295	238.05	0.672

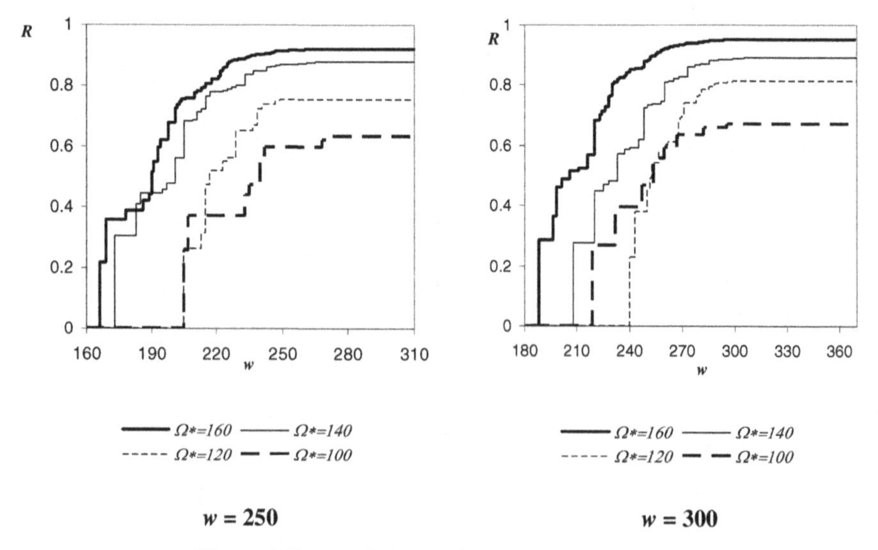

Figure 8.5. $R(w)$ functions for the solutions obtained

Observe that the greater the reliability level achieved, the greater the cost of further reliability improvement. The cost-reliability curves are presented in Figure 8.6. Each point on these curves corresponds to the best solution obtained by the GA.

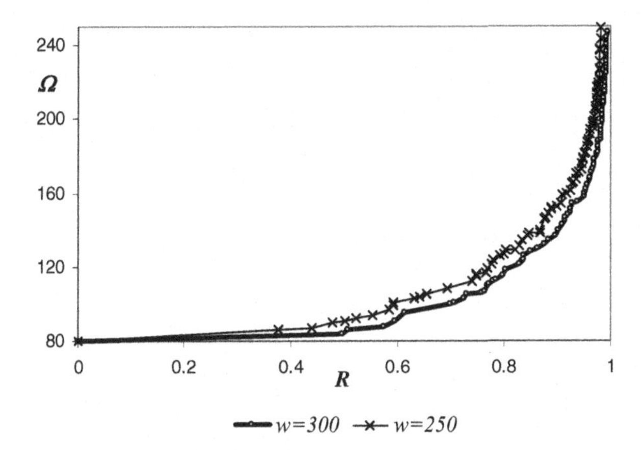

Figure 8.6. Reliability-cost curves for the best obtained solutions

References

1. Ushakov I. A universal generating function.Sov J Comput Syst Sci 1986;24:37-49.
2. Ushakov I. Optimal standby problem and a universal generating function. Sov J Comput Syst Sci 1987;25:61-73.
3. Grimmett G, Stirzaker D. Probability and random processes. Second edition. Clarendon Press, Oxford, 1992.
4. Ross S. Introduction to Probability Models. Seventh Edition. Boston: Academic Press, 2000.
5. Holland J. Adaptation in Natural and Artificial Systems. The University of Michigan Press, Ann Arbor, Michigan, 1975.
6. Goldberg D. Genetic Algorithms in Search, Optimization and Machine Learning. Addison-Wesley, 1989.
7. Michalewicz Z. Genetic Algorithms + Data Structures = Evolution Programs. Third edition. Berlin: Springer-Verlag, 1996.
8. Kinnear K. Generality and Difficulty in Genetic Programming: Evolving a Sort. In Proceedings of the Fifth International Conference on Genetic Algorithms. Ed. Forrest S. Morgan Kaufmann. San Mateo, CA. 1993; 287-94.
9. Syswerda G. A study of reproduction in generational and steady state genetic algorithms. In Foundations of genetic algorithms, Morgan Kauffman, San Mateo, CA. 1991; 94-101.
10. Vavak F, Fogarty T. A comparative study of steady state and generational genetic algorithms for use in nonstationary environments. In Evolutionary computing (Lecture notes in computer science; 1143), Springer, Brighton, UK, 1996; 297-306.
11. Whitley D. The Genitor algorithm and selective pressure: Rank best allocation of reproductive trails is best. Proc. 3th International Conference on Genetic Algorithms. Ed. Schaffer D. Morgan Kaufmann, 1989; 116-121.
12. Kuo W, Zuo M. Optimal reliability modelling. New Jersey: John Wiley & Sons, 2003.
13. Kozlov B, Ushakov I. Reliability Handbook. New York; London: Holt, Rinehart and Winston: 1970 (Translation from Russian edition 1966).
14. Barlow R, Heidtmann K. Computing *k*-out-of-*n* system reliability. IEEE Trans Reliab 1984;33:322-3.
15. Chiang D, Niu S. Reliability of consecutive-*k*-out-of-*n*:*F* systems. IEEE Trans Reliab 1981;30: 87-9.
16. Bollinger R, Salvia A. Consecutive-*k*-out-of-*n*:*F* networks. IEEE Trans Reliab 1982;31:53-5.
17. Chao M, Lin G. Economical design of large consecutive-*k*-out-of-*n*:*F* systems. IEEE Trans Reliab 1984;33:411-3.
18. Hwang FK. Invariant permutations for consecutive-*k*-out-of-*n*:*F* cycles. IEEE Trans Reliab 1989;38:65-7.

19. Kuo W, Zhang W, Zuo M. A consecutive-k-out-of-n:G system: the mirror image of a consecutive-k-out-of-n:F system. IEEE Trans Reliab 1990;39: 244-53.
20. Zuo M, Kuo W. Design and performance analysis of consecutive-k-out-of-n structure. Naval Research Logistics 1990;37:203-30.
21. Kontoleon J. Reliability determination of a r-successive-out-of-n:F system. IEEE Trans Reliab 1980;29:437.
22. Derman C, Lieberman G, Ross S. On the consecutive-k-out-of-n:F system. IEEE Trans Reliab 1982;31:57-63.
23. Bollinger R. Direct computations for consecutive-k-out-of-n:F systems. IEEE Trans Reliab 1982;31:444-6.
24. Lambiris M. Papastavridis S. Exact reliability formulas for linear and circular consecutive-k-out-of-n:F systems. IEEE Trans Reliab 1985;34:124-6.
25. Goulden I. Generating functions and reliabilities for consecutive-k-out-of-n:F systems. Utilitas Mathematic 1987;32:141-7.
26. Hwang F. Simplified reliabilities for consecutive-k-out-of-n:F systems. SIAM Journal on Algebraic and Discrete Methods 1986;7:258-64.
27. Hwang F. Fast solutions for consecutive-k-out-of-n:F system. IEEE Trans Reliab 1982;31:447-8.
28. Hwang F, Yao Y. A direct argument for Kaplansky's theorem on cyclic arrangement and its generalization. Oper Res Lett 1991;10:241-3.
29. Du D, Hwang FK. A direct algorithm for computing reliability of a consecutive-k cycle. IEEE Trans Reliab 1988;37:70-2.
30. Chang G, Cui L, Hwang FK. Reliabilities of consecutive-k systems. Boston: Kluwer, 2000.
31. Wu J, Chen R. An $O(kn)$ algorithm for a circular consecutive-k-out-of-n:F system. IEEE Trans Reliab 1992;41:303-5.
32. Tong Y. Some new results on the reliability of circular consecutive-k-out-of-n:F system. In Reliability and quality control. Ed. Basu A. North-Holland: Elsevier, 1986; 395-400.
33. Antonopoulou I, Papastavridis S. Fast recursive algorithm to evaluate the reliability of circular consecutive-k-out-of-n:F system. IEEE Trans Reliab 1987;36:83-4.
34. Korczak E. New efficient algorithm for calculating dependability indices of a circular consecutive k-out-of-n:F systems. Prace Przemyslowego Instytutu Telekomunikacji, 1995;45:20-7 (in Polish).
35. Chang J, Chen R, Hwang F. A fast reliability-algorithm for the circular consecutive-weighted- k-out-of-n:F system. IEEE Trans Reliab 1998;47:472-4.
36. Griffith W. On consecutive k-out-of-n failure systems and their generalizations. In Reliability and quality control. Ed. Basu A. North-Holland : Elsevier, 1986; 157-65.
37. Tong Y. A rearrangement inequality for the longest run with an application in network reliability. J Appl Probab 1985; 22: 386-93.
38. Saperstain B. The generalized birthday problem. J Amer Statistical Assoc 1972: 67: 425-8.
39. Saperstain B. On the occurrence of n successes within N Bernoulli trails. Technometrics 1973;15:809-18.
40. Naus J. Probabilities for a generalized birthday problem. J Amer Statistical Assoc 1974;69:810-5.
41. Nelson J. Minimal-order models for false-alarm calculations on sliding windows. IEEE Trans Aerospace Electronic Systems 1978;14:351-63.
42. Sfakianakis M, Kounias S, Hillaris A. Reliability of consecutive-k-out-of-r-from-n:F systems. IEEE Trans Reliab 1992;41:442-7.

43. Papastavridis S, Sfakianakis M. Optimal-arrangement & Importance of the components in a consecutive-k-out-of-r-from-n:F system. IEEE Trans Reliab 1991;40:277-9.
44. Cai J. Reliability of a large consecutive-k-out-of-r-from-n:F system with unequal component-reliability. IEEE Trans Reliab 1994;43:107-11.
45. Psillakis Z. A simulation algorithm for computing failure probability of a consecutive-k-out-of-r-from-n:F system. IEEE Trans Reliab 1995;44:523-31.
46. Malinowski J. Preuss W. A recursive algorithm evaluating the exact reliability of a consecutive-k-within-m-out-of-n:F system. Microelectron Reliab 1995;35:1461-65.
47. Levitin G. Consecutive-k-out-of-r-from-n system with multiple failure criteria. IEEE Trans Reliab 2004;53:394-400.
48. Billinton R, Allan R. Reliability evaluation of power systems. New York: Plenum Press, 1996.
49. Murchland J, Fundamental concepts and relations for reliability analysis of Multistate systems. Reliability and Fault Tree Analysis. In: Theoretical and Applied Aspects of System Reliability, SIAM 1975,581-618.
50. El-Neveihi E, Proschan F, Setharaman J. Multistate coherent systems. J Appl Probab 1978;15:675-88.
51. Barlow R, Wu A. Coherent systems with multistate components. Mathematics of operations Research 1978;3:275-81.
52. Ross S. Multivalued state element systems. Annals of Probability 1979;7:379-83.
53. Griffith W. Multistate reliability models. J Appl Probab 1980;17:735-44.
54. Natvig B. Two suggestions of how to define a multistate coherent system. Adv. Applied Probab. 1982;14:434-55.
55. Hudson J, Kapur K. Reliability theory for multistate systems with multistate elements. Microelectron Reliab 1982;22:1-7.
56. Natvig, B. Multi-state coherent systems. In: Jonson N, Kotz S. editors. Encyclopedia of Statistical Sciences. NY: Wiley, 1984;5.
57. El-Neveihi E, Prochan F. Degradable systems: A survey of multi-state system theory. Comm. Statist. 1984;13(4):405-32.
58. Reinschke K, Ushakov I. Application of Graph Theory for Reliability Analysis. (in Russian). Moscow: Radio i Sviaz, 1988.
59. Lisnianski A, Levitin G. Multi-state system reliability. Assessment, optimization and applications. Singapore: World Scientific, 2003.
60. El-Neweihi E, Proschan F, Sethuraman J. Multistate coherent systems. J Appl Probab 1978;15:675-88.
61. Singh C. Forced frequency-balancing technique for discrete capacity systems. IEEE Trans Reliab 1983;32:350-3.
62. Rushdi A. Threshold systems and their reliability. Microelectron Reliab 1990;30(2):299-312.
63. Wu J, Chen R. An algorithm for computing the reliability of weighted-k-out-of-n systems. IEEE Trans Reliab 1994;43:327-8.
64. Huang J, Zuo M, Kuo W. Multi-state k-out-of-n systems. In: Pham H, editor. Handbook of Reliability Engineering, Springer, 2003;3-18.
65. Huang J, Zuo M, Wu Y. Generalized multi-state k-out-of-n:G systems. IEEE Trans Reliab 2000;49:105-11.
66. Mathur FP, de Sousa PT. Reliability models of NMR systems. IEEE Trans Reliab 1975;24:604-16.
67. Ben-Dov Y. Optimal reliability design of k-out-of-n systems subject to two kinds of failure. J Operations Research Society 1980;31:743-8.

68. Pham H, Pham M. Optimal design of (k,n−k+1) systems subject to two modes. IEEE Trans Reliab 1991;40:559-62.
69. Pham H. Optimal system size for k-out-of-n systems with competing failure modes. Math Comput Model 1991;15:77-82.
70. Pham H, Malon D. Optimal design of systems with competing failure modes. IEEE Trans Reliab 1994;43:251-4.
71. Satoh N, Sasaki M, Yuge T, Yanagi S. Reliability of 3-state device systems with simultaneous failures. IEEE Trans Reliab 1993;42:470-7.
72. Biernat J. The effect of compensating fault models on n-tuple modular redundant system reliability. IEEE Trans Reliab 1994;43:294-300.
73. Gifford DK. Weighted voting for replicated data. Proc. 7th ACM SIGOPS Symp. Operating System Principles, Pacific Grove, CA 1979:150-9.
74. Nordmann L, Pham H. Weighted voting systems. IEEE Trans Reliab 1999;48:42-9.
75. Levitin G, Lisnianski A. Reliability optimization for weighted voting system, Reliab Engng Syst Safety 2001;71:131-8.
76. Pham H. Reliability analysis of digital communication systems with imperfect voters. Mathematical and Computer Modeling Journal 1997;26:103-112.
77. Shanthikumar J. A recursive algorithm to evaluate the reliability of a consecutive-k-out-of-n:F system. IEEE Trans Reliab 1982;31:442-3.
78. Shanthikumar J. Reliability of systems with consecutive minimal cutsets. IEEE Trans Reliab 1987;36:546-50.
79. Hwang F, Yao Y. Multistate consecutively-connected systems. IEEE Trans Reliab 1989;38:472-4.
80. Kossow A, Preuss W. Reliability of linear consecutively-connected systems with multistate components. IEEE Trans Reliab 1995;44: 518-22.
81. Zuo M, Liang M. Reliability of multistate consecutively-connected systems, Reliab Engng Syst Safety 1994; 44:173-6.
82. Levitin G. Reliability evaluation for linear consecutively-connected systems with multistate elements and retransmission delays. Qual Reliab Eng Int 2001;17:373-8.
83. Malinowski J, Preuss W. Reliability of circular consecutively-connected systems with multistate components. IEEE Trans Reliab 1995;44:532-4.
84. Doulliez P, Jamoulle E. Transportation networks with random arc capacities. RAIRO 1972;3:45-60.
85. Evans J. Maximum flow in probabilistic graphs: The discrete case. Networks 1976;6:161-83.
86. Somers J. Maximum flow in networks with small number of random arc capacities. Networks 1982;12:241-53.
87. Alexopoulos C, Fishman G. Characterizing stochastic flow networks using the Monte Carlo method. Networks 1991;21:775-798.
88. Alexopoulos C, Fishman G. Sensitivity analysis in stochastic flow networks using the Monte Carlo method. Networks 1993;23:605-21.
89. Alexopoulos C. A note on state-space decomposition methods for analyzing stochastic flow networks. IEEE Trans Reliab 1995;44:354-7.
90. Lin Y. A simple algorithm for reliability evaluation of a stochastic-flow network with node failure. Comput Oper Res 2001;28:1277-85.
91. Lin Y. Two-commodity reliability evaluation for a stochastic-flow network with node failure. Comput Oper Res 2002;29:1927-34.
92. Lin Y. Using minimal cuts to evaluate the system reliability of a stochastic-flow network with failures at nodes and arcs. Reliab Engng Syst Safety 2002;75:41-6.
93. Lin Y. Study on the system capacity for a multicommodity stochastic-flow network with node failure. Reliab Engng Syst Safety 2002;78:57-62.

94. Lin Y. Extend the quickest path problem to the system reliability evaluation for a stochastic-flow network. Comput Oper Res 2003;30:567-75.
95. Lin Y. Using minimal cuts to study the system capacity for a stochastic-flow network in two-commodity case. Comput Oper Res 2003;30:1595-607.
96. Levitin G. Reliability evaluation for acyclic consecutively connected networks with multistate elements. Reliab Engng Syst Safety 2001;73:137-43.
97. Levitin G. Reliability evaluation for acyclic transmission networks of multi-state elements with delays. IEEE Trans Reliab 2003;52:231-7.
98. Yeh WC. An evaluation of the multi-state node networks reliability using the traditional binary-state networks reliability algorithm. Reliab Engng Syst Safety 2003;81:1-7.
99. Yeh WC. Multistate-node acyclic networks reliability evaluation based on MC. Reliab Engng Syst Safety 2003;81:225-31.
100. Yeh WC. Search for all *d*-Mincuts of a limited-flow network. Comput Oper Res 2002;29:1843-58.
101. Yeh W. Multistate-node acyclic network reliability evaluation. Reliab Engng Syst Safety 2002;78:123-9.
102. Malinowski J, Preuss W. Reliability evaluation for tree-structured systems with multistate components. Microelectron Reliab 1996; 36:9-17.
103. Malinowski J, Preuss W. Reliability of reverse-tree-structured systems with multistate components. Microelectron Reliab 1996;36:1-7.
104. Teng X, Pham H. Software fault tolerance. In: Pham H, editor. Reliability Engineering Handbook. New Jersey: Springer 2003;585-611.
105. Levitin G. Redundancy optimization for multi-state system with fixed resource requirements and unreliable sources. IEEE Trans Reliab 2001;50:52-9.
106. Levitin G. Universal generating function approach for analysis of multi-state systems with dependent elements. Reliab Engng Syst Safety 2004;84:285-92.
107. Apostolakis GE. The effect of a certain class of potential common mode failures on the reliability of redundant systems. Nuclear Engineering and Design 1976;36:123-33.
108. Dhillon BS, Anude OC. Common-cause failures in engineering systems: a review. Int J Reliability, Quality and Safety Eng 1994;1:103-29.
109. Vaurio JK. An implicit method for incorporating common-cause failures in system analysis. IEEE Trans Reliab 1998;47:173-80.
110. Chae KC, Clark GM. System reliability in the presence of common-cause failures. IEEE Trans Reliab 1986;35:32-5.
111. Levitin G. Incorporating common-cause failures into nonrepairable multi-state series-parallel system analysis. IEEE Trans Reliab 2001;50:380-8.
112. Levitin G. Lisnianski A. Optimal separation of elements in vulnerable multi-state systems. Reliab Engng Syst Safety 2001;73:55-66.
113. Fleming K, Silady F. A risk informed defense-in-depth framework for existing and advanced reactors. Reliab Engng Syst Safety 2002;78:205-25.
114. Birnbaum L. On the importance of different elements in a multi-element system. Multivariate analysis, 2. New York: Academic Press; 1969.
115. Fussell J. How to calculate system reliability and safety characteristics. IEEE Trans Reliab 1975;24:169-74.
116. Meng F. Comparing the importance of system elements by some structural characteristics. IEEE Trans Reliab 1996;45:59-65.
117. Meng F. Some further results on ranking the importance of system elements. Reliab Engng Syst Safety 1995;47:97-101.
118. Elsayed E. Reliability Engineering: Addison Wesley Longman; 1996.

119. Høyland A, Rausand M. System reliability Theory: models and Statistical methods. New York: Wiley; 1994.
120. Hong J, Lie C. Joint reliability-importance of two edges in an undirected network, IEEE Trans Reliab 1993;42:17-23.
121. Armstrong M. Joint reliability-importance of elements. IEEE Trans Reliab 1995;44:408-12.
122. Cheok MC, Parry GW, Sherry RR. Use of importance measures in risk informed applications. Reliab Engng Syst Safety 1998; 60:213-26.
123. Vasseur D, Llory M. International survey on PSA figures of merit. Reliab Engng Syst Safety 1999;66:261-74.
124. Van der Borst M, Shoonakker H. An overview of PSA importance measures. Reliab Engng Syst Safety 2001;72:241-5.
125. Borgonovo E, Apostolakis GE. A new importance measure for risk-informed decision making. Reliab Engng Syst Safety 2001;72:193-212.
126. Griffith W. Multistate reliability models, J Appl Probab 1980:17:735-44.
127. Wu S, Chan L. Performance utility-analysis of multi-state systems, IEEE Trans Reliab 2003;52:14-20.
128. Block H, Savits T. Continuous multi-state structure functions. Oper Res 1984;32:703-14.
129. Baxter L. Continuum structures I. J Appl Probab 1984;21:802-15.
130. Baxter L. Continuum structures II. Math Proc Cambridge Philos Soc 1986;99:331-8.
131. Brunelle R, Kapur K. Continuous-state system reliability: an interpolation approach. IEEE Trans Reliab 1998;47:181-7.
132. Lisnianski A. Estimation of boundary points for continuum-state system reliability measures. Reliab Engng Syst Safety 2001;74:81-8.
133. Gnedenko B, Ushakov I. Probabilistic reliability engineering. New York: John Wiley and Sons, 1995.
134. Levitin G, Lisnianski A, Ben-Haim H, Elmakis D. Redundancy optimization for series-parallel multi-state systems. IEEE Trans Reliab 1998;47:165-72.
135. Levitin G, Lisnianski A, Elmakis D. Structure optimization of power system with different redundant elements. Electr Pow Syst Res 1997;43:19-27.
136. Levitin G. Multi-state series-parallel system expansion scheduling subject to reliability constraints. IEEE Trans Reliab 2000;49:71-9.
137. Barbacci M. Survivability in the age of vulnerable systems. Computer 1996;29:8.
138. Levitin G, Lisnianski A. Optimizing survivability of vulnerable series-parallel multi-state systems. Reliab Engng Syst Safety 2003;79:319-31.
139. Levitin G. Optimal multilevel protection in series-parallel systems. Reliab Engng Syst Safety 2003;81:93-102.
140. Levitin G, Dai Y, Xie M, Poh KL. Optimizing survivability of multi-state systems with multi-level protection by multi-processor genetic algorithm. Reliab Engng Syst Safety 2003;82:93-104.
141. Kuo W, Prasad V. An annotated overview of system-reliability optimization. IEEE Trans Reliab 2000;40:176-87.
142. Quigley J, Walls L. Measuring the effectiveness of reliability growth testing. Qual Reliab Eng Int 1999;15:87-93.
143. Abdou M. A volumetric neutron source for fusion nuclear technology testing and development. Fusion Eng Des 1995; 27:111-53.
144. Zunzanyika X. Simultaneous development and qualification in the fast-changing 3.5" hard-disk-drive technology. In Proc. of the Annual IEEE Reliability and Maintainability Symposium 1995:27.

145. Perera U. Reliability of mobile phones, Proc. of the Annual IEEE Reliability and Maintainability Symposium 1995:33.
146. Cushing M, Krolewski J, Stadterman T, Hum B. The impact of Army reliability standardization improvement on reliability testing, RAC Journal 1996:4.
147. Popentiu F, Boros D. Software reliability growth supermodels. Microelectron Reliab 1996;36:485-91.
148. Lloyd DK. Forecasting reliability growth. Quality and Reliability Engineering Journal 1986;2:19-23.
149. Crow L. Reliability growth projections with applications to integrated testing. In Proceedings of the 41st Technical Meeting, Institute of Environmental Sciences, IES 1995:93-103.
150. Rajgopal J, Mazumdar M. A system based component test plan for series system with type-II sensoring. IEEE Trans Reliab 1996;45:375-8.
151. Coit D. Economic allocation of test times for subsystem-level reliability growth testing. IIE Trans 1998;30:1143-51.
152. Levitin G. Allocation of test times in multi-state systems for reliability growth testing. IIE Trans 2002;34:551-8.
153. Kececioglu D. Reliability Engineering Handbook. Part I and II. Englewood Cliffs, New Jersey: Prentice Hall, 1991.
154. Munoz A, Martorell S, Serdarell V. Genetic algorithms in optimizing surveillance and maintenance of components. Reliab Engng Syst Safety 1997;57:107-20.
155. Monga A, Zuo M. Optimal system design considering maintenance and warranty. Comput Oper Res 1998;25:691-705.
156. Martorell S, Sanchez A, Serdarell V. Age-dependent reliability model considering effects of maintenance and working conditions. Reliab Engng Syst Safety 1999;64:19-31.
157. Van Noortwijk J, Dekker R, Cooke R, Mazzuchi T. Expert judgment in maintenance optimization. IEEE Trans Reliab 1992;41:427-32.
158. Levitin G, Lisnianski A. A new approach to solving problems of multi-state system reliability optimization. Qual Reliab Eng Int 2001;47:93-104.
159. Levitin G, Lisnianski A. Joint redundancy and maintenance optimization for multistate series-parallel systems. Reliab Engng Syst Safety 1998;64:33-42.
160. Levitin G, Lisnianski A. Optimal multistage modernization of power system subject to reliability and capacity requirements. Electr Pow Syst Res 1999;50:183-90.
161. Nakagawa T. Sequential imperfect preventive maintenance policies. IEEE Trans Reliab 1988;37:295-8.
162. Boland P. Periodic replacement when minimal repair costs vary with time. Naval Research Logistics 1982;29:541-6.
163. Levitin G, Lisnianski A. Optimization of imperfect preventive maintenance for multi-state systems. Reliab Engng Syst Safety 2000;67:193-203.
164. Levitin G, Lisnianski A. Structure optimization of power system with bridge topology. Electr Pow Syst Res 1998;45:201-8.
165. Lisnianski A, Levitin G. Structure optimization of multi-state system with time redundancy. Reliab Engng Syst Safety 2000;67:103-12.
166. Levitin G, Lisnianski A. Survivability maximization for vulnerable multi-state systems with bridge topology. Reliab Engng Syst Safety 2000;70: 125-40.
167. Moore E, Shannon C. Reliable circuits using less reliable relays. J. Franklin Inst. 1956;9:191-208.
168. Creveling C. Increasing the reliability of electronic equipment by the use of redundant circuits. Proc. IRE 1956;44:509-15.

169. Lipp J. Topology of switching elements vs. reliability. IRE Trans. Reliability Qual. Contr. 1957;6:35-9.
170. Dhillon B. Design reliability. Fundamentals and applications. CRC Press, 1999; 396.
171. Barlow R. Engineering reliability. Philadelphia: SIAM, 1998; 199.
172. Yokota T, Gen M, Ida K. System reliability optimization problems with several failure modes by genetic algorithm. Japanese Journal of Fuzzy Theory and Systems 1995;7:119-32.
173. Page L, Perry J. Optimal "series-parallel" networks of 3-state devices. IEEE Trans Reliab 1988;37:388-94.
174. Barlow R, Hunter L, Proschan F. Optimum redundancy when components subject to two kinds of failure. J. Indust. Appl. Math. 1963;II:64-73.
175. Kolesar P. Linear programming and the reliability of multicomponent systems. Naval Research Logistics 1967;14:317-28.
176. Levitin G, Lisnianski A. Structure optimization of multi-state system with two failure modes. Reliab Engng Syst Safety 2001;72:75-89.
177. Levitin G. Optimal series-parallel topology of multi-state system with two failure modes. Reliab Engng Syst Safety 2002;77:93-107.
178. Levitin G. Analysis and optimization of weighted voting systems consisting of voting units with limited availability. Reliab Engng Syst Safety 2001;73:91-100.
179. Levitin G. Maximizing survivability of vulnerable weighted voting systems. Reliab Engng Syst Safety 2004;83:17-26.
180. Levitin G. Asymmetric weighted voting systems. Reliab Engng Syst Safety 2002;76: 99-206.
181. Levitin G. Evaluating correct classification probability for weighted voting classifiers with plurality voting. Eur J Oper Res 2002;141:596-607.
182. Levitin G. Threshold optimization for weighted voting classifiers. Naval Research Logistics 2003;50:322-44.
183. Levitin G. Linear multi-state sliding window systems. IEEE Trans Reliab 2003;52:263-9.
184. Levitin G. Element availability importance in generalized *k*-out-of-*r*-from-*n* systems. IIE Trans 2003;35:1125-31.
185. Papastavridis S, Sfakianakis M. Optimal-arrangement & importance of the components in a consecutive-*k*-out-of-*r*-from-*n*:*F* system. IEEE Trans Reliab 1991;40:277-9.
186. Levitin G. Optimal allocation of elements in linear multi-state sliding window system. Reliab Engng Syst Safety 2002;76:247-55.
187. Levitin G. Uneven allocation of elements in linear multi-state sliding window system. Eur J Oper Res 2004;163:418-33.
188. Levitin G. Common supply failures in linear multi-state sliding window systems. Reliab Engng Syst Safety 2003;82:55-62.
189. Malinowski J, Preuss W. Reliability increase of consecutive-*k*-out-of-*n*:*F* and related systems through components' rearrangement. Microelectron Reliab 1996;36:1417-23.
190. Levitin G. Optimal allocation of multi-state elements in a linear consecutively-connected system. IEEE Trans Reliab 2003;52:192-9.
191. Levitin G. Optimal allocation of multi-state elements in linear consecutively-connected systems with delays. Int J Reliab Qual Saf Eng 2002;9:89-108.
192. Levitin G. Optimal allocation of multi-state elements in linear consecutively-connected systems with vulnerable nodes. Eur J Oper Res 2003;150:406-19.
193. Levitin G. Reliability of acyclic transmission networks with constant transmission characteristics of lines. Reliab Engng Syst Safety 2002;78:297-305.

194. Levitin G. Maximizing survivability of acyclic transmission networks with multi-state retransmitters and vulnerable nodes. Reliab Engng Syst Safety 2002;77:189-99.
195. Levitin G. Optimal reliability enhancement for multi-state transmission networks with fixed transmission times. Reliab Engng Syst Safety 2002;76:287-99.
196. Belli F, Jedrzejowicz P. Fault-tolerant programs and their reliability. IEEE Trans Reliab 1990;39:184-92.
197. Ashrafi N, Berman O, Cutler M. Optimal design of large software-systems using *N*-version programming. IEEE Trans Reliab 1994;43:344-50.
198. Tai A, Meyer J, Avizienis A. Performability enhancement of fault-tolerant software, IEEE Trans Reliab 1993;42:227-37.
199. Meyer JF. On evaluating the performability of degradable computing systems, IEEE Trans Computers 1980;29:720-31.
200. Grassi V, Donatiello L, Iazeolla G. Performability evaluation of multicomponent fault-tolerant systems. IEEE Trans Reliab 1988;37:216-22.
201. Levitin G. Reliability and performance analysis for fault-tolerant programs consisting of versions with different characteristics. Reliab Engng Syst Safety 2004;86:75-81.
202. Levitin G. Optimal version sequencing in fault-tolerant programs. Asia-Pacific J Oper Res 2005;22:1-18.
203. Levitin G. Optimal structure of fault-tolerant software systems. Reliab Engng Syst Safety 2005;89:286-95.

Index